Lecture Notes in Computer Science   10270

Commenced Publication in 1973
Founding and Former Series Editors:
Gerhard Goos, Juris Hartmanis, and Jan van Leeuwen

More information about this series at http://www.springer.com/series/7412

Puneet Sharma · Filippo Maria Bianchi (Eds.)

Image Analysis

20th Scandinavian Conference, SCIA 2017
Tromsø, Norway, June 12–14, 2017
Proceedings, Part II

 Springer

Editors
Puneet Sharma ⓘ
University of Tromsø
Tromsø
Norway

Filippo Maria Bianchi ⓘ
University of Tromsø
Tromsø
Norway

ISSN 0302-9743   ISSN 1611-3349   (electronic)
Lecture Notes in Computer Science
ISBN 978-3-319-59128-5   ISBN 978-3-319-59129-2   (eBook)
DOI 10.1007/978-3-319-59129-2

Library of Congress Control Number: 2017940836

LNCS Sublibrary: SL6 – Image Processing, Computer Vision, Pattern Recognition, and Graphics

Printed on acid-free paper

This Springer imprint is published by Springer Nature
The registered company is Springer International Publishing AG
The registered company address is: Gewerbestrasse 11, 6330 Cham, Switzerland

Preface

This book constitutes the refereed proceedings of the 20th Scandinavian Conference on Image Analysis, SCIA 2017, held in Tromsø, Norway, in June 2017.

The 87 revised papers presented were carefully reviewed and selected from 133 submissions. The 87 accepted articles are organized in two volumes, i.e., volumes 1 and 2. Volume 1 comprises topical sections on the history of SCIA, motion analysis and 3D vision, pattern detection and recognition, machine learning, and image processing and applications. Volume 2 is structured in topical sections on remote sensing, medical and biomedical image analysis, feature extraction and segmentation, and face, gesture, and multispectral analysis.

June 2017

Puneet Sharma
Filippo Maria Bianchi

Organizers

Sponsors

Organization

General Chair

Robert Jenssen   University of Tromsø - The Arctic University
of Norway

Program Chairs

Puneet Sharma   University of Tromsø - The Arctic University
of Norway

Filippo Maria Bianchi   University of Tromsø - The Arctic University
of Norway

Program Co-chairs

Arnt Børre Salberg   Norwegian Computing Center, Norway
Jon Yngve Hardeberg   Norwegian University of Science and Technology,
Norway
Trym Haavardsholm   Norwegian Defence Research Establishment, Norway

Program Committee

Adrien Bartoli	ISIT – CENTI, France
Anders Heyden	Lund University, Sweden
Anne H. Schistad Solberg	University of Oslo, Norway
Arnt-Børre	Norsk Regnesentral, Norway
Atsuto Maki	Kungliga Tekniska Högskolan, Sweden
Cristina Soguero Ruiz	Rey Juan Carlos University, Spain
Daniele Nardi	Sapienza University, Italy
Domenico Daniele Bloisi	Sapienza University, Italy
Enrico Maiorino	Sapienza University, Italy
Erkki Oja	Aalto University, Finland
Fredrik Kahl	Lund University, Sweden
Gustau Camps-Valls	University of Valencia, Spain
Heikki Kälviäinen	Lappeenranta University of Technology, Finland
Helene Schulerud	Sintef, Norway
Ingela Nyström	Uppsala University, Sweden
Janne Heikkilä	University of Oulu, Finland
Jens Thielemann	SINTEF, Norway
Joni Kämäräinen	Tampere University of Technology, Finland
Karl Øyvind Mikalsen	University of Tromsø, Norway
Kjersti Engan	University of Stavanger, Norway

Contents – Part II

Feature Extraction and Segmentation

Simplification of Polygonal Chains by Enforcing Few Distinctive
Edge Directions 3
  Melanie Pohl, Jochen Meidow, and Dimitri Bulatov

Leaflet Free Edge Detection for the Automatic Analysis of Prosthetic
Heart Valve Opening and Closing Motion Patterns from High Speed
Video Recordings 15
  Maryam Alizadeh, Melissa Cote, and Alexandra Branzan Albu

Max-Margin Learning of Deep Structured Models
for Semantic Segmentation 28
  Måns Larsson, Jennifer Alvén, and Fredrik Kahl

Robust Abdominal Organ Segmentation Using Regional
Convolutional Neural Networks 41
  Måns Larsson, Yuhang Zhang, and Fredrik Kahl

Detecting Chest Compression Depth Using a Smartphone Camera
and Motion Segmentation 53
  Øyvind Meinich-Bache, Kjersti Engan, Trygve Eftestøl, and Ivar Austvoll

Feature Space Clustering for Trabecular Bone Segmentation 65
  Benjamin Klintström, Eva Klintström, Örjan Smedby,
  and Rodrigo Moreno

Airway-Tree Segmentation in Subjects with Acute Respiratory
Distress Syndrome 76
  Kristína Lidayová, Duván Alberto Gómez Betancur, Hans Frimmel,
  Marcela Hernández Hoyos, Maciej Orkisz, and Örjan Smedby

Context Aware Query Image Representation for Particular Object Retrieval . . . 88
  Zakaria Laskar and Juho Kannala

Granulometry-Based Trabecular Bone Segmentation 100
  Manish Chowdhury, Benjamin Klintström, Eva Klintström,
  Örjan Smedby, and Rodrigo Moreno

Automatic Segmentation of Abdominal Fat in MRI-Scans,
Using Graph-Cuts and Image Derived Energies. 109
   Anders Nymark Christensen, Christian Thode Larsen,
   Camilla Maria Mandrup, Martin Bæk Petersen, Rasmus Larsen,
   Knut Conradsen, and Vedrana Andersen Dahl

Remote Sensing

Two-Source Surface Reconstruction Using Polarisation 123
   Gary A. Atkinson

Synthetic Aperture Radar (SAR) Monitoring of Avalanche Activity:
An Automated Detection Scheme 136
   H. Vickers, M. Eckerstorfer, E. Malnes, and A. Doulgeris

Canonical Analysis of Sentinel-1 Radar and Sentinel-2 Optical Data 147
   Allan A. Nielsen and Rasmus Larsen

A Noncentral and Non-Gaussian Probability Model for SAR Data 159
   Anca Cristea, Anthony P. Doulgeris, and Torbjørn Eltoft

Unsupervised Multi-manifold Classification of Hyperspectral Remote
Sensing Images with Contractive Autoencoder 169
   Aidin Hassanzadeh, Arto Kaarna, and Tuomo Kauranne

A Clustering Approach to Heterogeneous Change Detection. 181
   Luigi Tommaso Luppino, Stian Normann Anfinsen, Gabriele Moser,
   Robert Jenssen, Filippo Maria Bianchi, Sebastiano Serpico,
   and Gregoire Mercier

Large-Scale Mapping of Small Roads in Lidar Images
Using Deep Convolutional Neural Networks. 193
   Arnt-Børre Salberg, Øivind Due Trier, and Michael Kampffmeyer

Physics-Aware Gaussian Processes for Earth Observation. 205
   Gustau Camps-Valls, Daniel H. Svendsen, Luca Martino,
   Jordi Muñoz-Marí, Valero Laparra, Manuel Campos-Taberner,
   and David Luengo

Medical and Biomedical Image Analysis

Automatic Segmentation of Bone Tissue from Computed Tomography
Using a Volumetric Local Binary Patterns Based Method. 221
   Jukka Kaipala, Miguel Bordallo López, Simo Saarakkala,
   and Jérôme Thevenot

Local Adaptive Wiener Filtering for Class Averaging in Single
Particle Reconstruction   233
    Ali Abdollahzadeh, Erman Acar, Sari Peltonen, and Ulla Ruotsalainen

Comparison of Concave Point Detection Methods for Overlapping Convex
Objects Segmentation   245
    *Sahar Zafari, Tuomas Eerola, Jouni Sampo, Heikki Kälviäinen,
    and Heikki Haario*

Decoding Gene Expression in 2D and 3D   257
    *Maxime Bombrun, Petter Ranefall, Joakim Lindblad, Amin Allalou,
    Gabriele Partel, Leslie Solorzano, Xiaoyan Qian, Mats Nilsson,
    and Carolina Wählby*

Estimation of Heartbeat Peak Locations and Heartbeat Rate
from Facial Video   269
    Mohammad A. Haque, Kamal Nasrollahi, and Thomas B. Moeslund

Segmentation of Multiple Structures in Chest Radiographs
Using Multi-task Fully Convolutional Networks   282
    Chunliang Wang

A Novel Method for Automatic Localization of Joint Area on Knee
Plain Radiographs   290
    Aleksei Tiulpin, Jerome Thevenot, Esa Rahtu, and Simo Saarakkala

Semi-automatic Method for Intervertebral Kinematics Measurement
in the Cervical Spine   302
    *Anne Krogh Nøhr, Louise Pedersen Pilgaard, Bolette Dybkjær Hansen,
    Rasmus Nedergaard, Heidi Haavik, Rene Lindstroem,
    Maciej Plocharski, and Lasse Riis Østergaard*

Memory Effects in Subjective Quality Assessment of X-Ray Images   314
    Victor Landre, Marius Pedersen, and Dag Waaler

Classification of Fingerprints Captured Using Optical
Coherence Tomography   326
    Ctirad Sousedik, Ralph Breithaupt, and Patrick Bours

Interpolation from Grid Lines: Linear, Transfinite and Weighted Method   338
    *Anne-Sofie Wessel Lindberg, Thomas Martini Jørgensen,
    and Vedrana Andersen Dahl*

Automated Pain Assessment in Neonates   350
    *Ghada Zamzmi, Chih-Yun Pai, Dmitry Goldgof, Rangachar Kasturi,
    Yu Sun, and Terri Ashmeade*

Enhancement of Cilia Sub-structures by Multiple Instance Registration
and Super-Resolution Reconstruction.   362
    Amit Suveer, Nataša Sladoje, Joakim Lindblad, Anca Dragomir,
    and Ida-Maria Sintorn

Faces, Gestures and Multispectral Analysis

Residual vs. Inception vs. Classical Networks for Low-Resolution
Face Recognition   377
    Christian Herrmann, Dieter Willersinn, and Jürgen Beyerer

Visual Language Identification from Facial Landmarks   389
    Radim Špetlík, Jan Čech, Vojtěch Franc, and Jiří Matas

HDR Imaging Pipeline for Spectral Filter Array Cameras.   401
    Jean-Baptiste Thomas, Pierre-Jean Lapray, and Pierre Gouton

Thistle Detection.   413
    Søren I. Olsen, Jon Nielsen, and Jesper Rasmussen

An Image-Based Method for Objectively Assessing Injection Moulded
Plastic Quality   426
    Morten Hannemose, Jannik Boll Nielsen, László Zsíros,
    and Henrik Aanæs

Creating Ultra Dense Point Correspondence Over the Entire Human Head . . .   438
    Rasmus R. Paulsen, Kasper Korsholm Marstal, Søren Laugesen,
    and Stine Harder

Collaborative Representation of Statistically Independent Filters' Response:
An Application to Face Recognition Under Illicit Drug Abuse Alterations . . .   448
    Raghavendra Ramachandra, Kiran Raja, Sushma Venkatesh,
    and Christoph Busch

Multispectral Constancy Based on Spectral Adaptation Transform.   459
    Haris Ahmad Khan, Jean Baptiste Thomas, and Jon Yngve Hardeberg

State Estimation of the Performance of Gravity Tables Using Multispectral
Image Analysis.   471
    Michael A.E. Hansen, Ananda S. Kannan, Jacob Lund, Peter Thorn,
    Srdjan Sasic, and Jens M. Carstensen

Author Index   481

Contents – Part I

History of SCIA

Image Processing and Its Hardware Support Analysis
vs Synthesis - Historical Trends 3
  Ewert Bengtsson

Motion Analysis and 3D Vision

Averaging Three-Dimensional Time-Varying Sequences of Rotations:
Application to Preprocessing of Motion Capture Data 17
  *Tomasz Hachaj, Marek R. Ogiela, Marcin Piekarczyk,
  and Katarzyna Koptyra*

Plane Refined Structure from Motion. 29
  Branislav Micusik and Horst Wildenauer

A Time-Efficient Optimisation Framework for Parameters
of Optical Flow Methods. 41
  Michael Stoll, Sebastian Volz, Daniel Maurer, and Andrés Bruhn

Subpixel-Precise Tracking of Rigid Objects in Real-Time 54
  Tobias Böttger, Markus Ulrich, and Carsten Steger

Wearable Gaze Trackers: Mapping Visual Attention in 3D. 66
  *Rasmus R. Jensen, Jonathan D. Stets, Seidi Suurmets, Jesper Clement,
  and Henrik Aanæs*

Image Processing of Leaf Movements in *Mimosa pudica* 77
  Vegard Brattland, Ivar Austvoll, Peter Ruoff, and Tormod Drengstig

Evaluation of Visual Tracking Algorithms for Embedded Devices. 88
  *Ville Lehtola, Heikki Huttunen, Francois Christophe,
  and Tommi Mikkonen*

Multimodal Neural Networks: RGB-D for Semantic Segmentation
and Object Detection. 98
  *Lukas Schneider, Manuel Jasch, Björn Fröhlich, Thomas Weber,
  Uwe Franke, Marc Pollefeys, and Matthias Rätsch*

Uncertainty Computation in Large 3D Reconstruction 110
  Michal Polic and Tomas Pajdla

Robust and Practical Depth Map Fusion for Time-of-Flight Cameras.   122
  Markus Ylimäki, Juho Kannala, and Janne Heikkilä

An Error Analysis of Structured Light Scanning of Biological Tissue   135
  Sebastian Nesgaard Jensen, Jakob Wilm, and Henrik Aanæs

Structure from Motion by Artificial Neural Networks   146
  *Julius Schöning, Thea Behrens, Patrick Faion, Peyman Kheiri,
  Gunther Heidemann, and Ulf Krumnack*

Pattern Detection and Recognition

Computer Aided Detection of Prostate Cancer on Biparametric MRI
Using a Quadratic Discriminant Model   161
  *Carina Jensen, Anne Sofie Korsager, Lars Boesen,
  Lasse Riis Østergaard, and Jesper Carl*

Pipette Hunter: Patch-Clamp Pipette Detection   172
  Krisztian Koos, József Molnár, and Peter Horvath

Non-reference Image Quality Assessment for Fingervein Presentation
Attack Detection.   184
  *Amrit Pal Singh Bhogal, Dominik Söllinger, Pauline Trung,
  Jutta Hämmerle-Uhl, and Andreas Uhl*

Framework for Machine Vision Based Traffic Sign Inventory.   197
  Petri Hienonen, Lasse Lensu, Markus Melander, and Heikki Kälviäinen

Copy-Move Forgery Detection Using the Segment Gradient Orientation
Histogram   209
  Ali Retha Hasoon Khayeat, Paul L. Rosin, and Xianfang Sun

BriefMatch: Dense Binary Feature Matching for Real-Time Optical
Flow Estimation   221
  Gabriel Eilertsen, Per-Erik Forssén, and Jonas Unger

Robust Data Whitening as an Iteratively Re-weighted Least
Squares Problem.   234
  Arun Mukundan, Giorgos Tolias, and Ondřej Chum

DEBC Detection with Deep Learning   248
  Ian E. Nordeng, Ahmad Hasan, Doug Olsen, and Jeremiah Neubert

Object Proposal Generation Applying the Distance Dependent Chinese
Restaurant Process   260
  Mikko Lauri and Simone Frintrop

Object Tracking via Pixel-Wise and Block-Wise Sparse Representation   273
 Pouria Navaei, Mohammad Eslami, and Farah Torkamani-Azar

Supervised Approaches for Function Prediction of Proteins Contact
Networks from Topological Structure Information   285
 Alessio Martino, Enrico Maiorino, Alessandro Giuliani,
 Mauro Giampieri, and Antonello Rizzi

Top-Down Deep Appearance Attention for Action Recognition.   297
 Rao Muhammad Anwer, Fahad Shahbaz Khan, Joost van de Weijer,
 and Jorma Laaksonen

Machine Learning

Soft Margin Bayes-Point-Machine Classification via Adaptive
Direction Sampling   313
 Karsten Vogt and Jörn Ostermann

ConvNet Regression for Fingerprint Orientations.   325
 Patrick Schuch, Simon-Daniel Schulz, and Christoph Busch

Domain Transfer for Delving into Deep Networks Capacity
to De-Abstract Art.   337
 Corneliu Florea, Mihai Badea, Laura Florea, and Constantin Vertan

Foreign Object Detection in Multispectral X-ray Images of Food Items
Using Sparse Discriminant Analysis   350
 Gudmundur Einarsson, Janus N. Jensen, Rasmus R. Paulsen,
 Hildur Einarsdottir, Bjarne K. Ersbøll, Anders B. Dahl,
 and Lars Bager Christensen

Sparse Approximation by Matching Pursuit
Using Shift-Invariant Dictionary   362
 Karl Skretting and Kjersti Engan

Diagnosis of Broiler Livers by Classifying Image Patches   374
 Anders Jørgensen, Jens Fagertun, and Thomas B. Moeslund

Historical Document Binarization Combining Semantic Labeling
and Graph Cuts.   386
 Kalyan Ram Ayyalasomayajula and Anders Brun

Convolutional Neural Networks for Segmentation and Object Detection
of Human Semen   397
 Malte S. Nissen, Oswin Krause, Kristian Almstrup, Søren Kjærulff,
 Torben T. Nielsen, and Mads Nielsen

Convolutional Neural Networks for False Positive Reduction
of Automatically Detected Cilia in Low Magnification TEM Images   407
  Anindya Gupta, Amit Suveer, Joakim Lindblad, Anca Dragomir,
  Ida-Maria Sintorn, and Nataša Sladoje

Deep Kernelized Autoencoders   419
  Michael Kampffmeyer, Sigurd Løkse, Filippo M. Bianchi,
  Robert Jenssen, and Lorenzo Livi

Spectral Clustering Using *PCKID* – A Probabilistic Cluster Kernel
for Incomplete Data   431
  Sigurd Løkse, Filippo M. Bianchi, Arnt-Børre Salberg,
  and Robert Jenssen

Automatic Emulation by Adaptive Relevance Vector Machines   443
  Luca Martino, Jorge Vicent, and Gustau Camps-Valls

Image Processing and Applications

Deep Learning for Polar Bear Detection   457
  Scott Sorensen, Wayne Treible, Leighanne Hsu, Xiaolong Wang,
  Andrew R. Mahoney, Daniel P. Zitterbart, and Chandra Kambhamettu

Crowd Counting Based on MMCNN in Still Images   468
  Tao Wang, Guohui Li, Jun Lei, Shuohao Li, and Shukui Xu

Generation and Authoring of Augmented Reality Terrains Through
Real-Time Analysis of Map Images   480
  Theodore Panagiotopoulos, Gerasimos Arvanitis,
  Konstantinos Moustakas, and Nikos Fakotakis

Solution of Pure Scattering Radiation Transport Equation (RTE)
Using Finite Difference Method (FDM)   492
  Hassan A. Khawaja

Optimised Anisotropic Poisson Denoising   502
  Georg Radow, Michael Breuß, Laurent Hoeltgen, and Thomas Fischer

Augmented Reality Interfaces for Additive Manufacturing   515
  Eythor R. Eiriksson, David B. Pedersen, Jeppe R. Frisvad,
  Linda Skovmand, Valentin Heun, Pattie Maes, and Henrik Aanæs

General Cramér-von Mises, a Helpful Ally for Transparent Object
Inspection Using Deflection Maps?   526
  Johannes Meyer, Thomas Längle, and Jürgen Beyerer

Dynamic Exploratory Search in Content-Based Image Retrieval   538
    Joel Pyykkö and Dorota Głowacka

Robust Anomaly Detection Using Reflectance Transformation Imaging
for Surface Quality Inspection   550
    Gilles Pitard, Gaëtan Le Goïc, Alamin Mansouri, Hugues Favrelière,
    Maurice Pillet, Sony George, and Jon Yngve Hardeberg

Block-Permutation-Based Encryption Scheme with Enhanced
Color Scrambling   562
    Shoko Imaizumi, Takeshi Ogasawara, and Hitoshi Kiya

Author Index   575

Feature Extraction and Segmentation

Simplification of Polygonal Chains by Enforcing Few Distinctive Edge Directions

Melanie Pohl[(✉)], Jochen Meidow, and Dimitri Bulatov

Fraunhofer IOSB, Ettlingen, Germany
{melanie.pohl,jochen.meidow,dimitri.bulatov}@iosb.fraunhofer.de

Abstract. Simplification of polygonal chains by reducing the number of vertices becomes challenging when additionally dominant edge expansions of the polygonal chains shall be exposed. Such simplifications are often sought in order to generalize polygonal chains representing borders or medial axes of man-made structures such as buildings or road networks, observed in aerial images.

In this paper, we present two methods that reduce the number of vertices in polygonal chains meanwhile featuring additional properties: First, the resulting polygonal chains are irrespective of coordinate axes as pixel-based approaches tend to produce. Second, the simplified chains keep the rough shape of the initial ones and emphasize dominating edge expansions. Optionally, detected perpendicularities may be enforced. Third, polygons with holes are supposed to exhibit parallel segments in interior and exterior polygonal chains. Our methods treat the associated polygonal chains simultaneously by emphasizing common distinctive directions.

Keywords: Polygonal chain simplification · Generalization · Distinctive/Dominant directions · Expectation maximization

1 Introduction

In object-based image analysis, vectorization of image data is the most essential concept to compress and abstract image information. It has numerous applications, e.g., digitizing calligraphy, creating topographic databases, or deriving semantic representations of image content. An important intermediate step of the latter application is classification of image regions [11,18]. Our main application area is creation of virtual cityscapes with buildings and roads as most important objects of recognition value. Vectorization of outlines or (medial) axes of these objects is the principal task in urban terrain reconstruction [4].

Preliminaries. A plausible boundary of a classified object would be the pixel conducted polygonal chain separating all *in-object* pixels from *outside* pixels. To outline exterior and interior contours, we utilize a modification of the Moore contour-tracing algorithm [14] on the mask of the detected object and on its

© Springer International Publishing AG 2017
P. Sharma and F.M. Bianchi (Eds.): SCIA 2017, Part II, LNCS 10270, pp. 3–14, 2017.
DOI: 10.1007/978-3-319-59129-2_1

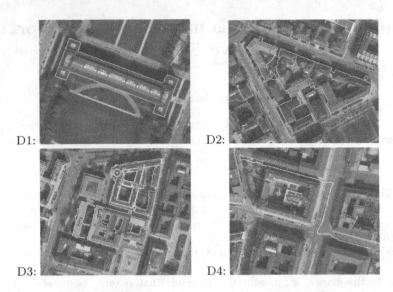

Fig. 1. Test data extracted from a test side of Munich, Southern Germany [15]. D1: single building outline with orthogonal directions. D2: building complex of several directions. D3: building complex of several directions and multiple atriums (holes). D4: medial axes of a road segment.

complement. Our methods are designed to generate outlines of buildings and medial axes of road courses, but all procedures may immediately be applied to other objects featuring dominating edge expansions. The contours are pre-simplified by means of Douglas-Peucker [6] with an extremely tight threshold (1 to 1.5 pixels) to eliminate only the axially parallel stair-shaped polygonal chains from contour tracing without changing the initial shape. The choice of this tight threshold avoids topological inconsistencies and dependencies of initial vertices [16] in closed polygonal chains. To identify the medial axes of objects like roads, we use thinning of the corresponding classification mask with a following vector-ization according to [17] instead. This is done to preserve the topology of junctions and branch points. Nevertheless, the final step is again pre-simplification with a small threshold value. The outlines of four classified objects – one building, two building complexes and a road course – are exemplary shown in Fig. 1.

Related Work. Complicated pipelines are designed to obtain polygonal chains which approximate a classification result, and are robust against data noise and outliers. Additionally, for many man-made structures, constraints of symmetry must be imposed. The resulting polygonal chains should capture the object's topological properties and have a small number of vertices. It has been pointed out in [12] that the generic meshes obtained by crust algorithms [1], ball-pivoting methods [2] or alpha-shapes [7] do not always satisfy these criteria.

To reduce the number of vertices, one can apply a simplification routine, such as the one presented in [6]. However, this approach depends on the choice of starting vertex and is liable to noise. Furthermore, decreasing the number of vertices does not necessarily decrease the number of included edge directions. To compensate this drawback, [10] searched for distinctive directions before polygonization. Straight lines within the images are used to add and to remove rectangles to form polygons [9]. Furthermore, [13] attempts post-processing of the directions obtained by concave hulls [12]. Given the hull of a building footprint, two directions should be derived from the histogram formed by edges of the polygonal chain weighted by their lengths. Additionally, it was suggested to work with orthogonal projections and to limit the search direction for the concave hull to multiples of $\pi/2$. If this range is too coarse, multiples of smaller values that are typical for man-made objects can be chosen. This approach is also transferable to one of the standard polygonization algorithms, such as [1] based on medial axes, or on the Moore neighbor-tracing approach [14].

Contribution. For a variety of reasons, segmentation results may be noisy or flawed. Hence, the pre-simplified polygonal chains may wriggle with many vertices and directions; e.g., straight segments of building outlines are not obtained properly. The task of this paper is to explicate methods to reduce the number of vertices in a polygonal chain while simultaneously enforcing dominant directions of the polygonal chain's edge expansion. These dominant directions are also called distinctive directions.

Although man-made objects feature curvy structures, too, we will focus on (complexes of) buildings and roads in metropolises, where streets are roughly laid out in piecewise straight-lined segments and where building outlines are mostly characterized by orthogonality or parallelism to several distinctive directions. Segments of distinctive directions distinguish themselves by multiple occurrences. To extract them we present and compare two approaches: The former, an intuitive way to solve the problem is to transfer the observed edge directions into a histogram weighted by the edges' lengths followed by analysis of histogram peaks and non-maxima suppression. Alternatively, distinctive directions can be determined by fitting parametric distributions to the edge directions. The latter can be done effectively by applying the expectation-maximization (EM) algorithm for mixture models with a usually a priori unknown number of components, i.e., distinctive directions.

We apply our methods also to a special issue: Some polygons formed by closed polygonal chains contain empty areas, called *holes*, which are for their part limited by an interior polygonal chain. Some holes are difficult to detect from image data or could be represented by noisy or imprecise observations. A simultaneous treatment of interior and exterior polygonal chains, on the one hand, stabilizes the detection of holes and, on the other hand, guarantees the outlines to be of similar or parallel shape. Furthermore, this method enforces parallelism of segments in exterior and interior polygonal chains.

We will show the usability of our procedures exemplary for outlines of buildings (or building complexes) and a road course.

2   Distinctive Directions and Simplification

Our task is to reduce the number of vertices in a polygonal chain while keeping the most prevalent directions of its edges. In case of building outlines, most European buildings feature two distinctive rectangular directions; however, there are many exceptions. Nevertheless, these man-made structures include just a few distinctive directions. Likewise, roads often feature few distinctive directions and sometimes even rectangular structures. There are definitions for what a road course is in [5]. For our purpose, it is sufficient to consider a road course as a connected sequence of road segments. An example is shown in Fig. 1(D4).

In this section, we present two different approaches for selecting distinctive directions in a polygonal chain with n vertices $\{v_1, v_2, \ldots, v_n\}$, $v_i = (x_i, y_i)^\top$ and corresponding edges vectors $e_i = v_{i+1} - v_i$. All methods rely on the enclosed counterclockwise angles α_i of the positive x-axis and e_i assessed by the edges' lengths ℓ_i. Each α_i represents an edge direction. Parallel edges in the polygonal chain are supposed to enhance the appearance of the same direction but are – due to polygonal sequence – often oriented oppositely. Hence, the edge direction α_i is set irrespectively of the orientation of e_i to $\alpha_i \leftarrow \alpha_i \bmod \pi$.

2.1   Hill-and-Valley Decomposition

Similarly to [13], we create a length-weighted histogram for each polygonal chain with $\pi/180$ bin discretization for α_i (abscissa) and $q_j = \sum_s \ell_s$ for all e_s assigned to bin j (ordinate). If the polygon contains holes, values of α_i and ℓ_i are collected over all polygonal chains.

Extracting Distinctive Directions. We assume a distinctive direction to occur often in the polygonal chain and therewith to correspond to a local maximum of the histogram. A local maximum, is either a histogram bin of larger value than the incident bins to both sides, or the medial bin of a *maximum plateau* that is a set of successive bins of equal and larger value than the closest different valued bins to both sides. If the number of bins belonging to a plateau is even, we chose the bin of smaller angle. (Accordingly, local minima are defined the other way around.)

In the initial histogram, many local maxima are present. In order to extract distinctive directions representing the shape of polygonal chains with adequate accuracy and to suppress weak local maxima, we smooth the initial circular histogram. As a Gaussian filter lends itself to (histogram) smoothing, we use the composition of its simplest approximation, the binomial filter $1/4 \begin{bmatrix} 1 & 2 & 1 \end{bmatrix}$, several times. If we know the number of smoothing steps $\#\mathcal{S}$, the filter might be pre-calculated as an autonomous approximation for a Gaussian filter. Hence, smoothing could take place in one step. Note, that bin width and amount of

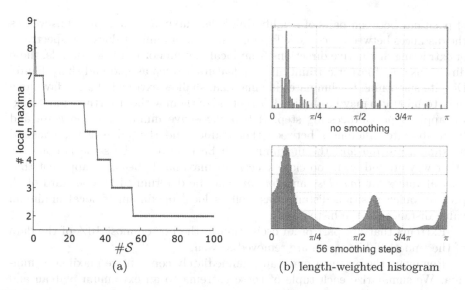

Fig. 2. Amount of local maxima of the length-weighted histogram during smoothing process (data set D4 in Fig. 1).

observations influences $\#S$. We found a bin discretization of $\pi/180$ useful for simplification of polygonal chains of roads or buildings. But in general, the bin width might be adapted according to the task of application and accuracy.

The challenge is to estimate $\#S$ to get the optimal result. It is hard to guess the ideal number of desired local maxima in the histogram without knowing the true number of distinctive directions in the polygonal chain. Since the binning is quite fine $\#S$ should be bounded away from zero ($\#S \geq 30$) but far less than 100. Plotting the number of local maxima as a function of smoothing steps, we obtain a decreasing graph with many plateaus as shown in Fig. 2a. There are some strategies to determine $\#S$ optimally, such as taking the closest point from all coordinate tuples initiating a plateau to the point of origin. But to the best of our knowledge, the optimal number is chosen best heuristically by taking all available prior information into account.

Following the assumption above, after smoothing, the remaining local maxima correspond to distinctive directions of the polygonal chain. This strategy is plausible and useful. To make the procedure more robust against the choice of bin width, we recommend a small modification. Enlarging the bin width may cause shifted local maxima because of coinciding bins. To compensate this effect, the distinctive directions φ_f are extracted from the centroid of a histogram *hill* – that means from all bins between two successive local minima (the *valleys*) – to cover most α_i, instead of its local maximum.

Elimination of Shallow Extrema. The certainty of distinctive directions depends on the varying severity of the *max-to-min ratio* including the distances of positions between pairs of neighboring local maxima (and minima) as well as the distances between successive local maxima and minima values. Irrespectively of extracting distinctive directions from local maxima or centroids of hills, these directions are of larger certainty if they feature a steep ascend and sharp figure. Despite smoothing to eliminate side maxima, shallow extrema of an unfavorable max-to-min ratio may be obtained. To get rid of them without further smoothing, we apply a post-processing step: If two successive minima are *too close* and both absolute differences between their values and the value of the enclosed maximum are *too low*, the maximum will be removed. The same is done the other way around with too close successive maxima. It becomes apparent that critical values for *too close* and *too low* may be determined by one third of all present valley position distances as well as local maximum to local minimum value distances in the histogram.

Extrema that are part of an oscillation in the bottommost 10% of the data of the smoothed histogram are removed as well.

Removing the extrema may cause immediately consecutive maxima or minima. We summarize each tuple of these extrema to an extremum plateau and compute from this set the medial bin as extremum.

Working with the modification, the distinctive directions can afterwards be extracted from the centroids of the remaining hills.

2.2 Parametric Distribution Analysis

A disadvantage of hill-and-valley decomposition is the large number of degrees of freedom. Varying the bin size influences $\#\mathcal{S}$. The choice of this number is rather heuristic and per-object. The more prior knowledge about the underlying object is available the more robust the simplification of the polygonal chain works. But since we expect a few distinctive directions and since the shape of the smoothed histogram indicates the presence of an underlying compound distribution, we pursue a statistical approach. The following method models distinctive directions and background noise with parametric distributions by means of expectation maximization.

Modeling. The statistical theory of directional data usually considers distributions on the unit circle. For the statistical analysis, we transform the observed chain directions α_i mod π by doubling them, estimate distinctive directions φ_f among other distribution parameters, and back-transform the results.

Besides the distinctive directions φ_f, we expect background clutter due to imperfect data. The parametric approach allows to model the background explicitly by means of the uniform distribution $p(\alpha) = 1/(2\pi)$, $0 \leq \alpha \leq 2\pi$. Thus, all directions α between 0 and 2π are equally likely. For the dominant directions, we utilize the *von Mises* distribution which is in many respect the "natural" analogue on the circle of the normal distribution on the real line [8]. The probability density function reads

$$p(\alpha|\varphi,\kappa) = \frac{1}{2\pi I_0(\kappa)} \exp\left\{\kappa\cos(\alpha - \varphi)\right\}, \qquad 0 \leq \varphi \leq 2\pi, \quad 0 \leq \kappa \leq \infty \quad (1)$$

where $I_0(\kappa)$ is the modified Bessel function of order zero, φ is the mean direction, and κ is the so-called concentration parameter. As $\kappa \to 0$, the distribution converges to the uniform distribution; as $\kappa \to \infty$, it tends to the point distribution concentrated in the direction φ. The maximum likelihood estimate $\hat{\kappa}$ for the concentration parameter κ can be seriously biased when the sample size n and R/n are small, whereas R is the length of the resultant $(\sum_i \cos \alpha_i, \sum_i \sin \alpha_i)$ [8]. In that case, $\hat{\kappa}$ can substantially over-estimate the true value of κ. Correction formulas are provided in [3].

For uncertainty analysis of each estimated mean direction φ_f, two measures are essential: Given the concentration κ_f for a direction φ_f, the circular standard deviation $\sigma_f = \sqrt{-2 \log \rho_f}$, $\rho_f = I_1(\kappa_f)/I_0(\kappa_f)$ can be computed, to measure the spread of data. For distinctive directions, we expect circular standard deviations of less than say $20°$. For testing the potential orthogonality of distinctive directions φ_f, we utilize the precision $\sigma_{\hat{\varphi}_f} = 1/\sqrt{R\hat{\kappa}_f}$ for each φ_f.

For the application at hand, we expect several distinctive directions and background clutter, modeled by a mixture of F von Mises distributions and a uniform distribution, whereas the number of components F is unknown a priori. For fitting this mixture, we apply the EM algorithm and determine the number of components by considering information theoretic criteria such as the Bayesian Information Criterion.

For the outlines of buildings, we expect the distinctive directions not to vary too much, i.e., the dispersion σ_f should be small for each direction φ_f. Thus, dominance is not only defined by frequently occurring directions, but also by directions featuring a small variation.

Enforcing Orthogonal Directions. To detect orthogonality, we perform statistical parametric tests for all pairs of estimated distinctive directions. The corresponding test statistic for a pair of distinctive directions (φ_g, φ_h) is

$$T_{gh} = \frac{\sin(|\hat{\varphi}_g - \hat{\varphi}_h| - \pi)}{\sigma_{\hat{\varphi}_g \hat{\varphi}_h}} \sim N(0, 1) \tag{2}$$

with the standard deviation $\sigma_{\hat{\varphi}_g \hat{\varphi}_h} = \sqrt{\sigma_{\hat{\varphi}_g}^2 + \sigma_{\hat{\varphi}_h}^2}$ of the estimated difference $\hat{\varphi}_g - \hat{\varphi}_h$. The values of the test statistic T_{gh} are standard normal distributed.

Once a pair of distinctive directions are in line for orthogonality, it can be enforced by applying the EM algorithm again with a bimodal von Mises distribution featuring the three parameters φ_g, κ_g and κ_h, whereby the second mean direction is given by $\varphi_h = \varphi_g + \frac{\pi}{2}$. Alternatively, the estimated directions $\hat{\varphi}_g$ and $\hat{\varphi}_h$ can be adjusted according to their uncertainty provided by the estimated variances $\sigma_{\hat{\varphi}_g}^2$ and $\sigma_{\hat{\varphi}_h}^2$. The statistical optimal corrections for the directions φ_g and φ_h are

$$\begin{pmatrix} \widehat{\Delta \varphi_g} \\ \widehat{\Delta \varphi_h} \end{pmatrix} = \frac{\hat{\varphi}_g - \hat{\varphi}_h + \pi/2}{\sigma_{\hat{\varphi}_g}^2 + \sigma_{\hat{\varphi}_h}^2} \begin{pmatrix} -\sigma_{\hat{\varphi}_g}^2 \\ \sigma_{\hat{\varphi}_h}^2 \end{pmatrix} \tag{3}$$

i.e., uncertain distinctive directions will be shifted more than certain directions to fulfill the orthogonality constraint.

2.3   Simplification of Polygonal Chains

The extracted distinctive directions are now used to simplify the initial polygonal chain as described in [13]. The polygon simplification is divided into two steps: The assignment of edges and their adjustment.

For *assignment*, we label each edge e_i according to $\widehat{f_i} = \arg\min_f(|\varphi_f - \alpha_i|)$.

For *adjustment*, all vertices belonging to successive edges of same label f are replaced by a line through the centroid of this group of vertices along the vector $(\cos\varphi_f, \sin\varphi_f)$. Each line is limited by intersection with neighboring lines. First and last line in open polygonal chains are limited to the "open" side by projecting the first vertex v_1 as well as the last vertex v_n onto the line.

3   Results and Discussion

We demonstrate the usability of our methods by means of four data sets D1 to D4, extracted from a test side of Munich, Southern Germany (Fig. 1, [15]). As closed polygonal chains we chose three building structures D1 to D3 and as an open polygonal chain a road example D4. One can see that D1 and D4 have a quite simple shape with only two perpendicular directions. To show the behavior of our algorithm on polygonal chains with multiple directions, we chose complexes of buildings D2 and D3. Note that D3 also highlights the usability on polygons with holes. All bordering polygonal chains have some influence in the computation of distinctive directions.

Results. The color coding of the edge labels of D1's outline resulting from hill-and-valley decomposition (Sect. 2.1) is shown in Fig. 3a. The initial polygonal chain includes 88 different edge directions and may be assigned to only two resulting distinctive directions (Table 1). Hence, it is possible to simplify the polygonal chain only with edges of these two directions. Using the parametric approach (Sect. 2.2), we obtain the estimates $\widehat{\varphi}_1 = 108.9°\,(30.9\%$ proportion$)$ and $\widehat{\varphi}_2 = 21.0°\,(69.1\%)$ which enclose an angle of $87.9°$. The circular standard deviations of the directions are in an expected magnitude of approximately $22°$.

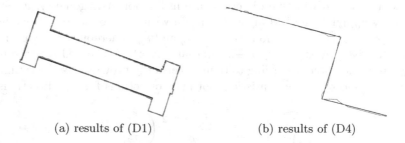

(a) results of (D1)   (b) results of (D4)

Fig. 3. Color coding and polygonal chain simplification (black underlay). The color coding results from both algorithms.

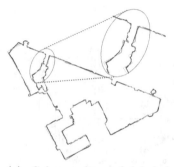

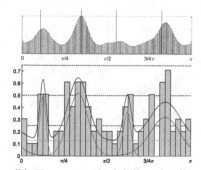

(a) Color coding of labeled edges; black: original polygonal chain; pink: hill-and-dale; blue: EM

(b) Top: smoothed hill-and-valley histogram with orange marked distinctive directions; bottom: histogram with EM components

Fig. 4. Color coding of edge labels and extraction of distinctive directions visualized on the histogram for test set D2. (Color figure online)

The hypothesis of a present right angle is not rejected, thus the distinctive directions can be adjusted proportionally to their estimated standard deviations according to (3), which are almost equal for this data.

The adapted boundaries of D2 are depicted in Fig. 4a. Both, hill-and-valley decomposition and EM-based algorithm lead to four extracted directions as can be seen in Fig. 4b and Table 1. The results of both approaches do not differ very much (Fig. 4b) but the determined directions vary enough to result in a different edge labeling and therewith a slightly differing fitting to the original outline (Fig. 4a). Three directions obtained by hill-and-valley decomposition are outlined as sharp hills in the upper histogram of Fig. 4b. The fourth maximum is blurred and of much lower value compared to the other ones. This indicates higher uncertainty. A harder constraint concerning the value differences between local maxima and minima as well as their positions (Sect. 2.1) could eliminate the intuitively redundant appearing maximum. However, the algorithms are developed to work on large data sets with hundreds of buildings for instance. The calculation rule for a constraint needs to be global enough to work for every building in the data set.

The polygon D3 includes holes. The color coding in Fig. 5a shows the assignment of hole edges parallel to the exterior polygonal chain and therewith enforcing the simplification to parallel edges. Beside some isolated "wrong labeled" edges in between, all main edge expansions in the polygonal chain are exposed: The green and gray edges coding for the nearly rectangular parts in the outline (green: "side" parts, gray: "bottom/top" parts) as well as the orange colored non-rectangular walls in the upper part of the building outline. By hill-and-valley decomposition we reduce the initial number of 251 edge directions to four (Fig. 5a). With the EM-based algorithm, it is even possible to reduce the number of main directions to three (Fig. 5b, Table 1).

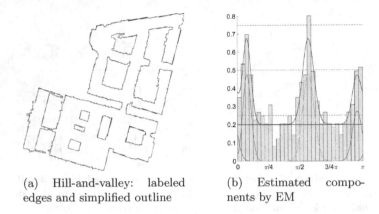

(a) Hill-and-valley: labeled edges and simplified outline

(b) Estimated components by EM

Fig. 5. Results for a data set of a building outline containing atriums (D3). The corresponding polygonal chains enclose a polygon with holes. (Color figure online)

Table 1. Numbers of directions present in the polygonal chains of D1–D4.

	D1	D2	D3	D4
Initial number of directions	88	165	251	61
Initial amount of local maxima	16	37	40	9
Number of hill-and-valley directions	2	4	4	2
Number of EM directions	2	4	3	2

As example for an open polygonal chain we consider the course of the road D4. The resulting labeling and simplified polygonal chain of the road course are shown in Fig. 3b. The results of both, hill-and-valley decomposition and EM, are very similar. Both approaches are able to detect both distinctive directions (out of initial 61, see Table 1) one intuitively assumes.

Discussion. Both introduced methods for detecting distinctive directions have different advantages over each other. Contrary to the EM-based algorithm, hill-and-valley decomposition is rather robust. It always results in at least one resulting direction even if the histogram is sparse or equally distributed. Prior knowledge about the number of distinctive directions supports the procedure and makes it quite robust. Otherwise, the approach has several important thresholds: It yields unsatisfactory results if the bin size or $\#S$ are chosen inappropriately or if the thresholds to eliminate shallow extrema (Sect. 2.1) do not fit the distribution. Enlarging the bin width may result in coincidence of neighboring bins and therewith a possibly tilted result. For compensation, we chose the average direction of hill as main direction instead of the local maximum. The EM-based approach works directly with the polygonal edges. Beside predefining the number of components no other parameters are required. The algorithm requires a clear mixed distribution. It is preferred if no or just few prior knowledge is available. But it has problems with lack of observations or major outliers, as well as nearly uniform distributions. So, the decision which algorithm is "best" for computation of distinctive directions depends on the background information, i.e., how many different observations are present, whether the number of free parameters can be decreased by prior knowledge, or whether the manifestation of several distinctive directions can be assumed.

Please note that the extraction of distinctive directions lying close to each other may lead to extrapolations in the simplification step if these directions are assigned to incident edges of the polygonal chain. Elimination of "isolated" edge labels will not solve that problem because it may lead to collapsing parts in the polygonal chain. This deficit should be avoided initially by adapting the number of smoothing steps (Sect. 2.1) or components (Sect. 2.2).

4   Conclusion and Future Work

In this article, we presented two methods to extract distinctive directions in order to simplify polygonal chains. The number of edges and their corresponding directions can be reduced as shown, i.e., for the example of road courses. Simultaneous evaluation of exterior and interior polygonal chains of building outlines lead to a simplification with the same determined directions for both, polygonal outline and hole outlines, and therewith to a parallelization of edges in segments of interior and exterior polygonal chains.

Simplified polygonal chains feature few distinctive directions and depict a shape similar to the initial polygonal chain. This has advantages in reducing data for memory reasons but also provides a more intuitively outline if it comes to man-made structures, e.g., straight walls or straight courses. Hence, both methods generate polygonal chains with less vertices but an appropriate fitting to the underlying classified image object. The choice of method depends on the number of edges in the polygonal chain and whether more background information about the underlying object is available. If 90° angles are expected, the detected directions may be shifted according to their certainty afterwards.

Our aspiration was to keep the algorithm flexible for any type of cityscape. If data sets are laid out in a grid pattern, it would be recommended to include the data of, e.g., all buildings at once into the procedure. Alternatively, determined distinctive directions may successively be included into a maximum a posteriori optimization.

For future work, the (un)certainty of distinctive directions may be used to suppress unnecessary directions. Further, the assignment of polygonal edges to a distinctive direction can be changed from a winner-takes-all strategy to a probability-based method. The assignment of edges close to multiple distinctive directions should be influenced by incident edge labels in order to advance edge propagation.

Acknowledgment. We want to thank Christian Böge for coming up always with helpful ideas.

References

1. Amenta, N., Choi, S., Kolluri, R.K.: The power crust, unions of balls, and the medial axis transform. Comput. Geom. **19**(2), 127–153 (2001)
2. Bernardini, F., Mittleman, J., Rushmeier, H., Silva, C., Taubin, G.: The ball-pivoting algorithm for surface reconstruction. IEEE Trans. Vis. Comput. Graph. **5**(4), 349–359 (1999)
3. Best, D.J., Fisher, N.I.: The BIAS of the maximum likelihood estimators of the von Mises-Fisher concentration parameters: the BIAS of the maximum likelihood estimators. Commun. Stat. Simul. Comput. **10**(5), 493–502 (1981)
4. Bulatov, D., Häufel, G., Meidow, J., Pohl, M., Solbrig, P., Wernerus, P.: Context-based automatic reconstruction and texturing of 3D urban terrain for quick-response tasks. ISPRS J. Photogrammetry Remote Sens. **93**, 157–170 (2014)
5. Bulatov, D., Häufel, G., Pohl, M.: Identification and correction of road courses by merging successive segments and using improved attributes. In: Proceedings SPIE Remote Sensing, p. 100080V. International Society for Optics and Photonics (2016)
6. Douglas, D., Peucker, T.: Algorithms for the reduction of the number of points required to represent a digitized line or its caricature. Can. Cartographer **10**(2), 112–122 (1973)
7. Edelsbrunner, H., Mücke, E.P.: Three-dimensional alpha shapes. ACM Trans. Graph. (TOG) **13**(1), 43–72 (1994)
8. Fisher, N.I.: Statistical Analysis of Circular Data. Cambridge University Press, Cambridge (1995)
9. Grigillo, D., Kanjir, U.: Urban object extraction from digital surface model and digital aerial images. In: Proceedings of XXII ISPRS Congress, Melbourne, Australia (2012)
10. Gross, H., Thoennessen, U., Hansen, W.V.: 3D modeling of urban structures. Int. Arch. Photogrammetry Remote Sens. **36**(Part 3), W24 (2005)
11. Lafarge, F., Mallet, C.: Creating large-scale city models from 3D-point clouds: a robust approach with hybrid representation. Int. J. Comput. Vis. **99**(1), 69–85 (2012)
12. Moreira, A., Santos, M.Y.: Concave hull: a k-nearest neighbours approach for the computation of the region occupied by a set of points (2007)
13. Pohl, M., Feldmann, D.: Generating straight outlines of 2D point sets and holes using dominant directions or orthogonal projections. In: Proceedings of the 11th Joint Conference on Computer Vision, Imaging and Computer Graphics Theory and Applications (VISIGRAPP 2016). GRAPP, Rome, Italy, 27–29 February 2016, vol. 1, pp. 59–71 (2016)
14. Pradhan, R., Kumar, S., Agarwal, R., Pradhan, M.P., Ghose, M.: Contour line tracing algorithm for digital topographic maps. Int. J. Image Process. (IJIP) **4**(2), 156–163 (2010)
15. Rothermel, M., Wenzel, K., Fritsch, D., Haala, N.: Sure: photogrammetric surface reconstruction from imagery. In: Proceedings LC3D Workshop, Berlin, vol. 8 (2012)
16. Saalfeld, A.: Topologically consistent line simplification with the Douglas-Peucker algorithm. Cartography Geogr. Inf. Sci. **26**(1), 7–18 (1999)
17. Steger, C.: An unbiased detector of curvilinear structures. IEEE Trans. Pattern Anal. Mach. Intell. **20**(2), 113–125 (1998)
18. Wegner, J.D., Montoya-Zegarra, J.A., Schindler, K.: Road networks as collections of minimum cost paths. ISPRS J. Photogrammetry Remote Sens. **108**, 128–137 (2015)

Leaflet Free Edge Detection for the Automatic Analysis of Prosthetic Heart Valve Opening and Closing Motion Patterns from High Speed Video Recordings

Maryam Alizadeh, Melissa Cote, and Alexandra Branzan Albu$^{(\boxtimes)}$

University of Victoria, Victoria, BC, Canada
{alizadeh,mcote,aalbu}@uvic.ca

Abstract. Prosthetic heart valves (PHVs) are routinely used in clinical settings to replace defective native heart valves in patients suffering from valvular heart disease. Although PHV designs must be rigorously tested using cardiovascular testing equipment to ensure their optimal characteristics and safe operation, visual data obtained during simulations are typically assessed manually, a tedious and error-prone task. The valve orifice area over time, which informs on the opening and closing motion patterns, constitutes a key quality metric for PHV assessment. In addition to the very fast motion of the valve's leaflets, a major issue lies in the orifice being partly occluded by the leaflets' inner side or inaccurately depicted due to its transparency, which is not addressed in the literature. In this paper, we propose a novel orifice segmentation approach for automatic PHV quantitative performance analysis, based on the detection of the leaflet free edges to accurately extract the actual orifice area. Utilizing video frames recorded with a high speed digital camera during in vitro simulations, an initial estimation of the orifice area is first obtained via active contouring and then refined to capture the leaflet free edges via a curve extension scheme based on brightness and smoothness criteria. Evaluation on three different PHVs demonstrated the effectiveness of our approach to detect valve leaflet free edges and extract the actual orifice area, significantly outperforming a baseline algorithm both in terms of valve design evaluation metrics and computer vision evaluation metrics.

Keywords: Active contours · Biomedical image analysis · Leaflet edge detection · Motion pattern · Orifice area segmentation · Prosthetic Heart Valves · Video analysis

1 Introduction

1.1 Context

More than 5 million people are diagnosed with valvular heart disease each year in the United States alone [1]. Prosthetic heart valves (PHVs) are routinely used

© Springer International Publishing AG 2017
P. Sharma and F.M. Bianchi (Eds.): SCIA 2017, Part II, LNCS 10270, pp. 15–27, 2017.
DOI: 10.1007/978-3-319-59129-2_2

in clinical settings to replace defective native heart valves in patients. Compared to mechanical valves, bioprosthetic valves typically behave more similarly to native valves, but have shorter life cycles and are prone to calcification, an issue that can be accelerated by the valve's design and consequent increased stress on the leaflets [2]. Researchers are constantly working on new designs for more reliable and durable biological PHVs. PHV designs must be rigorously tested using cardiovascular testing equipment, to assess their performance and ensure their optimal characteristics. In addition to flow and pressure measurements, it is also common to manually assess the performance of the PHVs from visual data (images/videos) that are collected during simulations. This manual assessment is a tedious and error-prone task, with potentially devastating consequences.

In this paper, we are interested in the automatic, accurate, and efficient quantitative performance analysis of biological PHVs from video footage acquired during in vitro testing, using computer vision methods. A quantitative analysis of videos also allows one to study additional quality parameters, such as the valves' orifice area over time, their opening and closing motion patterns, and their leaflet kinematics. Figure 1 shows sample frames of the opening motion of a biological PHV, acquired during in vitro simulations. One difficulty, computer vision-wise, comes from the fast motion of the valve's leaflets during the opening and closing phases, which results in blurry edges. However, the major issue that we address in this paper lies in the orifice area being partly occluded by the inner side of the leaflets or inaccurately depicted due to its transparency.

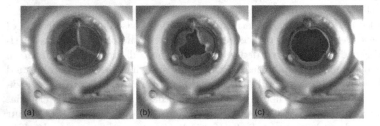

Fig. 1. Example frames from a test video, showing the opening motion of a tricuspid prosthetic valve: valve closed (a), partly open (b), and completely open (c).

1.2 Related Works

High-speed digital cameras offer an interesting medium for the quantitative evaluation of PHVs. Although they require direct optical access, unlike imaging modalities that are typically used for in vivo assessments [3], they are capable of recording fast movement such as the valve's leaflet opening and closing patterns with great detail. Earlier works in the literature utilizing high-speed cameras have typically relied on the addition of external mechanisms to help process the videos. Examples include particles added to the working fluid [4,5], manually placed markers on the leaflets [5–7], and laser light points projection

[8]. Advancements in the quality, availability, and affordability of high-speed digital cameras have allowed researchers to circumvent the need for these external mechanisms; recent works typically make use of computer vision and image processing methods, including digital kymography, thresholding, and deformable models, to focus exclusively on the valves' original pixel intensities.

A few works [9–11] have reported using digital kymography to analyze valve opening and closing procedures. While enabling a fast analysis of PHV recordings, digital kymograms, which consist of image lines projected along a time axis, focus on a very local region of the valve only.

Thresholding is arguably the simplest image segmentation method. Three variations have been proposed in [12] to segment the orifice area of trileaflet PHVs and assess the leaflet fluttering behavior: user-defined thresholding, Otsu's method, and finite mixture model-based thresholding. Thresholding typically fails due to the valve orifice not having a homogeneous gray-level representation.

Deformable models, such as active contours (snakes), have been utilized to address this issue of non-homogeneity of valve orifices. Wittenberg et al. [13] investigated valve movements via endoscopic high-speed recordings of native pig heart valves in an ex vivo (explanted heart) setting. They computed the orifice area over time using manually initialized snakes. Local constraints on three nodes, corresponding to the commissions between the valve leaflets, were also marked manually. Kondruweit et al. [11], in a similar setup, proposed a combination of digital kymograms calculated at different angles and snakes (based on [13]), to analyze the effective orifice area over time. Both kymogram lines and snakes were manually initialized. In an effort to automate the PHV assessment, Condurache et al. [14] proposed a method to segment the orifice area during in vitro mechanical simulations and compared automatic thresholding, similar to that of [12], with snakes. Their snake implementation included anchor-points-based attractors, automatically added to the frames by changing the intensity of strategic pixels, to better fit the leaflet boundaries. In a following paper [15], the authors tackled the automatic analysis of leaflet fluttering, but incorrectly assumed that the detected orifice boundaries were in line with leaflet borders.

1.3   Contributions

Of all the reviewed related works, [15] presented the most interesting computer vision-based approach for analyzing valve opening and closing motion patterns in an automated way. However, one of the difficulties that they failed to address was the fact that part of the orifice may be occluded by the valve's leaflets, or inaccurately depicted due to visible inner valve regions, making their method incapable of fully detecting and tracking the actual free edges of the leaflets. Figure 2 illustrates this last issue, showing two troublesome cases where part of the inner side of the leaflets is visible. In this paper, we address these issues by developing an automated method for tracking free edges separately, and not simply extracting them from the boundaries of the segmented orifice area.

Our contributions are two-fold. From a theoretical viewpoint, we propose a 2D curve extension scheme based on brightness and smoothness criteria that allows us to recover the true shape of the curve when deformable models typically fail due to visible inner regions of the underlying object. From a practical viewpoint, we apply this curve extension scheme to the problem of detecting and tracking the free edges of PHV leaflets and propose a novel approach for the automatic analysis of prosthetic heart valve opening and closing motion patterns from high-speed video recordings.

The paper is structured as follows. Section 2 details our approach, Sect. 3 discusses its experimental evaluation, and Sect. 4 presents concluding remarks.

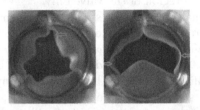

Fig. 2. Example frames in which the orifice area is partially visually bordered by the inner side of the leaflets (blue arrows) instead of the free edges of the leaflets. (Color figure online)

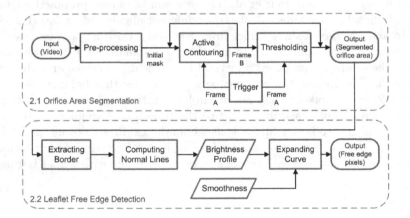

Fig. 3. Flow chart of the proposed method.

2   Proposed Method

Two main algorithms are proposed to extract information from the simulation videos that can be used for the quantitative assessment of PHV designs (Fig. 3):

1. segmentation of the orifice area, which utilizes active contours;
2. detection of the leaflet free edges, which allows us to refine the shape of the orifice and address cases for which the orifice is partially bordered by the inner side of the valve's leaflets (Fig. 2).

The extracted information can be used to determine several quality parameters, such as how fast the valve opens and closes, how long the valve stays open and its maximum opening, which constitute crucial data for assessing PHV designs. Each algorithm is further detailed in the remainder of this section.

2.1 Orifice Area Segmentation

Figure 3 (top) presents an overview of the orifice area segmentation algorithm. First, pre-processing is carried out to determine the region of interest (ROI) inside the frames and obtain an initial approximation (mask) of the orifice area. The orifice area is then segmented either via active contouring, initialized using the mask, or via thresholding, depending on where the valve is in its opening and closing cycle (trigger). This first algorithm builds upon the method proposed in [15], but differs on several levels. Although the general idea of the pre-processing steps is similar, the way to select the final ROI and the construction of the mask differ. The trigger part was also absent from [15].

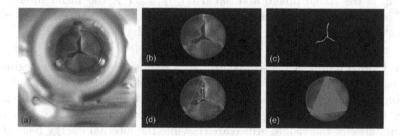

Fig. 4. Pre-processing: original image (a), circular outline of the valve (ROI) (b), thinned opening area (c), centerlines (in green) and corresponding anchor points (in red, on circular outline) (d), and triangular initial mask M_0 for active contouring (e). (Color figure online)

Fig. 5. Triggering active contouring: observation of the mean gray level value inside the maximal circle fitted inside the initial triangular mask, at different frames of the opening phase. The last circle on the right corresponds to Frame A.

Pre-processing. Knowing the actual radius of the imaged PHV, an ROI that corresponds to the inside of the valve's circular outline is first determined via the circular Hough transform [16] applied on a Canny edge map of the frame (Fig. 4a–b). A frame showing a small valve opening (akin to a star shape) is preferred. Then, an initial mask M_0 for active contouring is determined from the anchor/commission points as follows. The opening area is segmented via thresholding, then dilated and thinned to remove any artifacts on the area borders (Fig. 4c). The linear Hough transform is then applied on the thinned segmented area to find the three centerlines, at approximately 120° from each other (Fig. 4d). M_0 (Fig. 4e) is derived as the triangle connecting three anchor points (red 'x' on the circular outline in Fig. 4d) corresponding to the intersection of the centerlines with the circular outline. The anchor points are found by extending the lines from their end point closest to the circular outline.

Trigger. The trigger part determines when to activate (and deactivate) active contouring, i.e. when the valve is open enough for the active contour to work properly. The maximal circle $MaxC$ that can be fitted inside M_0 is used for observing the gray values of the orifice area. Figure 5 depicts how the orifice area is observed at different frames in the opening phase to determine the first frame to be segmented via active contouring (Frame A) as follows:

$$|\bar{u}_0(MaxC)_{FrameA} - \bar{u}_0(MaxC)_{FrameA-1}| > 0.1\bar{u}_0(MaxC)_{FrameA-1} \quad (1)$$

where u_0 is the frame image and $\bar{u}_0(MaxC)_{FrameA}$ is the mean intensity of $MaxC$ in Frame A. The orifice area in Frame A is used as a threshold value for deactivating active contouring at Frame B during the closing phase.

Active Contouring. An active contour is a parametric curve. The segmentation is carried out as a minimization of the energy of this curve, which is defined as a combination of internal and external energy terms. The external energy coming from the image data pushes the active contour to fit the data (here the boundaries of the visible orifice area), while the internal energy restricts its deformation to ensure its smoothness. We used the Chan-Vese model [17], which tries to segment the image based on intensities as opposed to edges, according to the following energy functional for curve C (2) and minimization problem (3):

$$F(c_1, c_2, C) = \mu L(C) + \nu A(in(C)) + \lambda_1 \int_{in(C)} |u_0(x, y) - c_1|^2 dxdy$$
$$+ \lambda_2 \int_{out(C)} |u_0(x, y) - c_2|^2 dxdy \quad (2)$$

$$\inf_{c_1, c_2, C} F(c_1, c_2, C) \quad (3)$$

where u_0 is the frame image, c_1 and c_2 are respectively the averages inside (in) and outside (out) C, L and A are the length and area, and $\mu \geq 0$, $\nu \geq 0$, $\lambda_1 > 0$, and $\lambda_2 > 0$ are fixed parameters (typically $\lambda_1 = \lambda_2 = 1$, $\nu = 0$, and μ is selected according to the dataset, here $\mu = 1$). At Frame A, the active contour is initialized using the contour of M_0. The final contour in a given frame, evolved until equilibrium, is used as the initial approximation for the next frame. This procedure continues up to Frame B. The segmented orifice areas (OA) at each frame are obtained from the pixels inside the evolved curve C. They may be smaller than the actual orifice areas due to the positioning of the leaflets (Fig. 2), and will be enlarged via the leaflet free edge detection algorithm.

Thresholding. For the few frames before Frame A and after Frame B, global thresholding (typically 0.2 over a maximal value of 1) is used to obtain OA.

2.2   Leaflet Free Edge Detection

Figure 3 (bottom) presents an overview of the proposed leaflet free edge detection algorithm. The goal is to detect the leaflet free edges in every frame in the valve cycle, by extending the curve derived from the segmented orifice area OA (Sect. 2.1) that could have stopped short of the actual free edges, thus addressing the issue of the visible inner side of the leaflets bordering the orifice area. Starting from the border pixels of OA, the curve is expanded along normal lines using a combination of brightness and smoothness criteria, yielding the set of free edge pixels as output. Figure 6 illustrates the various steps.

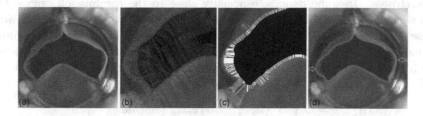

Fig. 6. Leaflet free edge detection: cropped region of initial frame (a), normal lines (blue) to each segmented orifice border pixels (red) and segmented orifice area (black) (b), expanded normal lines showing a jagged profile (white) when considering only the brightness criterion (c), and detected free edge pixels (red) when considering both brightness and smoothness criteria (d). Blue arrows in (d) point to key expansion locations; the actual free edges were successfully recovered. (Color figure online)

Extracting Border. The boundary pixels of OA, output by the first algorithm, are extracted to form the initial free edge border $FE = \{(x, y) \in u_0 \mid B_{OA}(x, y) = 1\}$, where B indicates the boundary. OAs are also superimposed on the original frames to change the corresponding intensities to 0 (Fig. 6b).

Computing Normal Lines. The next step is to compute a normal line to each border pixel. We find the orientation of the normal line at each border pixel $fe_i \in FE$ via the first derivative in the X and Y directions:

$$N_{orientation}(fe_i) = -dx/dy \qquad (4)$$

We then extend the normal lines locally from the border pixels inward and outward of OA, by 25 pixels (Fig. 6b, blue lines), thus obtaining two sets of pixels NFE_{in} and NFE_{out} for each border pixel fe_i. The length of the normal lines could also be determined dynamically as a percentage of the valve's radius.

Brightness Profile. Free edge pixels tend to be the brightest locally. We thus compute the intensity profile along each normal line, find the maximal value in the profile, then expand the orifice by incorporating the normal line going from the border pixel to the location of the maximum (Fig. 6c):

$$EOA_{MaxB} = OA \cup \{nfe_1, ..., nfe_{max} \mid u_0(max)$$
$$= \max_{j \in NFE_{in}, k \in NFE_{out}} (u_0(j), u_0(k))\} \quad (5)$$

where nfe_1 is the border pixel (first element of NFE_{in} and NFE_{out} for fe_i) and nfe_{max} is the pixel with the maximal value in either NFE_{in} or NFE_{out}. Morphological closing is then applied on EOA_{MaxB} to fill any gap between the added normal lines. A first set of free edge pixels $FE_{MaxB} = \{(x, y) \in u_0 \mid B_{EOA_{MaxB}}(x, y) = 1\}$ is derived, where B indicates the boundary.

Smoothness. In some regions, especially close to the commission points, the pixel with the maximal brightness may not correspond to a free edge and cause a false detection (Fig. 6c, lower left region). The smoothness of the final contour obtained from EOA is thus also considered. A second set of free edge pixels FE_{LocB} is found by selecting the local (instead of absolute) maximal brightness pixels in the normal line profiles that are closer to the segmented orifice area:

$$EOA_{LocB} = OA \cup \{nfe_1, ..., nfe_{max} \mid u_0(max) = peak,$$
$$dist(max, 1) = \min_{j \in Peaks} (dist(j, 1))\} \quad (6)$$

$$FE_{LocB} = \{(x, y) \in u_0 \mid B_{EOA_{LocB}}(x, y) = 1\} \quad (7)$$

where $Peaks$ represents the set of pixels along the normal lines NFE_{in} and NFE_{out} for which the intensity corresponds to a $peak$ value.

Expanding Curve. For each original border pixel, the two sets FE_{MaxB} and FE_{LocB} are compared; the corresponding pixel in either FE_{MaxB} or FE_{LocB} providing the smoother final contour, i.e. the smallest distance with the neighboring border pixel, is selected as the final free edge pixel:

$$FE_{Final} = \{\{fe_{i+1} \in FE_{MaxB} \mid dist(fe_{i+1}, fef_i) < dist(fe_{j+1}, fef_i)\}$$
$$\cup \{fe_{j+1} \in FE_{LocB} \mid dist(fe_{j+1}, fef_i) < dist(fe_{i+1}, fef_i)\}\} \quad (8)$$

where fef_i is already in FE_{Final}. The final expanded curve that represents the leaflet free edges is the collection of all final candidate free edge pixels (Fig. 6d).

3   Experimental Results

3.1   Experimental Setup and Video Dataset

As there are no public PHV video datasets available, we created our own with an experimental setup using tricuspid biological PHVs submitted to the Pulse

Duplicator system from ViVitro Labs Inc., the world's most widely cited and used in vitro cardiovascular hydrodynamic testing system [18]. The Pulse Duplicator simulates one half of the human heart and generates a pulsatile flow through the mounted PHV. Flow and pressure sensors, located on each side of the valve, measure and collect the flow data during each simulation.

Videos of the simulations were recorded with a Photron SA3 high speed digital camera with a 1000 fps frame rate, a 1/1000 s shutter speed, and a 300 frame test cycle, i.e. from the moment the valve opens until it closes completely, with resolutions between 400×400 and 1024×1024 pixels depending on the valve. Sample frames are shown in Fig. 1. The dataset comprises three videos taken from three different PHVs, for a total of 1745 frames. Ground truth data, defining the leaflet free edges, were obtained in a semi-automatic fashion.

3.2 Evaluation

We compare our proposed extension scheme for the detection of the leaflet free edges (Sect. 2.2) to the active contour-based orifice area segmentation (Sect. 2.1, based on the work in [15]), the latter acting as a baseline. We first compare the PHVs' orifice curves, which constitute key data for PHV performance evaluation. Figure 7 illustrates the orifice area over time for the three PHVs. For our proposed method, the orifice area is computed from the region bordered by FE_{Final}. The results are in very good agreement with the ground truth, both in terms of trend and magnitude, and show a significant improvement over the baseline, which systematically falls short of the actual orifice area. Table 1, which shows the root-mean-square error (RMSE) values for the three valves computed from the differences with the ground truth, confirms the accuracy of our proposed method versus that of the baseline. The information extracted from our proposed method can be used, for instance, to determine how fast the valve opens and closes, how long its stays open, as well as its maximum opening, which constitute crucial data for PHV design assessment.

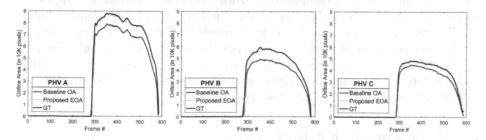

Fig. 7. Comparison of the segmented orifice areas over time for the baseline and proposed approaches with the ground truth data (GT).

Although the orifice curves are relevant from a valve evaluation perspective, they do not allow to properly evaluate the orifice shape accuracy. Therefore, we

Table 1. RMSE values of the segmented orifice areas for the baseline and proposed approaches, both absolute (in pixels) and relative (to the maximal orifice area).

Method	PHV A	PHV B	PHV C
Baseline OA	6996 (8.0%)	6242 (10.5%)	3215 (6.7%)
Proposed EOA	2409 (2.7%)	1898 (3.2%)	972 (2.0%)

Table 2. Average Hausdorff distance d_H ($\pm$ standard deviation) in pixels between the ground truth and the orifice boundaries for the baseline and proposed approaches.

Method	PHV A	PHV B	PHV C
Baseline OA	19.0 ± 4.3	32.8 ± 9.9	14.6 ± 6.0
Proposed EOA	8.7 ± 3.7	14.0 ± 12.2	11.5 ± 6.6

also evaluate our approach with a boundary matching metric, i.e. the Hausdorff distance, defined between curves A and B as follows [19]:

$$d_H(A, B) = \max \left(\max_{i}\{d(a_i, B)\}, \max_{j}\{d(b_j, A)\}\right) \tag{9}$$

$$d(a_i, B) = \min_{j}\|b_j - a_i\| \tag{10}$$

where a_i is a point on A and b_j a point on B. Table 2 shows the average d_H over all frames between the ground truth and the orifice boundaries extracted by the baseline and the proposed approaches, for the three PHVs. The smaller distances obtained for our method confirm its ability to recover the shape of the actual orifice area more accurately then the baseline. On a frame-by-frame basis, the largest distances typically occur when the valve is almost closed.

Figure 8 shows typical examples of the orifice boundaries extracted by the baseline and proposed approaches at various moments in the cycle, along with d_H values. The first three columns constitute good cases, in which our approach successfully tracked the leaflet free edges, whereas the baseline approach failed locally either due to the leaflet's visible inner side (2nd and 3rd columns), or due to the presence of small attached pieces (see 1st column, left region). The last column constitutes a bad case in which our approach yielded small errors due to local brightness maxima, typically occurring near the commission points. A stricter smoothness criterion could address this issue.

3.3   Comparison with Attractors

We also compare our free edge detection approach to that based on attractors [15]. In [15], artificial edges are added to the circular outline of the valve on both sides of the commission points to improve the active contour segmentation and detect the actual orifice area. However, this approach only works in cases where the free edges are located on the circular outline, which only covers a special case

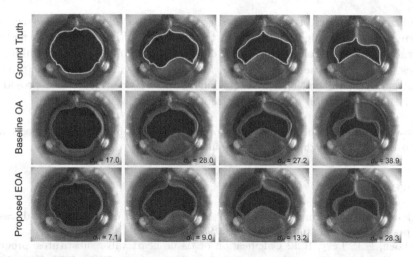

Fig. 8. Typical results and Hausdorff distances d_H, at different moments in the cycle, for the baseline and proposed approaches, along with the ground truth, for PHV A.

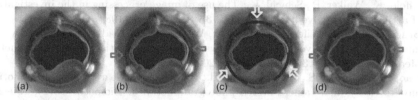

Fig. 9. Comparison with attractors (yellow arrows): original (cropped) frame (a), results (red) of [15] without (b) and with (c) attractors and of our proposed method (d). Blue arrows indicate problematic areas, where attractors fail and our method succeeds. (Color figure online)

of valves and motion. Figure 9 shows that the method in [15], contrary to ours, is unable to detect the free edges when the valve's motion is such that leaflet inner sides are visible and the free edges are away from the circular outline.

4   Conclusion

This paper presents an automatic approach for the quantitative performance analysis of bioprosthetic heart valves from high speed videos recorded during in vitro simulations of the opening and closing motions. The proposed approach addresses the issue of the valve orifice being partially bordered by the inner side of its leaflets, by tracking the actual leaflet free edges. An initial estimation of the orifice area is first obtained via active contouring and then refined via a curve extension scheme, allowing the method to successfully segment the orifice area for valve design evaluation purposes. Evaluation on videos of three valves demonstrated our approach's effectiveness to detect the leaflet free edges and thus obtain the actual orifice area during a complete cycle. Our approach out-performed a baseline algorithm both in terms of valve design evaluation metrics (accuracy of the temporal orifice area curves) and computer vision evaluation metrics (accuracy of the orifice shape). Future works will explore a setup using several cameras, mitigating the occlusion of the valve orifice while providing more visual information on the leaflet motion pattern.

Acknowledgments. This work was supported in part by NSERC Canada and ViVitro Labs Inc. through the Engage Grants Program.

References

1. Nkomo, V.T., Gardin, J.M., Skelton, T.N., et al.: Burden of valvular heart diseases: a population-based study. Lancet **368**(9540), 1005–1011 (2006)
2. Schoen, F.J., Levy, R.J.: Calcification of tissue heart valve substitutes: progress toward understanding and prevention. Ann. Thorac. Surg. **79**(3), 1072–1080 (2005)
3. Sucha, D., Symersky, P., Tanis, W., et al.: Multimodality imaging assessment of prosthetic heart valves. Circ. Cardiovasc. Imaging **8**(9), e003703 (2015)
4. Affeld, K., Walker, P., Schichl, K.: The use of image processing in the investigation of artificial heart valve flow. ASAIO J. **35**(3), 294–297 (1989)
5. Bruecker, C., Steinseifer, U., Schroeder, W., et al.: Unsteady flow through a new mechanical heart valve prosthesis analysed by digital particle image velocimetry. Meas. Sci. Technol. **13**(7), 1043–1049 (2002)
6. Gao, Z.B., Pandya, S., Hosein, N., et al.: Bioprosthetic heart valve leaflet motion monitored by dual camera stereo photogrammetry. J. Biomech. **33**(2), 199–207 (2000)
7. Lu, P.-C., Liu, J.-S., Huang, R.-H., et al.: The closing behavior of mechanical aortic heart valve prostheses. ASAIO J. **50**(4), 294–300 (2004)
8. Iyengar, A.K.S., Sugimoto, H., Smith, D.B., et al.: Dynamic in vitro quantification of bioprosthetic heart valve leaflet motion using structured light projection. Ann. Biomed. Eng. **29**(11), 963–973 (2001)
9. Friedl, S., Koenig, S., Kondruweit, M., et al.: Digital kymography for the analysis of the opening and closure intervals of heart valves. In: Handels, H., Ehrhardt, J., Deserno, T.M., Meinzer, H.-P., Tolxdorff, T. (eds.) Bildverarbeitung fur die Medizin 2011. Informatik aktuell, pp. 144–148. Springer, Heidelberg (2011)
10. Kondruweit, M., Friedl, S., Wittenberg, T., et al.: Description of a novel ex-vivo imaging and investigation technique to record, analyze and visualize heart valve motions under physiological conditions. Heart Lung Circ. **19**(Supp 2), S174 (2010)
11. Kondruweit, M., Friedl, S., Heim, C., et al.: A new ex vivo beating heart model to investigate the application of heart valve performance tools with a high-speed camera. ASAIO J. **60**(1), 38–43 (2014)
12. Hahn, T., Condurache, A.P., Aach, T., et al.: Automatic in-vitro orifice area determination and fluttering analysis for tricuspid heart valves. In: Handels, H., Ehrhardt, J., Horsch, A., Meinzer, H.P., Tolxdorff, T. (eds.) Bildverarbeitung fur die Medizin 2006, pp. 21–25. Springer, Heidelberg (2006)
13. Wittenberg, T., Cesnjevar, R., Rupp, S., et al.: High-speed-camera recordings and image sequence analysis of moving heart-valves: experiments and first results. In: Buzug, T.M., Holz, D., Bongartz, J., et al. (eds.) Advances in Medical Engineering, pp. 169–174. Springer, Berlin (2007)

14. Condurache, A.P., Hahn, T., Hofmann, U.G., et al.: Automatic measuring of quality criteria for heart valves. In: Pluim, J.P.W., Reinhardt, J.M. (eds.) Medical Imaging 2007: Image Processing. Proceedings SPIE, vol. 6512, p. 65122Q1–1. SPIE (2007)

15. Condurache, A.P., Hahn, T., Scharfschwerdt, M., et al.: Video-based measuring of quality parameters for tricuspid xenograft heart valve implants. IEEE Trans. Biomed. Eng. **56**(12), 2868–2878 (2009)

16. Atherton, T.J., Kerbyson, D.J.: Size invariant circle detection. Image Vis. Comput. **17**(11), 795–803 (1999)

17. Chan, T.F., Vese, L.A.: Active contours without edges. IEEE Trans. Image Process. **10**(2), 266–277 (2001)

18. Information about our pulse duplicator system. http://vivitrolabs.com/product/pulse-duplicator/

19. Chalana, V., Kim, Y.: A methodology for evaluation of boundary detection algorithms on medical images. IEEE Trans. Med. Imaging **16**(5), 642–652 (1997)

Max-Margin Learning of Deep Structured Models for Semantic Segmentation

Måns Larsson[1]([⊠]), Jennifer Alvén[1], and Fredrik Kahl[1,2]

[1] Chalmers University of Technology, Gothenburg, Sweden
{mans.larsson,alven,fredrik.kahl}@chalmers.se
[2] Centre for Mathematical Sciences, Lund University, Lund, Sweden

Abstract. During the last few years most work done on the task of image segmentation has been focused on deep learning and Convolutional Neural Networks (CNNs) in particular. CNNs are powerful for modeling complex connections between input and output data but lack the ability to directly model dependent output structures, for instance, enforcing properties such as smoothness and coherence. This drawback motivates the use of Conditional Random Fields (CRFs), widely applied as a post-processing step in semantic segmentation.

In this paper, we propose a learning framework that jointly trains the parameters of a CNN paired with a CRF. For this, we develop theoretical tools making it possible to optimize a max-margin objective with back-propagation. The max-margin loss function gives the model good generalization capabilities. Thus, the method is especially suitable for applications where labelled data is limited, for example, medical applications. This generalization capability is reflected in our results where we are able to show good performance on two relatively small medical datasets. The method is also evaluated on a public benchmark (frequently used for semantic segmentation) yielding results competitive to state-of-the-art. Overall, we demonstrate that end-to-end max-margin training is preferred over piecewise training when combining a CNN with a CRF.

Keywords: Segmentation · Convolutional Neural Networks · Markov random fields

1 Introduction

Convolutional Neural Networks (CNNs) have, during the last few years, been used with great success on a variety of computer vision problems such as image classification [12] and object detection [8]. The capability of CNNs to learn high-level abstraction of data makes them well suited for the task of image classification. Following this development, there have been several successful attempts to extend CNN based methods to tasks done on the pixel level such as semantic segmentation [9,16,17].

Electronic supplementary material The online version of this chapter (doi:10.1007/978-3-319-59129-2_3) contains supplementary material, which is available to authorized users.

P. Sharma and F.M. Bianchi (Eds.): SCIA 2017, Part II, LNCS 10270, pp. 28–40, 2017.
DOI: 10.1007/978-3-319-59129-2_3

A drawback of CNNs is that they do not have the ability to directly model statistical dependencies of output variables. Hence they cannot explicitly enforce smoothness constraints or encourage spatial consistency of the output, something that arguably is important for the task of semantic segmentation. To deal with this a Markov Random Field (MRF), or its variant Conditional Random Field (CRF), can be used as a refinement step. This was done by Chen *et al.* in [4] where they used CNNs to form the unary potential of the dense CRF model presented by Krähenbühl *et al.* in [11]. However, the CNN and the CRF models are trained separately in [4] meaning that the parameters of the CRF are learnt while holding the CNN weights fixed. In other words, the deep features are learnt disregarding statistical dependencies of the output variables. In reaction to this, several approaches for jointly training deep structured models, combining CNNs and CRFs, have recently been proposed [5,14,15,24,28]. In these approaches, as well as the one presented in this paper, the parameters of the CRF and the weights of the CNN can be trained jointly, enabling the possibility to learn deep image features taking dependencies of the output variables into account.

1.1 Contributions

What differentiates this paper from previous work done on learning deep structured models is mainly the joint learning algorithm. We apply a max-margin based learning framework inspired by [23]. This removes the need to calculate, or approximate, the partition function present in learning algorithms that try to maximize the log-likelihood. For instance, in [14,28], the inference step is approximately solved using a few iterations of the mean-field algorithm or gradient descent, respectively. Similarly, in [5], sampling techniques are used to approximate the partition function. In our learning framework, we can use standard graph cut methods to perform optimal inference in the CRF model. We also show how the CNN weights can be trained to optimize the max-margin criterion via standard back-propagation. To our knowledge, we are the first to present a method for jointly training deep structured models with a max-margin objective for semantic segmentation.

Our experiments show that training deep structured models using our method gives better results than piecewise training where the CNN and CRF models are trained separately. This proves that training deep structured models jointly enables the model to learn deep features that take output dependencies into account which in turn gives better segmentations. We tested our method on the Weizmann Horse dataset [2] for proof of concept. In addition we applied it to two medical datasets, one for heart ventricle segmentation in ultrasound images and one for pericardium segmentation in CTA slices.

1.2 Related Work

The concept of deep structured models has been examined extensively in recent work. In [20] Ning *et al.* combine a CNN with an energy based model, similar to a MRF, for segmentation of cell nuclei and in [4] a dense CRF with unary potentials from a CNN is used to achieve state-of-the-art results on several semantic segmentation benchmarks.

Methods for jointly training these deep structured models have also received a lot of attention lately. In [24] Tompson *et al.* present a single learning framework unifying a novel ConvNet Part-Detector and an MRF inspired Spatial-Model achieving state-of-the-art performance on the task of human body pose recognition. Further, Chen *et al.* present a more general framework for joint learning of deep structured models that they apply to image tagging and word from image problems in [5]. Zheng *et al.* [28] show that the mean-field inference algorithm with Gaussian pairwise potentials from [11] can be modeled as a Recurrent Neural Networks. This enabled them to train their model within a standard deep learning framework using a log-likelihood loss. In [15], they formulated a CRF model with CNNs for estimating the unary and pairwise potentials via piecewise training.

In the field of medical image analysis, methods based on CNNs have also received an increased interest during the last few years with promising results [6,19,22]. Recently, more intricate deep learning approaches have been proposed. Ronneberger *et al.* [21] proposed the U-Net, a network based on the idea of "fully convolutional networks" [16]. A similar network structure was also proposed by Brosch *et al.* in [3]. However, to our knowledge, methods utilizing end-to-end training of deep structured models have yet to be presented for medical image segmentation tasks.

2   A Deep Conditional Random Field Model

The deep structured model proposed in this paper consists of a CNN coupled with a CRF. This setup allows the model to learn deep features while still taking dependencies in the output data into account. Denote the set of input instances by $X = \{x^{(n)}\}_n$ and their corresponding labelings by $Y = \{y^{(n)}\}_n$. The input and output instances are images indexed for each pixel by $x^{(n)} = (x_1^{(n)}, \ldots, x_N^{(n)})$ and $y^{(n)} = (y_1^{(n)}, \ldots, y_N^{(n)})$ respectively. We only consider the binary labeling case, hence $y_i^{(n)} = \{0, 1\}$. Our deep structured model is described by a CRF of the form

$$P(Y|X; w, \theta) = \frac{1}{Z} e^{-\sum_n E(y^{(n)}, x^{(n)}; w, \theta)}, \tag{1}$$

where w are the weights of the CRF, θ are the weights of the CNN and Z is the partition function. The energy E considered decomposes over unary and pairwise terms according to the following form

$$E(y, x; w, \theta) = \sum_{i \in \mathcal{V}} \phi_i(y_i, x; w, \theta) + \sum_{(i,j) \in \mathcal{E}} \phi_{ij}(y_i, y_j, x; w), \tag{2}$$

where $\mathcal{V}$ is the set of nodes (*i.e.* pixels) and $\mathcal{E}$ is the set of edges connecting neighbouring pixels.

The unary term of the energy E has the following form

$$\phi_i(y_i, \boldsymbol{x}; \boldsymbol{w}, \boldsymbol{\theta}) = w_1 \log(\Phi_i(y_i, \boldsymbol{x}; \boldsymbol{\theta})), \tag{3}$$

where $\Phi_i(y_i, \boldsymbol{x}; \boldsymbol{\theta})$ denotes the output of the neural network for pixel i. There are no explicit requirements for the CNN except that it should output an estimate of the probability for each pixel being either foreground or background.

The pairwise term consists of two parts both penalizing two neighbouring pixels being labeled differently. The first part adds a constant cost while the other one adds a cost based on the contrast of the neighbouring pixels. If $\mathbb{1}_{y_i \neq y_j}$ denotes the indicator function equaling one if $y_i \neq y_j$, the pairwise term has the following form

$$\phi_{ij}(y_i, y_j, \boldsymbol{x}; \boldsymbol{w}) = \mathbb{1}_{y_i \neq y_j} \left(w_2 + w_3 \, e^{-\frac{(x_i - x_j)^2}{2}} \right). \tag{4}$$

Note that, given these unary and pairwise terms, the energy is linear with respect to the weights $\boldsymbol{w}$.

2.1   Inference

Given an input instance $\boldsymbol{x}$, the inference problem equates to finding the maximum a posteriori labeling $\boldsymbol{y}^*$ given the model in (1). This is equivalent to finding a minimizer of the energy E in (2):

$$\boldsymbol{y}^* = \arg \min_{\boldsymbol{y}} E(\boldsymbol{y}, \boldsymbol{x}; \boldsymbol{w}, \boldsymbol{\theta}). \tag{5}$$

For our deep structured model the inference is done in two steps. Firstly, an estimation of the probability of each pixel being either foreground or background is computed by a forward pass of the CNN. Secondly, problem (5) is solved. We add the constraints $w_i \geq 0$, $i = 2, 3$ when learning the weights to make the energy E submodular. This means that graph cut algorithm can be used to efficiently find a global optimum [10].

2.2   Max-Margin Learning

There are two sets of learnable parameters, the weights of the CRF $\boldsymbol{w}$ and the weights of the CNN $\boldsymbol{\theta}$. The method of learning is based on an algorithm proposed by Szummer et al. [23] where the goal is to find a set of parameters $\boldsymbol{w}, \boldsymbol{\theta}$ such that

$$E(\boldsymbol{y}^{(n)}, \boldsymbol{x}^{(n)}; \boldsymbol{w}, \boldsymbol{\theta}) \leq E(\boldsymbol{y}, \boldsymbol{x}^{(n)}; \boldsymbol{w}, \boldsymbol{\theta}) \quad \forall \boldsymbol{y} \neq \boldsymbol{y}^{(n)}, \tag{6}$$

i.e. we want to learn a set of weights that assign the ground truth labeling an equal or lower energy than any other labeling. Since this problem might have multiple or no solutions we introduce a margin ζ and try to maximize it according to

$$\max_{\boldsymbol{w}:|\boldsymbol{w}|=1} \zeta$$
$$s.t. \quad E(\boldsymbol{y}, \boldsymbol{x}^{(n)}; \boldsymbol{w}, \boldsymbol{\theta}) - E(\boldsymbol{y}^{(n)}, \boldsymbol{x}^{(n)}; \boldsymbol{w}, \boldsymbol{\theta}) \geq \zeta \quad \forall \boldsymbol{y} \neq \boldsymbol{y}^{(n)}. \tag{7}$$

Finding the set of parameters that provides the largest margin regularizes the problem and tends to give good generalization to unseen data. However, for the final objective we make a few changes suggested by Szummer *et al.* [23]. To start of, a slack variable for each training sample ξ_n is introduced to make the method more robust to noisy data. In addition, we use a rescaled margin, demanding a larger energy margin for labelings that differ a lot from the ground truth. Also, the program described in (7) includes an exponential amount of constraints which makes solving it intractable, we therefore perform the optimization over a much smaller set $S^{(n)}$. These changes, given the variable transformation $||\boldsymbol{w}|| \leftarrow 1/\zeta$, give rise to the following problem

$$\gamma = \min_{\boldsymbol{w}} \frac{1}{2}\|\boldsymbol{w}\|^2 + \frac{C}{N}\sum_n \xi_n \quad s.t. \quad \forall \boldsymbol{y} \in S^{(n)} \; \forall n$$

$$E(\boldsymbol{y}, \boldsymbol{x}^{(n)}; \boldsymbol{w}, \boldsymbol{\theta}) - E(\boldsymbol{y}^{(n)}, \boldsymbol{x}^{(n)}; \boldsymbol{w}, \boldsymbol{\theta}) \geq \Delta(\boldsymbol{y}^{(n)}, \boldsymbol{y}) - \xi_n \tag{8}$$

$$\xi_n \geq 0, w_2 \geq 0, w_3 \geq 0,$$

where N is the number of training samples, C is a hyperparameter regulating the slack penalty and $\Delta(\boldsymbol{y}^{(n)}, \boldsymbol{y})$ is the Hamming loss $\Delta(\boldsymbol{y}^{(n)}, \boldsymbol{y}) = \sum_i \delta(y_i^{(n)}, y_i)$.

The constraint set $S^{(n)}$ is iteratively grown by adding labelings that violate the constraints in (8) the most. For each iteration, the weights are then updated to satisfy the new, larger constraint set. This weight update is repeated until the weights no longer change. The complete learning algorithm is summarized in Algorithm 1.

Input: image-labeling pairs $\{(\boldsymbol{x}^{(n)}, \boldsymbol{y}^{(n)})\}$ in the training set
initialize $S^{(n)} = \emptyset$ for each training instance n and $\boldsymbol{w} = \boldsymbol{w}_0$
while $\boldsymbol{w}$ *not converged* **do**
    for *all training instances n* **do**
        find MAP labeling of instance n:
        $\boldsymbol{y}^* \leftarrow \arg\min_{\boldsymbol{y}} E(\boldsymbol{y}, \boldsymbol{x}^{(n)}; \boldsymbol{w}, \boldsymbol{\theta}) - \Delta(\boldsymbol{y}^{(n)}, \boldsymbol{y})$
        if $\boldsymbol{y}^* \neq \boldsymbol{y}^{(n)}$ **then**
          | add $\boldsymbol{y}^*$ to constraint set: $S^{(n)} \leftarrow S^{(n)} \cup \{\boldsymbol{y}^*\}$
        end
        update $\boldsymbol{w}$ to ensure ground truth has the lowest energy by solving
          program (8)
    end
end
Output: $\boldsymbol{w}$

Algorithm 1. Pseudocode for the CRF weight learning algorithm from [23].

2.3 Back-Propagation of Error Derivatives

In this section, we show how the max-margin objective from the previous section can be optimized for our coupled CNN and CRF model. Our main goal during learning is to maximize the margin, or equivalently, minimize the objective γ as defined in (8). To be able to perform a gradient based weight update we need to calculate the derivative of this objective with respect to the weights of the network

$$\frac{\partial \gamma}{\partial \theta_j} = \sum_n \sum_i \frac{\partial \gamma}{\partial \Phi_i} \frac{\partial \Phi_i}{\partial \theta_j}, \tag{9}$$

where the two sums are over the training instances, n, and the pixels, i. As previously, Φ_i is the output of the network. Given a well-defined network structure the term $\frac{\partial \Phi_i}{\partial \theta_j}$ can be easily calculated using standard back-propagation. Henceforth we will focus on calculating the term $\frac{\partial \gamma}{\partial \Phi_i}$. To simplify notation we will introduce z_i as the output of the network of pixel i being foreground, $z_i = \Phi_i(y_i = 1, \boldsymbol{x}; \boldsymbol{\theta})$. We start of by expressing (8) on the following compact form

$$\gamma(\boldsymbol{z}) = \min_{\boldsymbol{w}, \boldsymbol{\xi}} f(\boldsymbol{w}, \boldsymbol{\xi}),$$
$$\text{s.t.} \quad h_k(\boldsymbol{w}, \boldsymbol{\xi}, \boldsymbol{z}) \leq 0, \quad k = 1, \ldots, M, \tag{10}$$

where f is the objective function, h_k characterize the constraints and M is the total number of constraints. We will treat γ as a function depending on the network output, $\gamma(\boldsymbol{z})$.

In addition, the minimizers $\boldsymbol{w}^*$ and $\boldsymbol{\xi}^*$ can also be seen as functions of $\boldsymbol{z}$, that is, $\boldsymbol{w}^* = \boldsymbol{w}^*(\boldsymbol{z})$ and $\boldsymbol{\xi}^* = \boldsymbol{\xi}^*(\boldsymbol{z})$, which gives that

$$\gamma(\boldsymbol{z}) = f(\boldsymbol{w}^*(\boldsymbol{z}), \boldsymbol{\xi}^*(\boldsymbol{z})) = \frac{1}{2}\|\boldsymbol{w}^*\|^2 + \frac{C}{N}\sum_{n=1}^N \xi_n^*,$$

$$\frac{\partial \gamma}{\partial z_i} = \sum_{j=1}^D f_{w_j} \frac{\partial w_j}{\partial z_i} + \sum_{n=1}^N f_{\xi_n} \frac{\partial \xi_n}{\partial z_i} = \sum_{j=1}^D w_j \frac{\partial w_j}{\partial z_i} + \frac{C}{N}\sum_{n=1}^N \frac{\partial \xi_n}{\partial z_i}, \tag{11}$$

where D is the number of weights and N is the number of slack variables. To be able to calculate $\frac{\partial \gamma}{\partial z_i}$ we need $\frac{\partial w_j}{\partial z_i}$ and $\frac{\partial \xi_n}{\partial z_i}$. These derivatives are found by creating and solving a system of equations from the optimality conditions of the problem. The Lagrangian for the constrained minimization problem in (10) is

$$L(\boldsymbol{w}, \boldsymbol{\xi}, \boldsymbol{\lambda}) = f(\boldsymbol{w}, \boldsymbol{\xi}) + \sum_{k=1}^M \lambda_k h_k(\boldsymbol{w}, \boldsymbol{\xi}),$$

where $\boldsymbol{\lambda}$ is the vector of Langrangian multipliers with elements λ_k. At optimum, the first-order optimality conditions are satisfied:

$$\nabla_{\boldsymbol{w}} L = \boldsymbol{w} + \sum_{k=1}^M \lambda_k \nabla_{\boldsymbol{w}} h_k = \boldsymbol{0} \text{ and } \nabla_{\boldsymbol{\xi}} L = \frac{C}{N} + \sum_{k=1}^M \lambda_k \nabla_{\boldsymbol{\xi}} h_k = \boldsymbol{0}. \tag{12}$$

Now, the conditions for the implicit function theorem are satisfied and we also get that

$$\frac{\partial(\nabla_{\boldsymbol{w}}L)}{\partial z_i} = \frac{\partial \boldsymbol{w}}{\partial z_i} + \sum_{k=1}^{M}\left(\frac{\partial \lambda_k}{\partial z_i}\nabla_{\boldsymbol{w}}h_k + \lambda_k \frac{\partial \nabla_{\boldsymbol{w}}h_k}{\partial z_i}\right) = \boldsymbol{0}, \tag{13}$$

$$\frac{\partial(\nabla_{\boldsymbol{\xi}}L)}{\partial z_i} = \sum_{k=1}^{M}\left(\frac{\partial \lambda_k}{\partial z_i}\nabla_{\boldsymbol{\xi}}h_k + \lambda_k \frac{\partial \nabla_{\boldsymbol{\xi}}h_k}{\partial z_i}\right) = \boldsymbol{0}. \tag{14}$$

Note that λ_k is a function of $\boldsymbol{z}$. For the active constraints, where $h_k = 0$, it holds that $\frac{\partial h_k}{\partial z_i} = 0$. For the passive constraints, $h_k < 0$, we use the following identities:

$$\lambda_k = 0 \qquad \text{and} \qquad \frac{\lambda h_k}{\partial z_i} = 0. \tag{15}$$

The equations in (12) to (15) give a linear system of equations with the unknowns $\frac{\partial w_j}{\partial z_i}, \frac{\partial \xi_n}{\partial z_i}, \lambda_k$ and $\frac{\partial \lambda_k}{\partial z_i}$. Solving this enables us to calculate $\frac{\partial \gamma}{\partial z_i}$ from (11) and finally $\frac{\partial \gamma}{\partial \theta_j}$ according to (9). Having this derivative makes it possible to learn CNN weights that optimize the max-margin objective formulated in (8) using gradient based methods. For more details, see the supplementary material.

2.4   End-to-End Training in Batches

We have now derived all the theoretical tools needed to train our deep structured model in an end-to-end manner. The joint training is done in epochs, where all training samples are utilized in each epoch. In every training epoch, new CRF weights are computed and the CNN weights are updated using gradient descent: $\theta_j \leftarrow \theta_j + \eta \frac{\partial \gamma}{\partial \theta_j}$ for all j.

To facilitate the process of learning deep image features for the CNN we first pretrain the weights $\boldsymbol{\theta}$ on the dataset without the CRF part of the model. Note that the CNN we used is based on a network pretrained on the ImageNet dataset [7]. The pretraining is done using stochastic gradient descent with a standard pixelwise log-likelihood error function.

The original learning method involves the entire training set when computing the CRF weights. However, since the linear equation system that needs to be solved grows with the number of training instances the learning process quickly becomes impractical with an increasing number of images. Hence we propose a method to compute the derivatives in batches. In batch mode we apply the CRF learning method from Algorithm 1 for each batch separately, We also calculate $\frac{\partial \gamma_b}{\partial \theta_j}$ following the steps described in Sect. 2.3. Note that the objective γ_b that we actually minimize here is an approximation of the true objective since not all images are included. For each batch, the constraint set $S_b^{(n)}$ is saved. These are, at the end of the epoch, merged to a set $S^{(n)}$ containing the low-energy labelings for all training instances. Finally the optimization problem in (8) is solved with this $S^{(n)}$ to get the CRF weights. When solving for the CRF weights we also get the current value of our objective γ, which obviously should decrease during training. The algorithm is summarized in Algorithm 2.

Input: image-labeling pairs $\{(\boldsymbol{x}^{(n)}, \boldsymbol{y}^{(n)})\}$ in the training set.
initialize $\boldsymbol{w} = \boldsymbol{w}_0$ and $\boldsymbol{\theta} = \boldsymbol{\theta}_0$
for *number of epochs* **do**
    initialize $S^{(n)} = \emptyset$ for each training instance n
    for *each batch b* **do**
        CNN forward pass $\to \boldsymbol{z}$
        CRF learning by Algorithm 1
        add low-energy labelings to set $S^{(n)}$
        calculation of objective derivative $\to \frac{\partial \gamma_b}{\partial z_i}$, back-propagation $\to \frac{\partial \gamma_b}{\partial \theta_j}$
        update CNN weights, $\theta_j \leftarrow \theta_j + \eta \frac{\partial \gamma_b}{\partial \theta_j}$
    end
    update CRF weights by solving (8) $\to \boldsymbol{w}, \gamma$
end
Output: $\boldsymbol{w}, \boldsymbol{\theta}$

Algorithm 2. Pseudocode for joint learning of parameters in batches.

3 Experiments and Results

Now, we present the performance of our method on three different segmentation tasks including comparisons to two baselines. For the first baseline, "CNN (only)", the segmentation is created by thresholding the output of a pretrained CNN. For the second baseline, "CNN + CRF (piecewise)", a CNN coupled with a CRF is trained in a piecewise manner, meaning that the network weights are kept fixed while learning the CRF weights. The results for the joint learning is denoted "CNN + CRF (joint)". For all experiments the CNN had the same structure as the FCN-8 network introduced by Long *et al.* [16]. The parameter settings were the same for all three segmentation tasks (learning rate = 10^{-4}, batch size = 10 and $C = 1$). All routines for training and testing were implemented in MATLAB on top of MATCONVNET [25].

3.1 Weizmann Horse Dataset

The Weizmann Horse dataset [2] is widely used for benchmarking object segmentation algorithms. The dataset contains 328 images of horses in different environments, we divide these images into a training set of 150 images, a validation set of 50 images and a test set of 128 images.

Our algorithm is compared to the, to our knowledge, best previously published results on the data set; Reseg [26], CRF-Grad [14] and PatchCut [27]. There are a few variations of the Weizmann Horse dataset available, we used the same one as in PatchCut [27]. Our algorithm is also compared to the two baselines, "CNN (only)" and "CNN + CRF (piecewise)". Quantative results (mean Jaccard index) are shown in Table 1 for the test images. In Fig. 1 some qualitative results are presented.

Table 1. Mean Jaccard index for the Weizmann Horse dataset (test set).

Method	Jaccard (%)	Method	Jaccard (%)
PatchCut [27]	84.03	CNN (only)	79.97
ReSeg [26]	**91.60**	CNN + CRF (piecewise)	81.62
CRF-Grad [14]	83.98	CNN + CRF (joint)	84.54

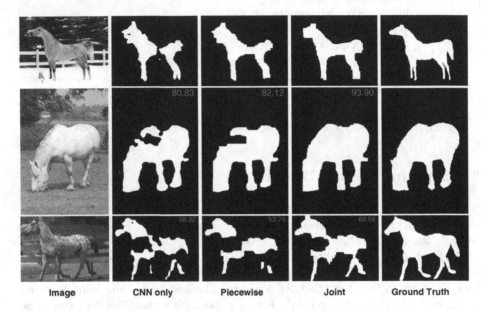

Image   CNN only   Piecewise   Joint   Ground Truth

Fig. 1. Qualitative results on the Weizmann Horse dataset. "Piecewise" denotes "CNN + CRF (piecewise)" and "Joint" denotes "CNN + CRF (joint)". The number shown in the upper right corner is the Jaccard index (%).

3.2 Cardiac Ultrasound Dataset

The second dataset we consider consists of 2D cardiac ultrasound images (2-chamber view, *i.e.* the left artrium and the left ventricle are visible). The ground truth consists of manual annotations of the left ventricle made by an experienced cardiologist according to the protocol in [13]. The dataset contains 66 images which are divided into a training set of 33 images, a validation set of 17 images and a test set of 16 images. See Fig. 2 and Table 2 for qualitative and quantitative results respectively.

3.3 Cardiac CTA Dataset

The third dataset we consider consists of 2D slices of cardiac CTA volumes originating from the SCAPIS pilot study [1]. The ground truth consists of slice-wise manual annotations of the pericardium made by a specialist in thoracic radiology and according to the protocol in [18]. The dataset includes in total 1500 2D

Table 2. Quantitative results for the Cardiac Ultrasound dataset (US) and the Cardiac CTA dataset (CTA). For the CTA dataset, the different types of slices are evaluated separately (ax - axial, cor - coronal and sag - sagittal). The mean Jaccard index (%) for the test sets are reported.

Method	US	CTA-ax	CTA-cor	CTA-sag
CNN (only)	82.28	81.40	77.11	75.96
CNN + CRF (piecewise)	85.79	81.84	77.12	75.83
CNN + CRF (joint)	**86.20**	**82.10**	**77.71**	**76.34**

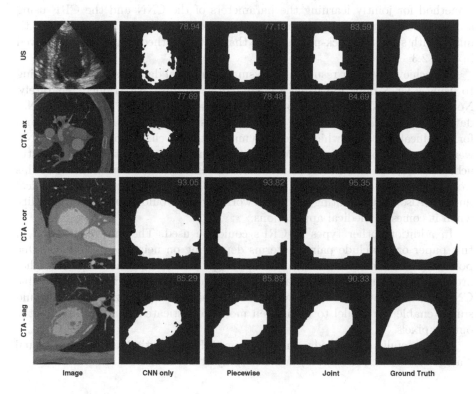

Fig. 2. Qualitative results on the Cardiac ultrasound dataset (US) and the Cardiac CTA dataset (CTA). For the CTA dataset, the different types of slices are evaluated separately (ax - axial, cor - coronal and sag - sagittal). "Piecewise" denotes "CNN + CRF (piecewise)" and "Joint" denotes "CNN + CRF (joint)". The red number shown in the upper right corner is the Jaccard index (%).

slices which are divided into three subsets of equal size to be evaluated separately representing three different views (*i.e.* axial, coronal and sagittal view). For each view the 2D slices were divided into a training set of 300 images, a validation set of 100 images and a test set of 100 images. Some of the 2D slices originate from regions where the pericardium is not visible. Thus, these images were excluded from the quantitative results since the Jaccard index is undefined if the ground truth and segmentation are both empty sets. Some qualitative results of the joint training process are visualized in Fig. 2 and quantitative results are presented in Table 2.

4   Conclusion and Future Work

In this paper, we have proposed a segmentation algorithm based on a deep structured model consisting of a CNN paired with a CRF. We also presented a method for jointly learning the parameters of the CNN and the CRF using a max-margin approach. Conveniently, the max-margin objective could be optimized with standard back-propagation thanks to the theoretical results derived in Sect. 2.3.

We achieve superior results on two smaller medical datasets when comparing to using a CNN only and using a CNN paired with a CRF trained separately. Note that the CNN we used is based on a network pretrained on the ImageNet dataset [7]. It has hence learnt image features for standard RGB images and for classification tasks, which of course makes it more challenging learning CNN weights well-adjusted for medical image segmentation. In spite of this, we still achieve good results on the two medical datasets. A future continuation of this work would be to combine the CRF with a CNN trained on a larger set of medical images. Also, implementing the framework for 3D would increase its usability when it comes to medical applications.

In addition, other types of CRFs could be used. The ones considered in this paper only include pairwise terms depending on neighbouring pixels. One possible extension would be to consider longer distance relationships or higher order energy terms. Also, the pairwise terms could be learned with a trainable CNN in the same way as for unary terms. A trainable regularization term would surely enable the model to learn even more sophisticated relationships for the output pixels.

We gratefully acknowledge funding from SSF (Semantic Mapping and Visual Navigation for Smart Robots) and VR (project no. 2016-04445).

References

1. Bergström, G., et al.: The Swedish CArdioPulmonary bioImage Study: objectives and design. J. Internal Med. **278**(6), 645–659 (2015)
2. Borenstein, E., Ullman, S.: Class-specific, top-down segmentation. In: Heyden, A., Sparr, G., Nielsen, M., Johansen, P. (eds.) ECCV 2002. LNCS, vol. 2351, pp. 109–122. Springer, Heidelberg (2002). doi:10.1007/3-540-47967-8_8
3. Brosch, T., Tang, L.Y.W., Yoo, Y., Li, D.K.B., Traboulsee, A., Tam, R.: Deep 3D convolutional encoder networks with shortcuts for multiscale feature integration applied to multiple sclerosis lesion segmentation. IEEE Trans. Med. Imag. **35**(5), 1229–1239 (2016)

4. Chen, L.-C., Papandreou, G., Kokkinos, I., Murphy, K., Yuille, A.L.: Semantic image segmentation with deep convolutional nets and fully connected CRFs. In: International Conference on Learning Representations (2015)
5. Chen, L.-C., Schwing, A.G., Yuille, A.L., Urtasun, R.: Learning deep structured models. In: International Conference on Machine Learning (2015)
6. Cireşan, D.C., Giusti, A., Gambardella, L.M., Schmidhuber, J.: Mitosis detection in breast cancer histology images with deep neural networks. In: Mori, K., Sakuma, I., Sato, Y., Barillot, C., Navab, N. (eds.) MICCAI 2013. LNCS, vol. 8150, pp. 411–418. Springer, Heidelberg (2013). doi:10.1007/978-3-642-40763-5_51
7. Deng, J., Dong, W., Socher, R., Li, L.-J., Li, K., Fei-Fei, L.: ImageNet: a large-scale hierarchical image database. In: Conference on Computer Vision and Pattern Recognition (2009)
8. Girshick, R., Donahue, J., Darrell, T., Malik, J.: Rich feature hierarchies for accurate object detection and semantic segmentation. In: Conference on Computer Vision and Pattern Recognition (2014)
9. Giusti, A., Cireşan, D.C., Masci, J., Gambardella, L.M., Schmidhuber, J.: Fast image scanning with deep max-pooling convolutional neural networks. In: International Conference on Image Processing (2013)
10. Kolmogorov, V., Zabin, R.: What energy functions can be minimized via graph cuts? IEEE Trans. Pattern Anal. Mach. Intell. $26(2)$, 147–159 (2004)
11. Krähenbühl, P., Koltun, V.: Efficient inference in fully connected CRFs with Gaussian edge potentials. In: Advances in Neural Information Processing Systems (2011)
12. Krizhevsky, A., Sutskever, I., Hinton, G.E.: ImageNet classification with deep convolutional neural networks. In: Advances in Neural Information Processing Systems (2012)
13. Lang, R.M., et al.: Recommendations for cardiac chamber quantification by echocardiography in adults: an update from the american society of echocardiography and the european association of cardiovascular imaging. J. Am. Soc. Echocardiogr. $28(1)$, 1–39 (2015)
14. Larsson, M., Arnab, A., Kahl, F., Zheng, S., Torr, P.: Learning arbitrary pairwise potentials in CRFs for semantic segmentation. arXiv preprint (2017)
15. Lin, G., Shen, C., van den Hengel, A., Reid, I.: Efficient piecewise training of deep structured models for semantic segmentation. In: Conference on Computer Vision and Pattern Recognition (2016)
16. Long, J., Shelhamer, E., Darrell, T.: Fully convolutional networks for semantic segmentation. In: Conference on Computer Vision and Pattern Recognition (2015)
17. Noh, H., Hong, S., Han, B.: Learning deconvolution network for semantic segmentation. In International Conference on Computer Vision (2015)
18. Norlén, A., Alvén, J., Molnar, D., Enqvist, O., Norrlund, R.R., Brandberg, J., Bergström, G., Kahl, F.: Automatic pericardium segmentation and quantification of epicardial fat from computed tomography angiography. J. Med. Imaging $3(3)$ (2016)
19. Prasoon, A., Petersen, K., Igel, C., Lauze, F., Dam, E., Nielsen, M.: Deep feature learning for knee cartilage segmentation using a triplanar convolutional neural network. In: Mori, K., Sakuma, I., Sato, Y., Barillot, C., Navab, N. (eds.) MICCAI 2013. LNCS, vol. 8150, pp. 246–253. Springer, Heidelberg (2013). doi:10.1007/978-3-642-40763-5_31
20. Ranzato, M., Taylor, P.E., House, J.M., Flagan, R.C., LeCun, Y., Perona, P.: Automatic recognition of biological particles in microscopic images. Pattern Recogn. Lett. $28(1)$, 31–39 (2007)

21. Ronneberger, O., Fischer, P., Brox, T.: U-Net: convolutional networks for biomedical image segmentation. In: Navab, N., Hornegger, J., Wells, W.M., Frangi, A.F. (eds.) MICCAI 2015. LNCS, vol. 9351, pp. 234–241. Springer, Cham (2015). doi:10.1007/978-3-319-24574-4_28

22. Roth, H.R., Lu, L., Farag, A., Shin, H.-C., Liu, J., Turkbey, E.B., Summers, R.M.: DeepOrgan: multi-level deep convolutional networks for automated pancreas segmentation. In: Navab, N., Hornegger, J., Wells, W.M., Frangi, A.F. (eds.) MICCAI 2015. LNCS, vol. 9349, pp. 556–564. Springer, Cham (2015). doi:10.1007/978-3-319-24553-9_68

23. Szummer, M., Kohli, P., Hoiem, D.: Learning CRFs using graph cuts. In: Forsyth, D., Torr, P., Zisserman, A. (eds.) ECCV 2008. LNCS, vol. 5303, pp. 582–595. Springer, Heidelberg (2008). doi:10.1007/978-3-540-88688-4_43

24. Tompson, J.J., Jain, A., LeCun, Y., Bregler, C.: Joint training of a convolutional network and a graphical model for human pose estimation. In: Advances in Neural Information Processing Systems (2014)

25. Vedaldi, A., Lenc, K.: MatConvNet: convolutional neural networks for MATLAB. In: International Conference on Multimedia (2015)

26. Visin, F., Ciccone, M., Romero, A., Kastner, K., Cho, K., Bengio, Y., Matteucci, M., Courville, A.: ReSeg: a recurrent neural network-based model for semantic segmentation. In: Conference on Computer Vision and Pattern Recognition Workshops (2016)

27. Yang, J., Price, B., Cohen, S., Lin, Z., Yang, M.-H.: PatchCut: Data-driven object segmentation via local shape transfer. In: Conference on Multimedia Computer Vision and Pattern Recognition (2015)

28. Zheng, S., Jayasumana, S., Romera-Paredes, B., Vineet, V., Su, Z., Du, D., Huang, C., Torr, P.H.S.: Conditional random fields as recurrent neural networks. In: International Conference on Computer Vision (2015)

Robust Abdominal Organ Segmentation Using Regional Convolutional Neural Networks

Måns Larsson[1]([✉]), Yuhang Zhang[1], and Fredrik Kahl[1,2]

[1] Chalmers University of Technology, Gothenburg, Sweden
{mans.larsson,zhangyu,fredrik.kahl}@chalmers.se
[2] Centre for Mathematical Sciences, Lund University, Lund, Sweden

Abstract. A fully automatic system for abdominal organ segmentation is presented. As a first step, an organ localization is obtained via a robust and efficient feature registration method where the center of the organ is estimated together with a region of interest surrounding the center. Then, a convolutional neural network performing voxelwise classification is applied. The convolutional neural network consists of several full 3D convolutional layers and takes both low and high resolution image data as input, which is designed to ensure both local and global consistency. Despite limited training data, our experimental results are on par with state-of-the-art approaches that have been developed over many years. More specifically the method is applied to the MICCAI2015 challenge "Multi-Atlas Labeling Beyond the Cranial Vault" in the free competition for organ segmentation in the abdomen. It achieved the best results for 3 out of the 13 organs with a total mean Dice coefficient of **0.757** for all organs. Top scores were achieved for the gallbladder, the aorta and the right adrenal gland.

Keywords: Medical image analysis · Segmentation · Convolutional neural networks

1 Introduction

Segmentation is a key problem in medical image analysis, and an automated method for organ segmentation can be crucial for numerous applications in medical research and clinical care such as computer aided diagnosis and surgery assistance. The high variability of the shape and position of abdominal organs makes segmentation a challenging task. Previous work done on segmentation of abdominal organs includes, among others, multi-atlas methods [18], patch-based methods [17], and methods based on a probabilistic atlas [2,12]. These techniques achieve great results for several abdominal organs but struggle with the segmentation of organs where the anatomical variability is large such as the gallbladder.

During the last few years, deep convolutional neural networks have shown great performance and achieved state of the art results in many computer vision applications [9,11]. This fact can be partly attributed to the constant increase

© Springer International Publishing AG 2017
P. Sharma and F.M. Bianchi (Eds.): SCIA 2017, Part II, LNCS 10270, pp. 41–52, 2017.
DOI: 10.1007/978-3-319-59129-2_4

in available computing power, most notably GPU computing solutions, and the availability of large annotated datasets. In the field of medical image analysis this development has led to an increased interest in methods based on deep convolutional neural networks with promising results [4, 14]. Recently, more intricate deep learning approaches have been proposed in the field of image analysis. Ronneberger et al. [13] proposed the U-Net, a network based on the idea of "fully convolutional networks" [11]. This work was extended to a network utilizing 3D convolutional filters by Çiçek et al. [3], a similar network structure where also proposed by Brosch et al. [1]. Another example is Kamnitsas et al. [8] that used an eleven layer deep 3D convolutional network for brain lesion segmentation with good results.

In this paper, instead of designing larger and more complicated network structures we propose a two step method, simplifying the task the convolutional neural network need to solve. In summary, an automatic system for the segmentation of abdominal organs in contrast enhanced CT images is presented. Our main contribution is to show that despite limited training data (compared to other successful deep learning approaches), it is possible to design a system that achieves on par with state-of-the-art for very challenging segmentation tasks with high anatomical variability. Another contribution is that we develop a computationally efficient framework that allows for a fully integrated 3D approach to multi-organ segmentation. To our knowledge, this is the first attempt on using 3D CNNs for multi-organ abdominal segmentation.

The presented method is trained and test on the MICCAI2015 challenge "Multi-Atlas Labeling Beyond the Cranial Vault" [19] where it achieved state of the art results in the free competition for organ segmentation in the abdomen. To this date, our method gives the best results for 3 out of the 13 organs.

2   Proposed Solution

Our system segments each organ independently and can be divided into three steps:

1. Localization of region of interest using a multi-atlas approach.
2. Voxelwise binary classification using a convolutional neural network.
3. Postprocessing by thresholding and removing all positive samples except the largest connected component.

Each step will now be described in detail.

2.1   Localization of Region of Interest

This part provides a robust initialization of the segmentation. The goal is to locate the center voxel of the organ in the target image. When this has been done a prediction mask is placed centered around the predicted organ center. The prediction mask later defines the region of interest where the convolutional neural network is initially applied. The use of an initialization method enables us to train more specialized, or regional, networks that only need to differentiate between a certain organ and the background. This means that the classification task that the network needs to perform is simplified and computationally less demanding networks can be applied.

The location of the organ center in the target image is obtained using a feature-based multi-atlas approach. Each atlas image is registered to the target using the method described in [7]. The method is computationally efficient compared to traditional intensity-based registration methods. More importantly though, this approach provides a robust and reliable estimate for organ locations which have been demonstrated for many different settings and modalities. In our framework, the registration is performed individually for each organ and atlas image. Affine transformations are computed and then used to transform each organ center point from an atlas image to the target image. The median of the transformed center points provides us with an estimate of the center point for the region of interest in the target image. The reason for using the median, and not for instance the mean operator, is that it provides a robust estimate of the center point, that is, it is not affected by a few, spurious outliers.

The prediction mask is estimated using the ground truth segmentations of the atlas images. Let the ground truth segmentations be represented by a binary image of the same dimension as the atlas image $G^{(l)}$, where l is the image id, and $G^{(l)}_{ijk} = 1$ if and only if voxel with index i, j, k in image with id l is foreground (or organ). Further, define $D^{(l)}$ as the binary image formed by dilating $G^{(l)}$ by a cube of size $25 \times 25 \times 25$ voxels and translating it so that the center of the organ is located at the center of the image. The prediction mask P is then defined as the binary image where each element P_{ijk} is

$$P_{ijk} = \begin{cases} 1 & \text{if } \frac{1}{N} \sum_{l=1}^{N} D^{(l)}_{ijk} \geq \delta \\ 0 & \text{otherwise} \end{cases} \tag{1}$$

where N is the number of atlas images and δ is a threshold set to $\delta = 0.5$ for the majority of the organs. Finally, the region of interest R is defined as the prediction mask centered around the estimated center point. An example of a localization of region of interest is shown in Fig. 1.

2.2 Voxel Classification Using a Convolutional Neural Network

The convolutional neural network is applied using a sliding window approach. For each voxel to be segmented two cubes of different resolutions centered around said voxel are extracted and used as input features to the network. The network in return outputs a probability, denoted p_{ijk}, of the voxel being organ.

To speed up the process the network is not applied to every voxel in the area that is being segmented, denoted S. Instead, it takes steps of three in each dimension over S. The probabilities output by the network are then interpolated to every voxel in S. All voxels in S that have been assigned an interpolated probability neither close to zero nor close to one (voxels with a probability between 0.1 and 0.9 to be specific) will be classified by the network once more.

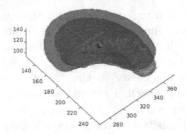

Fig. 1. Example of localization of region of interest for the Spleen. The green sphere is the estimated center point, the red mask escribes the ground truth and the blue mask describes the estimated region of interest. (Color figure online)

To reduce the dependency on the quality of the initial region of interest where the convolutional neural network is applied, a region growing algorithm is used. Call the set of voxels that should be segmented S. Further, call the set of voxels already classified by D and the set of voxels with an assigned probability larger than 0.5 by O. The region growing algorithm is then described by Algorithm 1. The use of a region growing algorithm means that even though the initial region may only cover part of the organ, a successful segmentation is still possible, see Fig. 3.

Initialize:
- S as the region of interest R
- D as $\varnothing$
- O as $\varnothing$.

while $S \neq \varnothing$ **do**

    – Classify voxels in S
    – Set $D = D \cup S$, and O as the set of voxels with an assigned probability larger than 0.5
    – Let O^+ be the set O dilated by a cube of size $12 \times 12 \times 4$ voxels
    – Set $S = O^+ \setminus D$

end
Output: O

Algorithm 1. Region growing algorithm for efficient classification.

Convolutional Neural Network Setup. The convolutional neural network used performs voxelwise binary classification. The input features for the network are two image cubes, one with a fine resolution similar to the original CT image and the other with a coarse resolution. The fine resolution input feature is meant to provide the network with local information ensuring local precision while the coarse resolution input feature is meant to ensure global spatial consistency. The inputs are processed in separate pipelines and the aggregated features from both pipelines are then merged for the last part of the network. A schematic of the convolutional neural network is shown in Fig. 2.

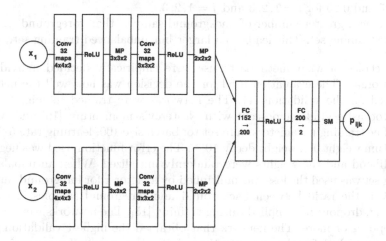

Fig. 2. Structure of the convolutional neural network. Both type and size of each layer are shown. The following abbreviations are used. Conv: Convolutional layer, ReLU: Rectified Linear Unit, MP: Max Pooling, FC: Fully Connected and SM: Soft Max. Both inputs are cubes containing $27 \times 27 \times 12$ voxels and are centered around the voxel being classified. Input x_1 has as high resolution with voxels of size $1 \times 1 \times 3$ mm^3, while input x_2 is downsampled by a factor of five in each dimension.

Implementation and Training. For the implementation of the convolutional neural network the framework Torch7 was used [5]. For each convolutional neural network the training and validation set were extracted from the region of interest calculated as described previously and the area around the part of the image describing the organ. This was done for each image in the training set. For the majority of the organs a balanced training set was used, meaning that there was an equal amount of foreground and background samples in the training set. However, since some of the organs are quite small this leads to a relatively small training set. Several methods, listed below, were used to deal with this problem.

1. For organs present in pairs, kidneys and adrenal glands, training samples from both the left and the right organ were used. Note that this does not pose a problem during inference since the initialization part of the method will separate the organs.
2. Expansion of the training set by adding slightly distorted CT images, transforming them using a random affine transformation similar to the identity transformation. The transformation matrix T was randomized as

$$T = \begin{pmatrix} 1+\delta_{11} & \delta_{12} & \delta_{13} & 0 \\ \delta_{21} & 1+\delta_{22} & \delta_{23} & 0 \\ \delta_{31} & \delta_{32} & 1+\delta_{33} & 0 \\ 0 & 0 & 0 & 1 \end{pmatrix}$$

where δ_{ij} are independently and uniformly randomized numbers between -0.25 and 0.25 for $i = 1, 2, 3$ and $j = 1, 2, 3$.

3. Including a greater number of background samples than foreground samples in the training set. This leads to a larger but unbalanced training set.

The choice of what methods to use were empirically decided individually for each organ. The evaluation used for the decision was how well the network performed on the validation set. The networks were trained in mini batches using stochastic gradient descent with Nesterov's momentum [16] and weight decay. The training parameters were set to: batch size 100, learning rate $5 \cdot 10^{-3}$, momentum weight 0.9, weight decay 10^{-5}. The error function used was negative log likelihood and the weights were randomly initialized. When an unbalanced training set was used the loss was multiplied by a factor k for foreground samples where k is the ratio between background and foreground samples. To avoid overfitting, dropout was applied during training [15]. The networks were trained for 10 epochs or more. The network that obtained the highest validation score was finally picked for the segmentation of the test images (see experimental section for different network and data settings).

2.3 Postprocessing

As a final step the probabilities from the convolutional neural network are thresholded, with a organ specific value estimated from data, in order to create a binary image. Everything but the largest connected component is set to zero producing the final segmentation.

3 Experimental Results

Our system was evaluated by submitting an entry to the MICCAI 2015 challenge "Multi-Atlas Labeling Beyond the Cranial Vault" in the free competition for organ segmentation in the abdomen [19]. In this challenge, there are 30 CT images coupled with manual segmentations of 13 organs, listed in Table 1. These 30 images and segmentations are available for method development and validation. Out of the 30 images 20 were used for training and 10 for validation.

In addition to these images training data from the VISCERAL challenge [6] was also used for training. The VISCERAL training data consists of 20 unenhanced whole body CT images and 20 contrast enhanced CT images over the abdomen and thorax. In these images organ ids 1, 2, 3, 4, 6, 8, 11, 12 and 13 were manually segmented. The unenhanced whole body CT images were excluded from the training set for organs with organ id 1, 2, 6, 8, 10, 11 and 12 since they differed too much from the enhanced CT images. All images were resampled to the same resolution of 1 mm × 1 mm × 3 mm. For the right kidney, a network trained on a training set formed by samples from both the right and the left kidney was used. For the stomach, the data set was expanded with distorted CT images and for the left adrenal gland an unbalanced data set was used with twice as many background samples as foreground samples.

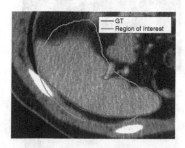 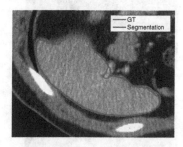

Fig. 3. Example of the resulting segmentation of the spleen for a CT slice. Note that even though the initial region of interest did not contain the entire organ the final result still does. This is due to region growing.

The test set of the MICCAI challenge consists of 20 CT images. The submitted segmentations are evaluated by calculating the Dice coefficient for each organ. The final results are given in Table 1 with the currently two best competitors:

- **IMI** - algorithm name: *IMI_deeds_SSC_jointCL* submitted by Mattias Heinrich at the Institute of Medical Informatics, Lübeck, Germany.
- **CLS** - algorithm name: *CLSIMPLEJLF_organwise* submitted by Zhoubing Xu at the Vanderbilt University, Nashville, TN, USA.
- **Other** - this column contains results from other competitors. The score is only shown if they are the highest for that organ.

For each organ the estimation of the organ center took about 20 s and the classification with CNN took between 30 s and 5 min, depending on organ size. The CNN computations were performed on a GeForce GTX TITAN X GPU with Maxwell architecture.

4 Discussion

In Table 2 a comparison of the validation score and the test score of our method is presented. As can be seen from the table there is a large difference between validation and testing scores for some organ. This means that our networks do not generalize well to the test set for these organs which might be an indication of overfitting and that the input features and structure of our network is not ideal to learn high order information that generalize to all other CT images. However,

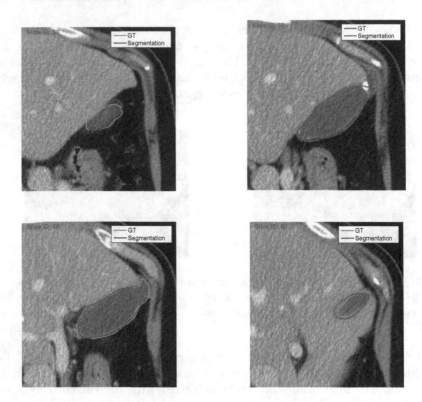

Fig. 4. Example of the resulting segmentation of the gallbladder. This is one of many images where our method achieved good result, for this image we got a Dice coefficient of 0.90.

since the validation data has not been used for the actual training, only for the decision on when to stop the training, these differences might not be only due to overfitting. Instead it might be due to the existence of anatomical variations in the test set that differs too much from anything seen in the training and validation images for the network to perform well. A specific example of where our method performed badly on the test data, for an organ with good validation result, is shown in Fig. 5. Here, the network has classified most of the right kidney correctly. However, it has also classified a lot of surrounding organs or tissue as right kidney as well.

The ideal solution to this problem would be to include more images in the training set. This however, requires more manually segmented CT images which are not always easy to acquire. Other approaches to solve this problem would be to train a network on several organs, and then fine tune the network weights for each specific organ. This could enable the network to learn higher order features that differentiates well between all organs in the CT image, not only between the organs located closest to the organ that is currently being segmented. Other future work would be to add a postprocessing step using a Conditional Random

Table 1. Final results measured in Dice metric for organ segmentation in CT images. Our approach gives the best results for 3 out of the 13 organs. Here '–' means that one of the specified methods achieved best result.

Organ	IMI	CLS	Other	Our
Spleen	0.919	0.911	**0.964**	0.930
Right kidney	**0.901**	0.893	–	0.866
Left kidney	0.914	0.901	**0.917**	0.911
Gallbladder	0.604	0.375	–	**0.624**
Esophagus	**0.692**	0.607	–	0.662
Liver	**0.948**	0.940	–	0.946
Stomach	**0.805**	0.704	–	0.775
Aorta	0.857	0.811	–	**0.860**
Inferior Vena Cava	**0.828**	0.760	–	0.776
Portal and Splenic Vein	0.754	0.649	**0.756**	0.567
Pancreas	**0.740**	0.643	–	0.602
Right adrenal gland	0.615	0.557	–	**0.631**
Left adrenal gland	**0.623**	0.582	–	0.583
Average	**0.790**	0.723	–	0.757

Table 2. Comparison of validation score and test score measured in Dice metric for organ segmentation in CT images.

Organ	Validation	Test
Spleen	0.944	0.930
Right kidney	0.940	0.866
Left kidney	0.928	0.911
Gallbladder	0.744	0.624
Esophagus	0.724	0.662
Liver	0.947	0.946
Stomach	0.823	0.775
Aorta	0.892	0.860
Inferior Vena Cava	0.823	0.776
Portal Vein and Splenic Vein	0.632	0.567
Pancreas	0.689	0.602
Right adrenal gland	0.600	0.631
Left adrenal gland	0.580	0.583
Average	0.790	0.757

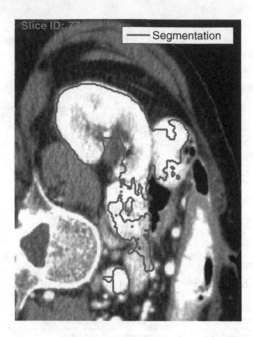

Fig. 5. Example of the resulting segmentation of the right kidney for a CT slice from the test set. The final segmentation is marked in blue. This segmentation was one of the examples where the method performed badly. (Color figure online)

Field (CRF). Recently, methods for training CRFs and CNNs jointly have been proposed for medical images [10]. In addition to this the type of CNN could be exchanged. For example, A "fully convolutional" type network mentioned in the introduction could be utilized to speed up training and inference. Note though that the memory requirements for these types of networks are larger since the input image cube and number of weights needed are usually larger.

5   Conclusion

In this paper, an efficient system for abdominal organ segmentation was presented. Our approach first uses a robust localization algorithm for finding the region of interest. As a second step a convolutional neural network is applied performing voxelwise classification. The network takes two different resolutions of image data as input to ensure both global and local consistency. The method was evaluated by submitting an entry to the MICCAI2015 challenge "Multi-Atlas Labeling Beyond the Cranial Vault" in the free competition for organ segmentation in the abdomen. The entry achieved on par with state-of-the-art for a majority of the organs with a mean Dice coefficient of 0.757.

We gratefully acknowledge funding from the Swedish Foundation for Strategic Research (Semantic Mapping and Visual Navigation for Smart Robots) and the Swedish Research Council (project no. 2016-04445).

References

1. Brosch, T., Tang, L.Y.W., Yoo, Y., Li, D.K.B., Traboulsee, A., Tam, R.: Deep 3D convolutional encoder networks with shortcuts for multiscale feature integration applied to multiple sclerosis lesion segmentation. IEEE Trans. Med. Imaging **35**(5), 1229–1239 (2016)
2. Chu, C., Oda, M., Kitasaka, T., Misawa, K., Fujiwara, M., Hayashi, Y., Nimura, Y., Rueckert, D., Mori, K.: Multi-organ segmentation based on spatially-divided probabilistic atlas from 3D abdominal CT images. In: Mori, K., Sakuma, I., Sato, Y., Barillot, C., Navab, N. (eds.) MICCAI 2013. LNCS, vol. 8150, pp. 165–172. Springer, Heidelberg (2013). doi:10.1007/978-3-642-40763-5_21
3. Çiçek, Ö., Abdulkadir, A., Lienkamp, S.S., Brox, T., Ronneberger, O.: 3D U-Net: Learning Dense Volumetric Segmentation from Sparse Annotation. Springer International Publishing, Cham (2016)
4. Cireşan, D.C., Giusti, A., Gambardella, L.M., Schmidhuber, J.: Mitosis detection in breast cancer histology images with deep neural networks. In: Mori, K., Sakuma, I., Sato, Y., Barillot, C., Navab, N. (eds.) MICCAI 2013. LNCS, vol. 8150, pp. 411–418. Springer, Heidelberg (2013). doi:10.1007/978-3-642-40763-5_51
5. Collobert, R., Kavukcuoglu, K., Farabet, C.: Torch7: a matlab-like environment for machine learning. In: BigLearn, NIPS Workshop (2011)
6. Hanbury, A., Müller, H., Langs, G., Menze, B.H.: Cloud–based evaluation framework for big data. In: Galis, A., Gavras, A. (eds.) FIA 2013. LNCS, vol. 7858, pp. 104–114. Springer, Heidelberg (2013). doi:10.1007/978-3-642-38082-2_9
7. Kahl, F., Alvén, J., Enqvist, O., Fejne, F., Ulén, J., Fredriksson, J., Landgren, M., Larsson, V.: Good features for reliable registration in multi-atlas segmentation. In: VISCERAL Anatomy3 Segmentation Challenge, pp. 12–17 (2015)
8. Kamnitsas, K., Chen, L., Ledig, C., Rueckert, D., Glocker, B.: Multi-scale 3D convolutional neural networks for lesion segmentation in brain MRI. In: Ischemic Stroke Lesion Segmentation, p. 13 (2015)
9. Krizhevsky, A., Sutskever, I., Hinton, G.E.: Imagenet classification with deep convolutional neural networks. In: Advances in Neural Information Processing Systems, vol. 25, pp. 1097–1105. Curran Associates Inc. (2012)
10. Larsson, M., Alvén, J., Kahl, F.: Max-margin learning of deep structured models for semantic segmentation. In: 20th Scandinavian Conference Image Analysis (SCIA 2017). Springer International Publishing (2017)
11. Long, J., Shelhamer, E., Darrell, T.: Fully convolutional networks for semantic segmentation. In: Proceedings of the IEEE Conference on Computer Vision and Pattern Recognition, pp. 3431–3440 (2015)
12. Park, H., Bland, P.H., Meyer, C.R.: Construction of an abdominal probabilistic atlas and its application in segmentation. IEEE Trans. Med. Imaging **22**(4), 483–492 (2003)
13. Ronneberger, O., Fischer, P., Brox, T.: U-Net: convolutional networks for biomedical image segmentation. In: Navab, N., Hornegger, J., Wells, W.M., Frangi, A.F. (eds.) MICCAI 2015. LNCS, vol. 9351, pp. 234–241. Springer, Cham (2015). doi:10.1007/978-3-319-24574-4_28. arXiv:1505.04597 [cs.CV])
14. Roth, H.R., Lu, L., Farag, A., Shin, H.-C., Liu, J., Turkbey, E.B., Summers, R.M.: DeepOrgan: multi-level deep convolutional networks for automated pancreas segmentation. In: Navab, N., Hornegger, J., Wells, W.M., Frangi, A.F. (eds.) MICCAI 2015. LNCS, vol. 9349, pp. 556–564. Springer, Cham (2015). doi:10.1007/978-3-319-24553-9_68

15. Sivastava, N., Hinton, G., Krizhevsky, A., Sutskever, I., Salakhutdinov, R.: Dropout: a simple way to prevent neural networks from overfitting. J. Mach. Learn. Res. **15**, 1929–1958 (2014)
16. Sutskever, I., Martens, J., Dahl, G., Hinton, G.: On the importance of initialization and momentum in deep learning. In: Proceedings of the 30th International Conference on Machine Learning (ICML-13), pp. 1139–1147 (2013)
17. Wang, Z., Bhatia, K.K., Glocker, B., Marvao, A., Dawes, T., Misawa, K., Mori, K., Rueckert, D.: Geodesic patch-based segmentation. In: Golland, P., Hata, N., Barillot, C., Hornegger, J., Howe, R. (eds.) MICCAI 2014. LNCS, vol. 8673, pp. 666–673. Springer, Cham (2014). doi:10.1007/978-3-319-10404-1_83
18. Wolz, R., Chu, C., Misawa, K., Fujiwara, M., Mori, K., Rueckert, D.: Automated abdominal multi-organ segmentation with subject-specific atlas generation. IEEE Trans. Med. Imaging **32**(9), 1723–1730 (2013)
19. Xu, Z.: Multi-atlas labeling beyond the cranial vault - workshop and challenge (2016). Accessed 10 Jan 2017

Detecting Chest Compression Depth Using a Smartphone Camera and Motion Segmentation

Øyvind Meinich-Bache[(✉)], Kjersti Engan, Trygve Eftestøl, and Ivar Austvoll

Department of Electrical Engineering and Computer Science,
University of Stavanger,
Kjell Arholmsgate 41, 4036 Stavanger, Norway
{oyvind.meinich-bache,kjersti.engan}@uis.no

Abstract. Telephone assisted guidance between dispatcher and bystander providing cardiopulmonary resuscitation (CPR) can improve the quality of the CPR provided to patients suffering from cardiac arrest. Our research group has earlier proposed a system for communication and feedback of the compression rate to the dispatcher through a smartphone application. In this paper we have investigated the possibilities of providing the dispatcher with more information by also detecting the compression depth. Our method involves detection of bystander's position in the image frame and detection of compression depth by generating Accumulative Difference Images (ADIs). The method shows promising results and give reason to further develop a general and robust solution to be embedded in the smartphone application.

Keywords: Video detection · Motion segmentation · CPR

1 Introduction

In Europe there are 370,000–740,000 out-of-hospital cardiac arrests every year with a survival rate as low as 7.6% [1]. Many are witnessed by a bystander and the bystander might not be skilled in cardiopulmonary resuscitation (CPR), thus there is a need for guided assistance to ensure the provision of quality CPR. The importance of quality CPR has been confirmed in many publications [2–4].

Smartphone applications for communication with the emergency unit and sending GPS location already exists in solution like *Hjelp 113-GPS* App by the Norwegian air ambulance[1]. Our group (Engan et al.) has earlier proposed an application for dispatcher communication which detects the compression rate [5]. Another important CPR quality metric is the compression depth which is crucial for generating sufficient circulation [6], thus providing the dispatcher with depth information can improve CPR quality and possibly save lives.

---

[1] https://www.itunes.apple.com/no/app/hjelp-113-gps/id363739748?l=no\&mt=8.

© Springer International Publishing AG 2017
P. Sharma and F.M. Bianchi (Eds.): SCIA 2017, Part II, LNCS 10270, pp. 53–64, 2017.
DOI: 10.1007/978-3-319-59129-2_5

Previously an accelerometer has been used to estimate the compression depth with the purpose of providing feedback in emergency or in training situations [7–9]. This requires the smartphone to be held in the hand of the bystander or at the chest of the patient during CPR. Since it is very important to maintain the phone connection between the bystander and the dispatcher we believe that placing the smartphone next to the patient and using the camera to perform the measurements would be more suited for emergency situations. This ensures that the microphone and loud speaker is not covered and that the phone connection is not interrupted by accidentally pressing a button. To our knowledge there has been made no attempt to estimate the compression depth from a smartphone camera with the attention to provide information to the dispatcher in an emergency situation. In this paper we have investigated this problem and propose a system that uses the front camera on a smartphone to estimate the compression depth. Figure 1 gives an overview of the proposed system, using generated Accumulative Difference Images (ADIs) [10] for motion segmentation to both detect the bystander position in the frame and to estimate the compression depth. These steps will be further explained in Sect. 3.

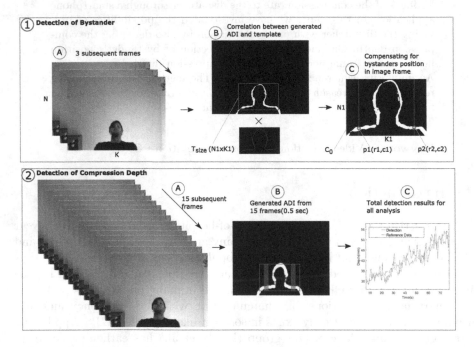

Fig. 1. Proposed system for detection of compression depth. Top: detecting bystander and regions of interest (ROIs). Bottom: detection of compression depth. (Color figure online)

2 Modelling of Scene

Modelling of the scene is necessary in order to estimate both the bystander's position in world coordinates and to compensate for the camera angle and position relative to the bystander.

2.1 Image to World Coordinates

We can find a model for the connection between world coordinates and image coordinates by calibration of the camera. By using camera coordinates for the world points it is sufficient to use the internal camera matrix K. The radial distortion must also be found and compensated for. Then we have

$$\lambda \begin{bmatrix} x_c \\ y_c \\ 1 \end{bmatrix} = KP_0 \begin{bmatrix} x_w \\ y_w \\ z_w \\ 1 \end{bmatrix} = \begin{bmatrix} \alpha & 0 & x_0 \\ 0 & \beta & y_0 \\ 0 & 0 & 1 \end{bmatrix} \begin{bmatrix} 1 & 0 & 0 & 0 \\ 0 & 1 & 0 & 0 \\ 0 & 0 & 1 & 0 \end{bmatrix} \begin{bmatrix} x_w \\ y_w \\ z_w \\ 1 \end{bmatrix} \tag{1}$$

where $\lambda = z_w$, P_0 a projection matrix, α and β the focal length of the camera and x_0 and y_0 the principal point offset in pixels. The distance, z_w, can be expressed $z_w = z_{w0} + \Delta z$ where z_{w0} is the distance between shoulders and ground and Δz is the compression depth in z-direction. A derivation of Eq. (1) for $\Delta z << z_{w0}$ gives the two expressions, approximated to be linear:

$$(y_c - y_0) = \beta \frac{y_w}{z_w} = \beta \frac{y_w}{z_{w0} + \Delta z} = \beta \frac{y_w}{z_{w0}} \frac{1}{1 + \frac{\Delta z}{z_{w0}}} \approx \beta \frac{y_w}{z_{w0}} (1 - \frac{\Delta z}{z_{w0}}) \tag{2}$$

$$(x_c - x_0) = \alpha \frac{x_w}{z_w} = \alpha \frac{x_w}{z_{w0} + \Delta z} = \alpha \frac{x_w}{z_{w0}} \frac{1}{1 + \frac{\Delta z}{z_{w0}}} \approx \alpha \frac{x_w}{z_{w0}} (1 - \frac{\Delta z}{z_{w0}}) \tag{3}$$

Figure 2 shows a model of the scene. Ellipsoid $1, 2$ and 3 illustrates the shoulder positions of the bystander. For illustration purpose ellipsoid 2 and 3 are scaled relative to ellipsoid 1 according to the camera enlargement model for approaching objects. $p.A, p.B, p.C$ and $p.D$ are camera positions along the positive y-axis where position $p.D$ defines the limit for camera positions where the bystander's shoulders are visible in the camera's field of view (FOV) and is a function of the distance between ground and shoulders along the z-axis given by $\frac{z_{w0}}{2}$. $L1$ and $L2$ represents motion vectors for the observed object enlargement in the image frame due to compression motions. The pink box is a zoomed in area of C illustrating the observed motion band in different camera positions.

The position of the ellipsoid marked as 1 illustrates the bystanders starting position, and 2 illustrates the new position if the compression motion is strictly in z-direction and the compression depth, Δz, is 50 mm. The enlargement for approaching objects for different z_{w0} is found from Eqs. (2) and (3) and is illustrated by using a 45 mm approaching object in Fig. 3. Since our method for

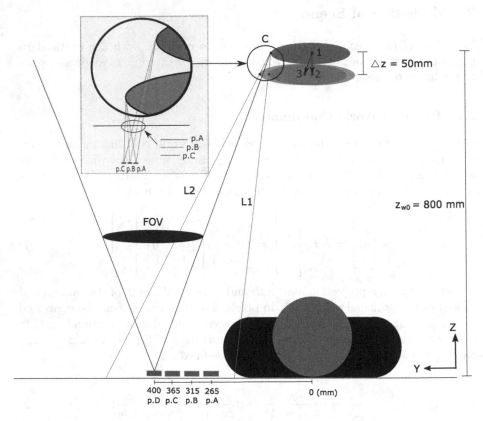

Fig. 2. Model of scene. Ellipsoid in position 1,2, and 3 illustrates the shoulder positions when compressing 50 mm. L1 and L2 illustrates the *blind spot* problem as a consequence of the different motions. *p.A*, *p.B*, *p.C* and *p.D* shows the possible camera positions for detections. The pink box shows the observed motion bands in the camera positions *p.A*, *p.B* and *p.C*. (Color figure online)

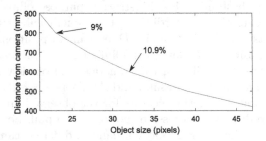

Fig. 3. Enlargement model for moving objects. The x-axis shows the observed size of the 45 mm square object in pixels and the y-axis show the distance between the object and the camera. Enlargement in % for object approaching 50 mm at 800 and 600 mm are marked.

detecting motion only captures changes in the contour of the bystander, a movement from shoulder position 1 to 2 and a camera positioned where $L1$ meets the ground floor line, would be represented by the same values for x_c and y_c. Thus, we would not be able to detect the change in the generated ADI and this position is further referred to as the *blind spot* and must be taken into account.

As shown in Fig. 2 a camera positioned where $L1$ meets the ground line is not possible since the camera would be placed underneath the patients shoulder. Camera positions $p.A$, $p.B$ and $p.C$ should therefore have no problem avoiding the *blind spot* problem. Positions where y-value $> p.C$ needs to be avoided since the bystander's shoulders no longer is guaranteed to be a part of the image frame. If the compression motion was strictly in z-direction the detected motion band should increase for each displacement along positive y-axis. This is not the case and it turns out that a compression motion will vary but are typically slightly positive along the y-axis, illustrated by the red ellipsoid at position 3 where line $L2$ indicates an approximation to a typical motion vector. This causes the *blind spot* line to move to the other side of the indicated camera positions $p.A$, $p.B$ and $p.C$. As a consequence, the detected motion band will shrink instead of increase as the camera is placed further along the positive y-axis. Since the y-value for $L2 > p.D$, the *blind spot* is not a problem, this is also true for a smaller bystander with $z_{w0} < 800$. Equation (1) and (2), as well as Fig. 3 shows that the linear model will change with z_{w0}, which is bystander and patient dependent (length of arms, size of torso).

2.2   Camera Angle Model

The camera angle problem is illustrated in the zoomed in area of circle C in Fig. 2 (pink box). Although the distance from the camera to the shoulders changes relatively little between positions $p.A$, $p.B$ and $p.C$, the displacements causes big variations in observed motion band. Since the compression movements will have small variations, the compensating model for displacement in y-direction is estimated by observing detected motion bands in given positions and at given compression depths. As the red, green and blue line in the pink box shows, this reduction of detected motion band is approximately linear which was also the case when studying the different detection results. The compensating model for the displacement in y-direction in the area between position $p.A$ and $p.C$ is estimated to be:

$$ang_{corr} = 1 + 0.0026(act_{pos} - p.A) \qquad (4)$$

where ang_{corr} is the compensating factor for displacement along positive y-axis and act_{pos} is the calculated position on the y-axis based on image to world conversion from Eq. (2). The model implies that a displacement from position $p.A$ to $p.C$ would mean a 26 % decrease in detected motion band. If the camera is positioned closer to the patient than position $p.A$ the observed motion band would increase and the model would scale down the detections. This will not be an issue here since the optimal position $p.A$ is next to the patient.

3 Proposed System

In Fig. 1 the system for detection of compression depth are shown step by step. The figure is divided into two main sections; detection of bystander and regions of interest (ROIs) (top), and detection of compression depth (bottom). ADIs [10] are used to carry out both sections. ADI is a well known method for motion segmentation and has earlier been used in many applications such as object tracking [11], vehicle surveillance systems [12] and smoke detection [13].

3.1 Detection of Bystander by Motion Segmentation

In the following let f indicate an $N \times K$ video frame where N is number of rows and K is number of columns, and $f(r, c, k)$ corresponds to row, r, and column, c, in frame number k.

From experiments we found that using three subsequent frames from the middle section of the sequences were enough to generate an ADI that revealed the position of the bystander. Spatial de-noising is done by Gaussian smoothing and the images are corrected for lens distortion [14] prior to ADI generation. The ADI is initialized by generating a $N \times K$ sized frame of zeros. Furthermore first of the three frames, k_0, is the reference frame and the ADI, $A(r, c)$, is found as:

$$A_k(r, c) = \begin{cases} A_{k-1}(r, c) + 1 & \text{if } |f(r, c, k_0) - f(r, c, k_0 + i)| > T \\ A_{k-1}(r, c) & \text{otherwise} \end{cases} \quad (5)$$

where T is a threshold value and i is an index for the subsequent frames. The resulting ADI used in detection of bystander will then consist of values from 0 to 2.

The generated absolute ADI is further correlated with templates to find the position of the bystander. This is illustrated in 1.B and 1.C in Fig. 1. The templates used are scaled and resized versions of a template of a person's head and shoulder contour created from an example sequence. To avoid higher correlation caused by thicker lines when the scale factor is above 1, a morphological *skeletonization* or *thinning* [15] of the scaled template is performed. The template position of the best match indicates the position of the bystander.

3.2 Position Compensation

In the detection of compression depth the information of the motion band in the shoulder areas are used. The desired camera position is when the bystander is centred in the image frame and the camera is placed close to the patient's arm. If the camera is positioned elsewhere compensation is needed. When compensating for position the bystander's shoulder points has to be detected. By starting in the first column, c_0, in the template match square marked T_{size} in Fig. 1(1C), the columns for the detection center points are found as follows:

$$c1 = c_0 + (\frac{1}{6} \cdot K1), \qquad c2 = c_0 + (\frac{5}{6} \cdot K1) \quad (6)$$

where $K1$ indicates the number of columns (width) of the matched template. Further the row number where the motion band starts is found by:

$$r_i = \min_r(A(r, c_i) \geqslant 1) \tag{7}$$

where $i = 1, 2$ indicates the two ROIs and r the row elements in the column c_i. Together with c1 and c2 these rows define the detection center points $p_1(c1, r1)$ and $p_2(c2, r2)$. The points are marked with a red circle in Fig. 1(1C). $p_1(c1, r1)$ and $p_2(c2, r2)$ are then converted from image to world coordinates, $w_1(x, y)$ and $w_2(x, y)$ by solving Eqs. (2) and (3) for $w_1(x, y)$ and $w_2(x, y)$. The actual distance, $d_{act,i}$, between the bystander and the camera is found by:

$$d_{act,i} = \sqrt{w_i(x)^2 + w_i(y)^2 + z_{w0}^2} \tag{8}$$

for $i = 1, 2$ which represents the two detections points and z_{w0} is illustrated in Fig. 2. The scaling factors for actual distance, $dist_{corr}$, for each detection point is found by:

$$dist_{corr,i} = \frac{d_{act,i}}{z_{w0}} \tag{9}$$

Further the compensating factor, ang_{corr}, for the camera angle is found by using the model given in Eq. (4). The same compensating factor is used for both $p_1(c1, r1)$ and $p_2(c2, r2)$ since these points lie approximately on the same horizontal line in the image frame.

3.3   Detection of Compression Depth

For the dispatcher-bystander communication to be efficient, the dispatcher should guide one problem at a time, thus the compression rate should first be guided to the desired range (100–120 cpm). Detection of compression rate is described in [5]. Knowing that the compression rate is in the desired range also makes the compression motion more predictable and furthermore the compression depth estimation less complicated.

The steps in detection of compression depth are shown in Fig. 1(2) and the compression depth is estimated every half second. Consider a videostream with 30 fps, providing $\frac{30}{2} = 15$ non-overlapping video frames in each compression depth estimation, $I(r, c, l_s)$, where l is the estimation number and s is a index for image number in this estimation. First, the images are spatially de-noised by Gaussian smoothing and corrected for lens distortion. Furthermore $I(r, c, l_1)$ is used as the reference frame and the other 14 frames to generate an ADI as shown in Eq. (5) and in Fig. 1(2A). For each new estimation the ADI is first set to zero before generating the ADI for the next estimation.

A reasonable width for the ROIs is found to be $M_{ROI} = 21$ columns when using image frame size of $N \times K = 480 \times 640$. The vertical motion band along the head/arms is then avoided but we still use enough columns to get a good average measurement of the motion band. An example is shown in Fig. 1(2B) where the ROIs is marked with red. Motion band vectors, $m_{band,i}$, for motion band size in columns, j, in the ROIs $i = 1, 2$ are found by:

$$m_{band,i}(o) = \sum_{q=1}^{N} A(q,j) > 1 \tag{10}$$

where o is a vector index for the columns used and q represents the row number.

Further the mean of these vectors are multiplied with their two compensating factors - position in image frame and camera angle, providing the corrected pixel size of the motion bands, $m_{mean,i}$:

$$m_{mean,i} = \frac{1}{M_{ROI}} \sum_{o=1}^{M_{ROI}} m_{band,i}(o) \cdot dist_{corr,i} \cdot ang_{corr} \tag{11}$$

used to find the combined detected motion band, m_{tot}, for this estimation, l:

$$m_{tot}(l) = \frac{1}{2}(m_{mean,1} + m_{mean,2}) \tag{12}$$

The last step is to filter the detections with a 3 coefficient weighted FIR filter to remove some of the noise caused by random movements from the bystander. The filter is selected from experimenting with different filter order and coefficient values to best suppress rapid changes without loosing important compression depth change information. $CD_{det}(l)$ represent the compression depth detection for estimation l and are found by:

$$CD_{det}(l) = 0.3 \cdot m_{tot}(l) + 0.35 \cdot m_{tot}(l-1) + 0.35 \cdot m_{tot}(l-2) \tag{13}$$

4   Experiments and Datasets

All compressions are performed on *Resusci Anne QCPR*[2] by the same bystander with $z_{w0} = 800$. Resusci Anne QCPR measures, among other things, the compression depth with an accuracy of ± 15 % and these data are used as reference data in development and verification testing of the proposed system. The smartphone used for the recordings is a *Xperia Z5 Compact (Sony, Japan)*.

The results are presented with Average error: $\mu_E = \frac{1}{L} \sum_{l=1}^{L} |CD_{det}(l) - CD_{true}(l)|$ where L is number of estimations and $CD_{true}(l)$ is the reference signal, and Performance, P, defined as percentage of the time where the $|CD_{det}(l) - CD_{true}(l)| < 10$ [mm]. According to the European Resuscitation Council Guidelines 2015 [16] 50–60 mm is the appropriate compression depth. A study of Stiell et al. [6] found that compression depth in the interval 40.3 to 55.3 mm provided maximum survival rate and the peak was found at 45.6 mm. Thus, the limit for accepted detection depths when calculating the P is here chosen to be ± 10 mm.

---

[2] http://www.laerdal.com/gb/ResusciAnne.

Each test starts with a target compression depth of approximately 20 mm and the target depth is gradually increased to 60 mm (maximum compression depth on Resusci Anne QCPR doll) during the 80–90 s recordings. The compression rate is in the desired range (100–120 cpm) for all tests. The detection of the bystander and the corresponding shoulder areas is performed once, and thereafter used throughout the sequence. Two different ways of finding the bystander's position are used; completely automatic using the method described in Sect. 3.1, and manually by a visual inspection.

The camera is calibrated with the procedure described in [14], which is based on [17,18]. The threshold used in generation of ADI is set to 50 and in the preprocessing of the images a Gaussian filter mask of size $N = 13$ with $\sigma = 3$ is used to reduce noise.

Modelling Experiment, Dataset 1

Equation (2) provides a theoretical conversion between pixels and mm. An experiment has been carried out to design a model for this conversion since a person performing compressions have larger movements than the actual compression depth itself. Dataset 1, D1, consist of 6 recordings where the phone for each recording is picked up and replaced at a point somewhere near the target of the optimal phone placement. The linear regression model for converting motion band in pixels to compression depth in mm is found to be:

$$CD_{conv}(l) = 2.7285 \cdot CD_{det}(l) - 13.9692 \qquad (14)$$

The data spread for D1 and the linear conversion model is shown in Fig. 4a.

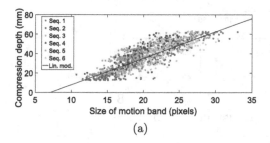
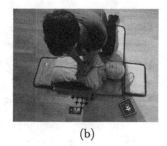

(a)   (b)

Fig. 4. (a): The spread of D1 and the connection between detected motion band in pixels and the actual compression depth at that time. Linear regression model is shown in purple. Different colors correspond to different recordings. (b): Scene for recording D2. The triangular system of black X's marks the phone position for each recording. (Color figure online)

Verification Test, Dataset 2

Dataset 2, D2, consists of 9 recordings, each with the phone placed at a different position marked with black X in Fig. 4b. If we define the desired position as $(0, p)$ where p represent position $p.A$ in Fig. 2, these positions corresponds to (−

100, p), (–50, p), (0, p), (50, p), (100, p), (–50, p+50), (0, p+50), (50, p+50) and (0, p+100). The values of the coordinates are given in millimetres. As shown on the smartphone in the figure, the (0, p+100) position is close to the limit of where the shoulders are included in the image frame, and is therefore the furthest distance from the bystander used in the recordings of D2. The y-coordinates chosen for D2 positions corresponds to position $p.A$, $p.B$ and $p.C$ in Fig. 2.

5 Results and Discussion

Table 1 shows the result from the proposed system, where the model found from D1 is tested on D2. The results from automatic detection of bystander shows poor results for position 2 and partly for position 1. By manually choosing the ROIs we get better results for position 1–4, but poorer results for position 5–9. The standard deviation given in parenthesis reveals little or no significant difference between the two methods for each position. Figure 5 also shows the results for each of the 9 positions in D2 arranged in the triangular form for the positions as in Fig. 4b. The reference data are shown in blue, the automatic bystander detection results in orange and when the bystander is manual detected in red. It can clearly be seen that the detection points chosen in automatic detection of bystander's shoulder points for position 2 provides poor detection results.

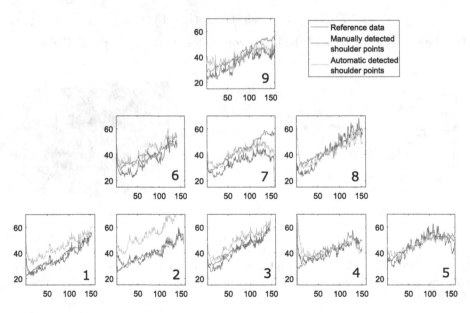

Fig. 5. Results for verification test, arranged in the same triangular form as seen in Fig. 4b. Blue graphs represent the reference data, orange the results with automatic detection of bystanders shoulders and red with manual detection of bystanders shoulders. The x-axis shows the estimation number (estimation each 0.5 s) and the y-axis shows the depth in millimetres. (Color figure online)

Table 1. Detection result for verification test performed on D2. Results are given as Average error, μ_E, with σ given in parentheses and Performance, P. Columns to the left, automatic detection of bystander's shoulder points. To the right, manually detection of bystander's shoulder points.

Position	Auto. detect. of bystander		Man. detect. of bystander	
	μ_E (mm)	P (%)	μ_E (mm)	P (%)
1	7.4 (3.8)	77.6	2.6 (3.3)	96.2
2	15.4 (4.1)	1.9	2.8 (3.7)	95.6
3	6.4 (3.1)	90.1	2.8 (3.1)	97.4
4	5.0 (7.5)	90.3	4.2 (6.0)	92.9
5	2.5 (3.4)	96.4	3.8 (5.0)	94.0
6	4.1 (3.5)	94.5	5.7 (3.4)	92.4
7	4.3 (7.3)	83.9	8.6 (6.6)	64.0
8	4.9 (7.6)	91.7	5.1 (5.0)	96.2
9	4.3 (5.0)	92.7	6.4 (3.7)	81.2
Mean	6.1	79.9	4.7	90.0
σ	3.8	29.8	2.0	10.9

The overall results indicates that as a consequence of determining the ROIs only once we might not have found suiting ROIs for the whole sequence, and that the detection results depend largely on the detection points chosen.

6 Conclusion and Future Work

The proposed system shows promising results for detection of compression depth by the use of a smartphone camera under the circumstances investigated in this paper. Although all tests are performed by only a single bystander with known distance between ground and shoulders, the model could be adapted for different distances.

In future work we will test the system for different bystander with known size/arm-length, as well as estimating the distance to the bystander when the distance is unknown. The latter is expected to be challenging since a small bystander would be similar to a big bystander further away.

Since the system is planned to be a part of an existing application for dispatcher feedback [5], the user could possibly type in some user information (height weight, age) when downloading and installing the app. This information would not only be useful for estimating distance, but would also be information relevant for the dispatcher. The system must also be able to track the bystander and to update the ROIs every 5 s or so during detection. Templates used to detect the bystander can here be developed from previous analysed ADIs. It could also be useful to use more of the information in the detected motion band when deciding the compression depth.

References

1. Bossaert, L.L., Perkins, G.D., Askitopoulou, H., Raffay, V.I., Greif, R., Haywood, K.L., Mentzelopoulos, S.D., Nolan, J.P., Van de Voorde, P., Xanthos, T.T., et al.: European resuscitation council guidelines for resuscitation 2015: section 11. The ethics of resuscitation and end-of-life decisions (2015)
2. Kern, K.B., Hilwig, R.W., Berg, R.A., Sanders, A.B., Ewy, G.A.: Importance of continuous chest compressions during cardiopulmonary resuscitation. Circulation **105**(5), 645–649 (2002)
3. Steen, S., Liao, Q., Pierre, L., Paskevicius, A., Sjöberg, T.: The critical importance of minimal delay between chest compressions and subsequent defibrillation: a haemodynamic explanation. Resuscitation **58**(3), 249–258 (2003)
4. Meaney, P.A., Bobrow, B.J., Mancini, M.E., Christenson, J., De Caen, A.R., Bhanji, F., Abella, B.S., Kleinman, M.E., Edelson, D.P., Berg, R.A., et al.: Cardiopulmonary resuscitation quality: improving cardiac resuscitation outcomes both inside and outside the hospital. Circulation **128**(4), 417–435 (2013)
5. Engan, K., Hinna, T., Ryen, T., Birkenes, T.S., Myklebust, H.: Chest compression rate measurement from smartphone video. Biomedical Eng. Online **15**(1), 95 (2016)
6. Stiell, I.G., Brown, S.P., Nichol, G., Cheskes, S., Vaillancourt, C., Callaway, C.W., Morrison, L.J., Christenson, J., Aufderheide, T.P., Davis, D.P., et al.: What is the optimal chest compression depth during out-of-hospital cardiac arrest resuscitation of adult patients? Circulation **130**(22), 1962–1970 (2014)
7. Gupta, N.K., Dantu, V., Dantu, R.: Effective CPR procedure with real time evaluation and feedback using smartphones. IEEE J. Transl. Eng. Health Med. **2**, 1–11 (2014)
8. Amemiya, T., Maeda, T.: Poster: depth and rate estimation for chest compression CPR with smartphone. In: 2013 IEEE Symposium on 3D User Interfaces (3DUI), pp. 125–126. IEEE (2013)
9. Song, Y., Oh, J., Chee, Y.: A new chest compression depth feedback algorithm for high-quality CPR based on smartphone. Telemedicine e-Health **21**(1), 36–41 (2015)
10. Gonzalez, R.C., Woods, R.E.: Digital Image Processing, 3rd edn. Pearson, Upper Saddle River (2008)
11. Yang, Y.H., Levine, M.D.: The background primal sketch: an approach for tracking moving objects. Mach. Vis. Appl. **5**(1), 17–34 (1992)
12. Koller, D., Weber, J., Malik, J.: Robust multiple car tracking with occlusion reasoning. In: Eklundh, J.-O. (ed.) ECCV 1994. LNCS, vol. 800, pp. 189–196. Springer, Heidelberg (1994). doi:10.1007/3-540-57956-7_22
13. Yuan, F.: A fast accumulative motion orientation model based on integral image for video smoke detection. Pattern Recogn. Lett. **29**(7), 925–932 (2008)
14. Bouguet, J.-Y.: Camera calibration toolbox for matlab (2004)
15. MATLAB, Image Processing Toolbox Morphological Operations, R.A.: The Mathworks Inc., Natick, Massachusetts, United States (2006)
16. Monsieurs, K.G., Zideman, D.A., Alfonzo, A., Arntz, H.R., Askitopoulou, H., Bellou, A., Beygui, F., Biarent, D., Bingham, R., et al.: European resuscitation council guidelines for resuscitation 2015: section 1. Executive summary (2015)
17. Zhang, Z.: A flexible new technique for camera calibration. IEEE Trans. Pattern Anal. Mach. Intell. **22**(11), 1330–1334 (2000)
18. Heikkila, J., Silven, O.: A four-step camera calibration procedure with implicit image correction. In: Proceedings of CVPR, IEEE Computer Society Conference on Computer Vision and Pattern Recognition, pp. 1106–1112. IEEE (1997)

Feature Space Clustering for Trabecular Bone Segmentation

Benjamin Klintström[1,2(✉)], Eva Klintström[1,3], Örjan Smedby[1,2,3],
and Rodrigo Moreno[2]

[1] Center for Medical Image Science and Visualization,
Linköping University, Linköping, Sweden
Benklint@gmail.com
[2] KTH Royal Institute of Technology, School of Technology and Health,
Huddinge, Stockholm, Sweden
[3] Department of Medical and Health Science, Division of Radiology,
Linköping University, Linköping, Sweden

Abstract. Trabecular bone structure has been shown to impact bone strength and fracture risk. *In vitro*, this structure can be measured by micro-computed tomography (micro-CT). For clinical use, it would be valuable if multi-slice computed tomography (MSCT) could be used to analyse trabecular bone structure. One important step in the analysis is image volume segmentation. Previous segmentation techniques have either been computer resource intensive or produced suboptimal results when used on MSCT data. This paper proposes a new segmentation method that tries to balance good results against computational complexity.

Material. Fourteen human radius specimens where scanned with MSCT and segmented using the proposed method as well as two segmentation methods previously used to segment trabecular bone (Otsu and Automated Region Growing (ARG)). The proposed method (named FCH) uses a combination of feature space clustering, edge detection and hysteresis thresholding. For evaluation, we computed correlations with the reference method micro-CT for 7 structure parameters and measured segmentation time.

Results. Correlations with micro-CT were highest for FCH in 3 cases, highest for ARG in 3 cases, and in general lower for Otsu. Both FCH and ARG had correlations higher than 0.80 for all parameters, except for trabecular thickness and trabecular termini. FCH was 60 times slower than Otsu, but 5 times faster than ARG.

Discussion. The high correlations with micro-CT suggest that with a suitable segmentation method it might be possible to analyse trabecular bone structure using MSCT-machines. The proposed segmentation method may represent a useful balance between speed and accuracy.

Keywords: Feature-space · Clustering · Segmentation · Trabecular bone

1 Introduction

Besides overall bone mass, the structure of the trabecular bone network has been shown to greatly influence bone strength and the risk of fractures [1–3]. Skeletal structures that mainly consist of trabecular bone are the vertebrae and they are also commonly affected by fractures among the elderly [4].

© Springer International Publishing AG 2017
P. Sharma and F.M. Bianchi (Eds.): SCIA 2017, Part II, LNCS 10270, pp. 65–75, 2017.
DOI: 10.1007/978-3-319-59129-2_6

In vitro, trabecular bone structure can be measured by micro-computed tomography (micro-CT) and by histomorphometry of bone biopsies and these two methods show good agreement [5]. *In vivo,* trabecular bone structure can be measured by high-resolution peripheral quantitative CT (HR-pQCT) [6]. Another clinically available device that shows promise for trabecular structure analysis is cone-beam computed tomography (CBCT), for which *in vitro* studies show good agreement to micro-CT data [7]. The method for imaging the vertebrae in clinical practise is multi-slice computed tomography (MSCT). Previous studies on MSCT have shown good agreement with micro-CT regarding parameters like bone volume over total volume (BV/TV), while other parameters like trabecular nodes (Tb.Nd) and trabecular separation (Tb.Sp) have not been as strongly correlated [8, 9]. Due to its clinical use for imaging the vertebrae, it would be interesting if data from MSCT machines could be used to measure trabecular bone structures parameters.

One important step in calculating trabecular bone structure parameters is the segmentation algorithm used to separate bone from other tissues in the imaged volume [10]. Segmentation can be performed in different ways. For example, an automatic or manual intensity threshold can be effective for high resolution images. Segmenting images acquired with HR-pQCT, CBCT and MSCT is more challenging. Our group has previously shown good results for segmenting CBCT data by using the automated region growing method (ARG) proposed in [11]. There are also segmentation methods using clustering-algorithms [12], but, to the best of our knowledge, they have not been applied for segmenting trabecular bone so far.

The aim of this paper is to suggest a new way of segmenting trabecular bone images using a combination of edge-detection, clustering and thresholding as well as to investigate the possibility of using MSCT data to measure trabecular bone structure parameters.

2   Materials and Methods

2.1   Material

Fourteen human radius specimens from cadavers were used for the analysis. The radius specimens were donated for medical research in accordance with the ethical guidelines regulating such donations at the University of California, San Francisco. The specimens have been used in previous studies from our group [13–16]. The specimens are almost cubic with a side of 12–15 mm and they all include a slab of cortical bone, facilitating orientation.

2.2   Image Acquisition

The images were acquired using two machines:

- One Multi-Slice Computed Tomography, the Somatom Force (Siemens AG, Erlangen Germany). The acquisition was performed using the protocol for imaging the middle ear bones with a tube voltage of 120 kVp and a tube current 230 mAs.

The slice thickness was 400 μm, the increment between slices 200 μm, the pitch 0.8 and the FOV 50 mm, resulting in an intra-slice-resolution of 98 × 98 μm. This machine will be referred to as MSCT.

- One Micro-Computed Tomography, the Skyscan 1176 (Bruker micro-CT, Kontich Belgium). The images were acquired using a tube voltage of 65 kV and a tube current of 385 μA. The total exposure time was approximately 2 h. The resulting volumes had an isotropic resolution of 8.67 μm. This machine will be referred to as micro-CT.

The comparisons were made with the micro-CT data segmented using Otsu-thresholding as a gold standard [17]. The MSCT data were interpolated to an isotropic resolution of 98 μm using spline-interpolation in MATLAB (Mathworks).

2.3 Specimen Preparation and Imaging

Before imaging, the bone samples were de-fatted and then placed in test tubes filled with water. During imaging with the MSCT the test tubes were then placed in the centre of a paraffin cylinder with a diameter of 100 mm. To simulate measurements *in vivo*, a paraffin cylinder was used to simulate soft tissue.

2.4 Image Pre-processing

After imaging and interpolation, the MSCT-volumes were manually registered using the micro-CT volumes as references to ensure as close a match as possible between the processed Volumes of Interests (VOI:s). This registration was performed in two steps using the manual registration tools in MeVisLab (MeVis Medical Solutions AG). This resulted in digitally extracted cubes consisting only of trabecular bones with sides of approximately 8 mm.

2.5 Image Segmentation Using Feature-Space Clustering

The proposed segmentation uses a combination of feature-space clustering with k-means and hysteresis thresholding [18, 19].

This algorithm starts with a pre-filtering step to reduce noise in the data and produce a smoother final segmentation. First with a median filter with a kernel size of approximately twice the mean thickness of trabecular bone, which has been determined from the micro-CT-data, and then with a Gaussian filter with a standard deviation equal to the thickness of trabecular bone.

To facilitate later calculations, the grayscale is then normalized to be centred at 0 with a fixed standard deviation. For this algorithm the standard deviation was set to 500 as previous testing from our group has found that it approximately equals the average for multiple different types of CT-machines (CBCT, HR-pQCT and MSCT). However, the exact value chosen has limited impact on the final segmentation. The intensity values after this pre-processing will be referred to as α.

The next step in the algorithm is to use a combination of Canny and Sobel edge detection in 3D to find the edges of the trabecular structure as well as the direction and magnitude of those edges [20, 21]. The information on the edges was then used to give each voxel a value based on whether it was inside or outside of the edges as well as the magnitude of the edges around it, voxels inside of a strong edge were assigned a large positive value and voxels outside of a strong edge were instead assigned a large negative value. Voxels inside or outside of weak edges were instead assigned a smaller positive or negative value and the values assigned by this function were stored in a matrix. These values will later be referred to as β. Notice that this idea is somehow similar to level sets where a signed distance transform is defined and the edges are defined as zero crossings of such a function. The main advantage of using Canny and Sobel is that the computational complexity is largely reduced compared to level sets.

The next part of the algorithm performs a local normalization of α and β. The average of the six-neighbours and their six-neighbours of every voxel is subtracted from its original intensity value α and β independently. The normalized images are referred to as δ and γ, respectively for α and β. This normalization procedure is performed in order to improve the performance of the method on regions with locally varying grayscales, which is typical in trabecular bone.

At this stage a four-dimensional feature space is created, where each voxel gets its position based on the α-, β-, δ- and γ-values assigned to it. After this, k-means is used in order to split that feature space into two clusters, which should correspond to bone and background. The centre of these two clusters is then used to calculate the Euclidian distance d of each voxel p from those centres in the feature-space:

$$d_{(P,C)} = \sqrt[2]{(\alpha_p - \alpha_c)^2 + (\beta_p - \beta_c)^2 + (\delta_p - \delta_c)^2 + (\gamma_p - \gamma_c)^2} \qquad (1)$$

where c is the centre of the cluster.

The cluster whose centre has the lower α value is assumed to represent the background (in our case water) and the one with the higher α value the foreground (which should be trabecular bone). Each voxel is then assigned the value of the quotient $\lambda(p)$ between the distance from the centre of the background cluster and the distance from the centre of the bone cluster:

$$\lambda_{(p)} = \frac{d_{(p,c_{background})}}{d_{(p,c_{bone})}} \qquad (2)$$

where $c_{background}$ and, c_{bone} are the two cluster centres.

After λ has been calculated, hysteresis thresholding is applied to λ in 3D. First, an upper threshold is used to segment the image, and the voxels exceeding this threshold are assigned a value of 1 while zero is assigned to the remaining voxel. The voxel is also segmented using a lower threshold. The segmentation using the upper threshold is then iteratively expanded into its six-neighbourhood. Voxels not belonging to the lowest segmentation are filtered out. This procedure is then repeated with the resulting matrix until the result no longer changes between two iterations. Figure 1 shows the

```
Function hysteresis
    new = λ >5;
    low = λ >0.9;
    old = low;

    while new != old
        old = new;
        new = dilate(new,6-neighbours);
        new = new & low;
    end
    return new;
```

Fig. 1. Pseudo-algorithm for the hysteresis procedure

pseudo-code of the hysteresis procedure. In the experiments, we fixed the two thresholds to 5 and 0.9, respectively, based on some preliminary tests.

The resulting image after the last iteration is considered the final segmentation and is the output of the algorithm. This segmentation method that uses a combination of feature-space clustering and a hysteresis threshold will from now on be referred to as FCH.

2.6 Parameter Calculation

The final segmentations are used to calculate the following 7 different bone structure parameters:

1. Trabecular nodes (Tb.Nd) which is defined as the number of trabecular intersections per mm^3
2. Trabecular termini (Tb.Tm) which is defined as the number of free ends of trabeculae per mm^3
3. Trabecular separation (Tb.Sp) which is defined as the thickness of the spaces between the trabeculae in mm
4. Trabecular spacing (Tb.Sc) which is defined as the distance between the midlines of the trabeculae in mm
5. Trabecular number (Tb.N) which is defined as the number of trabeculae per mm
6. Trabecular thickness (Tb.Th) which is defined as the thickness of trabecular structures in mm
7. Bone volume over total volume (BV/TV) which is calculated by dividing the number of voxels defined as bone with the total number of voxels in the analysed volume.

All of the above parameters were calculated in 3D. Four parameters (Tb.Nd, Tb.Tm, Tb.Sc and Tb.N) were calculated after skeletonizing the binary image to one-voxel-wide lines using the method described in [22] without any pruning of the resulting skeleton.

Besides the above-mentioned structural parameters, the contrast to noise ratio (CNR) of the resulting segmentations were calculated and the time each segmentation method took to run was measured. In order to describe the agreement between different segmentation methods, the Dice coefficient was calculated, which is given by:

$$\text{Dice}_{(S1,S2)} = \frac{2 \times |S1 \cap S2|}{|S1| + |S2|} \tag{3}$$

where S1 and S2 are two segmentations and |.| is the number of voxels defined as bone by the segmentation.

The segmentation and calculation of the parameters were performed on a personal computer (PC) with an Intel Core i5-4460 (Intel Santa Clara, CA) at 3.2 GHz, 16 GB of random access memory (RAM) and a 64-bit system and was implemented in MATLAB.

2.7 Statistical Methods

Results from the calculation of the parameters are presented as mean values with standard deviations. The correlation coefficients that are presented are Pearson correlations with 95% confidence intervals. Statistics were calculated using MS Excel, the Dice coefficient was calculated in MATLAB, and all graphs and tables were created in MS Excel. Images were created in MATLAB and MeVisLab.

3 Experimental Results

The MSCT volumes were segmented using Otsu thresholding, our implementation of the ARG algorithm [11] and FCH. Our implementation of ARG has previously been used to segment HR-pQCT and CBCT trabecular volumes [7, 9, 13]. Micro-CT data were segmented using Otsu's threshold and used as a gold standard. Figure 2 shows the result of different segmentation for MSCT as well as raw-image slices for both MSCT and micro-CT.

Table 1 shows the mean and standard deviation of different histomorphometry parameters computed on the segmented images of both micro-CT and MSCT. As shown, all three segmentation methods used on the MSCT resulted in an underestimation of Tb.Nd and Tb.Tm by a factor of 65 to 140 and Tb.N was underestimated by a factor of about 2. Tb.Th and BV/TV were instead overestimated by a factor of about 4 and Tb.Sc was overestimated by a factor of approximately 2. Tb.Sp produced similar results for the micro-CT and MSCT data. Overall the three different segmentation methods used for the MSCT data produced results of similar magnitude.

Table 2 shows Pearson correlation coefficients for the different parameters using micro-CT as reference. All segmentation methods showed correlation coefficients of 0.70 or higher for all structure parameters except for Tb.Th where Otsu and FCH showed correlations approximately equal to 0 and ARG had a correlation coefficient of 0.61.

Micro-CT   MSCT   MSCT   MSCT   MSCT
Raw   Raw   Otsu   FCH   ARG

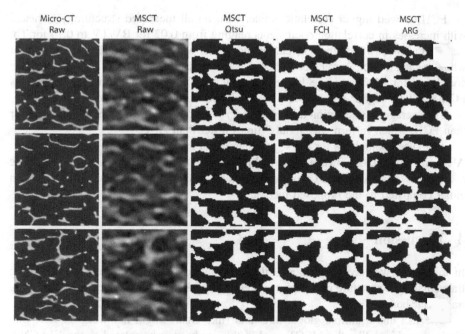

Fig. 2. Raw-image slices (left) for the micro-CT and the MSCT as well images from the MSCT segmented with the three different segmentation methods (right)

Table 1. Mean values of structural parameters, CNR and segmentation time in seconds ± standard deviations. $Tb.Nd$ [mm^{-3}], $Tb.Tm$ [mm^{-3}], $Tb.Sp$ [mm], $Tb.Sc$ [mm], $Tb.N$ [mm^{-1}], $Tb.Th$ [mm] and BV/TV [%]

Mean values ± SD		Tb.Nd	Tb.Tm	Tb.Sp	Tb.Sc	Tb.N	Tb.Th	BV/TV	CNR	Time
MSCT	Otsu	0.80 ± 0.22	0.67 ± 0.17	0.72 ± 0.09	1.23 ± 0.09	0.81 ± 0.06	0.46 ± 0.02	0.30 ± 0.08	5.15 ± 1.14	0.08 ± 0.02
	FCH	0.69 ± 0.16	0.61 ± 0.18	0.76 ± 0.10	1.40 ± 0.11	0.72 ± 0.06	0.58 ± 0.03	0.41 ± 0.10	4.74 ± 0.91	5.14 ± 1.36
	ARG	0.89 ± 0.16	0.75 ± 0.11	0.67 ± 0.06	1.20 ± 0.08	0.83 ± 0.05	0.48 ± 0.02	0.36 ± 0.07	5.68 ± 1.18	26.7 ± 8.16
micro-CT	Otsu	58.5 ± 15.6	86.3 ± 22.0	0.68 ± 0.10	0.74 ± 0.10	1.37 ± 0.18	0.13 ± 0.01	0.10 ± 0.03	29.4 ± 6.48	

Table 2. Correlation with micro-CT (reference method)

Pearson correlations coefficients (95% confidence interval)		Tb.Nd	Tb.Tm	Tb.Sp	Tb.Sc	Tb.N	Tb.Th	BV/TV
MSCT	Otsu	0.85	0.80	0.70	0.76	0.79	−0.04	0.85
		(0.58; 0.95)	(0.47; 0.93)	(0.27; 0.90)	(0.38; 0.92)	(0.45; 0.93)	(−0.56; 0.50)	(0.58; 0.95)
	FCH	0.92	0.87	0.80	0.90	0.88	−0.02	0.87
		(0.76; 0.97)	(0.63; 0.96)	(0.47; 0.93)	(0.71; 0.97)	(0.66; 0.96)	(−0.54; 0.52)	(0.63; 0.96)
	ARG	0.85	0.77	0.80	0.89	0.90	0.61	0.96
		(0.58; 0.95)	(0.40; 0.92)	(0.47; 0.93)	(0.68; 0.96)	(0.71; 0.97)	(0.12; 0.86)	(0.88; 0.99)

FCH showed higher correlations than Otsu on all measured structural parameters with increases in correlation coefficients ranging from 0.02 for BV/TV to 0.14 for Tb. Sc. FCH also showed higher correlation than ARG when measuring Tb.Nd and Tb.Tm while having lower correlation when measuring BV/TV and Tb.Th. The average correlation for ARG was 0.08 higher than that of FCH when including Tb.Th whereas FCH was 0.01 higher when Tb.Th was excluded from the comparison.

The average segmentation times (Table 1) show that Otsu was about 64 times faster than FCH, which in turn was about 5 times faster than ARG.

The Dice coefficients for the segmentations on the MSCT data were 0.93 between ARG and Otsu, 0.82 between ARG and FCH and 0.80 between Otsu and FCH. These results show that ARG and Otsu produces more similar results compared to each than either compared to the proposed method.

4   Discussion

In this study we have investigated the potential use of MSCT data for measuring trabecular bone structure parameters as well as proposing a new method for segmenting trabecular bone.

The fact that a simple global threshold (like Otsu) might not work so well with clinical machines like the MSCT used in this study may be related to the low CNR, compared to micro-CT images, when imaging trabecular bone. One reason for the lower CNR and another possible reason why a global threshold performs poorly could be the partial volume effect where the large image voxels of the MSCT, and other clinical machines, only partly contains bone. This causes a lot of voxels with intermediary intensity values that are both bone and background.

The MSCT data showed relatively high correlation with the micro-CT data with correlations above 0.80 for all structure parameters except Tb.Tm and Tb.Th when using either the ARG or FCH algorithm. This suggest that it might be possible to use MSCT:s for measuring and monitoring trabecular bone structure.

Otsu was by far the quickest segmentation method with an average segmentation-time of 0.08 s, but it also had the lowest average correlation with micro-CT data. In contrast, the ARG algorithm had the highest average correlation coefficient and was the slowest algorithm with an average segmentation-time of 26.7 s. FCH showed higher correlations with micro-CT than Otsu for all parameters and, except for Tb.Th, it showed correlations similar to ARG while at the same time being significantly faster with an average segmentation time of 5.14 s.

As in previous studies from our group on clinical machines, the MSCT data underestimated Tb.Nd and Tb.N while overestimating Tb.Th and BV/TV [7, 9]. As an over- or underestimation in absolute values can be corrected if the correlation is high, a higher correlation is more relevant for detecting true differences between individuals.

The finding that the Otsu algorithm is the fastest of the three compared algorithms was expected, due to the simple calculations that are performed on the histogram of the data. As the ARG algorithm is an iterative algorithm and our implementation performs 50 iterations of the region growing with different homogeneity thresholds, it is reasonable for it to take longer than Otsu to segment the image. It is also reasonable that

FCH takes longer to segment the image than Otsu as it uses more computer-intensive calculations to segment the image. Clinical workflows are often time-pressed which means that if a new algorithm is to be widely accepted it needs to not only give accurate and valid results, it has to do so in as short a time as possible. This makes FCH an interesting contribution to the arsenal for analysing trabecular bone as it results in correlations similar to ARG while taking approximately one fifth as long to compute.

The low correlation for Tb.Th for both Otsu and FCH when compared to ARG is interesting and would be interesting to investigate further. There could be multiple reasons for this result, but so far no satisfactory explanation has been found. As a previous study from our group [7] has shown that Tb.Th has little to no impact on the overall strength of the trabecular network, as measured by FEM, a lower performance on this parameter might be acceptable although better performance of course is preferred.

One interesting result is the fact that the Dice coefficient for ARG-FCH (0.82) and Otsu-FCH (0.80) is so much lower than the Dice coefficient for ARG-Otsu (0.93). One explanation for this could be the fact that while they use different types of thresholds (homogeneity vs. intensity) and a different process for segmenting the image (region growing vs. a global threshold), both ARG and Otsu only use intensity data. FCH, on the other hand, also processes information based on edge detection, which could introduce another dimension to base the segmentation on and result in less similar segmentations. If one looks at the segmented slices in Fig. 2, both ARG and FCH seems to be preserving more of the structure of the trabecular bone, which could explain their higher correlations with the micro-CT data regarding trabecular bone structure parameters. The visual similarity, but low Dice coefficient, between ARG and FCH is something that warrants further investigation as it might help with further development of the algorithms

One of the weaknesses of this study is the fact that it only used 14 bones, which means that the statistical power is relatively low. One of the ongoing projects for our team is to include more bone specimens to verify our results in a bigger sample. It would also have been of interest to study the agreement of each of the segmentation methods applied to MSCT data to the micro-CT segmentation results using the Dice coefficient. However, registration problems between the modalities made this comparison impractical. Another of our ongoing projects is to test other clinical protocols with varying scanning parameters on the MSCT and a clinical trial to evaluate the performance of the CBCT used in [13] when measuring trabecular bone structure parameters in osteoporotic individuals.

In conclusion, the proposed segmentation method shows a reasonable balance between performance and computational intensity while the MSCT shows potential for analysing trabecular bone structure if the correct segmentation method is used.

References

1. Kleerekoper, M., Villanueva, A.R., Stanciu, J., Rao, D.S., Parfitt, A.M.: The role of three-dimensional trabecular microstructure in the pathogenesis of vertebral compression fractures. Calcif. Tissue Int. **37**(6), 594–597 (1985)
2. Ulrich, D., van Rietbergen, B., Laib, A., Ruegsegger, P.: The ability of three-dimensional structural indices to reflect mechanical aspects of trabecular bone. Bone **25**(1), 55–60 (1999)
3. Parkinson, I., Badiei, A., Stauber, M., Codrington, J., Müller, R., Fazzalari, N.: Vertebral body bone strength: the contribution of individual trabecular element morphology. Osteoporos. Int. **23**(7), 1957–1965 (2012)
4. Mosekilde, L.: Vertebral structure and strength. In vivo and in vitro. Calcif. Tissue Int. **53**, S121–S126 (1993)
5. Thomsen, J.S., Laib, A., Koller, B., Prohaska, S., Mosekilde, L., Gowin, W.: Stereological measures of trabecular bone structure: comparison of 3D micro computed tomography with 2D histological sections in human proximal tibial bone biopsies. J. Microsc. **218**(Pt 2), 171–179 (2005)
6. Burghardt, A.J., Pialat, J.B., Kazakia, G.J., Boutroy, S., Engelke, K., Patsch, J.M., et al.: Multicenter precision of cortical and trabecular bone quality measures assessed by high-resolution peripheral quantitative computed tomography. J. Bone Mineral Res.: Official J Am. Soc. Bone Mineral Res. **28**(3), 524–536 (2013)
7. Klintström, E., Klintström, B., Moreno, R., Brismar, T.B., Pahr, D.H., Smedby, Ö.: Predicting trabecular bone stiffness from clinical cone-beam CT and HR-pQCT data; An in vitro study using finite element analysis. PLoS ONE **11**(8), e0161101 (2016)
8. Bauer, J.S., Link, T.M., Burghardt, A., Henning, T.D., Mueller, D., Majumdar, S., et al.: Analysis of trabecular bone structure with multidetector spiral computed tomography in a simulated soft-tissue environment. Calcif. Tissue Int. **80**(6), 366–373 (2007)
9. Klintström, E., Smedby, Ö., Moreno, R., Brismar, T.B.: Trabecular bone structure parameters from 3D image processing of clinical multi-slice and cone-beam computed tomography data. Skeletal Radiol. **43**(2), 197–204 (2014)
10. Bouxsein, M.L., Boyd, S.K., Christiansen, B.A., Guldberg, R.E., Jepsen, K.J., Muller, R.: Guidelines for assessment of bone microstructure in rodents using micro-computed tomography. J. Bone Mineral Res.: Official J. Am. Soc. Bone Mineral Res. **25**(7), 1468–1486 (2010)
11. Revol-Muller, C., Peyrin, F., Carrillon, Y., Odet, C.: Automated 3D region growing algorithm based on an assessment function. Pattern Recogn. Lett. **23**(1–3), 137–150 (2002)
12. Kettaf, F., Bi, D., de Beauville, J.A.: A comparison study of image segmentation by clustering techniques. In: 3rd International Conference on Signal Processing. IEEE (1996)
13. Klintström, E., Smedby, Ö., Klintström, B., Brismar, T.B., Moreno, R.: Trabecular bone histomorphometric measurements and contrast-to-noise ratio in CBCT. Dentomaxillofacial Radiol. **43**(8), 20140196 (2014)
14. Petersson, J., Brismar, T., Smedby, Ö.: Analysis of skeletal microstructure with clinical multislice CT. In: Larsen, R., Nielsen, M., Sporring, J. (eds.) MICCAI 2006. LNCS, vol. 4191, pp. 880–887. Springer, Heidelberg (2006). doi:10.1007/11866763_108
15. Moreno, R., Borga, M., Klintström, E., Brismar, T., Smedby, Ö.: Anisotropy estimation of trabecular bone in gray-scale: comparison between cone beam and micro computed tomography data. In: Tavares, J.M.R.S., Jorge, R.N. (eds.) Developments in Medical Image Processing and Computational Vision. LNCVB, vol. 19, pp. 207–220. Springer, Cham (2015). doi:10.1007/978-3-319-13407-9

16. Moreno, R., Borga, M., Klintström, E., Brismar, T., Smedby, Ö.: Correlations between fabric tensors computed on cone beam and micro computed tomography images. In: Computational Vision and Medical Image Processing (VIPIMAGE), pp. 393–398. CRC Press (2013)
17. Otsu, N.: Threshold selection method from gray-level histograms. IEEE Trans. Syst. Man Cybern. **9**(1), 62–66 (1979)
18. Hartigan, J.A.: Clustering algorithms (1975)
19. Hartigan, J.A., Wong, M.A.: Algorithm AS 136: a k-means clustering algorithm. J. Roy. Stat. Soc.: Ser. C (Appl. Stat.) **28**(1), 100–108 (1979)
20. Canny, J.: A computational approach to edge detection. IEEE Trans. Pattern Anal. Mach. Intell. **8**(6), 679–698 (1986)
21. Sobel, I., Feldman, G.: A 3 × 3 isotropic gradient operator for image processing. A talk at the Stanford Artificial Project. pp. 271–272 (1968)
22. Xie, W., Thompson, R.P., Perucchio, R.: A topology-preserving parallel 3D thinning algorithm for extracting the curve skeleton. Pattern Recogn. **36**(7), 1529–1544 (2003)

Airway-Tree Segmentation in Subjects with Acute Respiratory Distress Syndrome

Kristína Lidayová[1]([✉]), Duván Alberto Gómez Betancur[4], Hans Frimmel[2], Marcela Hernández Hoyos[4], Maciej Orkisz[5], and Örjan Smedby[3]

[1] Division of Visual Information and Interaction, Centre for Image Analysis, Uppsala University, Uppsala, Sweden
kristina.lidayova@it.uu.se

[2] Division of Scientific Computing, Department of Information Technology, Uppsala University, Uppsala, Sweden

[3] School of Technology and Health, KTH Royal Institute of Technology, Stockholm, Sweden

[4] Systems and Computing Engineering Department, School of Engineering, Universidad de Los Andes, Bogotá, Colombia

[5] CREATIS UMR 5220, U1206, CNRS, Inserm, INSA-Lyon, Université Claude Bernard Lyon 1, UJM-Saint Etienne, Univ. Lyon, 69621, Lyon, France

Abstract. Acute respiratory distress syndrome (ARDS) is associated with a high mortality rate in intensive care units. To lower the number of fatal cases, it is necessary to customize the mechanical ventilator parameters according to the patient's clinical condition. For this, lung segmentation is required to assess aeration and alveolar recruitment. Airway segmentation may be used to reach a more accurate lung segmentation. In this paper, we seek to improve lung segmentation results by proposing a novel automatic airway-tree segmentation that is able to address the heterogeneity of ARDS pathology by handling various lung intensities differently. The method detects a simplified airway skeleton, thereby obtains a set of seed points together with an approximate radius and intensity range related to each of the points. These seeds are the input for an onion-kernel region-growing segmentation algorithm where knowledge about radius and intensity range restricts the possible leakage in the parenchyma. The method was evaluated qualitatively on 70 thoracic Computed Tomography volumes of subjects with ARDS, acquired at significantly different mechanical ventilation conditions. It found a large proportion of airway branches including tiny poorly-aerated bronchi. Quantitative evaluation was performed indirectly and showed that the resulting airway segmentation provides important anatomic landmarks. Their correspondences are needed to help a registration-based segmentation of the lungs in difficult ARDS cases where the lung boundary contrast is completely missing. The proposed method takes an average time of 43 s to process a thoracic volume which is valuable for the clinical use.

Keywords: Airway segmentation · Airway-tree centerline detection · Thoracic CT · ARDS

© Springer International Publishing AG 2017
P. Sharma and F.M. Bianchi (Eds.): SCIA 2017, Part II, LNCS 10270, pp. 76–87, 2017.
DOI: 10.1007/978-3-319-59129-2_7

1   Introduction

Acute Respiratory Distress Syndrome (ARDS) is a life-threatening respiratory condition. This syndrome may occur as a consequence of a major injury or different pulmonary aggressions (bacteriological or chemical). As a result, the lungs are unable to fill with air and cannot provide enough oxygen into the bloodstream. The treatment requires the use of mechanical ventilation to pump air into the patient's lungs until the cause of the disease is detected and treated. Mechanical ventilator parameters (tidal volume and positive end-expiratory pressure - PEEP) need to be set carefully, taking into account the patient's clinical condition. Too large air volumes or too high PEEP may injure the lungs. On the other hand, lack of oxygen in the blood may lead to a multiple organ dysfunction syndrome. Both cases are with fatal consequences. Knowledge about the lung aeration is the key to helping prevent the injury. Since gray levels in Computed Tomography (CT) images are associated with the tissue density, thoracic CT scans are well suited to obtain this knowledge. Nevertheless, the accuracy of lung aeration quantification is hampered by two factors: the difficulty in delineating the outer boundary of the lungs (due to local lack of contrast), and the inclusion of internal structures not belonging to the parenchyma, such as the airway tree. To cope with both problems, airway segmentation can be useful. It can provide anatomic landmarks useful in lung delineation, and it can also serve for airway removal.

Airway segmentation is confronted with specific difficulties related to variable contrast in CT images from subjects with ARDS. In deflated lungs with low PEEP condition, the large opacities of non-aerated parenchyma make it very difficult, if at all possible, to see the lung boundary. Also, small bronchi are thinner in this condition. At the other extreme, when lungs are strongly inflated, the problem is mainly caused by a low contrast between the parenchyma and bronchi lumen. For smaller bronchi, the thickness of the wall may be below the scan resolution and can cause the segmentation algorithms to leak into the parenchyma. Bronchi filled with liquid or mucus seem to be disconnected on the CT scan, which presents another complication, mainly for segmentation algorithms based on a region-growing or wave propagation.

In this paper, we propose a novel airway-tree segmentation method that successfully deals with the challenges brought by ARDS. The method detects an approximate airway centerline tree and then applies the obtained intensity and distance information to restrict the onion-kernel region-growing segmentation and prevent it from leaking into the parenchyma. The method was evaluated on a series of thoracic CT images of subjects with ARDS, acquired at significantly different mechanical-ventilation conditions. The results show that the proposed method is able to find a large number of branches including tiny bronchi. This is important for achieving our overall goal - the improvement of the lung segmentation - especially in low PEEP conditions where the lung boundary contrast is missing. A possible approach is to use a self-atlas segmentation. It consists in segmenting the lungs in the most-contrasted volume and warping them towards the least-contrasted one by means of registration. In this work, we confirmed that using a hybrid registration that combines airway-tree landmark correspondences obtained from the segmented airway tree with gray-level information leads to an improvement in the lung segmentation. In addition, the proposed method is fast, which is valuable for clinical use.

2   Related Work

The segmentation of airway trees plays an important role in the analysis of various lung diseases. Since each disease brings its own challenges, specifications, and restrictions, there exist many different methods for airway-tree segmentation. An intuitive approach is based on a 3D region-growing [1–3] and relies on the contrast between the airway lumen and the airway wall that is usually relatively high. However, due to noise or partial volume effect, the contrast may be missing and the region-growing segmentation may leak into the parenchyma. Algorithms applying region-growing differ in the way they prevent the leakages. Many algorithms use region-growing only for large airways and implement additional procedures to identify smaller airways. A review work of Pu et al. [4] provides a division of other methods for airway segmentation into five categories based on their methodology as follows:

Morphological methods [5] use morphological reconstruction techniques to identify airways on CT slices and then reconstruct them as a connected 3D airway tree. Nevertheless, the reconstruction largely depends on how continuous the detected airway candidates are in space.

Knowledge-based methods [6] identify the airways by using various anatomical knowledge and rules. Since it is hard to list all the rules that characterize the airways, these methods are often used in combination with other approaches and remove false positives rather than detect the true positives.

Template matching methods [7] search for airway regions that match a set of predefined elliptical (2D) or tubular (3D) templates. Just like with the previous category, the main problem is to list all the templates that would fully describe the airways together with their size variability.

Machine learning methods [8] automatically learn specific airway characteristics during the training process. However, the diversity of the pre-labelled training data and the feature selection are critical for the final performance.

Geometric shape analysis methods [9] make use of Hessian-matrix based filters for tubular shape detection. However, these filters are sensitive to noise and irregularities caused by diseases. In addition, Gaussian convolution filter may blur small airways.

A summary and comparison of 15 different airway-tree segmentation methods can be found in the EXACT'09 challenge [10]. Each of the algorithms has its advantages and disadvantages. From the published results, it is evident that there is no one algorithm solving all application challenges. Therefore, we propose a method focused on subjects with ARDS. The method can be categorized as a knowledge-based method combined with 3D region-growing. A detailed description of the proposed method follows in the next section.

3   The Proposed Method

The proposed method takes a thoracic CT volume as an input. Since airways appear as elliptical regions on orthogonal CT image slices, our method uses knowledge-based filters to identify central voxels of these regions - the graph nodes. These nodes can represent bifurcating airways, as well as end or line points of the airway. There is no need to detect every single airway central voxel – a sparse detection is satisfactory. The distance between the central voxels (detected nodes) is, in general, longer for thicker airways and much shorter (only 1–2 voxels) for thinner ones which safely detects also curved branches. The nodes are then connected with straight edges into a tree graph structure representing the approximate centerline tree of the airway. The connection process is controlled by a set of criteria to ensure that the edges lie inside the airway tree and loops are avoided. The approximate centerline tree serves as a sufficient input for the subsequent segmentation algorithm. The complete centerline tree is voxelised, and for each voxel, information about its position, expected radius and an intensity range related to the position is stored. This information is important to prevent leakage into the parenchyma. Each voxel is then used as a seed point in a modified region-growing algorithm, which generates as output a binary segmentation of the airway tree. The pipeline of the whole algorithm is visualised in Fig. 1. In the following subsections, we describe the centerline tree extraction and segmentation algorithms more in detail.

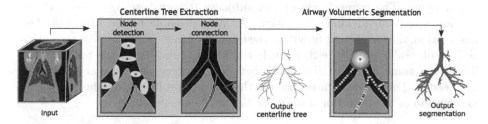

Fig. 1. Algorithmic overview. From input CT volume, a centerline tree is extracted in two steps, followed by airway segmentation based on region growing.

3.1   Centerline Tree Extraction Algorithm

The basic centerline tree extraction algorithm is based on an idea presented in [11]. However, it has been modified and extended for the needs of airway-tree centerline detection. Since airways are filled with air, potential graph nodes - voxels of interest - are expected to be aerated. Furthermore, the morphology of a lung suffering from the ARDS is very heterogeneous, it is meaningful to divide these potential voxels of interest into classes based on their physical density and to treat each of the classes differently. In literature, voxels in the parenchyma are divided based on their physical density into four aeration classes [12].

The denser the tissue inside the voxel, the higher the X-ray attenuation, which results in a higher gray level value (CT number). The gray level values are expressed in Hounsfield units (HU). Pure air voxels are assigned values of –1000 HU (no attenuation), whereas pure water voxels have values of 0 HU. Since tissues are mainly composed of water, their typical density is close to that of water. The four aeration classes then are:

- **Over-aerated voxels** [–1024, –900) HU may be found in over-distended parenchymal regions of mechanically ventilated lungs; in normal lungs, voxels with densities from this range are located mainly inside the largest airways.
- **Normally aerated voxels** [–900, –500) HU, in case of healthy subjects, are mostly located in parenchyma or can be located in thin airway bronchi, where the partial volume effect causes the rise in their intensities. In subjects suffering from ARDS, bronchi can be partly filled or completely blocked by mucus or liquid, which may increase the density even in the thick bronchi.
- **Poorly aerated voxels** [–500, –100) HU are usually observed in diseased parenchyma affected by ARDS, but gray levels from this range may also occur in tiny bronchi.
- **Non-aerated voxels** [–100, 100) HU, as the name of the class specifies, correspond to parenchymal regions that do not contain any air. These voxels may also correspond to structures like airway wall or vessels.

The potential graph nodes should be within the range [–1024, –100) HU which corresponds to the intensity ranges of the three aerated classes defined for the parenchyma. The parenchyma of sick subjects is very heterogeneous and can contain voxels of all kind of aeration. As it is possible to see from the class description, the voxels that fit within the intensity range of our interest are potential graph nodes present, either inside the airway tubes or in the parenchyma. Therefore, some filters to distinguish between the true and false airway voxels are needed. The rules are introduced in the following subsections where a further description of the detection and connection steps are provided.

Node Detection. The input volume is analysed through all slices in the three axis-oriented directions (axial, sagittal and coronal). In each slice, if a voxel has the intensity within the range [–1024, –100) HU, it is considered as a potential airway node. Such voxel needs to be examined further by looking into the intensities of surrounding voxels from the same slice. First, the position of the candidate voxel must be confirmed to be central within the lumen. This is done by casting rays in four main directions (up, down, left and right) starting from the candidate voxel position outwards until they reach a lumen border. The lumen border is defined as the first of n consecutive voxels of intensity higher than I, where I is either the upper bound of the aeration class where the candidate voxel belongs or intensity of the candidate voxel increased by 100 HU, whichever of them is higher. This offers an overlap if the candidate voxel intensity is similar to the higher bound intensity. If the two vertical distances and the two horizontal distances to the lumen border are pairwise equal, then these distances are saved as a major (a) and a minor (b) axis of a bounding ellipse. If the length of these two axes falls outside the specified minimum and maximum allowed radius, then the candidate voxel is discarded. In a subsequent step two auxiliary ellipses - inner and outer - are introduced. Both ellipses share the same center with the bounding ellipse and lie in the same slice, but have different axis lengths. They are shorter by x (inner ellipse) and longer by y (outer ellipse) pixels than the bounding ellipse axes. By subtracting the average intensity of voxels on the inner ellipse from the average intensity of voxels on the outer ellipse, we check if the region supposed to be the lumen is sufficiently darker than its wall. The difference, how much darker the inside should be, compared to the outside, is a function of a radius. The radius, $r = \sqrt{ab}$, is calculated from the major and minor axes as an area-preserving transformation from an ellipse to a circle. For bigger radii, the difference should be larger than for smaller radii since we expect that bigger bronchi are less affected by the partial-volume effect, as well as by physiological-liquid blockages, than smaller bronchi. The nodes that pass the complete set of filters are saved together with information about position, intensity and radius. Table 1 summarises the values for all the parameters.

Table 1. Parameters used in the algorithm listed with their corresponding values

Parameter name	[−1024, −900) HU	[−900, −500) HU	[−500, −100) HU
n number of border voxels	3	2	1
min_radius	1 mm	limited by the voxel size	limited by the voxel size
max_radius	20 mm	5 mm	3 mm
x voxels shorter	1	1	1
y voxels longer	2	1	0
max_edge_length	35 mm	10 mm	5 mm

Node Connection. In this step, the previously detected graph nodes are connected into a graph structure. Starting with each node considered as a separate graph, the graphs are gradually merged by adding links between them. A link between two nodes can be created only if the straight-line connection is shorter than *max_edge_length* (See Table 1), and lies within the airway lumen, *i.e.*, if all voxels intersected by the connection line have intensities within the range [−1024 HU, L), where L is the upper bound of the aeration class corresponding to the brighter of the two connected nodes. If the two nodes are already part of the same graph (connected via other nodes), or if there exists another node at a closer distance that fulfills the conditions, then the connection is not established. After all possible connections have been created, one or more approximate airway centerline trees represent the result. In situations when the bronchi contain mucus, the detected centerline trees are disconnected. The mucus consists mainly of liquid and therefore the corresponding intensity is outside the aerated range. Based only on the distance and the intensity, it is not possible to decide whether the connection passes through the mucus (and should be part of bronchi) or through the parenchyma (and should not be part of bronchi). Therefore, centerline trees of bronchi filled with liquid or mucus are left disconnected but not removed. If only one resulting tree is desired, all centerline trees not connected to the biggest centerline tree are removed.

3.2   Airway Segmentation

The approximate centerline tree is an important input for the modified onion-kernel region-growing algorithm, originally proposed for use in colon segmentation [13]. The centerline tree consists of nodes that contain information about their position, the airway radius and the intensity range in which the nodes have been detected. Now, the linear edge connections need to be voxelised and the node information interpolated for all the voxels on the line between two nodes. All these voxels become seeds for the region-growing algorithm. Since airway-tree segmentation is challenging mainly because the parenchyma has intensities very similar to those inside the airway, and the airway wall is often disconnected on several places, the region growing algorithm uses the information about the airway radius and intensity to stop the leakage. The region growing can not propagate further than twice the radius at the seed point. Furthermore, only voxels within the intensity range $[-1024$ HU, $L)$ - where L is again the upper bound of the aeration class corresponding to the brighter of the two connected nodes - should be included in the segmentation.

The onion-kernel region-growing algorithm itself is similar to classical region growing (26 neighbours). However, in the onion kernel case the growing is performed in layers from the seed point outwards (inverse onion peeling). First, the seed voxel is automatically included in the segmentation and is considered to be a zero layer. Then, all voxels from the first layer (1 voxel away from the seed) are processed. One voxel is included in the segmentation if it meets the intensity-based inclusion criterion and is 26-connected with at least one voxel from the previous layer that is already part of the segmentation. Afterward, all voxels which are 1 voxel further from the seed than the voxels considered in the previous step, are processed. This repeats until the distance from the seed point

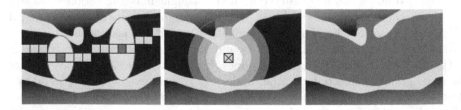

Fig. 2. Onion-kernel region-growing, (a) the input centerline tree is voxelised, (b) each centerline tree voxel becomes a seed; the segmentation is not allowed to grow backwards to fill the cavities (the dark green border delineates the region that would be filled otherwise, (c) the final segmentation. (Color figure online)

reaches twice the radius. The reasoning behind this inverse onion peeling is to achieve locally convex segmentation without segmentation "overhangs". Figure 2 shows the advantage of the inverse onion peeling propagation.

4   Evaluation and Results

We evaluated the proposed method, both qualitatively and quantitatively, using real 3D thoracic CT volumes of piglets with ARDS induced. The qualitative comparison was performed against a reference method based on region-growing and introduced by Mori et al. [1]. The quantitative evaluation tested if the goal of our work - to improve the lung segmentation by use of the airway-tree segmentation - is fulfilled.

4.1   Dataset

For each piglet, several 3D image pairs (end-inspiration/end-expiration) were acquired at various volume and pressure settings. For the qualitative evaluation two piglets were chosen, providing 35 image pairs, resulting in a total of 70 thoracic CT volumes that were processed. For the quantitative evaluation, large displacements and density changes between the analysed volumes are the most challenging and, therefore, three images acquired in end-inspiration at extreme and intermediate PEEP values – 20, 10, and 2 cmH_2O, with a constant tidal volume $V_t = 5$ ml/kg – were chosen.

4.2   Qualitative Evaluation

The reference segmentation method used for a visual comparison of the results is commonly referred to be a benchmark within the airway segmentation methods. It is simple and fast, and it avoids leakages into the parenchyma. The main idea of the algorithm is to gradually increase the threshold value of a region-growing until a sudden leakage appears. Afterward, the segmentation that was grown with the last threshold without leakage, is retained.

We have segmented the airway tree from 70 thoracic CT volumes using this method and the proposed one. In volumes acquired at high pressure, the two methods performed similarly, although the proposed method detected slightly longer branches. In images with lower PEEP, the pulmonary parenchyma has higher density and the contrast with the airway lumen may locally increase but leaving still leakages at various sites, therefore methods based on region growing don't perform better. Thanks to the underlying shape model the proposed method was able to better detect small airways, which resulted in more and longer branches compared to the standard method, as illustrated in Fig. 3.

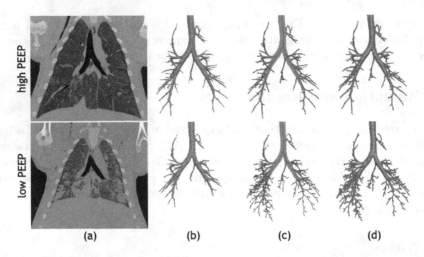

Fig. 3. Qualitative airway segmentation results, (a) Coronal slice of thoracic CT showing the differences in parenchyma intensities between high and low PEEP, (b) result of the reference method, (c) result of the reference method overlayed with a skeleton from the proposed method (d) result of the proposed method.

4.3   Quantitative Evaluation Pipeline

Quantitative evaluation was performed indirectly by comparing two registration-based lung-segmentation methods. The first one used an intensity-driven registration method while the second one used a hybrid registration where intensity information was enriched with airway-tree landmark correspondences. The goal was to test whether the bifurcation and end points obtained from the proposed airway-tree segmentation can improve the lung segmentation in situations where the lung boundary itself does not have sufficient contrast.

To assess registration accuracy, an expert in ARDS was asked to interactively segment the lungs and to mark several anatomical landmarks in the most challenging region of six CT volumes acquired at high, medium and low PEEP. The lung mask segmented by the expert in the image with the best lung boundary contrast (high PEEP), as well as the landmarks in this image, were warped using the transformation field resulting from the registration process between the original image with the highest PEEP and the original image with medium or low PEEP, where the lung is to be automatically segmented.

Registration accuracy was quantitatively evaluated in two ways: (1) the Dice score was calculated between the deformed lung mask of the image with the highest PEEP and the lung mask interactively segmented by the expert in the image with medium or low PEEP, and (2) the residual Euclidean distance was computed, after warping, between corresponding pairs of landmarks.

The landmark correspondences needed to drive the hybrid registration process were obtained from each airway segmentation by first extracting its graph representation [14] where nodes represent bifurcations or end points, and then by using a matching algorithm proposed in [15]. The latter provides a list of node pairs that match between two airway trees.

4.4   Quantitative Evaluation Results

The accuracy of the registration process is closely related to the number and location of landmark correspondences. Thus, it is important to have accurate airway segmentation with as many branches as possible and without leakages. Therefore, the visual assessment of the segmentation results motivated the use of the proposed method to improve the registration process.

The Dice similarity score for the lung mask segmentation was calculated for the intensity-driven and hybrid registration. For images acquired at high and low PEEP the Dice score was 0.81 for intensity-driven registration and 0.88 for the hybrid registration. Figure 4 visually illustrates these similarities on coronal slices from the posterior region, where the lack of contrast hampered the registration. The main difference is observed in the lowest part of the lung where the hybrid approach reached a better alignment than the intensity-based one. For images acquired at high and medium PEEP, the Dice score was 0.94 for intensity-driven and 0.96 for the hybrid registration.

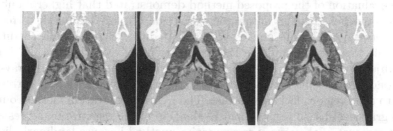

Fig. 4. Coronal slices of moving and fixed images, from the experiment between high PEEP and low PEEP, superposed: before registration (left), after intensity-driven registration (center), and after hybrid registration (right). Color areas correspond to regions where gray levels differ between the images due to lack of alignment and/or to density changes in diseased zones. (Color figure online)

The Euclidean distance between the pairs of manually placed landmarks was calculated between volumes acquired at high and medium PEEP. The average residual distance between the pairs of landmark locations before any transformation was 25.3 mm (range 21.9–29.7 mm). After applying the intensity-based registration and after applying the hybrid registration the average residual distance was 23.5 mm (range 20.4–26.3 mm) and 10.9 mm (range 0.71–29.3 mm), respectively.

Nevertheless, almost the same results (Dice scores and residual distances) were obtained using the reference airway-segmentation method (Mori et al. [1]), within the hybrid registration.

4.5 Computation Time

The proposed algorithm for airway-tree segmentation was implemented in C++. The average running time of a single-threaded implementation was 43 s per thoracic CT volume of mean size $512 \times 512 \times 441$ voxels. The centerline tree extraction took an average time of 41 s, while the onion-kernel region-growing algorithm took an average time of 2 s.

5 Discussion and Conclusions

The method is able to cope with the gray-level diversity in subjects with ARDS that changes with time, mechanical ventilation, and patient position, and it improved lung segmentation by providing landmarks for registration.

Qualitative assessment of the proposed method and the reference method showed that the proposed method detected a larger number of small branches, particularly in poorly-contrasted images acquired at low-pressure conditions. Occasionally, the resulting airway-tree segmentation contained small spurious bulges, caused by a leakage that was stopped by the distance criterion. The distance criterion was set to cover the local neighbourhood that is part of bronchi and at the same time stops the leakage growth already in an early stage. Quantitative evaluation of the proposed method demonstrated that higher number of detected branches yields to automatic extraction of more landmark correspondences between trees from images acquired at different ventilation conditions. This result was used to improve the registration between thoracic CT images from subjects with ARDS, by means of a hybrid approach that uses gray-level correspondences to align well-contrasted structures and landmark correspondences to align regions lacking contrast. The hybrid registration method outperformed gray-level-based registration. Evaluation of the registration serves as an indirect measure of the airway segmentation method because landmark distribution and location strongly depend on the airway-tree segmentation quality.

Although registration results using landmarks from both airways segmentation methods are comparable, this does not devalue the superiority of the proposed method but rather reflects a limitation of the hybrid registration, which would require landmarks in the least contrasted region, where the bronchi are not visible, because they are too thin, and none of the methods can segment them.

In future work, we would like to explore an anatomical evaluation of the approximate centerline tree, based on angles at the bifurcations, to be able to remove the *max_edge_length* criterion. This should allow establishing longer connections between the nodes and, at the same time, the anatomical evaluation will filter out possible spurious branches. The anatomical evaluation will be helpful also for connecting bronchi that are partially filled with mucus and appear disconnected in the CT volumes.

Acknowledgement. The authors thank Dr. Jean-Christophe Richard from team of Réanimation Médicale of the Hôpital de la Croix-Rousse, Lyon, France, for facilitating the images used in this work and helping with the segmentation of the same ones.

K. Lidayová, H. Frimmel, and Ö. Smedby have been supported by the Swedish Research Council (VR), grant no. 621-2014-6153. D. Gómez Betancur has been supported by Colciencias doctoral scholarships program. M. Hernández Hoyos, D. Gómez Betancur, and M. Orkisz were also partly supported by the French-Colombian ECOS Nord grant no. C15M04.

References

1. Mori, K., Hasegawa, J.I., Toriwaki, J.I., Anno, H., Katada, K.: Recognition of bronchus in three-dimensional X-ray CT images with applications to virtualized bronchoscopy system, Proc. 13th Int. Conf. Pattern Recogn. **3**, 528–532 (1996)
2. Wiemker, R., Bülow, T., Lorenz, C.: A simple centricity-based region growing algorithm for the extraction of airways. In: Proceedings of 2nd International Workshop on Pulmonary Image Analysis (MICCAI), pp. 309–314 (2009)
3. van Rikxoort, E.M., van Ginneken, B.: Automated segmentation of pulmonary structures in thoracic computed tomography scans: a review. Phys. Med. Biol. **58**, 187–220 (2013)
4. Pu, J., Gu, S., Liu, S., Zhu, S., Wilson, D., et al.: CT based computerized identification and analysis of human airways: a review. Med. Phys. **39**, 2603–2616 (2012)
5. Aykac, D., Hoffman, E.A., McLennan, G., Reinhardt, J.M.: Segmentation and analysis of the human airway tree from three-dimensional X-ray CT images. IEEE Trans. Med. Imag. **22**(8), 940–950 (2003)
6. Sonka, M., Park, W., Hoffman, E.A.: Rule-based detection of intrathoracic airway trees. IEEE Trans. Med. Imaging **15**(3), 314–326 (1996)
7. Bartz, D., Mayer, D., Fischer, J., Ley, S., del Rio, A., Thust, S., Straßr, W.: Hybrid segmentation and exploration of the human lungs. In: Visualization, pp. 177–184. IEEE (2003)
8. Lo, P., Sporring, J., Ashraf, H., Pedersen, J.J., de Bruijne, M.: Vessel-guided airway tree segmentation: a voxel classification approach. Med. Image Anal. **14**(4), 527–538 (2010)
9. Li, Q., Sone, S., Doi, K.: Selective enhancement filters for nodules, vessels, and airway walls in two-and three-dimensional CT scans. Med. Phys. **30**(8), 2040–2051 (2003)
10. Lo, P., van Ginneken, B., Reinhardt, J.M., Tarunashree, Y., et al.: Extraction of airways from CT (EXACT'09). IEEE Trans. Med. Imaging. **31**, 2093–2107 (2012)
11. Lidayová, K., Frimmel, H., Wang, C., Bengtsson, E., Smedby, Ö.: Fast vascular skeleton extraction algorithm. Pattern Recogn. Letters **76**, 67–75 (2016)
12. Gattinoni, L., Caironi, P., Pelosi, P., Goodman, L.R.: What has computed tomography taught us about the acute respiratory distress syndrome? Am. J. Respir. Crit. Care Med. **164**(9), 1701–1711 (2001)
13. Frimmel, H., Näppi, J., Yoshida, H.: Centerline-based colon segmentation for CT colonography. Med. Phys. **32**(8), 2665–2672 (2005)
14. Flórez-Valencia, L., Morales Pinzón, A., et al.: Simultaneous skeletonization and graph description of airway trees in 3D CT images. In: Proceedings of 25th GRETSI (2015)
15. Morales Pinzón, A., Hernández Hoyos, M., Richard, J.C., Flórez-Valencia, L., Orkisz, M.: A tree-matching algorithm: application to airways in CT images of subjects with the acute respiratory distress syndrome. Med. Image Anal. **35**, 101–115 (2017)

Context Aware Query Image Representation
for Particular Object Retrieval

Zakaria Laskar[✉] and Juho Kannala

Aalto University, Espoo, Finland
{zakaria.laskar,juho.kannala}@aalto.fi

Abstract. The current models of image representation based on Convolutional Neural Networks (CNN) have shown tremendous performance in image retrieval. Such models are inspired by the information flow along the visual pathway in the human visual cortex. We propose that in the field of particular object retrieval, the process of extracting CNN representations from query images with a given region of interest (ROI) can also be modelled by taking inspiration from human vision. Particularly, we show that by making the CNN pay attention on the ROI while extracting query image representation leads to significant improvement over the baseline methods on challenging Oxford5k and Paris6k datasets. Furthermore, we propose an extension to a recently introduced encoding method for CNN representations, regional maximum activations of convolutions (R-MAC). The proposed extension weights the regional representations using a novel saliency measure prior to aggregation. This leads to further improvement in retrieval accuracy.

Keyword: Image retrieval

1 Introduction

With the introduction of scale invariant local features, such as SIFT [20], the field of image retrieval has benefited tremendously over the last decade, extending its popularity and applicability to other fields of research like loop closure in robotics [19], and structure from motion [3]. In particular, the extension of bag-of-words model from text retrieval to the case of particular object retrieval in videos by Sivic and Zisserman [33] has been seminal for the developments in image retrieval [8,24]. One of the advantages of such local features is that it allows to follow-up the initial retrieval results with a costly but more accurate spatial verification step [31]. The initial retrieval results are obtained by matching local descriptors using selective matching kernels [34]. The issue of scalability in terms of memory requirements and computational cost associated with pairwise descriptor matching for large scale image retrieval databases was addressed by encoding the local descriptors into a single compact global image representation. Popular techniques are Fisher vectors [23], and VLAD [5].

The increase in computational capacity of GPUs and the generation of large datasets, such as ImageNet [29] have made Convolutional Neural Networks

© Springer International Publishing AG 2017
P. Sharma and F.M. Bianchi (Eds.): SCIA 2017, Part II, LNCS 10270, pp. 88–99, 2017.
DOI: 10.1007/978-3-319-59129-2_8

(CNN) the popular choice for a broad spectrum of computer vision tasks like image classification [18], object detection [9], and camera pose estimation [17]. As training a CNN from scratch requires a large amount of data, using activations from different layers of a CNN, trained on a large dataset like [29], as off-the-shelf image representation has bridged the applicability of CNN to different domains [9,30]. In case of limited amount of data, the parameters of a CNN pre-trained on ImageNet or other large datasets can be used to initialize the network parameters before training the CNN on the target dataset. Such a process is known as fine-tuning and has leveraged the use of CNN to other domains [22,38]. For the case of image retrieval, several works [7,10,27,35] propose the use of activations from a pre-trained CNN as image descriptors. As the pre-trained CNNs employed for such instance-level retrieval tasks were generally trained to suppress intra-class variations (observed in generic computer vision problems, like object detection), the performance of CNN based descriptors lagged behind the conventional local descriptors. Babenko et al. [7] first demonstrated that fine-tuning a pre-trained CNN on a Landmark dataset [7] significantly improves the retrieval accuracy on standard benchmark datasets of landmarks, such as Oxford5k [24] and Paris6k [22]. Arandjelovic et al. [4] used a similar paradigm of learning the image representations from a large dataset of geo-tagged images. However, the training was done using a ranking loss instead of the classification loss used in [7]. Radenovic et al. [26] showed the importance of hard positive and hard negative mining using unsupervised methods in improving the retrieval accuracy. Post training, the final representation of an image is encoded/computed using regional maximum activations of convolutions (R-MAC) [26,31]. R-MAC aggregates maximum activations over multiple regions into a compact image representation. The regions are generated using a fixed grid, which is designed to make the final image representation robust to scale and translation variation. Gordo et al. [11] proposed learning R-MAC representation using a Region Proposal Network (RPN) [28]. The R-MAC and the RPN were trained in an end-to-end manner resulting in powerful image representation that obtain the existing state-of-the-art performance on benchmark retrieval datasets.

In this paper, we focus on particular object retrieval, which is a special case of image retrieval, whereby, a query image(s) is given along with a region of interest (ROI) containing an object of interest. The retrieval engine then returns a ranked list of the database images, such that the images containing the object of interest are ranked higher. Traditionally, image retrieval methods encode the query image using feature representation extracted only from the ROI. This serves two purposes: reduced interference from background clutter, and, suppression of distractive patterns outside the ROI, which, sometimes maybe more salient than the ROI. On the other hand, regions outside the ROI can add contextual information to the ROI representation that can facilitate improvement in retrieval performance. Thus it can be stated that the suppression and encoding of information from regions outside the ROI is a tightly coupled problem. To address this issue, we propose the extension of computational model of hippocampus spatial attention, introduced by Mozer et al. in 1998 [21], to the problem of particular object retrieval. In particular, we show that by partial suppression of intermediate CNN representations (representations from intermediate layers of the CNN) of regions in the query image outside the ROI, we can obtain a proper trade-off between the two problems stated above, and thereby, obtain state-of-the-art retrieval performance on standard benchmark datasets like Oxford [24] and Paris [22].

Our second contribution is that we propose an extension to the conventional R-MAC encoding technique. The standard R-MAC suffers from the drawback of assigning uniform weightage to all the regions generated by the pre-defined grid prior to aggregation. As the regions are generated independent of the image content, responses from background clutter can cause negative interference due to equal weightage. Gordo et al. [11] proposed the use of RPN to generate image content dependent regions to circumvent this problem. However, this has certain challenges: (i) the number of regions from RPN is 3–10 times more than R-MAC [12], and (ii) additionally, the RPN model parameters need to be trained for the given task separately. Instead, we show that by using a simple saliency measure, obtained from the existing representation technique, one can weight the regional representations of R-MAC before aggregating. The saliency measure assigns higher weight to landmark type regions and lower weight to the background clutter.

Using the proposed modifications, we are able to achieve state-of-the-art retrieval results in standard object retrieval datasets. Our work can be seen as an extension to the work of [11,12] since we use off the shelf CNN representations from their trained network.

2   Background

In this section, we provide the reader with a brief background with the methods and terminologies encountered in CNN based literature.

2.1   CNN

When using a pre-trained CNN network like VGGNet (VGG) [32] or Residual Network (ResNet) [14], the network is often cropped at the last convolutional or pooling layer [11,22]. For example, conv5/pool5 layer in VGG, or res5c in ResNet-101. As such in the remainder of the paper, the term CNN will be associated with such cropped networks. Now, consider a CNN with L layers. Given an input image $I \in \mathbb{R}^{W_I \times H_I \times 3}$, the response obtained at the output of the layer $l \in L$ is a 3D tensor $\boldsymbol{X}^l \in \mathbb{R}^{W \times H \times K}$. K is the number of channels and $W \times H$ the spatial dimensions of the output feature map. The spatial resolution of the feature map depends on the network architecture and size of the input image, while the number of output channels K equals the number of filters in layer l. Additionally, it is assumed the feature map $\boldsymbol{X}^l$ is passed through a rectified linear unit (ReLU) activation function to ensure the non-negativeness of the activations.

The feature map $\boldsymbol{X}^l$ can be denoted as a set of K 2D feature maps $\boldsymbol{X}^l = \{X_k^l\}, k = 1...K$. Alternatively, each feature map $\boldsymbol{X}^l$ can be said to be a set of K dimensional feature representations for $W \times H$ activations. Instead of the term 'pixels', the term 'activation' is used in CNN feature space. The activation at spatial location $p \in \mathbb{R}^2$ in feature map X_k^l is represented by $X_{k,p}^l$. The set of all such locations p in a feature map be represented by $\boldsymbol{S} = [1, W] \times [1, H]$. As the layers are arranged in an hierarchical order, each layer computes a higher level abstraction from the feature representations of the previous layers.

2.2 R-MAC

Although the feature maps represent high quality abstraction of the image, they are very high dimensional. Typical dimensionality of feature maps extracted at the output of res5c layer in ResNet-101 network is $23 \times 13 \times 2048$ for an image of spatial resolution 800×600. As such, these high dimensional representations are encoded to fixed length global representations using techniques [4,6,16,35]. Among the various encoding methods, R-MAC has shown highest performance [35] and is a widely popular choice of encoding CNN representations.

As R-MAC can be used to encode feature map from any layer, l, we continue with the same set of notations introduced in Sect. 2.1, and, do not introduce any layer specific notations. For a given feature map, $\boldsymbol{X}^l$, R-MAC first generates a set of rectangular regions $\boldsymbol{R} = \{R_i\}, i = 1...N$, where $R_i \subseteq \boldsymbol{S}$ and N is the number of regions that depends on the size of feature map. For each region, the maximum activations of convolutions (MAC) [27] is computed by spatial max-pooling across the K dimensions resulting in a $1 \times K$ dimensional feature vector $\mathbf{f}_{R_i}$ per region R_i, where

$$\mathbf{f}_{R_i} = [f_{R_i,1}...f_{R_i,k}...f_{R_i,K}],$$
$$\text{such that } f_{R_i,k} = \max_{p \in R_i} X_{k,p}^l \tag{1}$$

Each region vector $\mathbf{f}_{R_i}$ is l_2 normalized, followed by whitening with PCA and l_2 normalized again. The regional feature vectors are sum aggregated to get the final image representation $\mathbf{f}$:

$$\mathbf{f} = \sum_{i=1}^{N} \mathbf{f}_{R_i} \tag{2}$$

As a result of sum aggregation, the final feature vector still retains the $1 \times K$ dimensionality. The final image representation $\mathbf{f}$ is again l_2 normalized, such that a simple dot product can be used to compute image similarity.

3 Contextual Information

In the general setting of particular object retrieval, contextual information can be viewed as the information outside the ROI, $\mathcal{R}$ in the query image. Such information can be facilitatory or inhibitory in the retrieval process. In other words, the contextual information along with ROI can increase or decrease the distinctiveness of the query image. Traditional models for extracting query image CNN representations include:

Full Query (FQ): The full query image is fed to a CNN [6], and the resulting feature representation is encoded to fixed length feature vector.

Cropped ROI (RQ): The ROI in the query image is cropped [11,26] and fed into a CNN, followed by encoding the resulting feature map.

Cropped Activation (AQ): The feature map representation is first obtained by feed-forwarding the whole query image through a CNN. Projection of the ROI on the feature map is computed, which is represented by the set of activations that have their center of receptive fields inside the ROI as shown in Fig. 1. The representations within the projected ROI are then encoded using standard encoding methods [4,6,26].

Methods that use **FQ** models are expected to perform better than **RQ** based methods when the contextual information is facilitatory. However, the facilitatory nature of context almost cannot be known apriori. On the other hand, methods that extract query image representation using **AQ** [4,6,26] are able to encode certain amount of contextual information. As can be seen from the Fig. 1, the total receptive field, $\tilde{\mathcal{R}}_L$ [2](dotted red colored box) of the activations in the projected ROI, $\mathcal{R}_L$ extend beyond the ROI, $\mathcal{R}$. Thus the extended ROI ($\tilde{\mathcal{R}}_L$ in Fig. 1) encodes certain amount of context. Note that the receptive field of layers $l < L$, $\tilde{\mathcal{R}}_l$ will have a smaller area than $\tilde{\mathcal{R}}_L$.

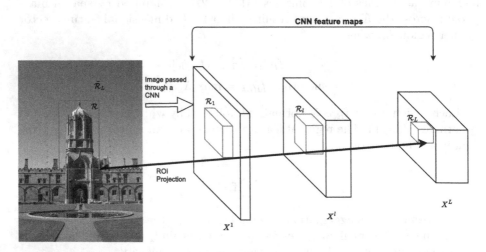

Fig. 1. An overview of feature extraction using a CNN. The network has l layers, $l=1..L$, where each layer computes a feature map $\boldsymbol{X}^l$ using the representations from the previous layers. The representations ($\mathcal{R}_1..\mathcal{R}_l..\mathcal{R}_L$) denote projections of the ROI, $\mathcal{R}$, on feature maps from different layers, l (bold black line) [13]. $\tilde{\mathcal{R}}_L$ is the cumulative receptive field of the activations inside $\mathcal{R}_L$ (Color figure online).

However, due to limited reach of the receptive field of the boundary activations, a large amount of contextual information is discarded. We propose to
combine the advantages of the three models mentioned above, by leveraging
the computational model of spatial attention observed in the hippocampus to
extract full query image CNN representation. The attention model is presented
next.

3.1 Computational Model of Spatial Attention

The main advantage of **RQ** and **AQ** models is that the representation from the
ROI has the highest response in the final query representation. Particularly, in
RQ, ROI representation has the sole representation, while, in **AQ**, it has higher
representation than the context. For **AQ**, this is based on the assumption that
only boundary activations are affected by regions outside the ROI. On the other
hand, the disadvantage of **FQ** based methods is that the ROI ceases to have a
higher prominence in the final representation.

Thus, an ideal model should not only encode information from regions beyond
the ROI, but, in the process should also maintain higher response from the
ROI in the final representation. Such a constraint can be modelled using the
computational model of spatial attention observed in hippocampus [21].

The attention model initiates with a saliency or attention map $A \in \mathbb{R}^{W_a \times H_a}$
defined over a feature map $\boldsymbol{X}^l$, such that $W_a = W$ and $H_a = H$. We do not
introduce any layer specific notations as the mask can be applied to feature
map at the output of any layer, l, of a CNN architecture. The same attention
mask is applied across all the channels, K, of the feature map $\boldsymbol{X}^l$ (see Sect. 2.1).
Therefore, each element p in the mask A, $A_p \in [0,1]$, affects the activations
occurring at the spatial location p across the K channels, $X^l_{1:K,p}$. The activity
levels are defined as follows:

$$A_p = \begin{cases} 1, & \text{if } p \in \mathcal{R}_l \\ M_p, & \text{if } p \notin \mathcal{R}_l, \end{cases} \tag{3}$$

where $\mathcal{R}_l$ is the projection of the ROI, $\mathcal{R}$, onto the feature map $\boldsymbol{X}^l$ [13]
(see Fig. 1), and M is the saliency map introduced in Sect. 4.1. Note that the
attention mask is specific to the spatial location and independent of the channel
dimension. So, identifying the activations, across different channels, with their
spatial location suffices, i.e. the notation $p \in \mathcal{R}_l$ represents all activations occur-
ring at spatial location p in the feature map $\boldsymbol{X}^l$ and lying within $\mathcal{R}_l$, $X^l_{1:K,p} \in \mathcal{R}_l$.
Now, activations occurring at position p in each feature map $X^l_k \in \boldsymbol{X}^l, k = 1...K$,
are modulated by the attention mask as follows:

$$\tilde{X}^l_{k,p} = \begin{cases} A_p X^l_{k,p}, & \text{if } p \in \mathcal{R}_l \\ g(A_p) X^l_{k,p}, & \text{if } p \notin \mathcal{R}_l, \end{cases} \tag{4}$$

$g(.)$ is a monotonic function [21]:

$$g(a) = \lambda_1 + \lambda_2 a^{\phi} \tag{5}$$

The constants λ_1, $\lambda_2 \in (0,1)$ are so chosen such that the function $g(.)$ always maintains a value less than one i.e. $g(.) < 1$. Additionally, the function $g(.)$ has a lower bound at λ_1, that defines the maximum attenuation which can be applied to any activation outside the projected ROI, $\mathcal{R}_l$. In all our experiments we set $\lambda_1 = 0.5$, and, $\lambda_2 = 0.4$. The constant ϕ suppresses activations with weak attention levels (less salient). As in [21], we set $\phi = 4$ for all experiments.

The modulated feature map representations $\tilde{X}$ are feed-forwarded through the remaining layers of the CNN network, or, directly encoded into a fixed length global image representation (detailed in Sect. 5).

4 Weighted R-MAC

The feature map is now encoded to a lower dimensional feature representation. Recent state-of-the-art use R-MAC for this operation [11,26]. However, one of the criticality with the standard R-MAC is the equal weighting of each region vector $\mathbf{f}_{R_i}$ while aggregating in Eq. 2. This implies that responses generating from image clutter and background will negatively affect the retrieval process due to assignment of uniform weightage. As one does not have the prior information about the location of the object of interest in the database images, increasing the number of regions ensures higher coverage, but, it also increases interference from irrelevant regions.

Instead, we propose to use a weighted version of the standard R-MAC (WR-MAC) such that the final image representation is a weighted combination of representations from each region $R_i \in \mathbf{R}$. In general, Eq. 2 is modified to

$$\mathbf{f} = \sum_{i=1}^{N} w_i \mathbf{f}_{R_i} \tag{6}$$

The weights $w_i \in \mathbb{R}$ are generated using a saliency measure (Sect. 4.1) such that landmark type regions have higher weights than background clutter.

4.1 Saliency Map

We use a simple, yet effective saliency measure [36,37]. Given the feature map $\mathbf{X}^l$, the saliency function maps the 3D tensor $\mathbf{X}^l$ to a 2D tensor M by sum aggregating the feature map $\mathbf{X}^l$ over the channel dimensions. Mathematically, the function can be defined as $\psi : \mathbb{R}^{W \times H \times K} \rightarrow \mathbb{R}^{W \times H}$ such that $\psi(\mathbf{X}^l) = M$, where $M = \sum_{k=1}^{K} X_k^l$. The map is additionally max-normalized such that each element p has a range, $M_p \in [0,1]$.

In Sect. 2.2 where we introduced R-MAC, we observe that the regions $\mathbf{R}$ generated by the rigid grid depend on the spatial dimension $\mathbf{S}$ of the feature map $\mathbf{X}^l$. As the spatial dimension of feature map $\mathbf{X}^l$ is retained in the saliency map M, we can define the same set of regions $\mathbf{R}$ over the map M. Now, for each region R_i we compute MAC to obtain the weight w_i. That is

$$w_i = \max_{p \in R_i} M_p \tag{7}$$

5   Experiments

5.1   Network and Datasets

For all our experiments we use the ResNet-101 [14] CNN network architecture. The network has a very deep architecture and attains the state-of-the-art results in a variety of computer vision problems [14]. The original network, pretrained on ImageNet, is cropped at the layer res5c_ReLU and fine-tuned using the method proposed in [12]. The network parameters of the fine-tuned model (ResNet-IR) are also publicly available [1]. Additionally, the model ResNet-IR has additional layers on top of res5c_ReLU which performs PCA whitening on the region vectors obtained from res5c_ReLU representations using R-MAC grid and performs the final R-MAC aggregations. The additional layers were trained end-to-end during fine-tuning [12]. We use the learnt PCA parameters from [12] in our experiments.

For evaluation, we use the standard and publicly available particular object retrieval datasets: Oxford Buildings [24] and Paris Buildings [25]. Each dataset contains several thousands of images. Within these images, 55 queries are defined and a ROI containing the precise location of the landmark to be queried is defined. The retrieval performance is computed using mean average precision (mAP), where the mean is taken over all queries.

5.2   WR-MAC

As mentioned in Sect. 2.2, for a given image, R-MAC can be used to encode representations from any given CNN layer, but, similar to [12] we use the res5c_ReLU representations for WR-MAC. A saliency map is computed using the res5c_ReLU representations which is then normalized. The normalized map is used to generate weights. The weights are only applied at the time of aggregating the regional representations generated from the R-MAC grid. Note that prior to aggregation, these representations are l_2 normalized, whitened with PCA and l_2 normalized again. Similar to [12], we extract R-MAC and WR-MAC representations from 3 different scales of the given image: 550, 800, and, 1050 pixels on the largest side and maintaining the original aspect ratio. The representations from the 3 scales are aggregated and l_2 normalized to obtain the final image representation.

5.3   Attention Model

The only parameter in the attention model which needs to be defined is the layer number, l, i.e. the CNN layer on which the attention model is to be applied. As at each layer, feature representations are computed using the representations from the previous layers, application of attention model at a certain layer affects the representations of all the layers above. As the number of layers is high for ResNet-IR network, we choose the following layers: (i) res2c_ReLU, (ii) : res4b15_ReLU, and, (iii) : res5c_ReLU and evaluate their performance as described below.

Given a query image and a ResNet-IR layer, l' (sampled as mentioned above), the feature representations at the output of layer l' are used to compute the normalized saliency map. Thus, it is to be noted that the saliency map used in the experiments vary, i.e. for WR-MAC the saliency map is defined over the res5c_ReLU representations, while for attention model it is defined over the layer, l'. The modulated representations of the layer, l' are then used as an input to the layer, $l' + 1$, and feed-forwarded through the remaining layers of ResNet-IR to obtain the final res5c_ReLU representations. The res5c_ReLU representations are then encoded as discussed in Sect. 5.2.

5.4   Results

In order to better interpret the results, the following points need to be noted: (i) the spatial attention model is only applied to the query side, and (ii) the proposed encoding method, WR-MAC, is applied to both the query and database images.

The motivation for evaluating the performance of attention model across different CNN layers is to observe where in the CNN should the contextual representations be suppressed. In other words, should it be suppressed as early as lower level layers (res2c_ReLU), or, late at high level layers (res5c_ReLU), or, somewhere in between at mid level (res4b15_ReLU). From the results in Table 1, it can be observed that performance across all the layers (considered in Table 1) improve over the baseline (Table 2). However, the performance of res4b15_ReLU layer representations is more consistent across the datasets. As such we consider spatial attention model (**SA**) applied on res4b15_ReLU layer representations as our proposed model and compare with the existing approaches in Table 2. Similar to Sect. 5.3, for all the models in Table 2, the res5c_ReLU representations are obtained and then encoded as discussed in Sect. 5.2. All the models **FQ**, **AQ**, and **SA** perform better than the baseline **RQ** due to the addition of contextual information in the final image representation. However, the **SA** model outperforms the existing models across both the datasets.

The comparison between the performance of **FQ** and **AQ** models gives an idea of the effect of the contextual information in image retrieval. The model **AQ**, which encodes limited contextual information, outperforms **FQ** in Oxford dataset, but, is outperformed by **FQ** in Paris dataset. Our assumption is that in Paris dataset, the contextual information has a facilitatory effect, while, in Oxford dataset, it has a certain degree of inhibitory effect on the retrieval performance.

Table 1. Effect of applying attention model to different CNN layers, measured in terms of mAP

Layers	Paris6k		Oxford5k	
	R-MAC	WR-MAC	R-MAC	WR-MAC
res2c_ReLU	95.44	95.68	87.90	88.49
res4b15_ReLU	95.49	95.80	89.45	90.20
res5c_ReLU	95.53	95.84	89.04	89.37

Table 2. Performance comparison of our proposed model of extracting query image representation (Spatial Attention, **SA**) and encoding using weighted R-MAC (WR-MAC), with existing approaches. Previous baseline [12] is marked as †. Re-ranking stages were not applied.

Models	Paris6k		Oxford5k	
	R-MAC	WR-MAC	R-MAC	WR-MAC
Cropped ROI (**RQ**)	94.5†	94.8	86.1†	86.7
Cropped Activation (**AQ**)	94.81	95.13	88.77	89.03
Full Query (**FQ**)	95.47	95.76	87.30	87.86
Spatial Attention (**SA**)	95.49	**95.80**	89.45	**90.20**

An in-depth analysis is left for future work. Although, the performance of **SA** model is constrained by the requirement of ROI on the query side, that does not diminish the scope of the proposed encoding method, which shows consistent improvement even without the spatial attention model.

6 Conclusion

In this paper we address the different models of extracting and encoding image representation using CNN for the task of image retrieval. The current models of extracting CNN representations from query images cannot fully exploit the advantages of the provided ROI in the query image. We propose an extension of an attention model to this problem and demonstrate increase in retrieval accuracy on standard particular object retrieval datasets. Additionally, we also propose an extension to a recently introduced popular encoding method for CNN representations, R-MAC and show further improvements in retrieval performance. To the best of our knowledge our results are the best among global image representations reported so far for these datasets. Using efficient re-ranking strategies [8,15] can lead to further improvements in retrieval accuracy.

References

1. Deep Image Retrieval with Residual Network. http://www.xrce.xerox.com/Our-Research/Computer-Vision/Learning-Visual-Representations/Deep-Image-Retrieval/
2. Receptive Field. http://cs231n.github.io/convolutional-networks/
3. Agarwal, S., Furukawa, Y., Snavely, N., Simon, I., Curless, B., Seitz, S.M., Szeliski, R.: Building rome in a day. Commun. ACM **54**(10), 105–112 (2011)
4. Arandjelovic, R., Gronat, P., Torii, A., Pajdla, T., Sivic, J.: Netvlad: CNN architecture for weakly supervised place recognition. In: Proceedings of the IEEE Conference on Computer Vision and Pattern Recognition, pp. 5297–5307 (2016)
5. Arandjelovic, R., Zisserman, A.: All about vlad. In: Proceedings of the IEEE Conference on Computer Vision and Pattern Recognition, pp. 1578–1585 (2013)

6. Babenko, A., Lempitsky, V.: Aggregating local deep features for image retrieval. In: Proceedings of the IEEE International Conference on Computer Vision, pp. 1269–1277 (2015)
7. Babenko, A., Slesarev, A., Chigorin, A., Lempitsky, V.: Neural codes for image retrieval. In: Fleet, D., Pajdla, T., Schiele, B., Tuytelaars, T. (eds.) ECCV 2014. LNCS, vol. 8689, pp. 584–599. Springer, Cham (2014). doi:10.1007/978-3-319-10590-1_38
8. Chum, O., Mikulik, A., Perdoch, M., Matas, J.: Total recall ii: query expansion revisited. In: 2011 IEEE Conference on Computer Vision and Pattern Recognition (CVPR), pp. 889–896. IEEE (2011)
9. Girshick, R., Donahue, J., Darrell, T., Malik, J.: Rich feature hierarchies for accurate object detection and semantic segmentation. In: Proceedings of the IEEE Conference on Computer Vision and Pattern Recognition, pp. 580–587 (2014)
10. Gong, Y., Wang, L., Guo, R., Lazebnik, S.: Multi-scale orderless pooling of deep convolutional activation features. In: Fleet, D., Pajdla, T., Schiele, B., Tuytelaars, T. (eds.) ECCV 2014. LNCS, vol. 8695, pp. 392–407. Springer, Cham (2014). doi:10.1007/978-3-319-10584-0_26
11. Gordo, A., Almazán, J., Revaud, J., Larlus, D.: Deep image retrieval: learning global representations for image search. In: Leibe, B., Matas, J., Sebe, N., Welling, M. (eds.) ECCV 2016. LNCS, vol. 9910, pp. 241–257. Springer, Cham (2016). doi:10.1007/978-3-319-46466-4_15
12. Gordo, A., Almazan, J., Revaud, J., Larlus, D.: End-to-end learning of deep visual representations for image retrieval (2016). arXiv:1610.07940
13. He, K., Zhang, X., Ren, S., Sun, J.: Spatial pyramid pooling in deep convolutional networks for visual recognition. In: Fleet, D., Pajdla, T., Schiele, B., Tuytelaars, T. (eds.) ECCV 2014. LNCS, vol. 8691, pp. 346–361. Springer, Cham (2014). doi:10.1007/978-3-319-10578-9_23
14. He, K., Zhang, X., Ren, S., Sun, J.: Deep residual learning for image recognition. In: Proceedings of the IEEE Conference on Computer Vision and Pattern Recognition, pp. 770–778 (2016)
15. Iscen, A., Tolias, G., Avrithis, Y., Furon, T., Chum, O.: Efficient diffusion on region manifolds: recovering small objects with compact CNN representations (2016). arXiv:1611.05113
16. Kalantidis, Y., Mellina, C., Osindero, S.: Cross-dimensional weighting for aggregated deep convolutional features. In: Hua, G., Jégou, H. (eds.) ECCV 2016. LNCS, vol. 9913, pp. 685–701. Springer, Cham (2016). doi:10.1007/978-3-319-46604-0_48
17. Kendall, A., Grimes, M., Cipolla, R.: Posenet: a convolutional network for real-time 6-dof camera relocalization. In: Proceedings of the IEEE International Conference on Computer Vision, pp. 2938–2946 (2015)
18. Krizhevsky, A., Sutskever, I., Hinton, G.E.: Imagenet classification with deep convolutional neural networks. In: Advances in Neural Information Processing Systems, pp. 1097–1105 (2012)
19. Laskar, Z., Huttunen, S., Herrera, D., Rahtu, E., Kannala, J.: Robust loop closures for scene reconstruction by combining odometry and visual correspondences. In: 2016 IEEE International Conference on Image Processing (ICIP), pp. 2603–2607. IEEE (2016)
20. Lowe, D.G.: Distinctive image features from scale-invariant keypoints. Int. J. Comput. Vis. **60**(2), 91–110 (2004)
21. Mozer, M.C., Sitton, M.: Computational modeling of spatial attention. In: Pashler, H. (ed.) Attention, vol. 9, pp. 341–393. UCL Press, London (1998)

22. Oquab, M., Bottou, L., Laptev, I., Sivic, J.: Learning and transferring mid-level image representations using convolutional neural networks. In: Proceedings of the IEEE Conference on Computer Vision and Pattern Recognition, pp. 1717–1724 (2014)
23. Perronnin, F., Liu, Y., Sánchez, J., Poirier, H.: Large-scale image retrieval with compressed fisher vectors. In: 2010 IEEE Conference on Computer Vision and Pattern Recognition (CVPR), pp. 3384–3391. IEEE (2010)
24. Philbin, J., Chum, O., Isard, M., Sivic, J., Zisserman, A.: Object retrieval with large vocabularies and fast spatial matching. In: IEEE Conference on Computer Vision and Pattern Recognition, CVPR 2007, pp. 1–8. IEEE (2007)
25. Philbin, J., Chum, O., Isard, M., Sivic, J., Zisserman, A.: Lost in quantization: improving particular object retrieval in large scale image databases. In: IEEE Conference on Computer Vision and Pattern Recognition, CVPR 2008, pp. 1–8. IEEE (2008)
26. Radenović, F., Tolias, G., Chum, O.: CNN image retrieval learns from BoW: unsupervised fine-tuning with hard examples. In: Leibe, B., Matas, J., Sebe, N., Welling, M. (eds.) ECCV 2016. LNCS, vol. 9905, pp. 3–20. Springer, Cham (2016). doi:10. 1007/978-3-319-46448-0_1
27. Razavian, A.S., Sullivan, J., Carlsson, S., Maki, A.: Visual instance retrieval with deep convolutional networks (2014). arXiv:1412.6574
28. Ren, S., He, K., Girshick, R., Sun, J.: Faster R-CNN: towards real-time object detection with region proposal networks. In: Advances in Neural Information Processing Systems, pp. 91–99 (2015)
29. Russakovsky, O., Deng, J., Su, H., Krause, J., Satheesh, S., Ma, S., Huang, Z., Karpathy, A., Khosla, A., Bernstein, M., et al.: Imagenet large scale visual recognition challenge. Int. J. Comput. Vis. 115(3), 211–252 (2015)
30. Sharif Razavian, A., Azizpour, H., Sullivan, J., Carlsson, S.: CNN features off-the-shelf: an astounding baseline for recognition. In: Proceedings of the IEEE Conference on Computer Vision and Pattern Recognition Workshops, pp. 806–813 (2014)
31. Shen, X., Lin, Z., Brandt, J., Wu, Y.: Spatially-constrained similarity measure for large-scale object retrieval. IEEE Trans. Pattern Anal. Mach. Intell. 36(6), 1229–1241 (2014)
32. Simonyan, K., Zisserman, A.: Very deep convolutional networks for large-scale image recognition (2014). arXiv:1409.1556
33. Sivic, J., Zisserman, A., et al.: Video google: a text retrieval approach to object matching in video
34. Tolias, G., Avrithis, Y., Jégou, H.: To aggregate or not to aggregate: selective match kernels for image search. In: Proceedings of the IEEE International Conference on Computer Vision, pp. 1401–1408 (2013)
35. Tolias, G., Sicre, R., Jégou, H.: Particular object retrieval with integral max-pooling of CNN activations (2015). arXiv:1511.05879
36. Wei, X.S., Luo, J.H., Wu, J.: Selective convolutional descriptor aggregation for fine-grained image retrieval (2016). arXiv:1604.04994
37. Zagoruyko, S., Komodakis, N.: Paying more attention to attention: Improving the performance of convolutional neural networks via attention transfer (2016). arXiv:1612.03928
38. Zhang, N., Donahue, J., Girshick, R., Darrell, T.: Part-based R-CNNs for fine-grained category detection. In: Fleet, D., Pajdla, T., Schiele, B., Tuytelaars, T. (eds.) ECCV 2014. LNCS, vol. 8689, pp. 834–849. Springer, Cham (2014). doi:10. 1007/978-3-319-10590-1_54

Granulometry-Based Trabecular Bone Segmentation

Manish Chowdhury[1], Benjamin Klintström[1,2], Eva Klintström[2],
Örjan Smedby[1,2], and Rodrigo Moreno[1(✉)]

[1] KTH, School of Technology and Health,
Hälsovägen 11c, 14157 Huddinge, Sweden
{manish.chowdhury,orjan.smedby,rodrigo.moreno}@sth.kth.se,
benklint@gmail.com
[2] Center for Medical Image Science and Visualization, Linköping University,
Linköping, Sweden
evaklintstrom@gmail.com

Abstract. The accuracy of the analyses for studying the three dimensional trabecular bone microstructure rely on the quality of the segmentation between trabecular bone and bone marrow. Such segmentation is challenging for images from computed tomography modalities that can be used in vivo due to their low contrast and resolution. For this purpose, we propose in this paper a granulometry-based segmentation method. In a first step, the trabecular thickness is estimated by using the granulometry in gray scale, which is generated by applying the opening morphological operation with ball-shaped structuring elements of different diameters. This process mimics the traditional sphere-fitting method used for estimating trabecular thickness in segmented images. The residual obtained after computing the granulometry is compared to the original gray scale value in order to obtain a measurement of how likely a voxel belongs to trabecular bone. A threshold is applied to obtain the final segmentation. Six histomorphometric parameters were computed on 14 segmented bone specimens imaged with cone-beam computed tomography (CBCT), considering micro-computed tomography (micro-CT) as the ground truth. Otsu's thresholding and Automated Region Growing (ARG) segmentation methods were used for comparison. For three parameters (Tb.N, Tb.Th and BV/TV), the proposed segmentation algorithm yielded the highest correlations with micro-CT, while for the remaining three (Tb.Nd, Tb.Tm and Tb.Sp), its performance was comparable to ARG. The method also yielded the strongest average correlation (0.89). When Tb.Th was computed directly from the gray scale images, the correlation was superior to the binary-based methods. The results suggest that the proposed algorithm can be used for studying trabecular bone in vivo through CBCT.

Keywords: Cone beam computed tomography · Segmentation · Granulometry · Trabecular bone

© Springer International Publishing AG 2017
P. Sharma and F.M. Bianchi (Eds.): SCIA 2017, Part II, LNCS 10270, pp. 100–108, 2017.
DOI: 10.1007/978-3-319-59129-2_9

1   Introduction

The analysis of bone micro-architecture from 3D medical imaging techniques is of medical interest due to the clinical importance of osteoporosis [7]. For this purpose, high resolution peripheral quantitative computed tomography (HR-pQCT), micro computed tomography (micro-CT), cone beam computed tomography (CBCT), multi-slice computed tomography (MSCT), and magnetic resonance imaging (MRI) have been considered [8,9,15,19]. Among these, micro-CT is usually considered as the gold standard in preclinical studies thanks to its high resolution and contrast. However, micro-CT cannot be used in clinical settings, since it requires a high radiation dose [3]. For quantification, the computation of histomorphometric parameters plays a very important role [1]. Usually, these parameters are calculated from a binary image after segmenting trabecular bone from the background (bone marrow). Thus, the choice of a segmentation method may play an important role in structure measurements, but little attention is generally given to this fact [10].

Due to the variability in medical imaging techniques, there is no segmentation method that is ideal for all types of bone images. In histology images from prepared specimens, with high contrast and extremely high resolution, a simple thresholding is often sufficient for segmentation. In micro-CT, the resolution is usually high enough for such a simple approach. Unfortunately, for imaging methods applicable in vivo, such as CBCT and HR-pQCT, segmentation methods are more prone to errors. The most common segmentation methods for CT data are either based on adaptive or on double thresholding [2]. Similarly, Engelke et al. [5] have used thresholding based on the neighboring pixels in the micro-CT images of bone obtained through dual-energy CT. In our group, Petersson et al. [14] evaluated bone microstructure from clinical CT using the Automated Region Growing (ARG) algorithm [16].

CBCT is a 3D imaging modality that can be used in vivo, which might be considered for diagnosing bone-related diseases. The wide accessibility of the machines, the radiation dose, cost-effectiveness, and short scanning time, make this scanners appealing for evaluating the trabecular bone structure in clinical settings [17]. Still, trabecular bone characterization with CBCT has not yet been properly investigated, in particular when it comes to the importance of the selection of the segmentation method. In this study, we propose a new segmentation method based on granulometry, a concept from mathematical morphology that allows us to estimate thickness at a local scale [12]. The method is tested on CBCT data, using micro-CT images as a reference.

2   Segmentation Using Granulometries

In [12], we showed that granulometry in binary mathematical morphology is equivalent to the sphere-fitting method proposed in [6] for computing trabecular bone thickness. Such equivalence allowed us to create a method for estimating trabecular thickness on gray scale image data by replacing binary with gray scale mathematical morphology. In this paper, we show that the method can be adapted to generate the segmentation of the image that can be used for extracting additional histomorphometric parameters.

The theory of granulometries was originally introduced by Matheron [11] to compute size distributions of grains in digital images. Intuitively, granulometry is analog to sieving the image with filters that emulate sieves of different hole sizes, where the size of a grain is determined by the minimum size of the hole that sifts the grain. A detailed mathematical description is found in [12].

The main procedure for computing trabecular thickness in gray scale is as follows. First, a granulometry is generated by applying the opening operation of mathematical morphology to the image with spherical structuring elements of different sizes E_1, E_2, ..., E_n. The openings are given by:

$$O_i = I \circ E_i = (I \ominus E_i) \oplus E_i, \tag{1}$$

where I is the image, $\ominus$ and $\oplus$ are the erosion and dilation morphological operations respectively. The thickness at a specific location x is related to the scale where the maximum change between consecutive openings is attained:

$$Th(x) = 2 \arg\max_t (O_t(x) - O_{t-1}(x)). \tag{2}$$

This procedure can be applied using binary or gray scale mathematical morphology. In [12], we discussed implementation issues that have to be considered for accurate estimation of thickness.

In an ideal noiseless scenario, using structuring elements E_i of sizes larger than the thickest trabecula will not have any effect on the results. Indeed, opening a noiseless image with a big enough structuring element will result in an image where all voxels have an intensity value of zero. However, this is not the case for noisy images. Thus, for efficiency purposes, it is necessary to stop the computation of openings at a scale larger than the thickest expected trabecula in the image. In the experiments on CBCT, considering that the thickest trabecula is expected to be of around 1 mm, we stopped the computations at scale 10, which corresponds to 1.5 mm.

The main effect of the opening operation on the image is that the gray scale dynamic range of values is reduced. Thus, the main hypothesis of our segmentation method is that after the final opening with E_n, the gray scale at trabecular bone is largely reduced compared to that reduction at bone marrow. The pipeline of the method is shown in Fig. 1.

The difference between the original and the last opening operation (cf. Fig. 1C) is higher in trabecular bone than in marrow. Since the contrast to noise ratio of this image is enhanced, a threshold can be used to get a final segmentation (cf. Fig. 1D). Thus the segmented image is computed as:

$$S = T(I - (I \circ E_n)) \oslash I, Th), \tag{3}$$

where I is the CBCT 3D image, S is the segmented image, E_n is the largest structuring element used for computing thickness, $\oslash$ is the Hadamard division

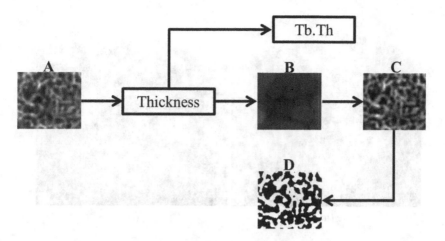

Fig. 1. Segmentation steps for the proposed method. A: original image, B: output of the last opening operation (residual), C: percentage of local gray scale reduction, D: final result after thresholding.

and T performs the threshold with Th. Although Th can be estimated adaptively using Otsu's threshold [13], we obtained good results with a fixed one (0.16) for our CBCT data in the experiments of Sect. 4.

3   Experimental Evaluation

We have studied our proposed granulometry-based segmentation technique on 14 human radius specimens [7]. The samples were donated for medical research in accordance with the ethical recommendations at the University of California, San Francisco. The specimens are almost cubic with a side of 12 to 15 mm and all include slabs of cortical bone, facilitating orientation. Two imaging techniques have been used in this study (Fig. 2):

- CBCT 3D images were acquired using the NewTom 5G (QR Verona, Verona, Italy) using a peak tube voltage of 110 kV, a tube current of 4.2–4.6 mA, and a field of view of 60 mm. After initial reconstruction with an isotropic resolution of 125 μm, the image was resampled by the scanner software to an isotropic voxel size of 75 μm.
- Micro-CT data were acquired with the Skyscan 1176 (Bruker micro-CT, Kontich Belgium) with a tube voltage of 65 kV, a tube current of 385 μA and an isotropic resolution of 8.67 μm.

The parameters were measured and calculated using MATLAB (MathWorks, Natick, MA). The code was developed in-house and calculated on a personal computer (PC) with Intel Core i7 (Intel Santa Clara, CA) at 2.60 GHz, 32 GB random access memory (RAM) and 64-bit operating system. We have compared

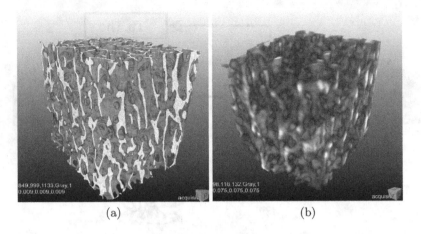

(a)   (b)

Fig. 2. 3D view (a) MicroCT and (b) CBCT.

our algorithm with the Automatic Region Growing (ARG) algorithm [16] and Otsu's threshold [13], in the following referred to as Otsu.

Segmenting micro-CT data is not an issue thanks to its high resolution and contrast. Thus, in general, all segmentation methods perform well for micro-CT, including a basic thresholding. In this study, the micro-CT data were segmented with Otsu in order to avoid the use of parameters. We have studied six histomorphometric parameters: trabecular node density (Tb.Nd), trabecular termini density (Tb.Tm), trabecular separation (Tb.Sp), trabecular number (Tb.N), trabecular thickness (Tb.Th), and bone volume over total volume (BV/TV). The details of these parameters are given in [1]. These parameters were computed in 3D.

4   Results

Figure 3 shows qualitative results in the form of selected slices from CBCT. Judging from these images, Otsu and ARG segmentation algorithms give visually similar results. It is observed that granulometry-based segmentation gives results that are visually similar to those of ARG. In order to assess these differences, we measured the pairwise spatial overlap using the Dice coefficients [4]. The Dice coefficients between Otsu - ARG, Otsu - Granulometry and ARG - Granulometry are 0.9784, 0.8793 and 0.8763 respectively, which confirms the visual assessment.

Table 1 shows the value of the mean and standard deviation of the histomorphometric parameters extracted from the binary images obtained from different segmentation techniques. This table also includes the Tb.Th computed in gray scale as we proposed in [12]. As shown, both BV/TV and Tb.Th were overestimated by approximately 4 times. However, Tb.Sp and Tb.N were underestimated by a small amount. On the other hand, Tb.Nd and Tb.Tm were highly underestimated. We think this finding can be explained by spurious branches generated by the used skeletonization algorithm [18] on the micro-CT data. Moreover, these

Original Micro-CT	Original CBCT	Otsu	ARG	Proposed

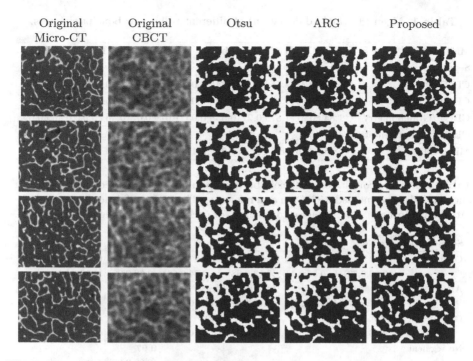

Fig. 3. Visual Segmentation Results using Otsu, ARG and Granulometry (proposed) for some selected slices.

parameters reflect the connectivity/topology of the network. The estimation of Tb.Th in gray scale is largely overestimated in this dataset.

In order to assess the performance of the proposed method, we computed the Pearson correlation coefficient of different parameters in Table 2, considering micro-CT images segmented with Otsu as our reference model.

Table 2 also reports the 95% confidence intervals for these differences. If a confidence interval does not include zero, the corresponding difference is statistically significant. As shown, for four parameters (Tb.Nd, Tb.N, Tb.Th and BV/TV), the proposed method reached larger correlations than 0.90, whereas Tb.Tm and Tb.Sp have correlations of 0.80 and 0.75, respectively. For three parameters (Tb.Nd, Tb.Tm and Tb.Sp), the strongest correlations with micro-CT images were found with ARG, and for three parameters with the new method (Tb.N, Tb.Th and BV/TV). As a way to compare the global performance of the methods, we computed the mean of the correlation coefficients. On average, our proposed granulometry-based segmentation yielded a correlation of 0.89 with micro-CT, which is stronger than for Otsu and ARG. Notice that the correlation between Tb.Th from micro-CT and Tb.Th in gray scale is the strongest. The reason of this is that this measurement is not affected by the segmentation algorithm and could be more appropriate for estimations in vivo. Unfortunately, except for Tb.Th, methods for estimating other parameters in gray scale are not currently available.

Table 1. Mean (± standard deviation) of different trabecular bone parameters.

Machine/ Segmentation	Tb.Nd (mm^{-3})	Tb.Tm (mm^{-3})	Tb.Sp (mm)	Tb.N (mm^{-1})	Tb.Th (mm)	BV/TV
CBCT/ARG	1.77±0.27	1.75±0.18	0.54±0.04	1.03±0.07	0.40±0.03	0.37±0.05
CBCT/ Otsu	2.66±0.65	1.68±0.42	0.46±0.09	1.31±0.21	0.25±0.02	0.24±0.04
CBCT/Granulometry	1.84±0.27	1.82±0.27	0.53±0.05	1.01±0.07	0.45±0.04	0.46±0.09
CBCT/No segmentation	-	-	-	-	0.82±0.25	-
Micro-CT/Otsu	58.5±15.6	86.3±21.9	0.68±0.10	1.37±0.18	0.13±0.01	0.10±0.02

Table 2. Pearson correlation coefficient between parameters computed on CBCT using micro-CT data segmented with Otsu as a reference. The 95% confidence limits are indicated in parenthesis. The strongest correlation for each parameter is highlighted in bold.

Segmentation	Tb.Nd	Tb.Tm	Tb.Sp	Tb.N	Tb.Th	BV/TV	Mean
Otsu	0.49 (0.00,0.79)	−0.07 (−0.55,0.43)	0.59 (0.13,0.83)	0.29 (−0.24,0.68)	0.89 (0.69,0.96)	0.59 (0.13,0.84)	0.45
ARG	**0.93** (0.81,0.97)	**0.81** (0.53,0.93)	**0.77** (0.45,0.91)	0.93 (0.80,0.97)	0.86 (0.64,0.95)	0.92 (0.76,0.97)	0.87
Granulometry	0.91 (0.76,0.97)	0.80 (0.49,0.92)	0.75 (0.40,0.90)	**0.97** (0.90,0.98)	0.91 (0.75,0.96)	**0.94** (0.83,0.97)	**0.89**
No segment	-	-	-	-	**0.97** (0.90,0.98)	-	-

5 Discussion

This paper has presented a method that uses the differences between the original and the residual images after granulometry analysis to increase the contrast of images acquired through CBCT. Such increase allowed us to use a threshold to segment the images.

The accuracy of the method was tested by comparing histomorphometric parameters computed in CBCT with respect to the ones obtained by micro-CT. Although gross systematic errors of the measurements on CBCT were found, the correlations were high for ARG and the proposed method. The proposed method yielded slightly better correlations compared to ARG.

Notice that the large systematic errors are more related to the low resolution and contrast of the CBCT images rather than to the segmentation algorithms. There are two strategies for handling these large systematic errors: (a) correct them using linear regression in order to give closer results to micro-CT or (b) use them as surrogates of the parameters estimated from micro-CT. In both cases, the measurements could be used for follow-up of treatments due to the strong correlations with micro-CT.

It is important to point out that, unlike ARG, the proposed method also estimates Tb.Th in gray scale. We found that such estimation of Tb.Th had the strongest correlation to micro-CT data, suggesting that it is advantageous to compute histomorphometric parameters that do not require segmentation.

The method is especially useful for images acquired with modalities that can be used in vivo, such as CBCT, MSCT and HR-pQCT, where noise, resolution and contrast are relevant issues. Our ongoing research includes testing this method in these modalities. While the method can also be applied to micro-CT, the result is not different to the one from less elaborate strategies such as Otsu, due to the high resolution and contrast of these images. Moreover, the proposed method can be used for computing the Tb.Th in gray scale as well as for performing the segmentation at the same time, something that is not possible with ARG or Otsu. In this study, we used the skeletonization method proposed in [18] for performing the analysis of segmented images both in micro-CT and CBCT. We plan to test other skeletonization methods in order to assess the sensitivity of the estimations of histomorphometric parameters with respect to the skeletonization method.

To summarize, the results from this paper suggest that the combination of CBCT and granulometry-based segmentation can be used for monitoring changes in the microarchitecture of trabecular bone in a clinical environment. From the results of this study, we argue that our granulometry-based technique seems more promising than ARG for CBCT data. A limitation of the present study is the low number of specimens available. In the future, we will apply the method to larger CBCT materials as well as to HR-pQCT data.

Acknowledgments. This research has been supported by Eurostars, grant no. E9126, and by the Swedish Council for Research (VR), grants no- 2012-3512 and 2014-6153. The authors are grateful to Britt-Marie Andersson at Uppsala University for performing the micro-CT imaging and to Sharmila Majumdar at University of California for kindly providing the specimens. Per-Magnus Johansson at Maxillofacial radiology in Växjö helped us with the NewTom 5G imaging.

References

1. Bouxsein, M.L., Boyd, S.K., Christiansen, B.A., Guldberg, R.E., Jepsen, K.J., Müller, R.: Guidelines for assessment of bone microstructure in rodents using micro-computed tomography. J. Bone Min. Res. **25**(7), 1468–1486 (2010)
2. Chevalier, F., Laval-Jeantet, A., Laval-Jeantet, M., Bergot, C.: CT image analysis of the vertebral trabecular network in vivo. Calcif. Tissue Int. **51**(1), 8–13 (1992)
3. Christiansen, B.A.: Effect of micro-computed tomography voxel size and segmentation method on trabecular bone microstructure measures in mice. Bone Rep. **5**, 136–140 (2016)
4. Dice, L.R.: Measures of the amount of ecologic association between species. Ecology **26**, 297–302 (1945)
5. Engelke, K., Graeff, W., Meiss, L., Hahn, M., Delling, G.: High spatial resolution imaging of bone mineral using computed microtomography: comparison with microradiography and undecalcified histologic sections. Invest. Radiol. **28**(4), 341–349 (1993)

6. Hildebrand, T., Rüegsegger, P.: A new method for the model-independent assessment of thickness in three-dimensional images. J. Microsc. **185**(1), 67–75 (1997)
7. Klintström, E., Klintström, B., Moreno, R., Brismar, T.B., Pahr, D.H., Smedby, Ö.: Predicting trabecular bone stiffness from clinical cone-beam CT and HR-pQCT data; an in vitro study using finite element analysis. PloS ONE **11**(8), e0161101 (2016)
8. Laib, A., Rüegsegger, P.: Comparison of structure extraction methods for in vivo trabecular bone measurements. Comput. Med. Imaging Graph. **23**(2), 69–74 (1999)
9. Majumdar, S., Gies, A., Newitt, D., Osman, D., Chiu, E., Truong, V., Genant, H., Lotz, J., Kinney, J.: Assessment of trabecular bone structure using magnetic resonance imaging and X-ray tomographic microscopy. Osteoporos. Int. **6**, 376–385 (1996)
10. Majumdar, S., Newitt, D., Jergas, M., Gies, A., Chiu, E., Osman, D., Keltner, J., Keyak, K., Genant, H.: Evaluation of technical factors affecting the quantification of trabecular bone structure using magnetic resonance imaging. Bone **17**(4), 417–430 (1995)
11. Matheron, G.: Random Sets and Integral Geometry. Wiley, New York (1975)
12. Moreno, R., Borga, M., Smedby, Ö.: Estimation of trabecular thickness in grayscale images through granulometric analysis. In: Proceedings of SPIE - Medical Imaging, vol. 8314, p. 831451 (2012)
13. Otsu, N.: Threshold selection method from gray-level histograms. IEEE Trans. Syst. Man Cybern. **9**(1), 62–66 (1979)
14. Petersson, J., Brismar, T., Smedby, Ö.: Analysis of skeletal microstructure with clinical multislice CT. In: Larsen, R., Nielsen, M., Sporring, J. (eds.) MICCAI 2006. LNCS, vol. 4191, pp. 880–887. Springer, Heidelberg (2006). doi:10.1007/11866763_108
15. Peyrin, F., Houssard, J.P., Maurincomme, E., Peix, G., Goutte, R., Laval-Jeantet, A.M., Amiel, M.: 3D display of high resolution vertebral structure images. Comput. Med. Imaging Graph. **17**(4–5), 251–256 (1993)
16. Revol-Muller, C., Peyrin, F., Carrillon, Y., Odet, C.: Automated 3D region growing algorithm based on an assessment function. Pattern Recogn. Lett. **23**(1–3), 137–150 (2002)
17. Van Dessel, J., Huang, Y., Depypere, M., Rubira-Bullen, I., Maes, F., Jacobs, R.: A comparative evaluation of cone beam CT and micro-CT on trabecular bone structures in the human mandible. Dentomaxillofacial Radiol. **42**(8), 20130145 (2013)
18. Xie, W., Thompson, R.P., Perucchio, R.: A topology-preserving parallel 3D thinning algorithm for extracting the curve skeleton. Pattern Recogn. **36**(7), 1529–1544 (2003)
19. Zhou, B., Wang, J., Yu, Y.E., Zhang, Z., Nawathe, S., Nishiyama, K.K., Rosete, F.R., Keaveny, T.M., Shane, E., Guo, X.E.: High-resolution peripheral quantitative computed tomography (HR-pQCT) can assess microstructural and biomechanical properties of both human distal radius and tibia: ex vivo computational and experimental validations. Bone **86**, 58–67 (2016)

Automatic Segmentation of Abdominal Fat in MRI-Scans, Using Graph-Cuts and Image Derived Energies

Anders Nymark Christensen[1(✉)], Christian Thode Larsen[1],
Camilla Maria Mandrup[2], Martin Bæk Petersen[2], Rasmus Larsen[1],
Knut Conradsen[1], and Vedrana Andersen Dahl[1]

[1] Department of Applied Mathematics and Computer Science,
Technical University of Denmark, Kongens Lyngby, Denmark
anym@dtu.dk
[2] Section of Systems Biology Research, Copenhagen University,
Copenhagen, Denmark

Abstract. For many clinical studies changes in the abdominal distribution of fat is an important measure. However, the segmentation of abdominal fat in MRI scans is both difficult and time consuming using manual methods. We present here an automatic and flexible software package, that performs both bias field correction and segmentation of the fat into superficial and deep subcutaneous fat as well as visceral fat with the spinal compartment removed. Assessment when comparing to the gold standard - CT-scans - shows a correlation and bias comparable to manual segmentation. The method is flexible by tuning the image-derived energies used for the segmentation, allowing the method to be applied to other body parts, such as the thighs.

1 Introduction

A large number of studies have investigated the importance of abdominal fat to health, or used it as a co-variate to other factors. A correlation between abdominal fat, insulin resistance and other metabolic risk factors [5,14,16] has been shown. The longitudinal changes has also been studied, in e.g. children [18] and pre-/post menopausal women [4].

The gold standard for quantifying abdominal fat is Computed Tomography (CT) scanning [17]. The modality yields absolute values on the Hounsfiled Unit (HU) scale, which makes quantification relatively easy and the image acquisition is fast, which minimises effects of organ movement. However, CT has the drawback of ionising radiation which limits its use in healthy subjects. Another modality is Dual X-ray Absorptiometry (DEXA) [6], which is fast and cheap, but does not allow for differentiation of subcutaneous and visceral fat. Finally Magnetic Resonance Imaging (MRI) can be used. MRI allows not only for the differentiation of the subcutaneous and visceral fat, but also for splitting the subcutaneous fat into a superficial and deep compartment. The main drawbacks

© Springer International Publishing AG 2017
P. Sharma and F.M. Bianchi (Eds.): SCIA 2017, Part II, LNCS 10270, pp. 109–120, 2017.
DOI: 10.1007/978-3-319-59129-2_10

are that MRI is costly, scanning is time consuming and segmentation of the fat is difficult. The latter point is mainly due to the bias field image artefact, a consequence of magnetic field inhomogeneities, and the fact that the values in an anatomic MRI scan are not quantitative like CT.

For the task of segmentation several options exist. There is the option of manual segmentation, which is known to produce good results. However, it suffers from being time consuming and is subject to inter- and intra-observer variance. As an alternative automatic methods are attractive. Commercial software exist in the form of sliceOmatic (Tomovision, Inc., Magog, Canada) which handles the segmentation in a semi-supervised fashion. For the fully automatic case, several approaches have been tried. Zhou et al. [19], used chain-coding, a heuristic method that requires the scan to be centred on the L4-L5 vertebra. Leinhard et al. [11] proposed binary operations with some heuristics for the final segmentation. Knutsson et al. [9] used a morphon based approach, that – while yeilding good results – is very sparsely described with regards to implementation. Finally, Mosbech et al. [15] used active contours for automatic segmentation, and – as the only method – described a way to find the fascia of Scarpa, which divides the subcutaneous fat, using dynamic programming.

We present here a flexible package, available for download at GitHub[1], that can be used to segment a variety of MRI data. The initial step - bias field correction - uses the method presented by [10]. The segmentation is based on the graph-cut method proposed by Li et al. [12]. We have found a number of image based energies that can be combined to suit the problem at hand. A preliminary study of the method can be found here [2]. We compare the automatic segmentation against the gold standard CT [17], and obtain correlations and bias between the two modalities comparable to manual segmentation [8].

2   Materials and Methods

All code was written in MATLAB (The MathWorks Inc., Natick, Massachusetts, USA).

2.1   Bias Field Correction

The bias field is an image artefact present in all MRI-scans, and can be described as a low frequent noise over the image. To correct for the bias field, the method by Larsen et al. [10] was used, which assumes that the observed image originates from a generative model. The underlying 'true' and uncorrupted image is described using a mixture of Gaussians, and the bias field artifact, assumed to be multiplicative and smooth, is modeled using a linear combination of cubic b-splines with regularization on the bending energy. Model parameters are estimated using generalized expectation maximization (GEM). When model parameters have been estimated, the bias field can be computed and divided into the

---

[1] https://github.com/AndersNymarkChristensen/FatSegmentationInMRI.

observed image in order to obtain its correction. The hyper parameters used as default in our package have been determined by testing on T1, T1 with water suppression and DIXON [13] sequences. For the first two on both the abdomen and the thigh, while the latter has only been tested on the abdomen.

2.2 Segmentation

Li et al. [12] formulated the segmentation of layers in volumetric data as a graph-problem. Representing the data as (x,y,z) where x and y are the horizontal plane and z the height, each terrain-like layer can only pass through each (x,y)-column once. For connectivity and to incorporate smoothness constraints, intra-column displacement of the layers is limited. The framework generalises easily for multiple layers, which can the be constrained with a min/max distance between them. The energy used (i.e. the vertex weights) are set depending on the problem. To solve the graph-cut problem, we use the solver implemented by Boykov and Kolmogorov [1].

To represent the MRI-scan in this fashion we start by unrolling the images using radial sampling, centred on the centroid of the slice. The abdomen is not round but elongated, and the subcutaneous layer of fat changes distance from the centroid rapidly in the lateral direction. We thus sample more densely in these regions, see Fig. 1a. This unrolling yields an image - Fig. 1b that can segmented by the framework of Li et al. [12].

Subcutaneous Fat. The first step in the actual segmentation is the subcutaneous fat, defined by the inner and the outer surface. Those surfaces are found simultaneously, using a constraint imposed on the distance between the two layer. For thighs or other structures these settings can be varied. The segmentation employs a variety of image derived energies, that can be grouped in two cost-classes: gradient-based surface cost, and surface cost based on cumulative sums.

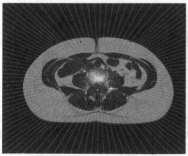

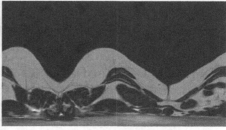

(a) Radial sampling of the image. Each radial ray passes through inner and outer surface once

(b) Unrolled image. In this representation, the inner and the outer surfaces are terrain-like and pass trough each image column once

Fig. 1. Radial sampling and the resulting unrolled image

With the slices along the z-direction, the surfaces are lying in the angular direction θ while the radial direction r takes the role of the height. The gradient-based surface energy is defined as

$$G = \text{sign}\left(\frac{\partial I}{\partial r}\right) \cdot \left(\left|\frac{\partial I}{\partial r}\right| + \left|\frac{\partial I}{\partial \theta}\right|\right) \tag{1}$$

where $\frac{\partial I}{\partial r}$ and $\frac{\partial I}{\partial \theta}$ are the gradients in the radial and angular direction.

We further define G_m the gradient applied to the median filtered image, G_g as applied to the Gaussian filtered image. For the outer surface we have G_o as the negative gradient and G_{os} as the negative gradient applied to the median and then Gaussian smoothed image. The costs for the inner surface are illustrated in Fig. 2, and the outer in Fig. 3.

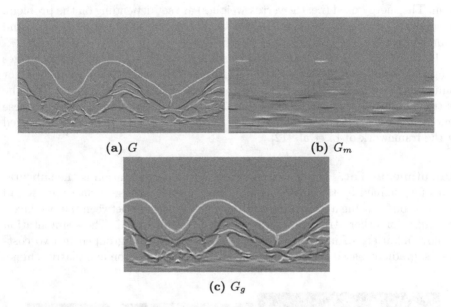

(a) G (b) G_m

(c) G_g

Fig. 2. The three gradient-based contributions to the cost of the inner surface. The cost is constructed such that it has low (dark) values where the surface is to be detected

The surface cost S_1 is derived by taking the G image and setting all values below zero to zero. A cumulative sum is then taken in the radial direction. For S_2 a cumulative sum is taken on the unfolded image. We define S_3 as the element wise product of S_1 and S_2 filtered with a median filter. For the outer layer we define a single energy S_o, which is the same as S_1 but applied to G_o. As illustrated by Fig. 4 these costs regulates how deep the layer can go in the segmentation.

The two surfaces can then be found by weighting these energies depending on the anatomical region and sequence used to record the images. A median filter is applied to the inner surface to get a smooth segmentation, see Fig. 5.

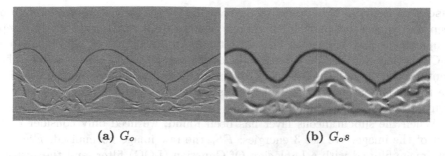

(a) G_o (b) $G_o s$

Fig. 3. The two gradient-based contributions to the cost of the outer surface

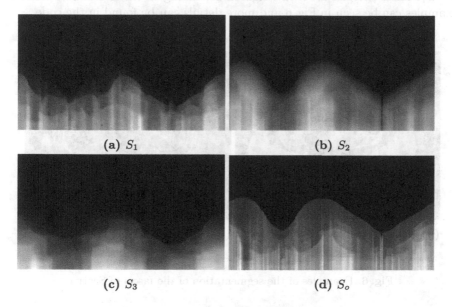

(a) S_1 (b) S_2

(c) S_3 (d) S_o

Fig. 4. The four surface costs based on cumulative sums

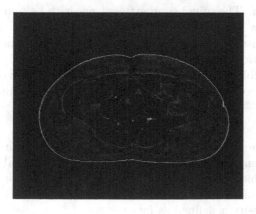

Fig. 5. Segmentation of the subcutaneous fat

Fascia of Scarpa. The fascia of Scarpa divides the subcutaneous fat into a deep and superficial part. In metabolic studies it is of interest, as the deep compartment may behave as visceral fat [3]. An advantage of MRI compared to CT is that the fascia is often visible and a segmentation thus possible. The identification of the fascia of Scarpa is optional, and would e.g. not be used on the thigh. The fascia is very fickle and is not always visible, in which case the segmentation is not meaningful.

When the subcutaneous layer has been found, we need only consider that part of the image. We use 3 energies: FS_r the raw image normalised, FS_l the raw image filtered with a Laplacian Of Gaussian (LOG) filter, and the surface cost S_1. These energies are then weighted depending on the anatomical position. Examples can be seen in Fig. 6 and more details can be found in the code.

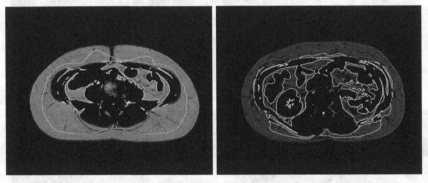

(a) Good segmentation of the fascia (green) in the posterior

(b) The segmentation fails, as the fascia (green) is not visible

Fig. 6. Examples of the segmentation of the fascia of Scarpa

Spine Extraction. The spinal compartment needs to be removed for accurate results. The marrow in the vertebra has comparable intensities to fat and is thus classified as such if not removed. The identification and removal of the spinal compartment is optional and would e.g. not be used on the thigh. The method is based on a new unfolding centred on the spine, with heuristic constraints added on the image derived energies. See the code for more details. An example is shown in Fig. 7.

Visceral Fat. The visceral compartment is defined by the inner layer, except for the compartment around the spinal column. The spinal compartment is removed - if available - before the segmentation of the fat. A k-means clustering using 5 groups is run and the median value is used as a threshold. Everything with a higher intensity is defined as fat.

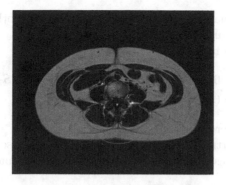

Fig. 7. Spinal compartment

2.3   Test Subjects

Data were acquired in connection with Copenhagen Women Study http://cws.
ku.dk. All participants received written and oral information about the study,
including risks and discomforts associated with participation, before they gave
their written consent to participate. The study was conducted according to the
Helsinki Declaration and approved by the ethical committee in the capital region
of Denmark, protocol nr. H-1-2012-150.

6 subjects - 3 overweight (ID 1 to 3) and 3 normal weight (ID 4 to 6) - were
chosen randomly among their weight groups. All subjects had a CT-scan performed on the abdomen. Within a mean/maximum of 11.7/24 days the subjects
also had a MRI scan of the abdomen. The CT were acquired on a Siemens
SOMATOM Definition scanner (Siemens, Erlangen, Germany) with settings
100 kVp and 60 mAs - the slice thickness 2 mm was and the pixel size was
1.5234 mm × 1.5234 mm. The MRI scans were acquired on a Siemens Avanto
1.5 T scanner (Siemens, Erlangen, Germany) using a T1 sequence with water
suppression - the slice thickness was 7.2 mm and the pixel size was 1.1719 mm
× 1.1719 mm. Before the MRI scan 20 mg of Hyoscinbutylbromide (Buscopan)
were administered intramuscular to minimise organ movements.

To compare consistent regions, slices were included from the first caudal slice
were the iliac crest were not visible, moving in the cranial direction until the last
slice before the liver becomes visible.

CT Segmentation. The CT scans were segmented using the method by Kim
et al. [7] with the modification that diagonal directions were included in the step
to find the visceral mask. All voxels with values between −150 HU and −50 HU
were classified as fat. All slices were inspected for major errors.

MRI Segmentation. The images were bias field corrected using the following settings: 'stepSize' was set to the voxel size, the 'desiredStepSize' was set

to 4, 'numberOfGaussians' 8, 'smoothingDistance' 25, 'smoothingRegularization' 5e-1, and 'mask' was an image of all positive voxels.

The subcutaneous fat was found with

$$\text{Energy}_{\text{inner}} = 0.5 \cdot G + 0.1 \cdot G_g + 0.1 \cdot G_m + 0.05 \cdot S_1 + 0.25 \cdot S_3 \qquad (2)$$

$$\text{Energy}_{\text{outer}} = 0.5 \cdot G_{os} + 0.5 \cdot S_o \qquad (3)$$

for the inner and outer layer respectively.

The spinal compartment was removed, and 5 clusters used for determination of the visceral fat threshold. All slices were inspected for major errors.

3   Results

The obtained volumes are given in Table 1 for the total volume i.e. the entire body, for the subcutaneous fat and for the visceral fat.

Table 1. Segmentation Volumes

Subject ID	Total volume			Subcutaneous fat			Visceral fat		
	CT	MRI	Diff. (%)	CT	MRI	Diff. (%)	CT	MRI	Diff. (%)
1	6277	6986	−709 (−11)	2992	3401	−409 (−14)	683	856	−173 (−25)
2	2835	3189	−354 (−12)	1038	1223	−185 (−18)	296	425	−129 (−44)
3	5662	3990	1672 (30)	1990	1413	576 (29)	1080	795	285 (26)
4	2737	2038	699 (26)	718	549	169 (24)	275	211	64 (23)
5	1653	1616	36 (2)	409	419	−10 (−3)	234	243	−10 (−4)
6	2183	2315	−132 (−6)	575	570	6 (1)	287	339	−53 (−18)
Average	3558	3356	202 (5.7)	1287	1262	24 (1.9)	476	478	−2 (−0.5)

Further, as the total volume deviates between the two modalities we normalise the subcutaneous and visceral with the total volume, to get the two fat measures as a fractional fat content. These are given in Table 2. The mean difference±standard deviation between CT and MRI was −0.005±0.0122 for the subcutaneous fat and −0.013±0.0088 for the visceral fat.

A correlation and a Bland-Altman (Tukey mean-difference) plot for the relative volumes are shown in Fig. 8.

Subject 3 has the largest deviation from CT. All the slices from the subject are shown in Fig. 9.

Table 2. Relative fat Volumes

Subject ID	Subcutaneous fat			Visceral fat		
	CT	MRI	Diff. (%)	CT	MRI	Diff. (%)
1	0.477	0.487	−0.010 (−2.134)	0.109	0.122	−0.014 (−12.545)
2	0.366	0.383	−0.017 (−4.712)	0.104	0.133	−0.029 (−27.582)
3	0.351	0.354	−0.003 (−0.809)	0.191	0.199	−0.008 (−4.426)
4	0.262	0.269	−0.007 (−2.701)	0.100	0.104	−0.003 (−3.280)
5	0.247	0.259	−0.012 (−4.854)	0.141	0.150	−0.009 (−6.544)
6	0.264	0.246	0.018 (6.676)	0.131	0.147	−0.015 (−11.606)
Average	0.328	0.333	−0.005 (−1.614)	0.130	0.143	−0.013 (−10.121)

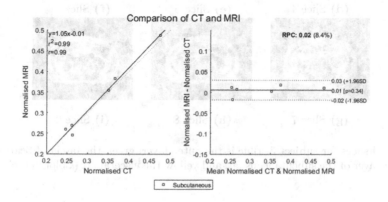

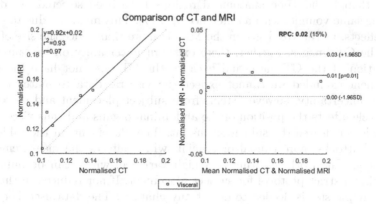

Fig. 8. Correlation and Bland-Altman plot for subcutaneous and visceral fat

4   Discussion

A software package for segmentation of abdominal fat in MRI has been implemented and validated by comparison with CT.

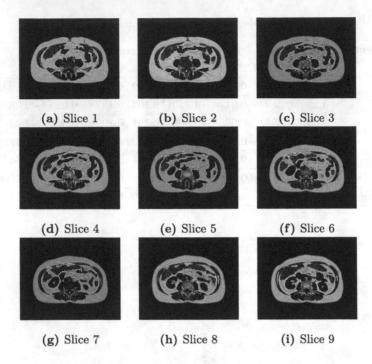

(a) Slice 1   **(b)** Slice 2   **(c)** Slice 3

(d) Slice 4   **(e)** Slice 5   **(f)** Slice 6

(g) Slice 7   **(h)** Slice 8   **(i)** Slice 9

Fig. 9. All slices for subject 3. Blue is the outer layer, green the fascia of Scarpa, red the inner layer of the subcutaneous fat, and yellow the visceral fat (Color figure online)

Even though the same anatomical regions have been selected, we do not obtain the same volume, with a difference of 5.7%. This might be due to partial volume effects, as the voxel size in the MRI is more than twice the size of what it is in the CT images. Further, there exist an uncertainty on the automatic segmentation of the CT data, and because the CT does not have the spinal compartment excluded we cannot expect the visceral data to match exactly. The main uncertainty, however, stems from subject placement and movement. Breath cycle affects the position of the abdominal organs and thus visceral fat, and to a lesser degree the subcutaneous fat. The placement on the table may push the buttocks more caudal or cranial, which affects the measurement of subcutaneous fat. Both resulting in a high variance, which can be minimised by following a strict protocol for subject placement. If not reduced in this way, a larger sample size is needed to detect any changes. The data used here were acquired with different aims, and do not follow the same protocol. With these caveats the mean difference of 1.9% for subcutaneous and -0.5% for visceral fat is acceptable.

When comparing the relative amount of fat present in each modality, we get very good correlations, better than with manual segmentation. For visceral fat manual/automatic r=0.89/r=0.97, subcutaneous fat manual/automatic r=0.92/r=0.99 [8]. This may partly be caused by the fewer number of subjects

(6 vs. 27) included in our study. Our bias is also comparable to or better than for manual segmentation. For visceral fat we get -0.013 ± 0.0088 compared to -0.029 ± 0.019, and for subcutaneous fat -0.005 ± 0.0122 compared to 0.004 ± 0.026 [8]. This is further confirmed by the Bland-Altman plots where all points are within the limits of agreement. The subject with the largest absolute error is shown. When inspected by an experienced medical doctor only minor errors were found.

Although not tried here, examples of the method applied to other sequences can be found in [2], where it also performed good as judged by inspections. The package is thus flexible across T1, T1 with water suppression and DIXON sequences. It should be noted that the data quality is very important to the quality of the segmentations. Artefacts from organ or breathing motion can easily lead to incorrect segmentations. Further, parameter tuning is often necessary, as the images can vary widely depending on the field strength of the MR-scanner and the sequence employed.

5 Conclusion

A flexible software package for automatic segmentation of abdominal fat has been implemented. Both the bias and variance is comparable to or smaller than for manual segmentation.

References

1. Boykov, Y., Kolmogorov, V.: An experimental comparison of min-cut/max-flow algorithms for energy minimization in vision. IEEE Trans. Pattern Anal. Mach. Intell. **26**(9), 1124–1137 (2004)
2. Christensen, A.N., Conradsen, K., Larsen, R.: Data Analysis of Medical Images: CT, MRI, Phase Contrast X-ray and PET. Ph.D. thesis (2016)
3. Enevoldsen, L.H., Simonsen, L., Stallknecht, B., Galbo, H., Bülow, J.: In vivo human lipolytic activity in preperitoneal and subdivisions of subcutaneous abdominal adipose tissue. Am. J. Physiol. Endocrinol. Metab. **281**(5), E1110–E1114 (2001)
4. Franklin, R.M., Ploutz-Snyder, L., Kanaley, J.A.: Longitudinal changes in abdominal fat distribution with menopause. Metab. Clin. Exp. **58**(3), 311–315 (2009)
5. Fujioka, S., Matsuzawa, Y., Tokunaga, K., Tarui, S.: Contribution of intra-abdominal fat accumulation to the impairment of glucose and lipid metabolism in human obesity. Metab. Clin. Exp. **36**(1), 54–59 (1987)
6. Haarbo, J., Gotfredsen, A., Hassager, C., Christiansen, C.: Validation of body composition by dual energy X-ray absorptiometry (DEXA). Clin. Physiol. Funct. Imaging **11**, 331–341 (1991). Oxford, England
7. Kim, Y.J., Lee, S.H., Kim, T.Y., Park, J.Y., Choi, S.H., Kim, K.G.: Body fat assessment method using CT images with separation mask algorithm. J. Digit. Imaging **26**(2), 155–162 (2013)
8. Klopfenstein, B.J., Kim, M.S., Krisky, C.M., Szumowski, J., Rooney, W.D., Purnell, J.Q.: Comparison of 3 T MRI and CT for the measurement of visceral and subcutaneous adipose tissue in humans. Br. J. Radiol. **85**(1018), e826–e830 (2012)

9. Knutsson, H., Andersson, M.: Morphons: segmentation using elastic canvas and paint on priors. In: Proceedings - International Conference on Image Processing ICIP, **2**, pp. 1226–1229 (2005)

10. Larsen, C.T., Iglesias, J.E., Leemput, K.: N3 bias field correction explained as a Bayesian modeling method. In: Cardoso, M.J., Simpson, I., Arbel, T., Precup, D., Ribbens, A. (eds.) BAMBI 2014. LNCS, vol. 8677, pp. 1–12. Springer, Cham (2014). doi:10.1007/978-3-319-12289-2_1

11. Leinhard, O., Johansson, A., Rydell, J., Smedby, Ö., Nyström, F., Lundberg, P., Borga, M.: Quantitative abdominal fat estimation using MRI. In: 19th International Conference on Pattern Recognition, ICPR 2008, pp. 1–4 (2008)

12. Li, K., Wu, X., Chen, D.Z., Sonka, M.: Optimal surface segmentation in volumetric images - a graph-theoretic approach. IEEE Trans. Pattern Anal. Mach. Intell. **28**(1), 119–134 (2006)

13. Ma, J.: Dixon techniques for water and fat imaging. J. Magn. Reson. Imaging **28**(3), 543–558 (2008)

14. Miyazaki, Y., Glass, L., Triplitt, C., Wajcberg, E., Mandarino, L.J., DeFronzo, R.A.: Abdominal fat distribution and peripheral and hepatic insulin resistance in type 2 diabetes mellitus. Am. J. Physiol. Endocrinol. Metab. **283**(6), E1135–E1143 (2002)

15. Mosbech, T.H., Pilgaard, K., Vaag, A., Larsen, R.: Automatic segmentation of abdominal adipose tissue in MRI. In: Heyden, A., Kahl, F. (eds.) SCIA 2011. LNCS, vol. 6688, pp. 501–511. Springer, Heidelberg (2011). doi:10.1007/978-3-642-21227-7_47

16. Rosito, G.A., Massaro, J.M., Hoffmann, U., Ruberg, F.L., Mahabadi, A.A., Vasan, R.S., O'Donnell, C.J., Fox, C.S.: Pericardial fat, visceral abdominal fat, cardiovascular disease risk factors, and vascular calcification in a community-based sample: the Framingham heart study. Circulation **117**(5), 605–613 (2008)

17. Seidell, J.C., Bakker, C.J., van der Kooy, K.: Imaging techniques for measuring adipose-tissue distribution-a comparison between computed tomography and 1.5-T magnetic resonance. Am. J. Clin. Nutr. **51**(6), 953–957 (1990)

18. Tinggaard, J., Hagen, C., Mouritsen, A., Mieritz, M., Wohlfahrt-Veje, C., Fallentin, E., Larsen, R., Christensen, A., Jensen, R., Juula, A., Maina, K.: Abdominal fat distribution measured by magnetic resonance imaging in 197 children aged 10–15 years – correlation to anthropometry and dual X-ray absorptiometry. In: ESPE Abstracts, pp. 84 P–2–330 (2015)

19. Zhou, A., Murillo, H., Peng, Q.: Novel segmentation method for abdominal fat quantification by MRI. J. Magn. Reson. Imaging **34**(4), 852–860 (2011)

Remote Sensing

Two-Source Surface Reconstruction Using Polarisation

Gary A. Atkinson[✉]

Bristol Robotics Laboratory, University of the West of England, Bristol, UK
gary.atkinson@uwe.ac.uk
http://www.cems.uwe.ac.uk/~gaatkins/

Abstract. Polarisation vision aims to capture and interpret the polarisation state of incoming light as part of a computer vision system. In this paper, polarisation information is fused with two-source photometric stereo (PS) in order to determine the three-dimensional geometry of surfaces in the presence of both specular and diffuse reflection. In addition to the primary benefit of applying to both reflection types, the method lessens the effects of the Lambertian assumption, which is prevalent in many PS methods other than those requiring many light sources. Further, the method overcomes inherent ambiguities that are present in basic polarisation vision systems. The proposed method commences by using PS to deduce a constrained mapping of the surface normal at each point onto a 2D plane. Phase information from polarisation is used to deduce a mapping onto a different plane. The paper then shows how the full surface normal can be obtained from the two mappings. The results section of the paper demonstrates strong performance of the novel approach against the baseline methods for a range of real-world objects.

Keywords: Surface reconstruction · Polarisation · Photometric stereo

1 Introduction

This paper presents a novel surface reconstruction algorithm that uses polarisation information from two light sources. The method initially applies a form of two-source photometric stereo (PS) to estimate part of the surface normal before polarisation information is then used in order to fully constrain the normals at each pixel. The motivation for this approach is to overcome some of the weaknesses in related approaches. Firstly, the need for three or more light sources in PS is often debilitating for real world applications. Secondly, many polarisation-based methods suffer from a range of ambiguities due to the non-monotonic mapping between the polarisation information and surface normals. The proposed method overcomes this issue using intensity information. Finally, a novel region-growing approach is applied that allows for reconstruction in the presence of both specular and diffuse reflection.

Of the many approaches to 3D vision found in the literature, polarisation vision (i.e. the use of polarised light for computer vision) is one of the less-studied. The principle of most methods of polarisation vision is that light changes

© Springer International Publishing AG 2017
P. Sharma and F.M. Bianchi (Eds.): SCIA 2017, Part II, LNCS 10270, pp. 123–135, 2017.
DOI: 10.1007/978-3-319-59129-2_11

its polarisation state on reflection from surfaces [1]. This is typically from unpolarised to partially linearly polarised form.

Most previous computer vision research using polarisation relies on Fresnel theory applied to *specular* reflection [2]. The specific properties of the polarisation correlate to the relationship between the surface orientation and the viewing direction [1,3]. Unfortunately, there are inherent ambiguities present and the refractive index of the target is typically required. Further, different equations are required to model reflection if a *diffuse* component is present.

Miyazaki et al. [4], Atkinson and Hancock [5] and Berger et al. [6] all use multiple viewpoints to overcome the ambiguity issue. In [4], specular reflection is used on transparent objects – a class of material that causes difficulty for most computer vision methods. In [5], a patch matching approach is used for diffuse surfaces to find stereo correspondence and, hence, 3D data. In [6], an energy functional for a regularization-based stereo vision is applied. Taamazyan et al. [7] use a combination of multiple views and a physics-based reflection model to simultaneously separate specular and diffuse reflection and estimate shape.

Drbohlav and Šára [8,9] use PS (as in this paper) but with linearly polarized incident illumination. Atkinson and Hancock [10] also use PS but with more light sources and less resilience to inter-reflections than the method of this paper. Garcia et al. [11] use a circularly polarised light source to overcome the ambiguities while Morel et al. [12] extend the methods of polarisation to metallic surfaces by allowing for a complex index of refraction.

In an effort to overcome the need for the refractive index, Miyazaki et al. constrain the histogram of surface normals [13] while Rahmann and Canterakis [14,15] use multiple views and the orientation of the plane of polarisation for correspondence. Huynh et al. [16] actually estimate the refractive index using spectral information.

2   Polarisation Vision

This paper is based on the premise that light undergoes a partial polarisation process upon reflection from smooth surfaces. Consider the experimental arrangement shown in Fig. 1, which is used for all of this work. Fresnel reflectance theory [2] is able to quantify the polarisation process that occurs when initially unpolarised incident light is reflected towards the camera. The reflected light can be parametrised by three values. First is the intensity, I, of the light. Second is the phase angle, ϕ, which defines the principle angle of the electric field component of the light as shown in the figure. Finally, the degree of polarisation, ρ, indicates the level of polarisation from 0 (unpolarised) to 1 (completely linearly polarised) [17].

As explained elsewhere in the literature [1,17], for specularly reflected light the phase angle aligns perpendicularly to the projection of the surface normal onto the image plane. Assuming that the Wolff subsurface scattering model [18] applies, diffuse reflection by contrast causes parallel alignment. Since there is no distinction in phase angle shifts of π radians, there are therefore four possible surface azimuthal angles, α (defined in Fig. 1), for an unknown reflectance type:

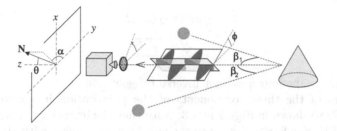

Fig. 1. Arrangement and definitions. Two polarisation images are acquired using the camera and rotating polariser. Each image has a different source illuminated. The surface normal, **N**, is defined by its zenith angle, θ, and azimuth angle, α.

$$\alpha \in \left[\phi, \phi + \frac{\pi}{2}, \phi + \pi, \phi + \frac{3\pi}{4}\right] \quad \text{where} \quad 0 \leq \phi < \pi \tag{1}$$

The degree of polarisation contains information principally related to the zenith angle of the surface. Unfortunately, the relationship between zenith angle and degree of polarisation varies substantially between specular and diffuse reflection and depends on the refractive index of the reflecting surface, which is typically unknown [17]. For this reason, its use for surface normal estimation is avoided in this paper. The method does however use the degree of polarisation as a measure of reliability of the phase data: any phase data with a corresponding degree of polarisation below a threshold (fixed at 1%) is deemed unreliable due their correspondingly high associated noise levels.

The method described in the next section requires two polarisation images of an object for input (a single polarisation image comprises a separate intensity, phase and degree of polarisation value for each pixel). There are now several commercially available cameras that capture such data directly such as the Fraunhofer "POLKA" [19]. For this paper however, a simple method is adopted where a motorised polariser is placed in front of the camera and images taken at various orientations [17]. The camera was a Dalsa Genie HM1400 fitted with a Schneider KMP-IR Xenoplan 23/1,4-M30,5 lens. Images are taken at polariser angles of 0° (horizontal), 45°, 90° and 135°. If the intensities measured for these angles are called I_0, I_{45}, I_{90} and I_{135}, then the polarisation data for each cell is then calculated via the Stoke's parameters S_0, S_1 and S_2 (assuming no circular polarisation is present) [2]:

Stoke's parameters:

$$S_0 = \frac{I_0 + I_{45} + I_{90} + I_{135}}{2} \tag{2}$$

$$S_1 = I_0 - I_{90} \tag{3}$$

$$S_2 = I_{45} - I_{135} \tag{4}$$

Polarisation image data:

$$I = S_0 \tag{5}$$

$$\phi = \frac{1}{2}\arctan_2\left(S_2, S_1\right) \tag{6}$$

$$\rho = \frac{\sqrt{S_1^2 + S_2^2}}{S_0} \tag{7}$$

where $\arctan_2$ is the four quadrant inverse tangent [20].

Examples of the three components of the polarisation image for a white snooker ball are shown in Figs. 2 and 3. Note that the intensity is normalised to the range [0,1]. For Fig. 2, the capture conditions were ideal, with the ball illuminated by a single small white LED placed close to the camera and blackout curtains surrounding the ball to diminish all inter-reflections from the environment. The reflection is therefore of diffuse type throughout the surface and so the first or third solutions to (1) must be true for all pixels. For Fig. 3, a plain matte white board was placed behind the ball. A sudden phase shift is present near the occluding contours of the ball due to an inter-reflection from the white backing. Since this region is effectively undergoing a specular reflection, either the second or fourth solution to (1) is true. By contrast, the first or third solutions are true for the rest of the ball, which is exhibiting diffuse reflection. Note also, that the degree of polarisation is higher near the inter-reflection. This is also predicted by the reflection theory but is of less significance to this paper [17]. The noise-ridden background to Fig. 3 is due to the very low degree of polarisation for the backing and the shadow cast by the ball; neither of which are of significance here.

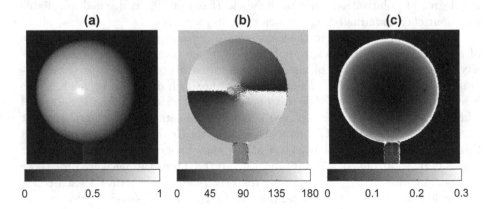

Fig. 2. Polarisation image of a white snooker ball with no inter-reflections and only one specularity. (a) Intensity I, (b) phase angle ϕ, (c) degree of polarisation, ρ.

3   Method

The approach can be broadly divided into the following steps assuming that the starting point is two full polarisation images corresponding to the two light source locations shown in Fig. 1.

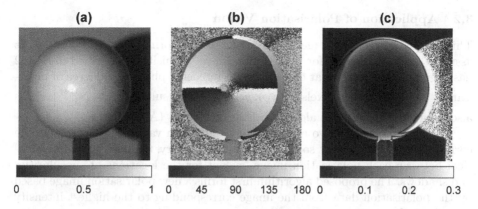

Fig. 3. Polarisation image of a white snooker ball with inter-reflections from a white sheet behind the ball. Otherwise, conditions were the same as in Fig. 2. Regions where $\rho > 0.3$ in (c) are shown white here to improve the clarity of the rest of the figure.

1. Extract estimates of 2D (y-z) surface normals from two-source PS.
2. (a) Extract estimates of 2D (x-y) surface normals from polarisation aided by PS.
   (b) Disambiguate spurious data or specular inter-reflections using region growing.
3. Combine data from each modality to obtain 3D (x-y-z) surface normals.

The first step essentially applies a 2D version of PS to obtain a 2D surface normal at each pixel. This is not merely a projection of the 3D normal onto the y-z plane but rather a vector representing one particular degree of freedom of the orientation of the surface at each point. The second step uses the polarisation phase angle of the incoming light and assumes diffuse reflection (for now) to estimate a 2D version of the surface normal in the x-y plane. PS data is applied to disambiguate most of these vectors but region growing is needed where specularities occur or where previous disambiguation is deemed incorrect. Finally, the information above is combined to form a 3D surface normal map.

3.1   Application of PS

The aim here is to use the method of PS to estimate a 2D normal in the y-z plane at each point. To do this, the original methodology of Woodham [21] is adapted to two images in order to obtain a set of n normals as follows:

$$\left\{ \mathbf{N}_i^{(\mathrm{ps})} = \begin{bmatrix} N_{y,i}^{(\mathrm{ps})} \\ N_{z,i}^{(\mathrm{ps})} \end{bmatrix} \right\}_{i=1}^{n} \leftarrow \begin{bmatrix} -\cos(\beta_R) & \cos(\beta_L) \\ \sin(\beta_R) & -\sin(\beta_L) \end{bmatrix} \begin{bmatrix} I_{L,i} \\ I_{R,i} \end{bmatrix} \forall i \qquad (8)$$

where the angles β_L and β_R are defined in Fig. 1 and the suffix "ps" is a reminder that these estimates are from photometric stereo. The "$\leftarrow$" symbol is used throughout this paper to refer to variable assignment.

3.2 Application of Polarisation Vision

This section describes the method to estimate 2D normals in the x-y plane. It assumes that the method for polarisation image acquisition described in Sect. 2 has been applied to arrive at corresponding intensity, phase and degree of polarisation values for all n pixels: $\{I_i, \phi_i, \rho_i\}_{i=1}^{n}$. The result of PS, $\left\{\mathbf{N}_i^{(ps)}\right\}_{i=1}^{n}$, is also used. This part of the algorithm is in two stages (Algorithm 1).

The raw data yields two estimates of polarisation values for each pixel (one corresponding to each light source direction). In theory, the data should be identical between each polarisation image, except for the locations of specularities and shadows. The proposed algorithm first forms a new polarisation image based on the polarisation data from the image corresponding to the highest intensity at each pixel.

The first three lines of Algorithm 1 are designed to obtain an initial azimuth angle estimation. The first line applies a sharpening operator to the phase data. The reason for this is that the subsequent specularity detection/disambiguation algorithm ("localAlign") is more reliable in the presence of sharp transitions between specular inter-reflections and diffuse regions.

The method makes the initial assumption that all surface normal projections onto the x-y (image) plane are aligned parallel to, or anti-parallel to, the phase angle, as predicted by the theory for diffuse reflection covered in Sect. 2. Line 2 sets the azimuthal angle, α, of each point accordingly. Whether said projections are parallel or anti-parallel depend on the best match to the estimated 2D normal from PS. Line 3 simply generates a set of 2D surface normals on the x-y plane using the calculated azimuth angles. The suffix "po" reminds us that these estimates are primarily from polarisation.

It is expected that most of the normal estimates will be correct at this point. However, it is likely that some regions of the image will be incorrect due to one of the following reasons:

Algorithm 1. Input: $\left\{\phi_i, \rho_i, \mathbf{N}_i^{(ps)}\right\}_{i=1}^{n}$ Output: $\left\{\mathbf{N}_i^{(po)}\right\}_{i=1}^{n}$

1: $\{\phi_i\} \leftarrow \text{sharpen}(\{\phi_i\})$

2: $\{\alpha_i\} \leftarrow \begin{cases} \phi_i & \forall\, i \mid \left(N_{y,i}^{(ps)} < 0 \wedge \phi_i > \frac{\pi}{2}\right) \vee \left(N_{y,i}^{(ps)} > 0 \wedge \phi_i < \frac{\pi}{2}\right) \\ \phi_i + \pi & \text{otherwise} \end{cases}$

3: $\left\{\mathbf{N}_i^{(po)} = \begin{bmatrix} N_{x,i}^{(po)} \\ N_{y,i}^{(po)} \end{bmatrix}\right\}_{i=1}^{n} \leftarrow \begin{bmatrix} -\sin\alpha_i \\ \cos\alpha_i \end{bmatrix} \forall\, i$

4: $\{R_i\}_{i=1}^{n} \leftarrow \begin{cases} 1 \; \forall\, i \mid \text{isBorderPixel}(i) \vee \rho_i < 0.01 \\ 0 \; \text{otherwise} \end{cases}$

5: **while** [More seed points?] **do**

6: $\quad j \leftarrow$ Get seed point

7: $\quad \left\{\mathbf{N}^{(po)}\right\} \leftarrow \text{localAlign}\left(\left\{\mathbf{N}^{(po)}, R\right\}, j\right)$

Algorithm 2. "localAlign". Input: $\left\{\mathbf{N}_i^{(\mathrm{po})}, R_i\right\}_{i=1}^n, j$ Output: $\left\{\mathbf{N}_i^{(\mathrm{po})}\right\}_{i=1}^n$

1: **if** $[R_j = 1]$ **then**

2:     **return** $\left\{\mathbf{N}^{(\mathrm{po})}, R\right\}$

3: $R_j \leftarrow 1$

4: **for** $[k \leftarrow \mathrm{neighboursOf}\,(j)]$ **do**

5:     $\Delta\alpha \leftarrow \arccos\left(\mathbf{N}_j^{(\mathrm{po})} \cdot \mathbf{N}_k^{(\mathrm{po})}\right)$

6:     **if** $[\mathrm{small}\,(\Delta\alpha)]$ **then**

7:         $\left\{\mathbf{N}^{(\mathrm{po})}\right\} \leftarrow \mathrm{localAlign}\left(\left\{\mathbf{N}^{(\mathrm{po})}, R\right\}, k\right)$

8:     **if** $[\mathrm{small}\,(\pi - \Delta\alpha)]$ **then**

9:         $\mathbf{N}_k^{(\mathrm{po})} \leftarrow -\mathbf{N}_k^{(\mathrm{po})}$

10:       $\left\{\mathbf{N}^{(\mathrm{po})}\right\} \leftarrow \mathrm{localAlign}\left(\left\{\mathbf{N}^{(\mathrm{po})}, R\right\}, k\right)$

11:     **if** $\left[\mathrm{small}\,\left(\frac{\pi}{2} - \Delta\alpha\right)\right]$ **then**

12:       $\mathbf{N}_{\mathrm{rot}} \leftarrow \begin{bmatrix} 0 & 1 \\ -1 & 0 \end{bmatrix} \mathbf{N}_k^{(\mathrm{po})}$

13:       **if** $\left[\arccos\left(\mathbf{N}_j^{(\mathrm{po})} \cdot \mathbf{N}_{\mathrm{rot}}\right) < \frac{\pi}{2}\right]$ **then**

14:         $\mathbf{N}_k^{(\mathrm{po})} \leftarrow \mathbf{N}_{\mathrm{rot}}$

15:       **else**

16:         $\mathbf{N}_k^{(\mathrm{po})} \leftarrow -\mathbf{N}_{\mathrm{rot}}$

17:       $\left\{\mathbf{N}^{(\mathrm{po})}\right\} \leftarrow \mathrm{localAlign}\left(\left\{\mathbf{N}^{(\mathrm{po})}, R\right\}, k\right)$

- Diffusely reflecting regions with incorrect disambiguation. This is typically where $N_y^{(\mathrm{ps})} \approx 0$, meaning the initial disambiguation is not robust. These regions have an azimuth error of π radians.
- Specular (direct or inter-reflective) regions. In these areas the azimuth angle error is $\pi/2$ radians, as predicted by the theory described in Sect. 2.

Lines 4 to 7 of Algorithm 1 are intended to address these possibilities, while enforcing (1). First, a set of pixels, $\{R_i\}_{i=1}^n$ are determined that should *not* be further considered by the algorithm. In the first instance, these pixels correspond to image border pixels and those with very low ($<1\%$) degree of polarisation.

Next, a seed point is chosen. This can easily be done either manually or randomly using a point of high confidence (i.e where the degree of polarisation is high and either $N_y^{(\mathrm{ps})} \ll 0$ or $N_y^{(\mathrm{ps})} \gg 0$). One weakness here however, is that this point must be of diffuse reflection and so may benefit from some heuristics-based selection in future work. Assume for now that only one seed point is needed, but note that the code permits more if necessary (e.g. for more complicated shapes). The rest of the process involves a recursive call to the region growing function, which progressively aligns spurious normals according to the constraints of (1).

The region growing works as follows (see Algorithm 2). First, the function checks whether the point in question j should be considered by reference to $\{R_i\}$ (lines 1 and 2). On occasions where $R_j = 1$, the function call terminates. Where this is not the case, point j is added to $\{R_i\}$, i.e. so it is not considered again. The remainder of the function is completed for each neighbour of the point (line 4). The neighbourhood is defined as simple 4-connected region for this paper.

Next, the angle between neighbouring x-y surface normals is calculated, $\Delta\alpha$ (line 5). There are four possibilities for $\Delta\alpha$:

- $\Delta\alpha$ is close to zero: assume both azimuth angles are correct, move from j to neighbouring point, k, and continue the process (lines 6 and 7).
- $\Delta\alpha$ is close to π: as above but assume azimuth disambiguation was incorrect so rotate by π (lines 8 to 10).
- $\Delta\alpha$ is close to $\pi/2$: rotate azimuth by $\pi/2$ as this region has specular reflection. The direction of rotation is chosen to minimise the angle between normals at j and k. Again, move from j to k and continue (lines 11 to 17).
- Otherwise: there could be a boundary of orientation so there is no reason to believe that the azimuth estimate is erroneous.

The definition of "close" for this recursive function remains open for now but only minor effects on results were observed as the threshold for closeness was varied between 5° and 10°.

3.3   Fusion of Data into Full 3D Surface Normals

The methods from the previous two sections give two sets of 2D surface normals: $\{\mathbf{N}^{(\mathrm{ps})}\}$ on the y-z plane and $\{\mathbf{N}^{(\mathrm{po})}\}$ on the x-y plane. These can be combined into a single 3D vector. First assume that the 2D vectors are normalised such that the following is true for each point:

$$N_x^{(\mathrm{po})^2} + N_y^{(\mathrm{po})^2} = 1 = N_y^{(\mathrm{ps})^2} + N_z^{(\mathrm{ps})^2} \tag{9}$$

where the subscripts i are omitted for the sake of brevity.

Denote the 3D surface normal at a particular point $\mathbf{N} = [N_x, N_y, N_z]^T$. Further, choose the components estimated by polarisation for x and y so there is only one unknown, N_z:

$$\mathbf{N} = [N_x, N_y, N_z]^T = \left[N_x^{(\mathrm{po})}, N_y^{(\mathrm{po})}, N_z\right]^T \tag{10}$$

Normalising this vector to unit length gives:

$$\mathbf{n} = \frac{\left[N_x^{(\mathrm{po})}, N_y^{(\mathrm{po})}, N_z\right]^T}{\sqrt{1 + N_z^2}} = \left[n_x^{(\mathrm{po})}, n_y^{(\mathrm{po})}, n_z\right]^T \tag{11}$$

This can then be re-normalised such that $n_y^{(po)^2} + n_z^2 = 1$, matching the form of the 2D estimate from PS as in (9). For the z component specifically:

$$\frac{n_z}{\sqrt{n_y^{(po)^2} + n_z^2}} = N_z^{(ps)} \tag{12}$$

Substituting the components of $\mathbf{n}$ from (11) and simplifying:

$$N_z^{(ps)} = \frac{N_z}{\sqrt{N_y^{(po)^2} + N_z^2}} \tag{13}$$

Rearranging:

$$N_z = \left| N_y^{(po)} \right| \left(\frac{1}{N_z^{(ps)^2}} - 1 \right)^{-1/2} \tag{14}$$

where the $|\cdot|$ sign is used since N_z must be positive. This means that the only unknown in (10) is resolved and so the surface normal is fully determined.

As stated earlier, the regions of the images where the degree of polarisation is less that 1% are not used in the algorithm to ensure robustness. For the results in the next section, bi-cubic interpolation was used to estimate the normals for these areas. It was also found that improvements could be made by interpolating over regions of very low N_y, although this was kept to a minimum. It is acknowledged that more sophisticated methods may be more appropriate for future work.

After completion of the surface normal calculations, the depth can be obtained by integration. For this paper, the well-established Frankot-Chellappa method [22] is used. The method is fast and highly robust to noise, but can suffer from over-smoothing.

4   Results

Consider Fig. 2, the polarisation image for a white snooker ball captured in ideal conditions as described in Sect. 2. The angle between the camera and light sources was $\beta_L = \beta_R = 19.6°$ (which was experimentally determined to be a reasonable trade-off between reconstruction accuracy and practicality). One hundred images were captured at a frame rate of 60 Hz at each of the four polariser angles (0, 45°, 90° and 135°) and the mean intensity used at each pixel to minimise noise. As expected, the phase angle directly relates to the surface azimuth up to a 180° ambiguity and the degree of polarisation is highest near the occluding contours [17].

As mentioned earlier in this paper and elsewhere in the literature, one of the issues facing polarisation vision algorithms is the set of complications caused by the simultaneous presence of specular and diffuse reflection. To test the robustness of the method proposed here, a second polarisation image was captured in exactly the same conditions as for Fig. 2 but with a large planar matte white surface placed 128 mm behind the target object. The resulting image is shown in Fig. 3, as also discussed in Sect. 2. The inter-reflection can only just be seen in the intensity image, yet the 90° phase shift and increased degree of polarisation make it clear in those two components of the polarisation image.

The results of applying the surface normal estimation algorithm to the polarisation images shown in Figs. 2 and 3 are shown in Fig. 4. The results are qualitatively good although the region of interpolation (described in Sect. 3.3) near the top-centre and bottom-centre of the normal maps is clearly apparent. Note that the data has been cropped here so only the object itself is being integrated.

Figure 5 shows a comparison of the profile of the reconstructions from Fig. 4 to ground truth data. To aid comparison, the profiles are aligned such that the tops of each curve are touching. Very similar results were found for pink and yellow balls, while blue and green results were slightly poorer due to their lower brightness causing higher noise levels.

One potential weakness of the method is the presence of high noise levels for areas of low degree of polarisation. Results so far in this paper overcame this in a rather expensive manner by capturing 100 images in very rapid succession for each of the four polariser angles. The effects of using fewer images is illustrated by Fig. 6 which shows the estimated surface normal (z-component only for compactness) using 1, 5, 20 and 100 images. At first sight, the quality of results is poor when only few images are used. However, the increased noise has a relatively low impact on surface height reconstruction, as shown in the figure. This is due to the facts that (1) the Frankot-Chellappa surface integrator

Fig. 4. Surface normals and depth estimated from the polarisation image shown in Fig. 2 (left) and Fig. 3 (right). Surface normals are encoded by colour (azimuth) and saturation (zenith). (Color figure online)

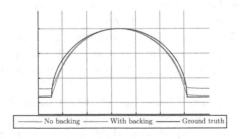

Fig. 5. Comparison between the estimated depth profile of the snooker ball and ground truth for the data represented in Fig. 4. (Color figure online)

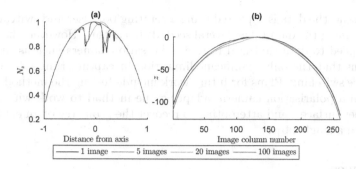

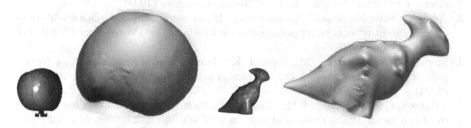

Fig. 6. Estimated (a) z-component of the surface normals and (b) surface height from the method using varying number of images per polariser angle.

Fig. 7. Surface reconstructions of an apple and a porcelain dinosaur model.

is smoothing over the noise, and (2) the higher noise regions correspond to areas of low zenith angle which always have smaller effects on the integration than those of larger zenith angle.

Tests on objects of more complicated geometry are shown in Fig. 7. The general geometry of the apple has clearly been reproduced. The model dinosaur also has reasonable shape, despite a few areas of incorrect disambiguation on the foot and head. It is hoped that such issues might be resolved in future work by optimising the location and number of seed points used.

5 Conclusion

This paper has demonstrated how PS and polarisation vision can be combined in order to overcome some of the weaknesses of each. The method may be useful, for example, in robotics where only two-source PS is feasible (it is often easy to add one light each side of a robot but difficult to place a third high above the ground). Results are comparable or better than baseline methods for PS or polarisation alone in the conditions considered. By way of example, the snooker ball in Fig. 2 yielded an ℓ_2-norm error of 0.066 for the proposed method compared to 0.089 for standard four-source PS. Use of polarisation alone yielded 0.070 but this required knowledge of refractive index and manual azimuth angle disambiguation [17].

At present, the data is captured using a rotating polariser and switching light sources meaning the data takes several seconds to capture. However, the method can be applied to data captured from a polarisation camera such as the Polka. This means that the only significant limitation in capture time relates to the light source switching. Plans for future work include testing the method at high-speed with a polarisation camera, adapting the method to work with metallic and rougher surfaces, and attempting to recover the geometry of an entire scene rather than a single object.

References

1. Wolff, L.B., Boult, T.E.: Constraining object features using a polarisation reflectance model. IEEE Trans. Pattern Anal. Mach. Intell. **13**, 635–657 (1991)
2. Hecht, E.: Optics, 3rd edn. Addison Wesley, Longman (1998)
3. Saito, M., Sato, Y., Ikeuchi, K., Kashiwagi, H.: Measurement of surface orientations of transparent objects using polarization in highlight. In: Proceedings CVPR, vol. 1, pp. 381–386 (1999)
4. Miyazaki, D., Kagesawa, M., Ikeuchi, K.: Transparent surface modelling from a pair of polarization images. IEEE Trans. Pattern Anal. Mach. Intell. **26**, 73–82 (2004)
5. Atkinson, G.A., Hancock, E.R.: Shape estimation using polarization and shading from two views. IEEE Trans. Pattern Anal. Mach. Intell. **29**, 2001–2017 (2007)
6. Berger, K., Voorhies, R., Matthies, L.: Incorporating polarization in stereo vision-based 3D perception of non-Lambertian scenes. In: Proceedings SPIE, vol. 9837 (2016)
7. Taamazyan, V., Kadambi, A., Raskar, R.: Shape from mixed polarization. arXiv preprint: arXiv:1605.02066 (2016)
8. Drbohlav, O., Šára, R.: Unambiguous determination of shape from photometric stereo with unknown light sources. In: Proceedings ICCV, pp. 581–586 (2001)
9. Drbohlav, O., Šára, R.: Specularities reduce ambiguity of uncalibrated photometric stereo. In: Heyden, A., Sparr, G., Nielsen, M., Johansen, P. (eds.) ECCV 2002. LNCS, vol. 2351, pp. 46–60. Springer, Heidelberg (2002). doi:10.1007/3-540-47967-8_4
10. Atkinson, G.A., Hancock, E.R.: Two-dimensional BRDF estimation from polarisation. Comput. Vis. Image Underst. **111**, 126–141 (2008)
11. Garcia, N.M., de Erausquin, I., Edmiston, C., Gruev, V.: Surface normal reconstruction using circularly polarized light. Opt. Express **23**, 14391–14406 (2015)
12. Morel, O., Stolz, C., Meriaudeau, F., Gorria, P.: Active lighting applied to three-dimensional reconstruction of specular metallic surfaces by polarization imaging. Appl. Optics **45**, 4062–4068 (2006)
13. Miyazaki, D., Tan, R.T., Hara, K., Ikeuchi, K.: Polarization-based inverse rendering from a single view. In: Proceedings ICCV, vol. 2, pp. 982–987 (2003)
14. Rahmann, S., Canterakis, N.: Reconstruction of specular surfaces using polarization imaging. In: Proceedings CVPR, pp. 149–155 (2001)
15. Rahmann, S.: Reconstruction of quadrics from two polarization views. In: Perales, F.J., Campilho, A.J.C., Blanca, N.P., Sanfeliu, A. (eds.) IbPRIA 2003. LNCS, vol. 2652, pp. 810–820. Springer, Heidelberg (2003). doi:10.1007/978-3-540-44871-6_94
16. Huynh, C.P., Robles-Kelly, A., Hancock, E.: Shape and refractive index recovery from single-view polarisation images. In: Proceedings CVPR, pp. 1229–1236 (2010)

17. Atkinson, G.A., Hancock, E.R.: Recovery of surface orientation from diffuse polarization. IEEE Trans. Image Proc. **15**, 1653–1664 (2006)
18. Wolff, L.B.: Diffuse-reflectance model for smooth dielectric surfaces. J. Opt. Soc. Am. A **11**, 2956–2968 (1994)
19. Fraunhofer Institute for Integrated Circuits. www.iis.fraunhofer.de/en/ff/bsy/tech/kameratechnik/polarisationskamera.html. Accessed 24 Mar 2017
20. Burger, W., Burge, M.J.: Principles of Digital Image Processing: Fundamental Techniques. Springer, Heidelberg (2010)
21. Woodham, R.J.: Photometric method for determining surface orientation from multiple images. Opt. Eng. **19**, 139–144 (1980)
22. Frankot, R.T., Chellappa, R.: A method for enforcing integrability in shape from shading algorithms. IEEE Trans. Pattern Anal. Mach. Intell. **10**, 439–451 (1988)

Synthetic Aperture Radar (SAR) Monitoring of Avalanche Activity: An Automated Detection Scheme

H. Vickers[1(✉)], M. Eckerstorfer[1], E. Malnes[1], and A. Doulgeris[2]

[1] Norut, Forskningsparken, P.O. Box 6434, 9294 Tromsø, Norway
hannah.vickers@norut.no
[2] CIRFA, University of Tromsø,
P.O. Box 6050 Langnes, 9037 Tromsø, Norway

Abstract. Snow avalanches are a recurring hazard during the winter months in mountainous regions such as Troms county. Monitoring of their occurrence has however, first become feasible through the launch of the Sentinel-1A and 1B Synthetic Aperture Radar (SAR) satellites which provide near-daily coverage of the area surrounding Tromsø. With the large areas covered by a single SAR image and the short times between repeat acquisitions, an enormous amount of data is now available, providing an ideal opportunity for operational monitoring of avalanche activity on a global scale. Such a system requires automated detection of avalanches since it is unrealistic to perform this task manually. A test version for an automatic avalanche detection algorithm based on change detection and K-means classification methods was developed and tested on the Tamokdalen area in Troms using Sentinel-1A images at 20 m resolution. However, the algorithm was not robust under variations in snow and weather conditions between acquisition dates and as such, we have revised the algorithm to address this. In the updated version we have retained several procedures of the first version, but the main difference being the replacement of the pre-classification change detection with a post-classification change detection scheme together with improved filtering techniques. Results are shown for examples acquired from the 2017 winter season and we summarize the main improvements and future requirements of the revised detection algorithm.

1 Introduction

Snow avalanches (hereafter called avalanches) are a real and frequent wintertime hazard in Troms county due to its mountainous terrain. They pose not only a risk to recreational users but also to property, major transportation networks and small communities that are exposed to avalanche runout paths. Knowledge of avalanche size and location is challenging to acquire. Nevertheless, such information is important as high avalanche activity indicates high avalanche danger. Traditional methods of avalanche activity monitoring involve field-based observations but these are frequently limited in temporal and spatial extent, particularly in remote mountain areas and during the period when the avalanche danger remains high and weather conditions are poor.

© Springer International Publishing AG 2017
P. Sharma and F.M. Bianchi (Eds.): SCIA 2017, Part II, LNCS 10270, pp. 136–146, 2017.
DOI: 10.1007/978-3-319-59129-2_12

Technological developments during the last few decades have allowed avalanche activity to be observed and mapped over large areas without involving human exposure to avalanche terrain. Both optical remote sensing [*Buhler et al.* 2009; *Larsen et al.* 2011; *Lato et al.* 2012] and Synthetic Aperture Radar (SAR) technologies [e.g. *Wiesmann et al.* 2001; *Martinez-Vasquez and Fortuny Guasch* 2008; *Malnes et al.* 2013; *Eckerstorfer and Malnes* 2015] have been exploited in the context of avalanche activity detection, with SAR having the advantage of being completely independent of light and weather conditions [*Eckerstorfer et al.* 2016]. Avalanche activity monitoring requires a fully operational system whereby avalanche activity can be automatically detected and mapped from an optical or SAR image. With the recent launches of both the Sentinel-1A and Sentinel-1B satellites, several parts of Troms county are currently imaged on a near-daily basis. Attempts to automate the process of avalanche mapping have used a variety of approaches based around simple thresholding methods, image segmentation and object based classification. However, to date none of these have been developed to a fully automated level of operation or tested extensively over large areas and over entire winter seasons.

In this work we have addressed issues associated with the earlier version of an avalanche detection method that was developed and applied to a test area in Troms county using Sentinel-1A images [*Vickers et al.* 2016]. The algorithm in this study combined a reference and avalanche activity image to form a backscatter change image, on which change detection and object classification was applied. This approach was satisfactory under favorable acquisition conditions, but performed poorly in cases where the reference image was affected by wet snow conditions. We have revised the algorithm in such a way that it can deal with the different image statistics characterized by variations in snow conditions within a single image and across images acquired on different days. This is often the case for an entire set of images acquired during a winter season. We also outline how the revised algorithm is superior to the earlier version and how it performs on the larger avalanche forecasting regions covered by the Norwegian avalanche forecasting service (NVE). Due to the near-impossible task of acquiring ground truth data for every single avalanche that can occur on the spatial and temporal scales in question, we quantify the accuracy of our automatic detections in terms of a comparison with avalanche debris that have been delineated manually from the same set of SAR images. An evaluation of the current strengths and weaknesses of the detection algorithm is made in terms of its suitability as an operational service.

2   Data Acquisition and Method

2.1   Sentinel-1A and 1B

Sentinel-1A was launched in April 2014 and has been delivering data since October 2014. Sentinel-1B meanwhile, was launched in April 2016. The interferometric wide swath mode (IW) covers a swath of approximately 250 km and is extended often up to 1000 km. This mode has 20 m × 5 m resolution (azimuth-ground range) and images are acquired in both co-polarization ("VV") and cross-polarization ("VH") bands. Both bands are utilized in the detection algorithm. Each image geometry has a repeat time of

12 days, but these paths often converge at high latitudes due to the polar orbit of the satellites, allowing some areas to be covered by multiple image geometries as often as every 2 to 3 days. The Sentinel-1 IW GRDH (Ground Range Detected High resolution) product, sampled with 10 m pixel spacing is geocoded using in-house 'GSAR' software [*Larsen et al.* 2005] to obtain the output radar backscatter image (σ^0). This is finally stored in GeoTIFF format using the UTM zone 33 N, WGS-84 projection and resampled to 20 m pixel spacing to ensure a high multi-look factor that helps to suppress most SAR speckle artifacts. During the geocoding process we also generate mask files for radar shadow and layover caused by the radar geometry and the terrain.

2.2   Automatic Detection Algorithm

The automatic avalanche detection algorithm builds on an earlier test version, as outlined in *Vickers et al.* [2016]. In that version, the key components that facilitated avalanche detection were the composition of a backscatter change image and the application of a K-means unsupervised clustering algorithm [*Theodoridis and Koutroumbas* 2006] which segmented the data into an 'avalanche' or 'not avalanche' class. As shown by the authors, the optimal threshold for the backscatter change image that produced the best balance between correct detections, missed detections and false alarms was variable across SAR scenes as well as within them. This was due to different snow and meteorological conditions during acquisition of the reference and activity images, made 12 days apart. The variable nature of the manual thresholding stage thus hinders a fully automated process for avalanche detection. As such we have explored alternative techniques for accurate characterization of avalanche debris in SAR images, which are also less dependent on manual intervention.

The revised version of the avalanche detection algorithm retains both the masking component and K-means clustering steps of the earlier version. The construction of a mask is necessary to eliminate areas where avalanche debris cannot be, or is unlikely to be detected due to the geometry of the SAR acquisition (radar layover and shadow) or due to slopes being too steep for debris to accumulate on (greater than 40°) and in areas where there exists dense forest cover or open water. The masking process not only speeds up the processing time but also allows us to discard data that do not contribute advantageously to the backscatter statistics that are used by the K-means algorithm. In this version, K-means clustering is utilized at an earlier stage in the algorithm to segment the two SAR images into 6 classes before change detection is implemented on the segmented images. Here, the pixels are assigned to the class whose mean value is closest to the value of the pixel. In all cases, we choose reference and activity images that correspond to repeat passes i.e. identical image geometry ensures that we do not need to account for terrain corrections and we always carry out coherent change detection. The choice of 6 classes for the segmentation is made to allow for satisfactory separability of the data across the range in backscatter values without compromising the processing time. Pixels are segmented according to their backscatter relative to the overall distribution of the image, which means that when meteorological changes cause the snow condition and backscatter properties of the snow to change, then pixels at comparable altitudes are affected in the same way. When the pixel class changes from one segmented image to another, this indicates a change in backscatter property due to factors other than those of a meteorological nature.

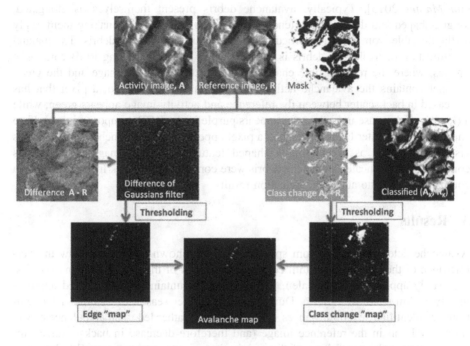

Fig. 1. Workflow illustration for the automatic avalanche detection algorithm

Avalanche debris are expected to produce the largest positive changes in backscatter and therefore class number since classes are assigned according to backscatter value. We therefore search for pixels whose class change lies in the tail of the class change distribution calculated for the segmented area. A class change threshold is currently set to lie at two standard deviations from the mean class change. Pixels exhibiting class change below this threshold are flagged as potential avalanche debris. An additional detection step is performed, where we construct a backscatter change image from the activity and reference image and apply a Difference of Gaussians filter in order to enhance spatial gradients in the backscatter change. This essentially enables edge detection since steep gradients in backscatter change correspond to edges of features in the image, for example those associated with avalanche debris. Similarly to the class change image, we set an automatically calculated threshold at two standard deviations from the mean for the Difference of Gaussians-filtered backscatter change image. Those pixels that satisfy the criteria imposed on both the class change image and Difference of Gaussians-filtered backscatter change image are retained in the final detection map. It should be noted that in both cases it is only the non-masked pixels that are included in calculating the statistical parameters required for determination of the threshold. Figure 1 illustrates more concisely the workflow of the detection algorithm.

2.3 Manual Identification

Using SAR backscatter images, we detect only the avalanche debris, which possesses increased surface roughness and hence increased scattering coefficient [*Eckerstorfer and Malnes* 2015]. Typically, avalanche debris present themselves as elongated, tongue-shaped and downslope-extending features in SAR images, making them easily distinguishable from the undisturbed snow cover surrounding the debris. The manual identification of avalanche debris is carried out by first constructing RGB composite images, where the red and blue channels contain the reference image and the green channel contains the avalanche activity image. Using this method, a pixel that has increased in backscatter between the reference and activity image appears green, while a (temporal) decrease in backscatter appears purple in the RGB composite image. If no change in backscatter takes place, then a pixel appears grey. Avalanche debris are thus visually identified as green, tongue-shaped features in RGB images. Vector files containing the delineated avalanche debris were converted into mask files to allow for a comparison with the automatic detection results.

3 Results

Avalanche detection results from specific cases are shown to illustrate how the performance of the revised algorithm compares with that of the earlier version. The first case study applies to Tamokdalen, a steep-sided mountainous area located approximately 80 km south of Tromsø. During the first winter season of Sentinel-1A operations we identified several instances where warmer weather led to an abundance of wet snow condition in the reference image (and therefore decrease in backscatter coefficient), whereas the acquisition of the activity image was made under cold, dry snow conditions (increase in backscatter coefficient relative to wet snow backscattering properties). The RGB composite image for this pair of images is shown in Fig. 2(a). Using the first version of the avalanche detection algorithm led to the detection of many 'false positives' as shown in Fig. 2(b). This was due to large parts of the area exhibiting a relatively high backscatter change that resulted from the difference in backscattering coefficient of wet snow and dry snow. Using the revised detection algorithm, we find that the overall number of detected avalanche pixels is substantially lower, as illustrated in Fig. 2(c) and corresponds better with the manual delineation of the avalanche debris (overlaid on Fig. 2(a) in yellow).

Additionally, we wish to test how the correct detection rate of the revised algorithm compares with that of the earlier version, when applied to the Tamokdalen area after a period of relatively high avalanche activity that occurred around New Year 2015. In this published example, using an activity image acquired on 6 January 2015 and a reference image from 13 December 2014, 62% of all manually delineated avalanche pixels were detected by the automatic detection algorithm while 38% were missed. The false alarm rate corresponding to this example was 46%. When compared feature-by-feature instead of in terms of number of pixels, the detection rate was almost unchanged, lying close to the 60% level. Figure 3 shows the detection results obtained using the revised algorithm. Visually, we see that the avalanche features detected are in good agreement with those detected manually, both in terms of location and size. However when expressed in terms

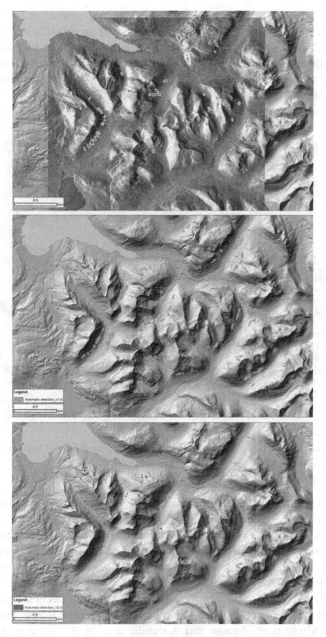

Fig. 2. (a) RGB composite image for Tamokdalen, constructed using a reference image obtained on 19th March 2015 and an activity image from 31st March 2015 with the manual delineations overlaid in yellow **(b)** Automatic detection results using the first version of the algorithm **(c)** Automatic detection results using the revised avalanche detection algorithm (Color figure online)

of number of pixels, the detection rate is 64%, indicating a similar number of correctly detected pixels compared with the earlier version. The main difference between the performances of the two algorithms for this case study lies in the proportion of false detections. Using the updated version of the algorithm, the false detection rate is reduced to 28%, which corresponds to an almost 20% reduction compared with the false alarm rate of 46% obtained using the earlier version.

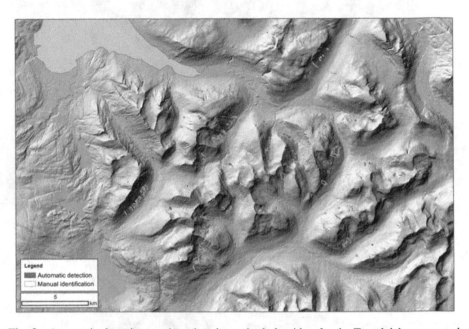

Fig. 3. Automatic detection results using the revised algorithm for the Tamokdalen case study from 6th January 2015, as presented in Vickers et al. 2016

Since the ultimate objective of our work is to develop an operational system for avalanche detection and monitoring, it is clear that the satisfactory success rate obtained for the Tamokdalen area must be replicable over much larger areas. With this in mind we have tested the revised algorithm on two larger subsections of a Sentinel-1A scene, though only results from one of these areas is presented in this paper. Results are obtained for a pair of S1A images with reference date 14th December 2016 and activity image with acquisition date 26th December 2016. During this 12-day interval, several days with high winds and warmer temperatures occurred leading to both transport of snow and destabilization of the snow pack, inevitably resulting in several large, naturally-released slab avalanches over a range of slope aspects. This is obvious when one inspects the RGB composite image constructed for the 'S1_Tamokdalen' region in Fig. 4(a). The areas that appear purple exhibit the presence of wet snow, since the backscatter coefficient is reduced here while areas of bright green features are evidence of avalanche debris.

Fig. 4. **(a)** RGB composite for 'S1_Tamokdalen' area using a reference image from 14th December 2016 and activity image from 26th December 2016. **(b)** Automatic detection results for the set of S1 images used Fig. 4(a). Automatically detected avalanche debris (light green) are overlaid on a terrain map and have been digitally enhanced to aid visibility (Color figure online)

This region covers an area of 75 km × 45 km, ranging from Lavangsdalen in the north to Tamokdalen in the south. The western edge corresponds to the southern part of the Malangen peninsula while the eastern edge covers the inner part of the Norwegian mainland known as Storfjord. The climates of the northern and western portions of this image can contrast greatly with the drier and colder inland climates of Tamokdalen and Storfjord. It is thus an appropriate test of the revised algorithm to determine whether it is capable of dealing with the variability of the snow conditions and hence, backscatter statistics within the entire image. Since the spatial area is much greater in comparison with only Tamokdalen, we break down the image processing into windows of 500 × 500 pixels and form a final detection map by combining the detection results from each individual window.

Figure 4(b) shows the detection map obtained for this larger area using this pair of images. In total, the algorithm detected 5588 pixels to be potential avalanche debris; of these, 62.4% were identified as correct detections while 37.6% of those identified manually were missed. The false alarm rate was calculated to be 75.7% assuming that the manual detections are a satisfactory representation of the ground truth. Of course we do not know exactly how accurate the manually identified avalanche debris compare with the true area covered by avalanche debris, since the manual detections are subject to user bias and still rely on an interpretation of a radar image which may not necessarily reveal all avalanche debris. For comparison, we have also obtained detections for the same area and dates using the earlier version of the algorithm, which yielded correct, missed and false detection rates of 38.8%, 61.2% and 94.6% respectively. This highlights the improvement in both the correct and false detection rate over the larger area, when compared with the performance of the earlier version.

4 Discussion and Conclusion

This paper presents the results of an automatic avalanche detection algorithm, which has undergone revision in order to achieve satisfactory detection rates when applied under typical operational monitoring circumstances. Specifically, the algorithm has been developed to be more robust towards variability in meteorological, and hence snow conditions across entire SAR images and between them. Compared with the earlier test version, we find that the updated version performs comparably with the earlier version when applied to high activity images, as was presented for a case study from January 2015 for the Tamokdalen area only. Under less favorable conditions, when the earlier version performed poorly and over-detected avalanche debris, the revised algorithm was more superior and produced detection results that were in better qualitative agreement with manual interpretations. On the larger test area, we obtain correct detection rates typically between 60 and 70% in cases of high avalanche activity. This is of the same order as was obtained on the small test region using the earlier algorithm. Even though we have not achieved significant improvement in the correct detection rate, when expressed in terms of the total number of pixels/area, we have demonstrated that the updated version is capable of performing consistently over large spatial areas where the local climate is expected to vary. The false alarm rate is also reduced over both the smaller test area and larger area when compared with that of the earlier algorithm.

The processing time is some 3 times longer on average, due to implementing K-means classification on the entire reference and activity images, rather than only once on the backscatter change image. The increased processing time is also due in part to the greater number of classes used to segment the images. Currently, the processing time can be considered acceptable on the local or regional scale, for example a single or pair of Sentinel-1 images covering the avalanche forecasting areas in Troms county. However, in order to offer such an operational monitoring service on a national and international scale, this processing time must be reduced substantially, and remains an area for further improvement and development.

There is the outstanding issue of improving the correct detection rate and reducing the number of false detections, especially if such an operational monitoring system is to be used to evaluate the accuracy of avalanche forecasts. There is also a degree of uncertainty around what types of avalanches are most or least detectable and why. Whether this is related to the avalanche debris lying close to areas of radar shadow/layover, or due to the spatial resolution of the radar image or even simply a result of the snow condition when the avalanche released, is a topic that should and will be more thoroughly investigated in order to reduce the number of avalanches missed by the algorithm. In the absence of a complete ground truth, we anticipate that comparisons with manually delineated avalanche debris will remain an important step in future development and evaluation of the automatic detection algorithm.

References

Bühler, Y., Hüni, A., Christen, M., Meister, R., Kellenberger, T.: Automated detection and mapping of avalanche debris using airborne optical remote sensing data. Cold Reg. Sci. Technol. **57**, 99–106 (2009)

Eckerstorfer, M., Malnes, E.: Manual detection of snow avalanche debris using high-resolution Radarsat-2 SAR images. Cold Reg. Sci. Technol. **120**, 205–218 (2015). doi:10.1016/j.coldregions.2015.08.016

Eckerstorfer, M., Bühler, Y., Frauenfelder, R., Malnes, E.: Remote sensing of snow avalanches: recent advances, potential, and limitations. Cold Reg. Sci. Technol. **121**, 126–140 (2016). doi:10.1016/j.coldregions.2015.11.001

Larsen, S.Ø., Salberg, A.B., Solberg, R.: Evaluation of automatic detection of avalanches in high-resolution optical satellite data. NR Note, Norwegian Computing Center, SAMBA/23/11 (2011)

Larsen, Y., Engen, G., Lauknes, T.R., Malnes, E., Høgda, K.A.: A generic differential interferometric SAR processing system, with applications to land subsidence and snow-water equivalent retrieval. In: Fringe ATSR Workshop 2005, Frascati, Italy (2005)

Lato, M.J., Frauenfelder, R., Bühler, Y.: Automated detection of snow avalanche deposits: segmentation and classification of optical remote sensing imagery. Nat. Hazards Earth Syst. Sci. **12**, 2893–2906 (2012)

Malnes, E., Eckerstorfer, M., Larsen, Y., Frauenfelder, R., Jonsson, A., Jaedicke, C., and Solbø, S.A.: Remote sensing of avalanches in northern Norway using synthetic aperture radar. In: Proceedings of the International Snow Science Workshop 2013, Grenoble–Chamonix, France, 9–13 October, pp. 955–959 (2013)

Martinez-Vasquez, A., Fortuny Guasch, J.: A GB-SAR processor for snow avalanche identification. IEEE Trans. Geosci. Remote Sens. **46**(11), 3948–3956 (2008)

Theodoridis, S., Koutroumbas, K.: Pattern Recognition. 3rd edn. Academic Press, Orlando (2006)

Vickers, H., Eckerstorfer, M., Malnes, E., Larsen, Y., Hindberg, H.: A method for automated snow avalanche debris detection through use of synthetic aperture radar (SAR) imaging. Earth Space Sci. **3** (2016). doi:10.1002/2016EA000168

Wiesmann, A., Wegmueller, U., Honikel, M., Strozzi, T., Werner, C.L.: Potential and methodology of satellite based SAR for hazard mapping. In: IGARSS 2001, Sydney, Australia, 9–13 July 2001

Canonical Analysis of Sentinel-1 Radar and Sentinel-2 Optical Data

Allan A. Nielsen[(✉)] and Rasmus Larsen

DTU Compute – Applied Mathematics and Computer Science,
Technical University of Denmark,
Richard Petersens Plads, Building 321, 2800 Lyngby, Denmark
{alan,rlar}@dtu.dk
http://people.compute.dtu.dk/alan

Abstract. This paper gives results from joint analyses of dual polarimety synthetic aperture radar data from the Sentinel-1 mission and optical data from the Sentinel-2 mission. The analyses are carried out by means of traditional canonical correlation analysis (CCA) and canonical information analysis (CIA). Where CCA is based on maximising correlation between linear combinations of the two data sets, CIA maximises mutual information between the two. CIA is a conceptually more pleasing method for the analysis of data with very different modalities such as radar and optical data. Although a little inconclusive as far as the change detection aspect is concerned, results show that CIA analysis gives conspicuously less noisy appearing images of canonical variates (CVs) than CCA. Also, the 2D histogram of the mutual information based leading CVs clearly reveals much more structure than the correlation based one. This gives promise for potentially better change detection results with CIA than can be obtained by means of CCA.
http://www.imm.dtu.dk/pubdb/p.php?6963.

Keywords: Canonical correlation analysis · Canonical information analysis

1 Introduction

In a preliminary investigation into change detection in data of different modalities this paper looks into canonical analysis of Sentinel-1[1] (S1) radar and Sentinel-2[2] (S2) optical remote sensing data. This kind of analysis is potentially important due to often occurring clouds in optical data and the all-weather acquisition capability of radar. The data are analysed both by means of traditional canonical correlation analysis (CCA) [4] and a more computer intensive method which we have named canonical information analysis (CIA) [13,14].

Earlier we have worked with CCA based change detection in bi-temporal optical data. We termed this method multivariate alteration detection (MAD)

---

[1] https://sentinel.esa.int/web/sentinel/missions/sentinel-1.
[2] https://sentinel.esa.int/web/sentinel/missions/sentinel-2.

© Springer International Publishing AG 2017
P. Sharma and F.M. Bianchi (Eds.): SCIA 2017, Part II, LNCS 10270, pp. 147–158, 2017.
DOI: 10.1007/978-3-319-59129-2_13

[10], and we also published an iterative version termed iteratively re-weighted MAD (IR-MAD or iMAD) [2,9]. The so-called MAD variates containing information on change are the differences between pairs of corresponding canonical variates.

CCA considers second order statistics of the involved variables only and as such it is ideal for Gaussian data. CCA is therefore not necessarily an obvious choice of analysis for data of such different modalities as here with radar and optical data (which have very different genesis and potentially follow very different statistical distributions), but since it is the basis of the widely used MAD methods we have included it here.

The idea in CIA is to replace correlation as the measure of association between variables with the more general information theoretical, entropy based measure mutual information (MI) [1,2,5,8]. Also in this case the change information is found in the differences between corresponding pairs of canonical variates. Since MI between variables is independent of their signs, this differencing is a little more tricky for MI based analysis than for correlation based analysis.

Other workers have dealt with canonical analysis based on mutual information [6,15]. Entropy and mutual information depend on the sample probability density functions of the involved variables and thus on higher order statistics.

In a situation with data of very different modalities the use of MI constitutes a conceptually much more pleasing way of measuring association between variables than the use of correlation.

For an illustrative (toy) example where CCA fails and CIA succeeds, (an example with RGB images covering a busy motorway for traffic surveillance), and an example of joint analysis of weather radar and Meteosat data, see [13].

2   Methods

Here we very briefly sketch the ideas in canonical correlation analysis, canonical information analysis, and basic information theory. In both CCA and CIA linear combinations $U = a^T X$ and $V = b^T Y$ of two sets of stochastic variables, k-dimensional X (here the Sentinel-1 data) and ℓ-dimensional Y (here the Sentinel-2 data), are determined.

2.1   Canonical Correlation Analysis

In canonical correlation analysis first published by Hotelling in 1936 [4] linear combinations which maximise correlation between U and V are found. Correlation considers second order statistics of the involved variables only. It is therefore ideal for Gaussian data. The canonical variates U and V are found by solving a generalised eigenvalue problem.

2.2   Canonical Information Analysis

Inspired by canonical correlation analysis, canonical information analysis is a method which replaces the maximisation of correlation with maximisation of mutual information between the linear combinations U and V. For further details see the next subsection and [11, 13, 14].

2.3   Basic Information Theory

In 1948 Shannon [12] published his now classical work on information theory. Below, we describe the information theoretical concepts entropy, relative entropy and mutual information for discrete stochastic variables, see also [1, 2, 5, 8].

Entropy. Consider a discrete stochastic variable X with probability density function (pdf) $p(X = x_i)$, $i = 1, \ldots, n$, i.e., the probability of observing a particular realization x_i of stochastic variable X, where n is the number of possible outcomes or the number of bins. Let us look for a measure of information content (or surprise if you like) $h(X = x_i)$ in obtaining that particular realization. If x_i is a very probable value, i.e., $p(X = x_i)$ is high, we receive little information by observing x_i. If on the other hand x_i is a very improbable value, i.e., $p(X = x_i)$ is low, we receive much information by observing x_i. The measure of information content should be a monotonically decreasing function of p. This can be obtained by choosing for example $h \propto 1/p$.

If we observe independent realizations x_i and x_j, i.e., the two-dimensional pdf $p(X = x_i, X = x_j)$ equals the product of the one-dimensional marginal pdfs $p(X = x_i)p(X = x_j)$, we would like the joint information content to equal the sum of the marginal information contents, i.e., $h(X = x_i, X = x_j) = h(X = x_i) + h(X = x_j)$. This can be obtained by transformation by means of the logarithm.

Thus the desired characteristics of the measure of information or surprise can be obtained if we define $h(X = x_i)$ as

$$h(X = x_i) = \ln \frac{1}{p(X = x_i)} = -\ln p(X = x_i). \tag{1}$$

The expectation $H(X)$ of the information measure, i.e., the average amount of information obtained by observing the stochastic variable X, is termed the entropy

$$H(X) = -\sum_{i=1}^{n} p(X = x_i) \ln p(X = x_i). \tag{2}$$

In the limit where p tends to zero and $\ln p$ tends to minus infinity, $-p \ln p$ tends to zero. $H(X) = -\mathrm{E}\{\ln p(X)\}$ is nonnegative. A discrete variable which takes on one value only has zero entropy; a uniform discrete variable has maximum entropy (equal to $\ln n$). For the joint entropy of two discrete stochastic variables X and Y we get

$$H(X, Y) = -\sum_{i,j} p(X = x_i, Y = y_j) \ln p(X = x_i, Y = y_j). \tag{3}$$

Relative Entropy. The relative entropy also known as the Kullback-Leibler divergence [7] between two pdfs $p(X = x_i)$ and $q(X = x_i)$ defined on the same set of outcomes (or bins) is

$$D_{KL}(p, q) = \sum_i p(X = x_i) \ln \frac{p(X = x_i)}{q(X = x_i)}. \tag{4}$$

This is the expectation of the logarithmic difference between p and q. $D_{KL} \geq 0$ with equality for $p(X = x_i) = q(X = x_i)$ only. The relative entropy is not symmetric in p and q (and therefore it is not a metric).

Mutual Information. The extent to which two discrete stochastic variables X and Y are not independent, which is a measure of their mutual information content, may be expressed as the relative entropy or the Kullback-Leibler divergence between the two-dimensional pdf $p(X = x_i, Y = y_j)$ and the product of the one-dimensional marginal pdfs $p(X = x_i)p(Y = y_j)$, i.e.,

$$D_{KL}(p(X, Y), p(X)p(Y)) = \tag{5}$$
$$\sum_{i,j} p(X = x_i, Y = y_j) \ln \frac{p(X = x_i, Y = y_j)}{p(X = x_i)p(Y = y_j)}.$$

This sum defines the mutual information $I(X, Y)$ of the stochastic variables X and Y. Mutual information equals the sum of the two marginal entropies minus the joint entropy

$$I(X, Y) = H(X) + H(Y) - H(X, Y). \tag{6}$$

Unlike the general Kullback-Leibler divergence in (4) this measure is symmetric. Mutual information is always nonnegative, it is zero for independent stochastic variables only.

Obviously we need to estimate marginal as well as joint pdfs to obtain the mutual information estimate in (6). We employ kernel density estimation, which uses N data samples to estimate these pdfs. Mutual information is subsequently estimated using the same N data points. This is possible in practice only due to a very fast estimation of pdfs, see [13].

3   S1 and S2 Data, Frankfurt Airport

Both the radar and the optical data cover the international airport in Frankfurt, Germany. The data are obtained from Google Earth Engine[3] (GEE) [3].

The Sentinel-1 data acquired in instrument Interferometric Wide Swath (IW) mode, is an S1 Ground Range Detected (GRD) scene, processed using the Sentinel-1 Toolbox[4] to generate a calibrated, ortho-corrected product.

---

[3] https://earthengine.google.com, https://developers.google.com/earth-engine.
[4] https://sentinel.esa.int/web/sentinel/toolboxes/sentinel-1.

Fig. 1. RGB image of Sentinel-1 C-band VV/VH data, VV as R and VH as G and B (i.e., cyan), 10 m pixels, 5 km north-south and 8 Km east-west, Franfurt Airport, Germany, acquired on 15 July 2016. (Color figure online)

Fig. 2. RGB image of Sentinel-2 MSI data, band 4 (near-infrared as R), band 3 (red as G), and band 2 (green as B), 10 m pixels, 5 km north-south and 8 Km east-west, Franfurt Airport, Germany, acquired on 12 Sep 2016. (Color figure online)

This processing includes thermal noise removal, radiometric calibration, and terrain correction using Shuttle Radar Topography Mission 30 m (SRTM 30) data. Finally it includes saturating the data (quoting GEE): "Values are then clamped to the 1st and 99th percentile to preserve the dynamic range against anomalous outliers, and quantised to 16 bits." This is to avoid excessive precision loss during conversion from floats to integers for storage. The outliers are usually due to strong reflections from sharp angles on antennas and other man-made objects. The spatial resolution is (range by azimuth) 20 m by 22 m and the pixel spacing is 10 m. The IW data are multi-looked, the number of looks is 5 by 1 and the equivalent number of looks is 4.9. We have 500 rows by 800 columns of 10 m pixels. Figure 1 shows an RGB image of Sentinel-1 C-band VV/VH data acquired on 15 July 2016, VV as red and VH as green and blue (i.e., cyan). The data are log transformed.

The Sentinel-2 data are the near-infrared, red, green and blue channels from the MultiSpectral Instrument (MSI), level-1C processed, 500 rows by 800 columns of 10 m pixels. Figure 2 shows an RGB image of bands 4, 3, and 2 (near-infrared, red, and green). These data are acquired on 12 Sep 2016. The saturation issues mentioned for the S1 data are not present for the S2 data.

Here k-dimensional X in the CCA/CIA analyses is the Sentinel-1 VV and HV data, $k = 2$, and the ℓ-dimensional Y is the Sentinel-2 VNIR data, $\ell = 4$.

4   Results

First: in a change detection setting with data of different modalities, the results obtained of course reflect both change over time and the different kinds of information contained in the different data modalities. It is very difficult to discriminate between the two.

A philosophical question is whether change detection between so different data sources/modalities can be carried out in a meaningful fashion at all. Maybe not, but it is always interesting to try to optimise the joint information content in the data irrespective of their origin. And this is exactly what the CIA method aims to do.

Figures 3 and 4 show the leading (CCA based) canonical variates (CV) for the S1 and the S2 data, respectively.

Figures 5 and 6 show the leading (CIA based) canonical variates for the S1 and the S2 data, respectively. Starting weights a and b are the values obtained by CCA.

As stated above CCA is not necessarily an obvious choice of analysis for data of quite different modalities. Neither is therefore the choice of starting values from CCA a good one. Figure 7 shows the leading (CIA based) canonical variate for the S2 data. Starting weights a and b are equal for each set. Table 1 shows correlation and mutual information for the three solutions, CCA, CIA with starting weights from CCA, and CIA with equal starting weights.

Fig. 3. Leading canonical variate from CCA of Sentinel-1 C-band VV/VH data.

Fig. 4. Leading canonical variate from CCA of Sentinel-2 VNIR data.

Fig. 5. Leading canonical variate from CIA of Sentinel-1 C-band VV/VH data, starting weights (*a* and *b*) from CCA.

Fig. 6. Leading canonical variate from CIA of Sentinel-2 VNIR data, starting weights (*a* and *b*) from CCA.

Fig. 7. Leading canonical variate from CIA of Sentinel-2 VNIR data, equal starting weights (a and b) for each set.

Table 1. Correlation and mutual information for the three solutions examined.

	CCA	CIA (cca)	CIA (equal)
Corr	0.4484	0.0305	−0.3297
MI	0.0453	0.2569	0.3183

Figures 8 and 9 show the development for mutual information and the weights (a and b) over the iterations for the CIA method with both starting conditions.

Figure 10 shows 2D histograms of the leading CVs for both CCA (left) and CIA with equal starting weights (right). CIA based CVs clearly reveal more structure in the data than the CCA CVs. The saturation issue mentioned above is clearly seen, especially in the CCA based S1 CV.

Although the weights in $a = [a_1 \, a_2]^T$ are not equal for the two different sets of starting values there is no visual difference between the two CIA based S1 CVs. Therefore only one of them is shown. On the other hand the two S2 CVs are very different. The solution based on equal weights have a higher MI and much more structure in both built-up areas in the town of Kelsterbach to the north of the airport and in wooded regions.

For both sets of starting values for a and b we observe the following:

- It is obvious, that the MI based solution is far less noisy for both the S1 and the S2 data and that it reveals more structure.
- The different appearance of the leading S1 CIA CV (compared to the S1 CCA CV) must be due mainly to the drastic change in a_1 associated with the S1 VV data. Apart from the conspicuously less noisy visual appearance,

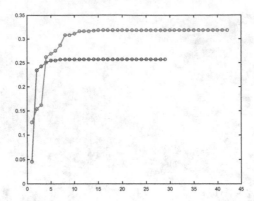

Fig. 8. Development of mutual information over the CIA iterations starting with the CCA solution (in blue) and starting with equal weights for each set (in red). (Color figure online)

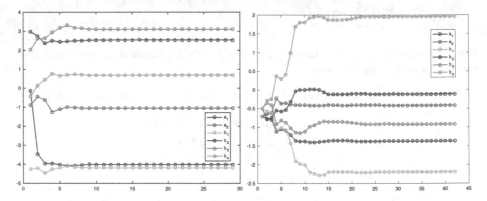

Fig. 9. Development of the weights $[a_1, a_2]^T$ for S1 data and $[b_1, \ldots, b_4]^T$ for S2 data over the CIA iterations starting with the CCA solution for each set (left), and equal weights for each set (right). For the left plot, especially coefficients b_1 and b_4 associated with the S2 blue and near-infrared bands change drastically and b_4 changes sign. Also in this case the coefficient a_1 for S1 VV changes drastically albeit less than b_1 and b_4. For the right plot, especially coefficients b_1 and b_4 associated with the S2 blue and near-infrared bands change drastically and b_4 changes sign. Also in this case the coefficient a_1 for S1 VV changes drastically albeit less than b_1 and b_4. (Color figure online)

we note that the leading S2 CIA CV looks very different from the CCA CV. This is especially true for most taxiways (but not for runways), aprons and other impervious regions, which have high values/appear bright in the CCA solution and have low values/appear dark in the two CIA solutions. Also the wooded areas surrounding the airport have low values/appear dark in the CCA solution whereas they have intermediate or high values/appear gray or bright in the two CIA solutions. This must be due to the change of sign in b_4 associated with the near-infrared channel of the S2 MSI data.

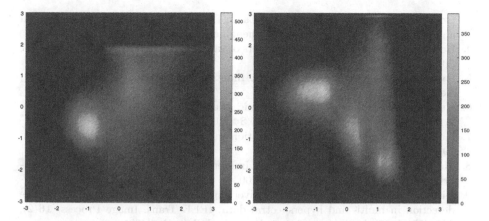

Fig. 10. 2D histograms for leading canonical variates, correlation based (left) and mutual information based with equal starting weights (right). The CVs all have mean value zero and variance one; the axes are stretched between minus and plus three standard deviations. The 2D histogram of the leading MI based CVs clearly reveals more structure in the data than the correlation based one. Also, we see that the MI based CVs are less sensitive to the saturation of the S1 data.

5 Conclusions

Although one may not be able to perform change detection between so different data sources/modalities in a meaningful fashion (here with radar and optical data), it is certainly interesting to try to optimise the joint information content in the data irrespective of their origin. And this is exactly what CIA aims to do.

Starting with equal weights for each set in the search for the CIA solution in our case is better than starting with weights from traditional CCA.

In this preliminary study of canonical analysis of data with different modalities, the results are a little inconclusive as far as the change detection aspect is concerned. Still, results show that CIA gives conspicuously less noisy appearing CV images than CCA. Also, a 2D histogram of the leading MI based CVs clearly reveals more structure in the data than the correlation based one. Finally, the MI based CVs are less sensitive to the saturation of the S1 data. Thus MI based canonical analysis reveals more signal and structure than correlation based analysis, and it gives promise for potentially better change detection results.

References

1. Bishop, C.M.: Pattern Recognition and Machine Learning. Springer, New York (2007)
2. Canty, M.J.: Image Analysis, Classification and Change Detection in Remote Sensing. With Algorithms for ENVI/IDL and Python, 3rd edn. Taylor & Francis, CRC Press (2014)

3. Google Earth Engine Team: Google Earth Engine: A planetary-scale geo-spatial analysis platform (2015). https://earthengine.google.com

4. Hotelling, H.: Relations between two sets of variates. Biometrika **XXVIII**, 321–377 (1936)

5. Hyvärinen, A., Karhunen, J., Oja, E.: Independent Component Analysis. Wiley, New York (2001)

6. Karasuyama, M., Sugiyama, M.: Canonical dependency analysis based on squared-loss mutual information. Neural Netw. **34**, 46–55 (2012)

7. Kullback, S., Leibler, R.A.: On information and sufficiency. Ann. Math. Stat. **22**(1), 79–86 (1951)

8. Mackay, D.J.C.: Information Theory, Inference and Learning Algorithms. Cambridge University Press, Cambridge (2003)

9. Nielsen, A.A.: The regularized iteratively reweighted MAD method for change detection in multi- and hyperspectral data. IEEE Trans. Image Process. **16**(2), 463–478 (2007). http://www2.imm.dtu.dk/pubdb/p.php?4695

10. Nielsen, A.A., Conradsen, K., Simpson, J.J.: Multivariate alteration detection (MAD) and MAF postprocessing in multispectral, bitemporal image data: New approaches to change detection studies. Remote Sens. Environ. **64**(1), 1–19 (1998). http://www2.imm.dtu.dk/pubdb/p.php?1220

11. Nielsen, A.A., Vestergaard, J.S.: Change detection in bi-temporal data by canonical information analysis. In: 8th International Workshop on the Analysis of Multitemporal Remote Sensing Images (MultiTemp), Annecy, France (2015). http://www.imm.dtu.dk/pubdb/p.php?6888, Matlab code at https://github.com/schackv/cia

12. Shannon, C.E.: A mathematical theory of communication. Bell Syst. Techn. J. **27**(3), 379–423, 623–656 (1948)

13. Vestergaard, J.S., Nielsen, A.A.: Canonical information analysis. ISPRS J. Photogrammetry Remote Sens. **101**, 1–9 (2015). http://authors.elsevier.com/a/1QAnN3I9x1EeMt, http://www.imm.dtu.dk/pubdb/p.php?6270, Matlab code at https://github.com/schackv/cia

14. Vestergaard, J.S.: Interpretation of images from intensity, texture and geometry. Ph.D. thesis, Technical University of Denmark (2015). http://orbit.dtu.dk

15. Yin, X.: Canonical correlation analysis based on information theory. J. Multivar. Anal. **91**, 161–176 (2004)

A Noncentral and Non-Gaussian Probability Model for SAR Data

Anca Cristea[(✉)], Anthony P. Doulgeris, and Torbjørn Eltoft

Earth Observation Laboratory, Department of Physics and Technology,
University of Tromsø, 9037 Tromsø, Norway
anca.cristea@uit.no

Abstract. A general compound statistical model for coherent imaging is developed and tested on single-channel Synthetic Aperture Radar (SAR) data. In this formulation, coherent scattering is taken into consideration and the texture is modeled using an Inverse Gaussian distribution. Parameter estimation is conducted via an Expectation Maximization (EM) scheme. A Maximum a Posteriori (MAP) speckle filter based on this model is also implemented. The filter shows good smoothing capabilities and preserves details in the selected scene, showing promise for target-detection applications.

Keywords: Speckle · Compound statistical model · Inverse Gaussian · MAP filter

1 Introduction

Coherent imaging systems such as radar, Synthetic Aperture Radar (SAR)/ Polarimetric Synthetic Aperture Radar (PolSAR), or medical ultrasound are subject to the effects of wave interference, which produces a granular texture called speckle in the image. Speckle is seen primarily as a type of noise that corrupts images and reduces the visibility of structures of interest. Moreover, being a type of multiplicative noise, it is especially difficult to remove. A number of strategies for despeckling filters have been implemented, mostly relying on statistical properties of the pixel values, thus including structural properties of the observed noisy image in the filtering mechanism. A comparative review of some of these techniques, including a MAP-type example, is presented in [1].

In SAR imaging, the received signal is complex and its amplitudes or intensities are often used for visual representation. The real and imaginary parts of the signal are considered as resulting from a random walk, which allows them to be modeled as random variables with distributions developed for different scattering regimes, depending on the characteristics of the imaged medium and system resolution. The simplest case to model is that of a random walk with a large number of steps, producing what is called fully developed speckle. The real and imaginary parts then follow a Gaussian distribution with zero mean

© Springer International Publishing AG 2017
P. Sharma and F.M. Bianchi (Eds.): SCIA 2017, Part II, LNCS 10270, pp. 159–168, 2017.
DOI: 10.1007/978-3-319-59129-2_14

and fixed variance, their amplitudes are Rayleigh distributed, and the intensities (squared amplitudes) are Chi-squared distributed. If a constant (coherent) component is added to the fully developed speckle, the mean of the Gaussian will be non-zero, the resulting signal amplitudes become Rice distributed, and the intensities Noncentral Chi-squared distributed. If the texture of the imaged object is taken into account as also varying, then this will result in a modulation of the mean and variance of the Gaussian by another random variable and the resulting model will be compound. Two examples of distributions often used for the texture are the Gamma and the Inverse Gaussian [2]. When the Gamma distribution is used, the K-distributions are derived. When the Inverse Gaussian is used, the Normal Inverse-Gaussian distribution is derived for the real and imaginary parts [3] and the Rician-Inverse Gaussian for the amplitudes [2].

We present a modified version of the model described in [2], in part because of the interesting filtering properties presented in [4]. Our primary purpose is to develop and implement a general model that can fit SAR data in any scattering conditions. A correct fit implies correct and robust parameter estimation, which can be challenging as the number of model parameters increases. Here, we use an EM-based estimator, as it fulfills the mentioned requirements. Moreover, due to the model formulation, one of the parameters is estimated directly from the data mean, thus requiring no further optimization. We also derive a closed-form MAP filter that uses this model to deliver an estimate of the texture values. The model derivation, the estimator, and the MAP filter are described in the following sections. An example of fitting and filtering applied on SAR data is also presented. At the end, we summarize the results and discuss other possible applications.

2   Model Derivation

The complex backscattered field has the form:

$$E = X + jY \tag{1}$$

where the real and imaginary parts X and Y are obtained by demodulating the signal received by the radar antenna and are assumed to be non-correlated. Each element of X or Y results from a sum of individual contributions from elementary scattering areas within the resolution cell (the minimum volume resolvable by the imaging system). The individual contributions cannot be measured, but each of them can be considered as a step in a random walk, and therefore a stochastic representation is required. In order to avoid redundance, the random variable associated with the distribution of either the real or the imaginary part of the complex scattered field will be noted here as X.

The distribution of X depends on the scattering suffered by electromagnetic waves when coming in contact with the imaged object. An object texture that is uniformly rough compared to the system resolution has a large number of scattering areas per resolution cell, which translates into a large number of steps in the random walk. In this case, the Central Limit Theorem applies and the distribution of X becomes Gaussian. If the scattering is exclusively diffuse, then the mean of the distribution is 0. If coherent scattering also occurs (in the presence of a strong reflector), then the mean is different from 0.

If the number of scattering areas is small relative to the resolution cell size, or if the texture is not uniformly rough (the number of scattering areas per resolution cell is variable over the entire object), then the conditions for the Central Limit Theorem may not respected locally and the distribution of X is non-Gaussian. The local variations in the number of scattering areas are modeled by another random variable (denoted Z), resulting in a modulation of the mean and variance of the Gaussian. X becomes a compound random variable called a normal variance-mean mixture:

$$X = \beta_x Z + \sqrt{Z}\mathcal{N}, \mathcal{N} \sim \mathcal{N}(0, \sigma). \tag{2}$$

In [5], the relation between this formulation and the geometric Brownian motion is established. In this context, $\beta_x = \beta cos x$ ($\beta_y = \beta sin y$ for Y) is a constant called drift coefficient, and Z is the random variable associated to the texture. The first term represents the coherent part, and the second term represents the diffuse part.

In the original formulation, $\sigma = 1$ and the distribution chosen for Z is a two-parameter (shape and scale) Inverse Gaussian. For the current model, we do not constrain the standard deviation of the speckle component σ, instead we set the mean of Z to 1, which leads to a one-parameter form for $p_Z(z)$:

$$p_Z(z) = \frac{1}{\sqrt{2\pi}}\delta exp(\delta^2)z^{-\frac{3}{2}}exp(-\frac{\delta^2}{2}(\frac{1}{z} + z)). \tag{3}$$

Figure 1 shows examples of Inverse Gaussian pdfs for different values of the shape parameter δ. The distribution is characterized by an asymmetric shape and a heavy tail for low δ values. As δ increases, the assymetry is reduced and the shape of the distribution approaches that of a Gaussian (Fig. 1). We consider it be Gaussian for a δ value of 10 and above (equivalent to $1/\delta < 0.1$), and non-Gaussian for lower values. In addition, the variance decreases with δ:

$$var[Z] = \frac{1}{\delta^2} \tag{4}$$

In order to derive the final form of $p_X(x)$, we must begin with the conditional probability distribution $p_{X|Z}(x|z)$, when z is fixed:

$$p_{X|Z}(x|z) = \frac{1}{\sqrt{2\pi\sigma^2 z}}e^{-\frac{(x-\beta_x z)^2}{2\sigma^2 z}}. \tag{5}$$

The probability distribution function for the normal-variance mean mixture can be obtained by integrating the conditional distribution over the entire span of Z values. The final form of the pdf is:

$$p_X(x) = \frac{1}{\pi}\frac{1}{\sigma}(\frac{\frac{\beta_x^2}{\sigma^2} + \delta^2}{\frac{x^2}{\sigma^2} + \delta^2})^{\frac{1}{2}}\delta exp(\delta^2)K_1(\sqrt{(\frac{\beta_x^2}{\sigma^2} + \delta^2)(\frac{x^2}{\sigma^2} + \delta^2)})exp(\frac{\beta_x x}{\sigma^2}) \tag{6}$$

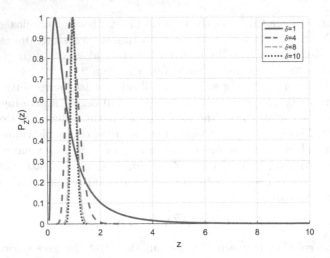

Fig. 1. Inverse Gaussian pdfs (normalized) for different values of δ. The domain of z is restricted for a better visibility (for $\delta = 1$, the tail extends up to z=20).

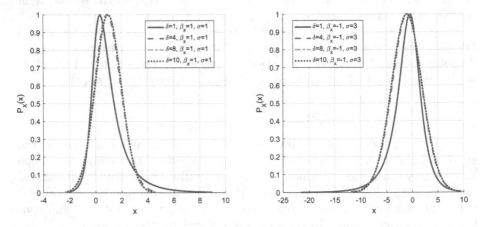

Fig. 2. Model pdfs (normalized) for different parameter values. The mean shifts with the value of β_x and σ acts as a scale parameter.

where $K_{\frac{3}{2}}(\cdot)$ is the modified Bessel function of the second kind and order $\frac{3}{2}$. The model can be considered as a re-parametrization of the Normal Inverse Gaussian (NIG) [5]. Examples of pdfs are shown in Fig. 2, where the influence of each parameter is illustrated: δ as shape parameter, σ as scale parameter and β_x (2) and $E[Z] = 1$).

2.1 Models for the Amplitude and Intensity of the Backscattered Field

If we wish to model the amplitudes and intensities of the complex backscattered field in (1), it is convenient to first develop the conditional distributions. In order to do this, it is necessary to look at the joint distribution of the variables X and Y. X and Y are assumed to be independent, therefore by simply multiplying the marginals we obtain the joint distribution, a bivariate Gaussian:

$$p_{XY|Z}(xy|z) = \frac{1}{2\pi\sigma^2 z}e^{-\frac{1}{2}(\frac{x-\beta_x z}{\sigma^2 z})^2(\frac{y-\beta_y z}{\sigma^2 z})^2}. \tag{7}$$

The amplitude is the equivalent of the ray (R) in polar coordinates, so a coordinate transformation is necessary. By setting $X = Rcos\theta$, $Y = Rsin\theta$, $\beta_x = \beta cos\phi$, $\beta_y = \beta sin\phi$, so that $R = \sqrt{X^2 + Y^2}$ and $\beta = \sqrt{\beta_X^2 + \beta_Y^2}$ and integrating the joint pdf $p_{R\Theta|Z}(r\theta|z)$ over the angles θ, we obtain a Rice conditional pdf for R:

$$p_{R|Z}(r|z) = \frac{x}{\sigma^2 z}e^{-\frac{x^2+\beta^2 z^2}{2\sigma^2 z}}I_0(\frac{x\beta}{\sigma^2}) \tag{8}$$

where $I_0(\cdot)$ is the modified Bessel function of the first kind and order 0. After performing the variable change $I = R^2$ and some additional manipulations, we derive the conditional pdf of the intensity:

$$p_{I|Z}(i|z) = \sigma^2 z p_{U|Z}(u|z) \tag{9}$$

where $U|Z \sim$ Non-central $\chi^2(2, \frac{\beta^2 z}{\sigma^2})$:

$$p_{U|Z}(u|z) = \frac{1}{2}exp(-\frac{1}{2}(u + \frac{\beta^2 z}{\sigma^2}))I_0(\sqrt{u\frac{\beta^2 z}{\sigma^2}}). \tag{10}$$

The form in (9) is particularly useful for parameter estimation, as the moments of the non-central chi-squared distribution have closed forms. After integrating over all the possible values for Z, the final probability distribution functions for the amplitude and intensity are:

$$p_R(r) = \sqrt{\frac{2}{\pi}}\frac{r}{\sigma^2}(\frac{\frac{\beta^2}{\sigma^2}+\delta^2}{\frac{r^2}{\sigma^2}+\delta^2})^{\frac{3}{4}}\delta exp(\delta^2)K_{\frac{3}{2}}(\sqrt{(\frac{\beta^2}{\sigma^2}+\delta^2)(\frac{r^2}{\sigma^2}+\delta^2)})I_0(\frac{\beta r}{\sigma^2}) \tag{11}$$

and, respectively:

$$p_I(i) = \frac{1}{\sqrt{2\pi}}\frac{1}{\sigma^2}(\frac{\frac{\beta^2}{\sigma^2}+\delta^2}{\frac{i}{\sigma^2}+\delta^2})^{\frac{3}{4}}\delta exp(\delta^2)K_{\frac{3}{2}}(\sqrt{(\frac{\beta^2}{\sigma^2}+\delta^2)(\frac{i}{\sigma^2}+\delta^2)})I_0(\frac{\beta\sqrt{i}}{\sigma^2}). \tag{12}$$

3   Parameter Estimation

Parameter estimation is conducted via an Expectation maximization (EM) scheme. Similar methods have been used in [6–8]. The complexity is reduced by the direct estimation of the drift coefficient as:

$$\beta_x = E[X]. \tag{13}$$

Once β_x is determined, the values of the remaining two parameters are initialized by computing the log-likelihood associated with the given data samples over a fixed grid, and optimized by using moment equations. We base the EM scheme on low-order moments, as they are reliable even for small sample sizes. A compromise between robustness and speed is desired. Therefore we choose moments with closed-form expressions, and try to minimize the implementation of numerical searches. The variance is an example of such a moment:

$$var[X] = \frac{\beta_x^2}{\delta^2} + \sigma^2. \tag{14}$$

A useful property of the NIG-derived model is the distribution of the posterior probability $p_{Z|X}(z|x)$, a Generalized Inverse Gaussian (GIG) [5] with parameters $GIG(-1, \sqrt{\frac{x^2}{\sigma^2} + \delta^2}, \sqrt{\frac{\beta_x^2}{\sigma^2} + \delta^2}))$, which has closed-form expressions for its moments [2]. The moment of order -1 is equal to:

$$E[Z^{-1}|X; \beta_x, \delta, \sigma] = (\frac{\delta_{GIG}}{\gamma_{GIG}})^{-1}(\frac{K_0(\gamma_{GIG}\delta_{GIG})}{K_1(\gamma_{GIG}\delta_{GIG})} + \frac{2}{\gamma_{GIG}\delta_{GIG}}) \tag{15}$$

where $\gamma_{GIG} = \sqrt{\frac{x^2}{\sigma^2} + \delta^2}, \delta_{GIG} = \sqrt{\frac{\beta_x^2}{\sigma^2} + \delta^2}$.

By averaging the moments computed using (15) over the entire value interval of the data samples, one can obtain moment estimates for the prior distribution $p_Z(z)$. Moreover, since the Inverse Gaussian is a particular case of the GIG, its moments also have closed-form expressions, so that:

$$E[Z^{-1}] = 1 + \frac{1}{\delta^2}. \tag{16}$$

The convergence condition can be chosen among a number of goodness-of-fit measures, for example the Kullback-Leibler Divergence:

$$D_{KL}(P||H) = \sum_i p(i)log\frac{p(i)}{H(i)} \tag{17}$$

between the histogram values $H(i)$ and the values of the fitted pdf $p(i)$.

Algorithm 1. EM for the SLC Model

---

1 **begin**
2 | $\beta_x = E[X]$
3 | Initialize δ_0 and σ_0 from maximum likelihood
4 | l=1
5 | **while** $D_{KL}(P(\beta_x, \delta_l, \sigma_l)||H) < D_{KL}(P(\beta_x, \delta_{l-1}, \sigma_{l-1})||H)$ **do**
6 | | $E[Z^{-1}] = E[E[Z^{-1}|X; \beta_x, \delta_{l-1}, \sigma_{l-1}]]$
7 | | $\delta_l = \sqrt{\frac{1}{E[Z^{-1}]-1}}$
8 | | $\sigma_l = \sqrt{Var[X] - \frac{\beta_x^2}{\delta_l^2}}$
9 | | $l \leftarrow l+1$
10 | **end**
11 **end**

---

4   MAP Filter

We propose a speckle filter associated with the developed model. The purpose is to remove the speckle and return a smooth image, with the important requirement that details of the scene are preserved and not over-smoothed. The model is integrated into a Maximum a Posteriori (MAP) estimation, similar to [4]. The posterior probability of the texture Z is developed in a Bayesian frame, using the prior $p_Z(z)$, the likelihood $p_{I|Z}(i|z)$ and the total probability of observing I, $p_I(i)$. The Intensity of the scattered field is used in order to include information from both the real and imaginary parts. The resulting posterior probability distribution also has the form of a GIG, similar to $p_{Z|X}(z|x)$, this time with parameters $(-\frac{3}{2}, \sqrt{\frac{\beta^2}{\sigma^2} + \delta^2}, \sqrt{\frac{i}{\sigma^2} + \delta^2})$. After the maximization of $p_{I|X}(i|x)$, the estimated texture value for each pixel is equivalent to:

$$\hat{z} = \frac{\sqrt{25 + 4(\frac{i}{\sigma^2} + \delta^2)(\frac{\beta^2}{\sigma^2} + \delta^2)} - 5}{2(\frac{\beta^2}{\sigma^2} + \delta^2)} \tag{18}$$

5   Fitting and Filtering SAR Data

The model and filter are tested on a PolSAR image obtained from the RADARSAT-2 EO Satellite working in Selective Single-Polarization Mode (Transmit H, receive H), at a resolution of 8 m (slant) × 5 m (azimuth). The scene is situated in the sea of Kattegat, off the coast of Denmark, and contains the small island Hesselø. In Fig. 3, the model is fitted to two different areas of the image representing the real part of the complex field. The difference in texture is obvious, but the relatively low values of the local mean, represented by the drift β_x, have a lower impact on the overall distribution.

The entire scene is represented in Fig. 4. The original data is presented in amplitude form for visibility. The high-intensity structure highlighted by the red square

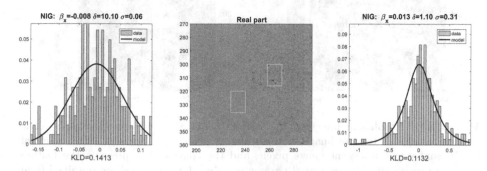

Fig. 3. Model fit and Kullback-Leibler Divergence on two areas of the test scene (yellow: sea, green: land). Each selection contains 220 pixels and is fitted using 50 bins. (Color figure online)

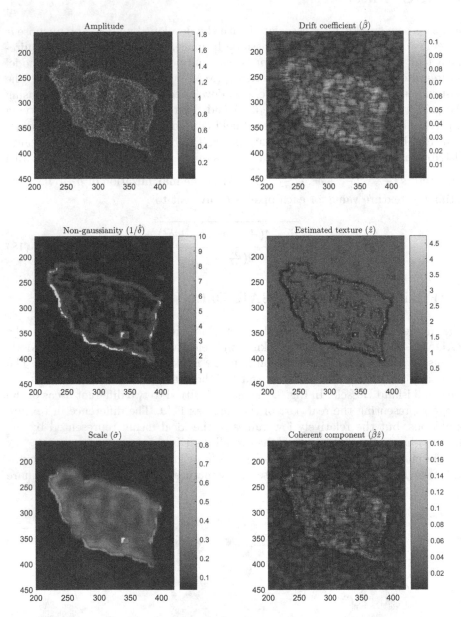

Fig. 4. SAR Amplitude Image of the island Hessel and corresponding parametric images, estimated texture image, and estimated coherent component. The values of the x- and y-axis represent image pixels, and the color scale represents the feature named in each title. The estimated parameter values for each pixel are obtained from neighborhoods of 11×11 pixels.

corresponds to a house in the real scene. Parameter estimation is performed on the real and imaginary parts, using the corresponding model (6) and Algorithm 1. β is obtained from its components β_x and β_y, and the other two parameters are averaged for their final values. The parametric images are computed for each pixel in the image, by applying the estimator on a 11×11 pixels neighborhood. The MAP filter is also applied on each pixel using (18), and finally an estimation of the purely coherent component of the amplitude $\hat{\beta}\hat{z}$ is given.

The estimated drift β has low values relative to the scale σ throughout the image, as expected for a mostly natural scene. The highest values are visible in an area with cliffs, respectively an area with an open field, suggesting the presence of coherent scattering. The non-Gaussianity (measured using $1/\delta$) shows small variations on land ($1/\delta < 1$ corresponding to $\delta > 1$) with extreme values present only at the boundaries between land and sea and in the house area. The house appears to have a strong brightness and associated σ value, but very low drift. The structure maintains its contrast and borders in the despeckled image, but is not visible on the image of the coherent component. The filtered image also highlights the boundaries between different areas, and a number of sharp maxima whose significance is less clear and requires further analysis.

6   Conclusions and Future Work

We have developed and tested a variation of the Normal-Inverse Gaussian distribution for the components of a complex SAR signal. Parameter estimation is conducted using an adapted, robust EM scheme. Once the model is fitted, it is possible to filter the data by applying a MAP filter which is also described here. Since this model accounts for a specular (coherent) component, however feeble, its detection makes it possible to make assumptions on the nature of the objects that it characterizes: they are likely to be large reflective surfaces. Further analysis of a variety of scenes is necessary for a clear interpretation of the obtained parameters and of the situations where they deliver unequivocal information on the nature of the surface. The main application that we wish to explore is target detection, on SAR images of different scales and resolutions, especially where the detection of the coherent part of the signal gives valuable information. It is also desired to assess the relevance of the new estimates for classification algorithms.

Other possible applications include an extension for Multilook Intensity, an extension to multichannel data (SAR Polarimetry), and the comparison with other texture models (for example the Gamma model).

References

1. Ozdarici, A., Akyurek, Z.: A comparison of SAR filtering techniques on agricultural area identification. In: ASPRS (2010)
2. Eltoft, T.: The Rician inverse Gaussian distribution: a new model for non-Rayleigh signal amplitude statistics. IEEE Trans. Image Process. **14**(11), 1722–1735 (2005)

3. Barndorff-Nielsen, O.: Normal inverse Gaussian distributions and stochastic volatility modelling. Scand. J. Stat. **24**(1), 1–13 (1997)
4. Eltoft, T.: Modeling the amplitude statistics of ultrasonic images. IEEE Med. Imaging **25**(2), 229–240 (2006)
5. Barndorff-Nielsen, O., Kent, J., Sørensen, M.: Normal variance-mean mixtures and Z distributions. Int. Stat. Rev. **50**(2), 145–159 (1982)
6. Karlis, D.: An EM type algorithm for maximum likelihood estimation of the normal-inverse Gaussian distribution. Stat. Probab. Lett. **57**(1), 43–52 (2002)
7. Oigard, T., Øigård, T., Hanssen, A.: The multivariate normal inverse Gaussian heavy-tailed distribution; simulation and estimation. In: EUSIPCO Proceedings, pp. 1489–1492 (2002)
8. Øigård, T., Hanssen, A., Hansen, R.: The multivariate normal inverse Gaussian distribution: EM-estimation and analysis of synthetic aperture sonar data. In: EUSIPCO Proceedings, no. 9037, pp. 1433–1436 (2004)

Unsupervised Multi-manifold Classification of Hyperspectral Remote Sensing Images with Contractive Autoencoder

Aidin Hassanzadeh[1](✉), Arto Kaarna[1], and Tuomo Kauranne[2]

[1] Machine Vision and Pattern Recognition Laboratory,
School of Engineering Science, Lappeenranta University of Technology,
Lappeenranta, Finland
{aidin.hassanzadeh,arto.kaarna}@lut.fi
[2] Mathematics Laboratory, School of Engineering Science,
Lappeenranta University of Technology, Lappeenranta, Finland
tuomo.kauranne@lut.fi

Abstract. Unsupervised classification is a crucial step in remote sensing hyperspectral image analysis where producing training labelled data is a laborious task. Hyperspectral imagery is basically of high-dimensions and indeed dimensionality reduction is considered a vital step in its pre-processing chain. A majority of conventional dimensionality reduction techniques rely on single global manifold assumptions and they can not handle data coming from a multi-manifold structure. In this paper, the unsupervised classification of hyperspectral imaging is addressed through a multi-manifold learning framework. To this end, this paper proposes a Contractive Autoencoder based multi-manifold spectral clustering algorithm for unsupervised classification of hyperspectral imagery. The proposed algorithm follows the same outline as the general multi-manifold clustering but exploits contractive autoencoder for tangent space estimation. We evaluate the proposed algorithm with two benchmark hyperspectral datasets, *Salinas* and *Pavia Center Scene*. The experimental results show the improvements made by the proposed method with respect to the conventional multi-manifold clustering based on local PCA and the basic autoencoder.

Keywords: Hyperspectral image · Multi-manifold learning · Spectral Clustering · Tangent space estimation · Contractive Autoencoder

1   Introduction

Unsupervised classification (clustering) is an indispensable technique in several modern data analysis tasks such as image segmentation, pattern recognition and data mining. Indeed, clustering plays a significant role in processing of hyperspectral remote sensing imagery, where labelled training samples are laborious to produce or often inadequate for application with supervised classification.

© Springer International Publishing AG 2017
P. Sharma and F.M. Bianchi (Eds.): SCIA 2017, Part II, LNCS 10270, pp. 169–180, 2017.
DOI: 10.1007/978-3-319-59129-2_15

A hyperspectral image (HSI) consists of hundreds of contiguous spectral bands that provide detailed information to distinguish spectrally similar materials. However, the high dimensionality of data introduces complexities, generally referred to the *curse of dimensionality* [1,2]. Thus, dimensionality reduction (DR) is of great importance in processing of HSI data for the unsupervised classification task.

Several DR techniques have been utilized in HSI analysis tasks. In this direction, the common strategies include linear DR approaches such as Principal Component Analysis (PCA), Multidimensional Scaling (MDS) and Independent Component Analysis (ICA). These linear techniques have been shown efficient in many scenarios, but as they rely on the linearity assumption and do not consider the potential non-linear dependencies they may be problematic in remote sensing HSI data usually coming with intrinsic non-linearities. The non-linearity in HSI data may be caused by non-linear scattering patterns, variations in local geometry of the sun-canopy-sensor triangle, nonuniform pixel composition, and the characteristics of transmission medium [3].

Non-linear dimensionality reduction (NLDR) approaches, mainly manifold learning algorithms, have been used to resolve both issues of high-dimensionality and the problems associated with non-linearity of HSI remote sensing data. The conventional non-linear manifold learning algorithms such as Isomap [4], kernel Principal Component Analysis (KPCA) [5], Locally Linear Embedding (LLE) [6], Local Tangent Space Alignment (LTSA) [7] and Laplacian Eigenmaps (LE) [8] are built on the single smooth manifold assumption. These approaches attempts to retrieve the structure of data while they assume that the data lie close to a low dimensional sub-manifold embedded in a uniformly distributed single topological manifold.

In many real world applications, data may lie on multiple manifolds. Manifold learning algorithms relying on a single global manifold will fail to achieve satisfactory results if the data is drawn from multiple manifolds with intersections. To this end, multi-manifold learning is an alternative line of research in NLDR that attempts to recover the structure of data with multi-manifold assumption. The multi-manifold assumption is about distribution of data where each data cluster is approximated by a specific sub-manifold that features distinct geometric structure being attributed to that data cluster.

There are a number of methods proposed in this context. A group of multi-manifold learning algorithms is about to complement the manifold learning algorithms based on the single manifold assumption to data with multiple manifolds. [9] proposes a multi-manifold clustering algorithm based on Multi Dimensional scaling (MDS). [10] extends ISOMAP to data of multiple manifolds. [11] proposes a multi-manifold learning algorithm that performs locality preserving projection (LPP) [12]. [13,14] are extensions of the well-know locally linear embedding (LLE) for data with multi-manifold structure.

Another group of the methods in multi-manifold learning incorporates the manifold structure into clustering problem. [15–17] incorporate dimension or density information of data into the clustering algorithm. [18] performs multi-manifold clustering via energy minimization. [19] proposes a sparse-coding based spectral clustering. [20], High Order Spectral Clustering (HOSC), performs spectral clustering based via higher order graph affinities. [21–26] propose spectral clustering utilizing local tangent space affinities.

In this paper, we present a new spectral multi-manifold clustering algorithm with application to HSI data. The proposed algorithm follows the same outline as the general local tangent space based spectral methods, but it adopts a variant of autoencoders, specifically Contractive Autoencoder (CAE) [27], for local tangent space estimation.

Autoencoder (AE) is a simple feed forward neural network that basically aims to reconstruct input data. The main purpose of AE is to perform dimensionality reduction or Sparse Coding (SC) through unsupervised learning. An autoencoder with the number of hidden units less than the number of input units (the original dimension of data) is seen as a dimensionality reduction technique. An AE with the number of hidden units larger than number of input units imposed by some constraint renders Sparse Coding. Indeed, PCA in this way may be pointed out as a special case of linear AE that includes rank deficient connection weight matrix [27,28].

Contractive Autoencoder is an improved variant of the basic autoencoder that directly attempts to produce robust latent representations that are less sensitive to infinitesimal variations in input data. CAE achieves this by balancing the reconstruction loss through an analytical regularizer term that penalizes the Frobenius norm of the Jacobian of the encoder. Through the manifold realization of data, Riafai et al. in [27–29] show that as the reconstruction objective in CAE attempts to produce accurate reconstructions, the contraction regularizer aims for the hidden representations that are also indifferent in all directions of input space. That is, the hidden representations in CAE prefer to change only along the directions of nearby training data points and accordingly tend to resemble a data manifold that is locally flat.

This observation leads to inspect the derivatives of the hidden representations with respect to the input data for estimation of the coordinates systems of the local tangent spaces as successfully exploited in [29]. Following this direction, we utilize CAE for local tangent space estimation required for a general spectral multi-manifold clustering applied for typical HSI unsupervised classification task.

In the proposed algorithm, we follow three sequential steps. First, we adopt and train an overcomplete single-layer CAE with unlabelled data. We compute the local tangent spaces at every point using the trained CAE. Second, we combine the obtained local tangent spaces with the proximity model of data and then we compute the affinity matrix. Third, we plug the affinity matrix into a standard spectral clustering algorithm and we classify data. We evaluate the performance of algorithm with two benchmark HSI data, namely *Salinas* and *Pavia Center Scene*.

The remainder of the paper is organized as follows. Section 2 reviews Contractive Autoencoder (CAE) and shows how it can be used for tangent space estimation. Section 3 presents the overall procedure of spectral multi-manifold clustering, presenting the main steps involved. Section 4 presents experimental results and evaluation of effectiveness of the proposed algorithm. Finally, Sect. 5 concludes the paper.

2 Tangent Space Estimation with Contractive Autoencoder

A basic autoencoder involves an encoder and a decoder. The encoder transforms data from the input space to the latent representations and the decoder transforms the latent representations back to the input space. Formally, a data vector from the original input space $\mathbf{x} \in \mathbb{R}^D$ is mapped to the latent representation $\mathbf{h} \in \mathbb{R}^{d_h}$ with encoder function $f(\mathbf{x})$ and the reconstruction of that input data $\hat{\mathbf{x}}$ is obtained from the latent representation with decoder function $g(\mathbf{h})$.

A basic autoencoder learns the encoder and decoder parameters Θ through minimizing the average reconstruction loss given by the following objective function:

$$\mathcal{J}_{AE} = \sum_{x \in X} L(\mathbf{x}, \hat{\mathbf{x}}) = \sum_{x \in X} L(\mathbf{x}, g(f(\mathbf{x}))), \tag{1}$$

where $L()$ is the reconstruction loss function that can be defined by the squared error for a linear decoder:

$$L(\mathbf{x}, \hat{\mathbf{x}}) = \| \mathbf{x} - \hat{\mathbf{x}} \|^2 \tag{2}$$

or can be defined by the Bernoulli cross entropy loss function for a logistic sigmoid decoder:

$$L(\mathbf{x}, \hat{\mathbf{x}}) = - \sum_{i=1}^{D} x_i \log(\hat{x}_i) + (1 - x_i) \log(1 - \hat{x}_i). \tag{3}$$

The encoder function $f(\mathbf{x})$ is of the form:

$$\mathbf{h} = s_h(\mathbf{W}_h \mathbf{x} + \mathbf{b}_h), \tag{4}$$

where $\mathbf{W}_h \in \mathbb{R}^{d_h \times D}$ and $\mathbf{b}_h \in \mathbb{R}^{d_h}$ are encoder network parameters, and $s_h(.)$ is a nonlinear element-wise function, commonly defined by logistic sigmoid $sigmoid(z) = (1 + \exp(-z))^{-1}$.

The decoder $g(\mathbf{h})$ has the following form:

$$\hat{\mathbf{x}} = s_r(\mathbf{W}_r \mathbf{h} + \mathbf{b}_r), \tag{5}$$

where $\mathbf{W}_r \in \mathbb{R}^{D \times d_h}$ and $\mathbf{b}_r \in \mathbb{R}^D$ are decoder network parameters. The typical choices for s_r are sigmoid and identity functions. To reduce the number of parameter to train and assist the training of autoencoder, the decoder may be tied to the encoder where the connection weight parameter are ($\mathbf{W}_h = \mathbf{W}$, $\mathbf{W}_r = \mathbf{W}^T$). In this way, the autoencoder network parameters set is defined by $\Theta = \{\mathbf{W}, \mathbf{b}_h, \mathbf{b}_r\}$.

Contractive Autoencoder is constructed by adding up an additional penalty term to the reconstruction objective cost function in Eq. 1. The role of added penalty term is to contract the reconstructions at training samples and make the reconstruction process robust to small variations in input data. CAE achieves this through the regularizer formed by the squared Frobenius norm of Jacobian of the encoder function, $\| J_f(x) \|_F^2$. The objective function in CAE is as follows:

$$\mathcal{J}_{CAE} = \sum_{x \in X} L(\mathbf{x}, g(f(\mathbf{x}))) + \lambda \| J_f(\mathbf{x}) \|_F^2, \tag{6}$$

where λ is a non-negative parameter controlling the strength of contraction regularizer. For an autoencoder with sigmoid encoder activation function the, jth row of Jacobian of the encoder, $J_f(\mathbf{x})_{j.}$, can be obtained as follows:

$$J_f(\mathbf{x})_{j.} = \frac{\partial f_j(\mathbf{x})}{\partial \mathbf{x}} = f_j(\mathbf{x})(1 - f_j(\mathbf{x}))\mathbf{W}_{j.}, \tag{7}$$

Recall the embedded submanifold assumption, the tangent space can be given by the image of Jacobian of submanifold embedding map (here in CAE the encoder function) which also equals to the column space of the Jacobian matrix. This motivates CAE to obtain the local tangent spaces of the points through the principal vectors of the transpose of the encoder Jacobian matrix. That is, CAE estimates the local tangent spaces by extracting the principal direction captured by the Jacobian matrix. In practice, at a given point $\mathbf{x}$, CAE estimates the coordinate axes of a local tangent space by the Singular Value Decomposition of the transpose of the Jacobian matrix:

$$J_f^T(\mathbf{x}) = \mathbf{USV}, \tag{8}$$

where $\mathbf{U}$ and $\mathbf{V}$ are left- and right-singular vectors, and $\mathbf{S}$ is the diagonal matrix containing singular values.

Spectral analysis of the Jacobian matrix shows that the majority of singular values of the Jacobian are zero (or close to zero) and only a few of them are reasonably large. Accordingly, the set of the orthonormal bases $\mathcal{B}_\mathbf{x}$ is given by the set of columns vectors of $\mathbf{U}$ corresponding to the dominant singular values

$$\mathcal{B}_\mathbf{x} = \{\mathbf{U}_{.k} | \mathbf{S}_{kk} > \epsilon, \} \tag{9}$$

which spans the tangent space at point $\mathbf{x}$, $T_\mathbf{x}$:

$$T_\mathbf{x} = \{\mathbf{x} + \mathbf{v} | \mathbf{v} \in \overline{Span}(\mathcal{B}_\mathbf{x})\}. \tag{10}$$

3 Multi-manifold Spectral Clustering

This section presents the main steps involved with the proposed multi-manifold spectral clustering (MMSC) algorithm. We briefly cover the methodologies used for estimating of local tangent spaces, computing of the affinities and spectral clustering.

Given the set of data points of the original input D-dimensional space $X = \{\mathbf{x}_i\}_{i=1}^N$, the MMSC assumes the data reside on m different embedded d-dimensional submanifolds. The proposed algorithm aims to cluster the data points to m different groups through a three-step procedure.

3.1   Estimation of Tangent Spaces

A submanifold embedded in Euclidean space or some abstract manifold can implicitly be approximated by its local tangent spaces. A MMSC based on local tangent space affinities models data manifolds via their local tangent spaces. Indeed, the quality of the estimated local tangent spaces has direct impacts on manifold sampling and the accuracy of data affinities.

The proposed MMSC algorithm exploits a single-layer contractive autoencoder for tangent space estimation, even though a multi-layer CAE can also be utilized. To this end, first a single layer contractive autoencoder is trained with unlabelled data. The network is set up with over-complete structure where $d_h > D$.

The trained CAE is then used to compute the encoder Jacobian and its SVD for each data point as formulated in Eqs. 7 and 8. The local tangent spaces of the point is extracted and stored as the set of column vectors of the left-singular vector of the Jacobian that corresponds to the dominant singular values, Eq. 9.

3.2   Estimation of Affinity Metrics

Following the smooth multi-manifold assumption, MMSC determines the affinity between a pair of points by the similarity between their local tangent spaces, captured by their principal angles.

For a given pair of points $\mathbf{x}_i$ and $\mathbf{x}_j$, let $T_{\mathbf{x}_i}$ and $T_{\mathbf{x}_j}$ be their tangent spaces respectively, and assume $dim(T_{\mathbf{x}_i}) = d_i$, $dim(T_{\mathbf{x}_j}) = d_j$ and $d \geq d_i \geq d_j \geq 1$. The principal angles θ_l between $T_{\mathbf{x}_i}$ and $T_{\mathbf{x}_j}$,

$$0 \leq \theta_1 \leq \ldots \theta_l \leq \ldots \leq \theta_b \leq \frac{\pi}{2} \tag{11}$$

are recursively given as follows:

$$cos(\theta_l) = \max_{\mathbf{p_k} \in T_{\mathbf{x}_i}, \mathbf{q_k} \in T_{\mathbf{x}_j}} \mathbf{p}_k^T \mathbf{q}_k , \quad k = 1, \ldots, l-1 \tag{12}$$

subject to

$$\| \mathbf{p}_k \| = \| \mathbf{q}_k \| = 1, \ \mathbf{p}^T \mathbf{p}_k = \mathbf{q}^T \mathbf{q}_k = 0, \tag{13}$$

where $\mathbf{p}_k$ and $\mathbf{q}_k$ are the set of the principal vectors of the pair tangent spaces.

Let $\mathcal{N}_k(\mathbf{x}_i)$ and $\mathcal{N}_k(\mathbf{x}_j)$ be the k-nearest neighbours sets that approximate the local proximity models at points $\mathbf{x}_i$ and $\mathbf{x}_j$. The MMSC computes the affinity as follows:

$$(A)_{ij} = \begin{cases} (\prod_{l=1}^{d_i} cos(\theta_l))^\alpha & \text{if } \mathbf{x}_i \in \mathcal{N}_k(\mathbf{x}_j), \mathbf{x}_j \in \mathcal{N}_k(\mathbf{x}_i), \\ \\ 0 & \text{otherwise} \end{cases} \tag{14}$$

Notice that here we use a similar affinity metric to the one defined in [22], but we employ CAE to compute the local tangent spaces.

3.3 Spectral Clustering

Given the affinity metrics, the spectral clustering algorithm is applied to the affinity matrix by which the data clusters are extracted.

4 Experiments

This section presents the qualitative and the quantitative results that evaluate the performance of the proposed algorithm. The proposed algorithm was applied to real world hyperspectral data and was evaluated by comparing with the clustering approaches based on the single manifold assumption and the MMSC algorithm based on local PCA. By this means, we would like to empirically analyse if the Contractive Autoencoder based multi-manifold clustering is an appropriate approach for a typical multi-manifold clustering of remote sensing hyperspectral imagery data.

4.1 Data

We evaluated our proposed algorithm with two different test hypercubes built on two benchmark hyperspectral remote sensing datasets: *Salinas* and *Pavia Center Scene*.

The *Salinas* dataset was collected by AVIRIS over Salinas valley in Southern California, USA. The hypercube data is characterized by 3.7 m pixel size and 224 bands (including the noisy bands). The data set consists of 16 different classes of land covers including: broccoli 1, broccoli 2, fallow, rough ploughed fallow, smooth fallow, stubble, celery, untrained grapes, soil/vineyard, senesced corn, Romaine lettuce 4wk, Romaine lettuce 5wk, Romaine lettuce 6wk, Romaine lettuce 7wk, untrained vineyard, vineyard with vertical trellis and shadow. Due to the close similarity between spectral features of the vineyard with vertical trellis and the untrained vineyard, the vineyard with vertical trellis class was excluded from the experiments. 900 samples from each land cover class were randomly chosen and for better visualization arranged in form of a rectangular cuboid. The RGB rendering and the ground truth of the test *Salinas* hypercube are shown in Fig. 1(a) and (b).

The *Pavia Center Scene* dataset was acquired by the reflective optics system imaging spectrometer (ROSIS) sensor over the city of Pavia, Italy. The hypercube data is characterized by 1.3 m pixel size and 102 bands with no noisy bands. The data set consists of 9 different classes of land cover including: water, trees, asphalt, self-blocking bricks, bitumen, tiles, shadows, meadows and bare soil. The shadow class was excluded from the experiments. 400 samples from each land cover class were randomly chosen and for better visualization arranged in form of a rectangular cuboid. The RGB rendering and the ground truth of the test *Pavia Center Scene* hypercube are shown in Fig. 2(a) and (b) respectively.

4.2 Experimental Set-Up

As described in Sect. 2, we computed the local tangent spaces using an overcomplete single-layer CAE. The architectures for CAEs used for *Pavia Center Scene* and *Salinas* test hypercubes are 102-150-102 and 204-250-204 respectively. The logistic sigmoid function was set for the encoder and the linear activation was set for the decoder. Since, the contraction coefficient, λ in Eq. 6, significantly affects the quality of CAE for local tangent estimation, we report CAE over different values of contraction coefficient λ.

We trained all the networks with all available samples over the squared error function with Stochastic Gradient Descent (SGD) driven by the momentum acceleration. The standard Spectral Clustering (SC) algorithm with the affinity metric defined in Sect. 3.2 is considered as the common baseline in all the experiments. Aiming for fair evaluation, all the SC experiments were performed with pre-known number of cluster and over 5 trials with random initialization. The performance of the clustering results were reported by the overall average accuracy (ACC) and the macro averaged Positive Predictive Value (PPV_m). We should notice that the architecture and the parameters utilized in the experiments are not optimal, the results may even be improved with some other architecture.

4.3 Results

The clustering results obtained by the proposed CAE based MMSC (MMSC+CAE) and the competing methods, Spectral Clustering with PCA (SC+PCA), MMSC with local PCA (MMSC+LPCA), MMSC with basic Autoencoder (MMSC+AE) are presented in Table 1. The results on the overall Average Accuracy (ACC) and the macro average Positive Predictive Value (PPV_m) show that MMSC+CAE outperforms all the other methods. In particular, in case of *Pavia Center Scene* test cube, CAE with the three reported contraction coefficients resulted higher clustering results compared to other methods.

Table 1. Highest overall clustering ACC and macro averaged PPV (in percentage) obtained by the considered algorithms with k nearest neighbours

Methods	Salinas			Pavia Center		
	k	ACC [%]	PPV_m [%]	k	ACC [%]	PPV_m [%]
SC+PCA	-	82.8	81.0	-	81.6	83.8
MMSC+LPCA	30	89.2	87.6	35	92.6	93.8
MMSC+AE	30	89.2	87.6	5	93.4	94.5
MMSC+CAE ($\lambda = 0.001$)	30	89.3	87.6	5	93.7	94.7
MMSC+CAE ($\lambda = 0.003$)	30	89.2	87.6	5	93.8	94.8
MMSC+CAE ($\lambda = 0.010$)	10	89.4	91.5	5	94.3	94.9

Figures 1 and 2 show the clustering maps obtained by the competing methods with *Salinas* and *Pavia center* test hypercubes respectively. The proposed MMSC+CAE with three different values of contraction coefficient are shown in Figs. f, g and h. The clustering maps show MMSC+CAE with reported coefficient values led to better clustering performance compared to the other methods. In particular, we observed that MMSC+CAE achieves better separation power

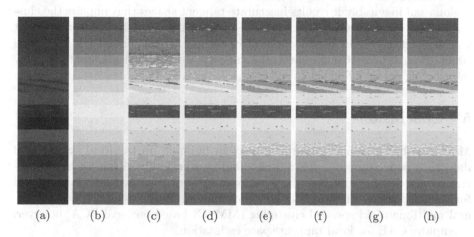

(a)   (b)   (c)   (d)   (e)   (f)   (g)   (h)

Fig. 1. Clustering performance comparison on *Salinas* test hypercube; RGB color rendering in (a), followed by the corresponding ground truth in (b). Clustering results obtained by (c) SC+PCA, (d) MMSC+LPCA, (e) MMSC+AE, (f) MMSC+CAE ($\lambda = 0.001$), (g) MMSC+CAE ($\lambda = 0.003$), (h) MMSC+CAE ($\lambda = 0.01$).

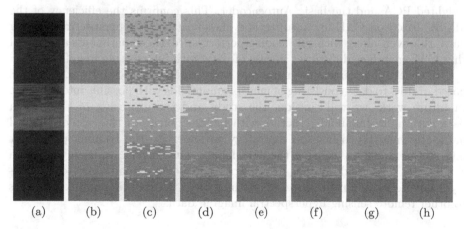

(a)   (b)   (c)   (d)   (e)   (f)   (g)   (h)

Fig. 2. Clustering performance comparison on *Pavia Center Scene* test hypercube; RGB color rendering in (a), followed by the corresponding ground truth in (b). Clustering results obtained by (c) SC+PCA, (d) MMSC+LPCA, (e) MMSC+AE, (f) MMSC+CAE ($\lambda = 0.001$), (g) MMSC+CAE ($\lambda = 0.003$), (h) MMSC+CAE ($\lambda = 0.010$).

on data classes of not very similar spectral classes, but it cannot fully resolve the data clusters with quite similar spectral features, e.g. in *Salinas* the spectral data from the untrained grape and the untrained vineyard classes, the first and the seventh blocks from above.

We suppose that this issue relates to the Jacobian of the encoder function in CAE used for tangent space estimation. To obtain tangent spaces the Jacobian can not differentiate the structures of data submanifolds within intersection regions and inevitably it results inaccurate tangent spaces that impairs the clustering performance. Overall, these observations indicate that to some extent CAE can effectively be used in MMSC for estimating tangent spaces but it will suffer from data with high spectral similarity, i.e. data with potential highly intersecting submanifolds.

5 Conclusions

Motivated by the ability of autoencoders in building up the local structure of data, this paper proposed a multi-manifold clustering algorithm for hyperspectral remote sensing imagery where a robust variant of autoencoder is employed for tangent space estimation. The proposed algorithm flows similar to the general multi-manifold spectral clustering (MMSC) based on local PCA, however, it employs CAE for local tangent space estimation.

We analysed the quality of the proposed algorithm with two test hypercubes extracted from two benchmark hyperspectral datasets: *Pavia Center Scene* and *Salinas*. The experiments show the outperformance of our proposed algorithm over the standard Spectral Clustering (SC) and the MMSC with local PCA, local weighted PCA and the basic Autoencoder. This confirms the efficiency of the MMSC based on CAE for unsupervised classification of hyperspectral imaging data. However, along with this leading performance, the experiments also reveal that CAE may face difficulties in case of data with intersecting manifold.

We believe that certain measures may be required to successfully extend CAE based MMSC to data consisting of multiple manifolds with possible intersections. In future work, we will consider the study and the extension of CAE for a data with multi-manifold structure with intersections. For the case of labelled data, this may be to develop a strategy to supervise the training process at intersection areas by utilizing the data labels. The current work only covers a single layer CAE and another future work will be to extend CAE to a multi-layer architecture. Furthermore, we are interested in extending our experiments to other remote sensing hyperspectral imaging data.

References

1. Lunga, D., Prasad, S., Crawford, M., Ersoy, O.: Manifold-learning-based feature extraction for classification of hyperspectral data: a review of advances in manifold learning. IEEE Sig. Process. Mag. **31**, 55–66 (2014)
2. Masulli, F., Rovetta, S.: Clustering high-dimensional data. In: Masulli, F., Petrosino, A., Rovetta, S. (eds.) CHDD 2012. LNCS, vol. 7627, pp. 1–13. Springer, Heidelberg (2015). doi:10.1007/978-3-662-48577-4_1
3. Bachmann, C., Ainsworth, T., Fusina, R.: Exploiting manifold geometry in hyperspectral imagery. IEEE Trans. Geosci. Remote Sens. **43**, 441–454 (2005)
4. Tenenbaum, J., Silva, V., Langford, J.: A global geometric framework for nonlinear dimensionality reduction. Science **290**, 2319–2323 (2000)
5. Schölkopf, B., Smola, A., Müller, K.: Nonlinear component analysis as a kernel eigenvalue problem. Neural Comput. **10**, 1299–1319 (1998)
6. Roweis, S., Saul, L.: Nonlinear dimensionality reduction by locally linear embedding. Science **290**, 2323–2326 (2000)
7. Zhang, Z., Zha, H.: Principal manifolds and nonlinear dimension reduction via local tangent space alignment. CoRR cs.LG/0212008 (2002)
8. Belkin, M., Niyogi, P.: Laplacian eigenmaps for dimensionality reduction and data representation. Neural Comput. **15**, 1373–1396 (2003)
9. Souvenir, R., Pless, R.: Manifold clustering. In: Tenth IEEE International Conference on Computer Vision, ICCV 2005, vol. 1, pp. 648–653. IEEE (2005)
10. Fan, M., Qiao, H., Zhang, B., Zhang, X.: Isometric multi-manifold learning for feature extraction. In: 2012 IEEE 12th International Conference on Data Mining, pp. 241–250 (2012)
11. Yang, W., Sun, C., Zhang, L.: A multi-manifold discriminant analysis method for image feature extraction. Pattern Recogn. **44**, 1649–1657 (2011)
12. He, X., Niyogi, P.: Locality Preserving Projections. Advances in Neural Information Processing Systems 16, pp. 153–160. MIT Press, Cambridge (2004)
13. Polito, M., Perona, P.: Grouping and Dimensionality Reduction by Locally Linear Embedding. Advances in Neural Information Processing Systems 14, pp. 1255–1262. MIT Press, Cambridge (2002)
14. Hettiarachchi, R., Peters, J.: Multi-manifold LLE learning in pattern recognition. Pattern Recogn. **48**, 2947–2960 (2015)
15. Barbará, D., Chen, P.: Using the fractal dimension to cluster datasets. In: Proceedings of the Sixth ACM SIGKDD International Conference on Knowledge Discovery and Data Mining, KDD 2000, pp. 260–264. ACM, New York (2000)
16. Gionis, A., Hinneburg, A., Papadimitriou, S., Tsaparas, P.: Dimension induced clustering. In: Proceedings of the Eleventh ACM SIGKDD International Conference on Knowledge Discovery in Data Mining, KDD 2005, pp. 51–60. ACM, New York (2005)
17. Haro, G., Randall, G., Guillermo, S.: Stratification learning: detecting mixed density and dimensionality in high dimensional point clouds. In: Advances in Neural Information Processing Systems 19, pp. 553–560. MIT Press (2007). http://papers.nips.cc/paper/3015-stratification-learning-detecting-mixed-density-and-dimensionality-in-high-dimensional-point-clouds.pdf
18. Guo, Q., Li, H., Chen, W., Shen, I., Parkkinen, J.: Manifold clustering via energy minimization. In: Sixth International Conference on Machine Learning and Applications, ICMLA 2007, pp. 375–380 (2007)

19. Elhamifar, E., Vidal, R.: Sparse manifold clustering and embedding. In: Shawe-Taylor, J., Zemel, R., Bartlett, P., Pereira, F., Weinberger, K. (eds.) Advances in Neural Information Processing Systems 24, pp. 55–63. Curran Associates, Inc. (2011)
20. Arias-Castro, E., Chen, G., Lerman, G.: Spectral clustering based on local linear approximations. Electron. J. Stat. **5**, 1537–1587 (2011)
21. Goldberg, A.B., Zhu, X., Singh, A., Xu, Z., Nowak, R.D.: Multi-manifold semi-supervised learning. In: AISTATS, pp. 169–176 (2009)
22. Wang, Y., Jiang, Y., Wu, Y., Zhou, Z.: Spectral clustering on multiple manifolds. IEEE Trans. Neural Netw. **22**, 1149–1161 (2011)
23. Gong, D., Zhao, S., Medioni, G.: Robust multiple manifolds structure learning. CoRR abs/1206.4624 (2012)
24. Arias-Castro, E., Lerman, G., Zhang, T.: Spectral clustering based on local PCA. arXiv preprint arXiv:1301.2007 (2013)
25. Deutsch, S., Medioni, G.: Intersecting manifolds: detection, segmentation, and labeling. In: Proceedings of the 24th International Conference on Artificial Intelligence, IJCAI 2015, pp. 3445–3452. AAAI Press (2015)
26. Hassanzadeh, A., Kauranne, T., Kaarna, A.: A multi-manifold clustering algorithm for hyperspectral remote sensing imagery. In: 2016 IEEE International Geoscience and Remote Sensing Symposium (IGARSS), pp. 3326–3329 (2016)
27. Rifai, S., Vincent, P., Muller, X., Glorot, X., Bengio, Y.: Contractive auto-encoders: explicit invariance during feature extraction. In: Proceedings of the 28th International Conference on Machine Learning (ICML-2011), pp. 833–840 (2011)
28. Rifai, S., Mesnil, G., Vincent, P., Muller, X., Bengio, Y., Dauphin, Y., Glorot, X.: Higher order contractive auto-encoder. In: Gunopulos, D., Hofmann, T., Malerba, D., Vazirgiannis, M. (eds.) ECML PKDD 2011. LNCS, vol. 6912, pp. 645–660. Springer, Heidelberg (2011). doi:10.1007/978-3-642-23783-6_41
29. Rifai, S., Dauphin, Y.N., Vincent, P., Bengio, Y., Muller, X.: The manifold tangent classifier. In Shawe-Taylor, J., Zemel, R.S., Bartlett, P.L., Pereira, F., Weinberger, K.Q. (eds.) Advances in Neural Information Processing Systems 24, pp. 2294–2302. Curran Associates, Inc. (2011)

A Clustering Approach to Heterogeneous Change Detection

Luigi Tommaso Luppino[1]([⊠]), Stian Normann Anfinsen[1], Gabriele Moser[2],
Robert Jenssen[1], Filippo Maria Bianchi[1] [iD], Sebastiano Serpico[2],
and Gregoire Mercier[3]

[1] Machine Learning Group, University of Tromsø, Tromsø, Norway
luigi.t.luppino@uit.no
[2] DITEN Department, University of Genoa, Genoa, Italy
[3] Dpt. Image et Traitement Information, Telecom Bretagne, Brest, France

Abstract. Change detection in heterogeneous multitemporal satellite
images is a challenging and still not much studied topic in remote sensing and earth observation. This paper focuses on comparison of image
pairs covering the same geographical area and acquired by two different sensors, one optical radiometer and one synthetic aperture radar, at
two different times. We propose a clustering-based technique to detect
changes, identified as clusters that split or merge in the different images.
To evaluate potentials and limitations of our method, we perform experiments on real data. Preliminary results confirm the relationship between
splits and merges of clusters and the occurrence of changes. However, it
becomes evident that it is necessary to incorporate prior, ancillary, or
application-specific information to improve the interpretation of clustering results and to identify unambiguously the areas of change.

Keywords: Domain adaptation · Heterogeneous image sources ·
Change detection · Clustering

1 Introduction

Change detection systems provide crucial information for damage assessment
after natural disasters such as floodings, earthquakes, landslides, or to detect
long-term trends in land usage, urban development, glacier dynamics, deforestation, and desertification [1–7]. In the last years, thanks to the development of heterogeneous or multimodal change detection methods, it was possible to relax the
assumption of homogeneous and co-calibrated measurements. However, despite
its undeniable potential, there is still a limited amount of research on heterogeneous change detection in the fields of computer vision, pattern recognition
and machine learning. In [8], copula theory is exploited to build local models
of dependence between unchanged areas in heterogeneous images and to link
their statistical distributions. In [9], joint distributions of heterogeneous images
are obtained by transforming their marginal densities in meta-Gaussian distributions, which provide simple and efficient models of multitemporal correlations.

© Springer International Publishing AG 2017
P. Sharma and F.M. Bianchi (Eds.): SCIA 2017, Part II, LNCS 10270, pp. 181–192, 2017.
DOI: 10.1007/978-3-319-59129-2_16

In [10,11], a method based on evidence theory is proposed, which fuses clustering maps of the individual heterogeneous images and then detects "change" and "no-change" classes, from the transition probabilities between clusters. In [12], the physical properties of the considered sensors and, especially, the associated measurement noise models and local joint distributions are exploited to define a "no-change" manifold.

The capability of processing data from heterogeneous sources in the same application opens for usage of a much larger amount of information. With respect to time series, the temporal resolution can be increased and the overall time window can be extended. Nonetheless, new issues arise. Different sensors are sensitive to distinct physical conditions and comparing their measurements may produce false detections, due to inconsistencies in sensor behaviour rather than actual changes in the monitored entities. As the complexity of the fused data set increases, there could be a requirement for more flexible and complicated statistical models, which are harder to fit on data, they may be characterized by larger uncertainty in the parameter estimation and a higher computational cost. Finally, detecting and characterizing changes in heterogeneous images is not as trivial as in the homogeneous case, where a change corresponds simply to a difference in the signal values.

In this work, we propose a novel cluster-based approach for change detection in heterogeneous data. We design an unsupervised method to be as general as possible, i.e. application-independent. The proposed method processes pairs of images, acquired at different times from different sensors. In particular, one image comes from an optical sensor, whereas the second is a synthetic aperture radar (SAR) image. The images must be co-registered by a pre-processing step, to avoid that spatial misalignment of the images is misclassified as a change. Moreover, a third type of images is considered, whose elements are obtained by stacking optical and SAR images. A clustering method is executed independently on each of the three data sets. Then, the clusters identified in the first two data sets are matched against the ones from the third data set, in order to determine if the clusters from the first image split or merge in the second image. We associate changes to the occurrence of such modifications.

In this preliminary study, the problem has been defined, a possible solution has been suggested and experiments have been performed to assess the capability of the proposed methodology. Making the whole process automatic is the following step, which will be treated in a further extension of this work.

2   Background

This work leverages on the information delivered by distance-based clustering analysis on image data. To select the proper distance measures, we first need to identify the correct statistical models to represent the data. Since we process optical and SAR images, we consider only models commonly used when dealing with these two specific data.

A simple probability distribution that describes well the optical images is the Gaussian distribution [13,14]. Specifically, a sensor with n channels yields feature vectors $\boldsymbol{x}_{opt} \in \mathbb{R}^n$, which are modelled by a multivariate Gaussian probability density function (pdf)

$$f(\boldsymbol{x}_{opt}|\boldsymbol{\mu}_i, \boldsymbol{\Sigma}_i) = \frac{1}{(2\pi)^{n/2}|\boldsymbol{\Sigma}_i|^{1/2}} \exp\left(-\frac{1}{2}(\boldsymbol{x}_{opt} - \boldsymbol{\mu}_i)^t \boldsymbol{\Sigma}^{-1}(\boldsymbol{x}_{opt} - \boldsymbol{\mu}_i)\right),$$

which compactly reads as $\boldsymbol{x}_{opt}|\omega_i \sim N(\boldsymbol{\mu}_i, \boldsymbol{\Sigma}_i)$. Here $\boldsymbol{\mu}_i$ and $\boldsymbol{\Sigma}_i$ are the mean vector and the covariance matrix associated to cluster ω_i, respectively.

Concerning SAR images in single polarisation, using the gamma distribution is a simplistic, yet effective option [15]:

$$f(x_{SAR}|\theta_i, L) = \frac{1}{\theta_i \Gamma(L)} \left(\frac{x_{SAR}}{\theta_i}\right)^{L-1} \exp\left(-\frac{x_{SAR}}{\theta_i}\right).$$

This is denoted by $x_{SAR}|\omega_i \sim \Gamma(\theta_i, L)$. $\Gamma(L)$ is the gamma function, while L and θ_i are the shape and the scale parameter, respectively. Since L (the number of looks) is the same for all the clusters, these can be fully characterised by their mean $\mu_i = L\theta_i$.

The log-normal distribution is an alternative to the gamma pdf. It fits data reasonably well under most circumstances and, contrarily to the gamma pdf, it allows to model heavy-tailed SAR intensity data [15]. A positive-valued random variable $X|\omega_i = e^Y$ follows a log-normal distribution if $Y|\omega_i = \log(X) \sim N(\mu_i, \sigma_i)$. The pdf reads

$$f(X|\mu_i, \sigma_i) = \frac{1}{X\sqrt{2\pi\sigma_i^2}} \exp\left(-\frac{(\log(X) - \mu_i)^2}{2\sigma_i^2}\right),$$

denoted by $X|\omega_i \sim logN(\mu_i, \sigma_i)$. The first two moments of random variables X and Y are related according to

$$\mu_{X|\omega_i} = \exp\left(\mu_i + \frac{\sigma_i^2}{2}\right), \quad \sigma_{X|\omega_i}^2 = \mu_i^2\left(e^{\sigma_i^2} - 1\right).$$

To conclude, if the statistical behaviour of a SAR image can be described by log-normal distributions, then a logarithmically transformed image can be modelled by a Gaussian distribution. This property will be useful to process the stacked data $\boldsymbol{x}_{st}$, which combines all features of the optical and the SAR image into one stacked feature vector, associated to each pixel.

As distance measures, we use Mahalanobis distance [16] for multivariate Gaussian distributed data and Hellinger distance [17] for gamma distributed data. A notorious drawback in cluster methods is the dependence of their results to initial conditions, such as initialization of cluster centers and ordering of the data. Additionally, the desired number of clusters or the scale parameter (used in methods such as hierarchical or density-based clustering) is often unknown. Ensemble clustering methods tackle these issues, by providing more stable results at the cost of higher computational complexity [18–20]. Ensemble methods can identify clusters of nontrivial shape and with different densities, handle noise and outliers, and they provide an estimate to the optimal number of clusters. In our case, such a number is unknown and, therefore, we perform cluster analysis with an ensemble approach based on Fuzzy C-Means (FCM) [21–23]. The ensemble procedure consists in repeating several times the FCM initialized with k different number of clusters, which each time is drawn from a uniform discrete distribution. FCM is implemented with the distance measures mentioned above. The FCM algorithm represents an iterative approach, where at each iteration a partition matrix U is returned as output. The membership values μ_{ij} contained in U are exploited to evaluate the covariance matrix of each cluster as:

$$\boldsymbol{\Sigma}_i = \frac{\sum\limits_{j=1}^{N} \mu_{ij}(\boldsymbol{x}_j - \boldsymbol{c}_i)(\boldsymbol{x}_j - \boldsymbol{c}_i)^t}{\sum\limits_{j=1}^{N} \mu_{ij}}, \quad i = 1, \ldots, k.$$

When multivariate Gaussian distributed data are involved, the Mahalanobis distance computed in the following iteration employs these updated covariance matrices. In a possible future development, we plan to examine the partition matrix to identify the most reliable clustering results, in order to improve the post-clustering analysis.

3   Recognition of Cluster Splits and Merges

Given two heterogeneous images of the same geographical area captured respectively at times t_1 and t_2, we want to detect if a change occurred during the time lapse. Each image is clustered by using the distance measures that captures its statistical properties. The clustering ensemble procedure on each image provides the partitions

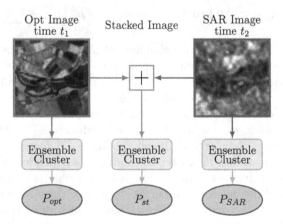

Fig. 1. First step of the proposed methodology: obtainment of the stacked image and of the three partitionings.

$$P_{opt} = c_{opt}^{(1)} \cup \quad \ldots \quad \cup c_{opt}^{(N_{opt})}$$

$$P_{SAR} = c_{SAR}^{(1)} \cup \quad \ldots \quad \cup c_{SAR}^{(N_{SAR})}$$

where $N_{opt} = |P_{opt}|$ and $N_{SAR} = |P_{SAR}|$ are the number of clusters in each partition. Then, if the SAR data x_{SAR} are assumed to follow a log-normal distribution, the logarithm of their intensities can be modelled by Gaussian pdfs. Since also the optical data x_{opt} are modelled by Gaussian pdfs, the stacked vector $x_{st} = [x_{opt}, \log(x_{SAR})]$ could be thought of a realization of a multivariate Gaussian random variable. Accordingly, we compute a third partition P_{st} on the stacked data, as shown in Fig. 1. Once the three partitions are obtained, we check whether a cluster from the image at time t_1 splits into two or more clusters in the stacked image, or whether two or more clusters from the stacked image may merge into one cluster of the image at time t_2. Instead of comparing directly the clusters from time t_1 and time t_2, with our method we leverage the information contained in the covariance matrix of the stacked image, which captures the cross-correlation between the original images. Moreover it may provide a regularization that filters out the effect of the speckle noise on the clustering results. The proposed methodology is depicted in Fig. 2.

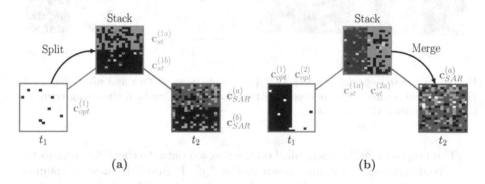

Fig. 2. Proposed methodology: it is possible to recognize changes as splits (a) and merges (b) by the comparison with the partitioning of the stacked image.

In Fig. 2(a), a region in the optical image at time t_1 is fully contained in a cluster $\mathbf{c}_{opt}^{(1)}$. The same region, is divided in two clusters, $\mathbf{c}_{st}^{(1a)}$ and $\mathbf{c}_{st}^{(1b)}$, in the stacked image. This means that in the SAR image at t_2 the region is split in two clusters as well, $\mathbf{c}_{SAR}^{(a)}$ and $\mathbf{c}_{SAR}^{(b)}$. This denotes that a change occurred in the time lapse $t_2 - t_1$. In Fig. 2(b) instead, we can see the region that in the stacked image corresponds to two clusters $\mathbf{c}_{st}^{(1a)}$ and $\mathbf{c}_{st}^{(2a)}$, merges into a single cluster $\mathbf{c}_{SAR}^{(a)}$ in the SAR image at time t_2. This indicates another type of change from t_1, where the region is characterized by two clusters $\mathbf{c}_{opt}^{(1)}$ and $\mathbf{c}_{opt}^{(2)}$ in the optical image.

4   Experiments and Results

In this section the proposed approach is applied, showing the potential and limitations of ensemble clustering and of an analysis of splits and merges. Consider the toy example in which two distinct ground cover classes are arranged as vertical stripes at time t_1 (Fig. 3(a)). At time t_2, after the occurrence of a change event, they are arranged as horizontal stripes (Fig. 3(b)). This case encompasses all the possible transitions between the two classes: (i) class 1 is unchanged (ii) Transition from class 2 to class 1 (iii) transition from class 1 to class 2 (iv) class 2 is unchanged.

The image in Fig. 3(a) emulates an optical acquisition with a single spectral channel. It is a field of uncorrelated Gaussian variables with mean $\mu_1 = 8$ for the left stripe, mean $\mu_2 = 12$ for the right stripe, and standard deviation $\sigma_1 = 1$ for both. The image in Fig. 3(b) is a plausible SAR acquisition generated as a field of uncorrelated gamma variables with mean $\mu_1 = 8$ for the upper stripe and mean $\mu_2 = 12$ for the lower stripe. The number of looks $L_1 = L_2 = 9$ for both clusters.

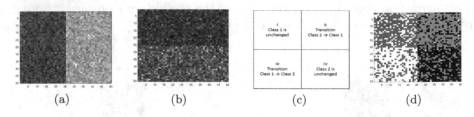

(a)   (b)   (c)   (d)

Fig. 3. Simulated optical (a) and SAR (b) acquisitions, before and after the event. Four possible combinations between the two classes (c). Common clustering result on the stacked data (d).

The proposed method is applied on the stacked data, to check if it reveals the same configuration as the one shown in Fig. 3(c). Figure 3(d) shows a common output obtained on this toy data set. For each corner of Fig. 3(d), most of the pixels are clustered together, whereas the ones clustered differently count as errors. The accuracy is calculated as $1 - \frac{\#\ of\ errors}{\#\ of\ pixels}$. Fifty experiment were done for each of the nine combinations of noise strength. The average of the accuracy is presented in Table 1.

Despite of the correlation with the state of the art, it is not possible to compare directly these results with the one found in the literature: first of all, this problem is faced in a different manner and in a different framework. Most importantly, the presented method is completely unsupervised, for which there is no similar work for comparison. However, what can be said is that the proposed approach is able to detect the behaviour of splits and merges between the clusters, when the noise remains under a reasonable level (e.g. $\sigma = 1$, $L = 9$).

The method is also applied to real satellite acquisitions. The images in Fig. 4 represent the countryside at the periphery of Gloucester, Gloucestershire, United

Table 1. Accuracy of the clustering result of the stacked data for different levels of noise in the two images.

	$L = 16$	$L = 9$	$L = 4$
$\sigma = 1$	78.0%	73.7%	67.3%
$\sigma = 2.5$	60.5%	57.6%	54.1%
$\sigma = 5$	50.7%	49.0%	46.5%

(a) Sept. the 5th, 1999   (b) Oct. the 21st, 2000   (c) Ground truth

Fig. 4. Gloucester before – optical image (a) – and after a flooding the event – SAR image (b). In (c), the ground truth of the change.

Kingdom, before and after a flood. Since the speckle noise affecting the latter was too strong, a 7-by-7 enhanced Lee filter [24] was applied to attenuate the noise, while preserving the details contained in heterogeneous areas. The analysis is carried out on the presented images by dividing them into smaller and non-overlapping windows of 50×50 pixels, and then by looking for changes inside them separately. In this way, it can be reasonably thought that pixels can be grouped into a limited number of clusters, making the clustering process easier and more accurate regardless of the spatial nonstationarity of the image data. Processing smaller windows also reduces the computational cost, which scales quadratically with the windows size and the number of clusters. The FCM algorithm has been iterated 20 times, drawing a different number of clusters each time from the uniform probability mass function U [4, 7].

4.1 First Experiment

The region selected for the first experiment is shown in Fig. 5. It contains some agricultural fields and a river in the lower part of it. As seen in Fig. 5(d), clusters relative to different parts of the image are very well separated.

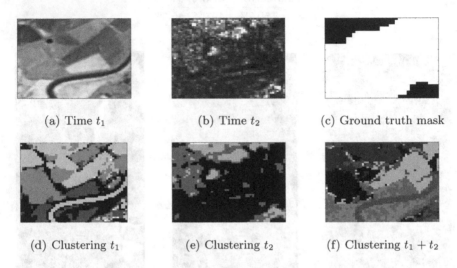

(a) Time t_1 (b) Time t_2 (c) Ground truth mask

(d) Clustering t_1 (e) Clustering t_2 (f) Clustering $t_1 + t_2$

Fig. 5. First experiment: (a) window of the optical image, (b) window of the SAR image, (c) window of the ground truth mask, (d) clustering result on image a, (e) clustering result on image b, (f) clustering result on the stacked image.

Concerning the SAR acquisition, from Fig. 5(b) we observe that the flooded area covers the majority of the window. Such area is correctly identified by the large black cluster in Fig. 5(e). Comparing the three clustering results, two clear examples of clusters merging and clusters splitting are spotted. The big cluster representing some fields, that from the upper part of the optical image goes down to the right, has split into two different clusters in the SAR image (the light grey one and the black one), and this is highlighted by the presence of the light grey and the dark grey clusters in Fig. 5(f). Then, the dominant cluster of Fig. 5(e) is the result of the merging of some clusters of the optical image, i.e. the white cluster (the river), the dark grey cluster (some fields close to the river), a good percentage of the black cluster (the boundaries around the river and the fields) and one part of the above mentioned big cluster which has split. All these clusters are visible in the result obtained with the stacked data, and they are respectively: the dark grey cluster (the river), the grey cluster close to it (the fields close to the river and the boundaries) and the light grey cluster (the part of the splitting).

4.2   Second Experiment

The region selected for the second experiment is displayed in Fig. 6. In this case, the different areas are not well separated (Fig. 6(d)), especially in the center and in the lower right corner of the window, mainly because these parts of the image present miscellaneous ground covers. For example, some of the central pixels in Fig. 6(a) look darker, so the clustering algorithm erroneously cluster them together with the ones belonging to the river, as it happened in the first experiment. Instead, the bare soil field presents some brighter pixels close to the river and some darker pixels far from it, and these two groups are divided. Moving on to the image in Fig. 6(b), it can be seen how it looks still noisy and muddled, even after being filtered. Consequently, the clustering in Fig. 6(e) does not yields the same quality of the first experiment. Making a comparison with Fig. 6(c), more accurate delineation of changed areas would have been emphasized if the grey and black classes were grouped together and, most importantly, some of the agricultural fields in the lower right corner were grouped differently. But this is not a fault of the ensemble clustering, as these last areas are very similar to the flooded portion of the region, due to the characteristics of the specific kind of field and its SAR signature. Recognizing the flooded area in Fig. 6(b) by visual inspection and without prior knowledge, is also very difficult. It is worth noting that the available ground truth itself is only partially accurate, because for example the sharp edge on the right side is unlikely, but it still gives an idea about the location of the affected areas.

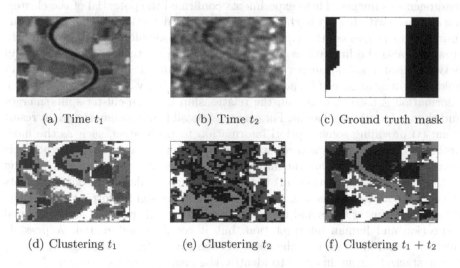

(a) Time t_1 (b) Time t_2 (c) Ground truth mask

(d) Clustering t_1 (e) Clustering t_2 (f) Clustering $t_1 + t_2$

Fig. 6. Second experiment: (a) window of the optical image, (b) window of the SAR image, (c) window of the ground truth mask, (d) clustering result on image a, (e) clustering result on image b, (f) clustering result on the stacked image.

The quality of the partitioning in Fig. 6(f) is heavily influenced by the speckle noise, which is a fundamental issue in the field of SAR data analysis. Under these conditions, it is not trivial to recognise splits and, most of all, merges, due to the amount of noise in the SAR image at time t_2. This case study highlights that an approach for change detection from an optical and a SAR image based on cluster splits and merges is limited by the clustering results. These latter are affected, in turn, by the characteristics of the input data (noise ratio, contrast, etc.), by the adopted clustering algorithm, and by the selection of its hyperparameters.

5   Conclusions and Future Works

In this paper, we studied the challenging problem of change detection in multitemporal and heterogeneous images, by means of a cluster-based techniques. Our study focused on the case of image pairs, relative to the same area, captured by heterogeneous sensors at different times. We evaluated to which extent a completely unsupervised approach could be successful in addressing change detection from heterogeneous image sources. The proposed idea is that changes on the ground can be related to clusters of the image at time t_1 splitting and/or merging into the clusters of the image at time t_2. The possibility to model SAR intensity as log-normally distributed and optical data as Gaussian allowed us to apply a multivariate Gaussian model in the joint domain of the optical channels and of the log-transformed SAR data. This allowed us to also apply the clustering algorithm on a stack of the images, to improve the chances of identify splits and merges. Experimental results were obtained on a toy example and on real satellite heterogeneous images. These experiments confirmed the potential of the clustering approach with respect to the problem of change detection from heterogeneous sources and suggested the effectiveness of the ensemble clustering approach. However, also the limitations were highlighted. In particular, the relationship between cluster splits/merges and changed/unchanged areas does not always hold. This limitation can be addressed if prior, ancillary, or application-specific information is used to constrain the relationship between cluster splits/merges and changed/unchanged areas. For example, possible improvements might result from: (i) providing some a priori information to the system, such as the most probable changing parts according to their position (ii) introducing hypothesis that the changing parts are the majority or the minority of the image, according to the particular application (iii) to indicate the particular class that represents the sought changed areas, e.g. water for floods, bare soil for forest res, etc. The detection of cluster splits and merges carried out in this work is based on visual inspection and human interpretation, but it could be automated. A possible solution would be to overlap the mask of each cluster from the image at time t_1 to the stacked image, in order to identify the areas where splits occur. Accordingly, this procedure could be applied to the clusters of the image at time t_2 to recognise merges. Once splits and merges are identified, one could rely on prior information (if available) to improve the accuracy of change detection. For example, if the location of the river in the image is provided, one could focus the search for flooded areas with in the clusters close to its position. Alternatively, one could leverage on the statistical characteristics of water in SAR images to identify the areas of interest. On one hand, this approach would provide an automatic tool to improve clusters interpretation and to identify relevant splits and merge, associated with changes of interest. On the other hand, the necessity of prior information confirms the extreme difficulty of performing automatic and unsupervised change detection in heterogeneous data.

A significant improvement is also expected if polarimetric SAR images are used instead of SAR acquisitions with only one polarisation. This is because they bring a lot more intrinsic information which would enhance the capability of clustering results to identify natural classes in feature spaces associated with SAR observations. Obviously, this would force to consider different and more complicated models and distance measures. The parallelisation of the proposed approach, which is favored by its window-based formulation and would benefit of current cluster or GPU-based architectures, represents another possible and interesting future development. Last, but not least, the research can be extended to the multitemporal case in which more than two images are considered, exploiting the proposed method for the analysis of long term trends such as deforestation, glacier dynamics, desertification, land use change and urban development.

References

1. Serpico, S.B., Dellepiane, S., Boni, G., Moser, G., Angiati, E., Rudari, R.: Information extraction from remote sensing images for flood monitoring and damage evaluation. Proc. IEEE **100**(10), 2946–2970 (2012)
2. Dell'Acqua, F., Gamba, P.: Remote sensing and earthquake damage assessment: experiences, limits, and perspectives. Proc. IEEE **100**(10), 2876–2890 (2012)
3. Guan, N., Yin, J., Li, C., Lei, M., Zhang, M.: Landslide recognition in remote sensing image based on fuzzy support vector machine. In: 2012 IEEE 12th International Conference on Computer and Information Technology (CIT), pp. 1103–1108. IEEE (2012)
4. Du, M., Cai, G.: Spatial and temporal dynamics of land cover in Beijing. In: 2012 IEEE International Geoscience and Remote Sensing Symposium (IGARSS), pp. 6333–6336. IEEE (2012)
5. Wu, X., Lu, A., Lihong, W., Pu, J., Huawei, Z., Haigang, T.: Glacier change along Wusun road in Chinese Tien Shan during 1973–2007 monitored by remote sensing. In: 2011 International Conference on Remote Sensing, Environment and Transportation Engineering (RSETE), pp. 1252–1256. IEEE (2011)
6. Latif, Z.A., Zaqwan, H.M., Saufi, M., Adnan, N.A., Omar, H.: Deforestation and carbon loss estimation at tropical forest using multispectral remote sensing: case study of Besul Tambahan permanent forest reserve. In: 2015 International Conference on Space Science and Communication (IconSpace), pp. 348–351. IEEE (2015)
7. Wu, J., Xia, H., Liu, Y.: Theory and methodology on monitoring and assessment of desertification by remote sensing. In: Proceedings of 2004 IEEE International Geoscience and Remote Sensing Symposium, IGARSS 2004, vol. 4, pp. 2302–2305. IEEE (2004)
8. Mercier, G., Moser, G., Serpico, S.: Conditional copula for change detection on heterogeneous SAR data. In: IEEE International Geoscience and Remote Sensing Symposium, IGARSS 2007, pp. 2394–2397. IEEE (2007)

9. Storvik, B., Storvik, G., Fjortoft, R.: On the combination of multisensor data using meta-gaussian distributions. IEEE Trans. Geosci. Remote Sens. **47**(7), 2372–2379 (2009)

10. Liu, Z., Dezert, J., Mercier, G., Pan, Q.: Dynamic evidential reasoning for change detection in remote sensing images. IEEE Trans. Geosci. Remote Sens. **50**(5), 1955–1967 (2012)

11. Liu, Z., Mercier, G., Dezert, J., Pan, Q.: Change detection in heterogeneous remote sensing images based on multidimensional evidential reasoning. IEEE Geosci. Remote Sens. Lett. **11**(1), 168–172 (2014)

12. Prendes, J., Chabert, M., Pascal, F., Giros, A., Tourneret, J.-Y.: A new multivariate statistical model for change detection in images acquired by homogeneous and heterogeneous sensors. IEEE Trans. Image Process. **24**(3), 799–812 (2015)

13. Bovolo, F., Bruzzone, L.: The time variable in data fusion: a change detection perspective. IEEE Geosci. Remote Sens. Mag. **3**(3), 8–26 (2015)

14. Goudail, F., Réfrégier, P.: Contrast definition for optical coherent polarimetric images. IEEE Trans. Pattern Anal. Mach. Intell. **26**(7), 947–951 (2004)

15. Oliver, C., Quegan, S.: Understanding Synthetic Aperture Radar Images. SciTech Publishing, USA (2004)

16. Basseville, M.: Distance measures for signal processing and pattern recognition. Sig. Process. **18**(4), 349–369 (1989)

17. Frery, A.C., Nascimento, A.D.C., Cintra, R.J.: Analytic expressions for stochastic distances between relaxed complex wishart distributions. IEEE Trans. Geosci. Remote Sens. **52**(2), 1213–1226 (2014)

18. Strehl, A., Ghosh, J.: Cluster ensembles-a knowledge reuse framework for combining multiple partitions. J. Mach. Learn. Res. **3**, 583–617 (2002)

19. Fred, A.L.N., Jain, A.K.: Combining multiple clusterings using evidence accumulation. IEEE Trans. Pattern Anal. Mach. Intell. **27**(6), 835–850 (2005)

20. Ghosh, J., Acharya, A.: Cluster ensembles. Wiley Interdisc. Rev.: Data Min. Knowl. Discov. **1**(4), 305–315 (2011)

21. Ghosh, S., Mishra, N.S., Ghosh, A.: Unsupervised change detection of remotely sensed images using fuzzy clustering. In: Seventh International Conference on Advances in Pattern Recognition, ICAPR 2009, pp. 385–388. IEEE (2009)

22. Singh, K.K., Mehrotra, A., Nigam, M.J., Pal, K.: Unsupervised change detection from remote sensing images using hybrid genetic FCM. In: 2013 Students Conference on Engineering and Systems (SCES), pp. 1–5. IEEE (2013)

23. Keller, J.M., Gray, M.R., Givens, J.A.: A fuzzy k-nearest neighbor algorithm. IEEE Trans. Syst. Man Cybern. **4**, 580–585 (1985)

24. Lopes, A., Touzi, R., Nezry, E.: Adaptive speckle filters and scene heterogeneity. IEEE Trans. Geosci. Remote Sens. **28**(6), 992–1000 (1990)

Large-Scale Mapping of Small Roads in Lidar Images Using Deep Convolutional Neural Networks

Arnt-Børre Salberg[1]([✉]), Øivind Due Trier[1], and Michael Kampffmeyer[2]

[1] Norwegian Computing Center, PO-Box 114, Blindern, 0314 Oslo, Norway
{salberg,trier}@nr.no
[2] UiT - The Arctic University of Norway, 9037 Tromsø, Norway
michael.c.kampffmeyer@uit.no

Abstract. Detailed and complete mapping of forest roads is important for the forest industry since they are used for timber transport by trucks with long trailers. This paper proposes a new automatic method for large-scale mapping forest roads from airborne laser scanning data. The method is based on a fully convolutional neural network that performs end-to-end segmentation. To train the network, a large set of image patches with corresponding road label information are applied. The final network is then applied to detect and map forest roads from lidar data covering the Etnedal municipality in Norway. The results show that we are able to map the forest roads with an overall accuracy of 97.2%. We conclude that the method has a strong potential for large-scale operational mapping of forest roads.

Keywords: Deep learning · Convolutional neural networks · Lidar · Remote sensing

1 Introduction

In 2015, Norway officially decided to collect airborne laser scanning (ALS) data for the entire land area below the timber line. The point density will be at least two first returns per square metre, with the main purpose to obtain a very detailed digital terrain model (DTM) of the entire country. For open areas above the tree line, i.e., in the mountains, the DTM will be based on automatic image matching from aerial photography.

The national coverage of ALS data provides large opportunities for new mapping products, e.g. maps of small roads like forest roads that are difficult to observe in optical remote sensing images. Forest roads are used for timber transport by trucks with long trailers, and due to the forest industry's demands for profitable management, accurate, detailed and complete mapping of forest roads is important.

Remote sensing imagery is often characterized by complex data properties in the form of heterogeneity and class imbalance, as well as overlapping class-conditional distributions [3]. Together, these aspects constitute severe challenges

© Springer International Publishing AG 2017
P. Sharma and F.M. Bianchi (Eds.): SCIA 2017, Part II, LNCS 10270, pp. 193–204, 2017.
DOI: 10.1007/978-3-319-59129-2_17

for creating land cover maps or detecting and localizing objects, producing a high degree of uncertainty in obtained results, even for the best performing model [13,16].

Automatic detection and mapping of road networks from remote sensing data has previously been studied by several authors [7,22], however, most of the work focus on optical data [22], and current state-of-the art algorithms fail to extract roads in optical images for cases where surrounding objects like water, buildings, trees, grass and cars occlude the road or cast shadows, especially with influence of spatial structures such as overpasses [22].

In recent years, deep convolutional neural networks (deep CNNs) have emerged as the leading modelling tools for image pixel classification and segmentation in general [8,14], and have had an increasing impact also in remote sensing [12,13,16,17,19]. This increasing interest is reflected for example in the ISPRS semantic segmentation challenge [11], where deep CNNs are dominating and are shown to provide the best performing models.

In practice, there are currently two main approaches to performing image segmentation using CNNs. The first one, which we refer to as patch-based, relies on predicting every pixel in the image by looking at the enclosing region of the pixel. This is commonly done by training a classifier on small image patches and then either classify all pixels using a sliding window approach, or more efficiently, converting the fully connected layers to convolutional layers [20], thereby avoiding overlapping computations. Further improvements may be achieved by using multi-scale approaches or by iteratively improving the results in a recurrent CNN [6,18].

The second approach is based on the idea of pixel-to-pixel semantic segmentation using end-to-end learning [14]. It uses the idea of a fully convolutional network (FCN), consisting of an encoder and a decoder. The encoder is responsible for mapping the image to a low resolution representation, whereas the decoder provides a mapping from the low resolution representation to the pixelwise predictions. Up-sampling is achieved using fractional-strided convolutions [14]. This approach has recently improved the state-of-the art performance on many image tasks and, due to the lack of fully-connected layers, allows pixel-wise predictions for arbitrary image sizes.

In this paper we build upon the work by Mnih and Hinton [15], who applied a neural network to detect roads in very high resolution optical remote sensing data, and hypothesize that we can train a deep convolutional neural network and apply it to perform automatic mapping of small roads in lidar data. To address the hypothesis we rely on state-of-the-art fully convolutional neural networks, tailored to perform semantic mapping [14], also with good results on remote sensing images [12].

2   Data

ALS data of the majority of Etnedal municipality, Oppland County, Norway, have been captured with an average of 6.5 ground hits per m^2. However, this varies from 0 (below dense canopies of deciduous trees) to $20/m^2$ at strip overlaps. For the entire Etnedal municipality, vector data in the form of ESRI shape files, containing the current official mapping of road centre lines and road area, are available.

2.1 Pre-processing

The ALS data consisted of point measurements (x, y, z) in UTM zone 32. Each point had a number of attributes, including:

1. Class (one of: ground, vegetation, building, other).
2. Return number, a number between 1 and 4 where 1 denotes the first return and 4 the last return.
3. Return intensity (uncalibrated radiance).

From these attributes, the following images were generated:

1. Digital terrain model (DTM) from all ALS points labelled as ground.
2. Elevation gradient of DTM, measured in degrees.
3. Digital surface model (DSM) from all ALS points labelled as first returns.

The pilot study [21] and further investigations indicated that the gradient image had the best potential for automatic detection of roads, compared with alternative representations of the ALS data. The other representations included: laser return intensity image, hill-shaded relief image, local relief image, aspect direction image, and the ALS point cloud of ground returns.

Two resolutions for the DTM, and thus, the gradient image, were evaluated. Although 0.2 m gives slightly better detail than 0.5 m whenever there are multiple ALS ground returns within a 0.5 m pixel, the increased data volume has a negative impact on the deep learning method. Smaller areas, measured in m^2, are input to the vision methods of the deep neural network, described below, meaning that the context of a road may be lost in the vision task. Also, with $6.5/m^2$ ALS ground point density on average, not much detail is lost (on average) when reducing the resolution from 25 pixels per m^2 to 4 pixels per m^2, corresponding to 0.2 m and 0.5 m pixel sizes, respectively. Another benefit is that the number of pixels is reduced by a factor of 6.25.

The gradient image of Etnedal municipality was divided into two sets, one for training and one for testing. Form each data set 256×256 image patches with 50% overlap were extracted. For the training and test dataset, only image patches that contain road segments were used. A total of 59004 images of size 256×256 pixels were available. This was divided into two equal sized datasets, one for training and one for test. 10% of the training images were used as validation data.

3 Automatic Detection of Roads

We applied the same FCN architecture as Kampffmeyer et al. [12], which allowed end-to-end learning of pixel-to-pixel semantic segmentation. The network was implemented on a graphical processing unit (GPU), in order to speed up computations. The network was trained in mini-batches on patches of 256 × 256 pixels. The patch size was chosen due to GPU memory considerations.

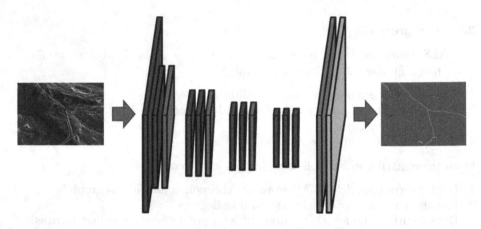

Fig. 1. Pixel-to-pixel architecture. Blue layers represent convolutional layers (including ReLU and batch-normalization layer), red layers represent pooling layers, the green layer represents the fractional-strided convolution layer and the yellow layer the softmax layer. (Color figure online)

Architecture. The CNN architecture of the FCN network (Fig. 1) consisted of four sets of two 3 × 3 convolutions (blue layers), each set separated by a 2 × 2 max pooling layer with stride 2 (red layers).

All convolution layers have a stride of 1, except the first one, which has a stride of 2. The change in the first convolution layer was a design choice, which was mainly made due to limits in GPU memory during test phase when considering large images. All convolutional layers were followed by a ReLU nonlinearity and a batch normalization layer [10]. Weights were initialized according to He et al. [9]. The final 3 × 3 convolution was followed by a 1 × 1 convolution, which consisted of one kernel for each class to produce class scores. The convolutional layers were followed by a fractional-strided convolution layer [14] (green layer in Fig. 1, sometimes also referred to as deconvolution layer), which learned to upsample the prediction back to the original image size, and a softmax layer (yellow layer in Fig. 1). The network was trained end-to-end using backpropagation.

Data Augmentation. The image patches were extracted from the input image with 50% overlap and were flipped (left to right) and rotated at 90 degree intervals, yielding 8 augmentations per image patch.

Median Frequency Balancing. Training of the FCN network was done using the cross-entropy loss function. However, as this loss was computed by summing over all the pixels, it did not account well for imbalanced classes. To take the imbalanced classes into account, the loss of the classes was weighted using median frequency balancing [1,5,12]. Median frequency balancing weights the class loss by the ratio of the median class frequency in the training set and the actual class frequency. The modified cross-entropy function is

$$L = -\frac{1}{N} \sum_{n=1}^{N} \sum_{c \in \mathcal{C}} \ell_c^{(n)} \log\left(\hat{p}_c^{(n)}\right),\tag{1}$$

where N is the number of samples in a mini-batch,

$$w_c = \frac{\text{median}\,(f_c | c \in \mathcal{C})}{f_c}\tag{2}$$

is the class weight for class c, f_c the frequency of pixels in class c, $\hat{p}_c^{(n)}$) is the softmax probability of sample n being in class c, $\ell_c^{(n)}$ corresponds to the label of sample n for class c when the label is given in one-hot encoding and $\mathcal{C}$ is the set of all classes.

3.1   Pre-processing and Post-processing

Merging Output Probabilities and Class Image. In test mode each 256×256 image was augmented by $90°$ rotation and flipping as described in Sect. 3, and sent through the CNN. The output of the CNN for a given image was a score map for each class. The score maps for each rotation and flip is rotated backwards, and merged by averaging. From the averaged score image, the class image was computed by, for each pixel, selecting the class with the largest score.

Merging of Classification Result. The neural network outputs classified images of size 256×256, based on input images of the same size. In order to avoid edge effects in the merged classification result, the input images are generated with 50% overlap between any two neighbouring images vertically or horizontally. In other words, subimages of size 256×256 pixels are generated with 128 pixels step size from the gradient image.

The classified images of size 256×256 pixels contain the values 1 (road) and 0 (background). These images are then cropped to size 128×128 pixels by removing pixels that are less than 64 pixels from the edge. These cropped images of size 128×128 pixels are then merged edge-to-edge to form the full classification map.

4   Results

The average classification accuracies of using the FCN approach to classify the validation dataset were

- Non-road: 97.2%.
- Road: 95.3%.
- Overall: 97.2%.

Please note that the overall accuracy is equal to the non-road accuracy. This is due to the high class imbalance between the road and non-road classes.

The automatic road detection method produced results that are not perfect. However, when comparing with the existing road centre lines, the automatic mapping often produced more accurate centre lines. For example, a gap in the existing tractor road centre line was closed by the automatic method (Fig. 2), the existing tractor road centre line ran outside of the tractor road at some curves, whereas the automatically generated centre lines stay inside the tractor road (Fig. 3), and there was no existing tractor road (or path) centre line at the detected location (Fig. 4).

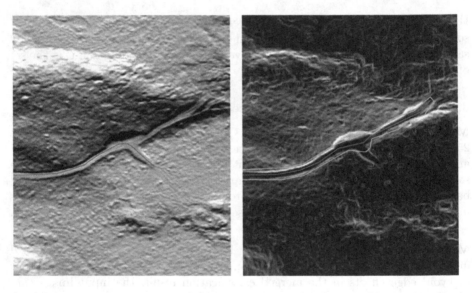

Fig. 2. There is a gap in the original centre line (left, orange) that is closed in the automatically detected centre line (right, orange) The cyan outline indicates the detected road area. (Color figure online)

There are also examples of situations where the automatic method had problems. For example, a road that is difficult to see in the gradient image (Fig. 5) may be missed. A road crossing a field (Fig. 6), may result in fragmented mapping. Some terrain features, e.g. two parallel ditches (Fig. 7) may result in a false road.

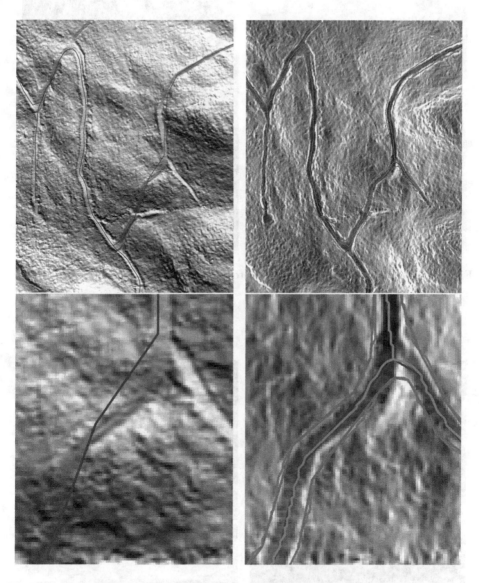

Fig. 3. Left: Existing tractor road centre line, with hill-shaded DTM. Right: road centre line and outline from automatic method, with gradient of DTM. Yellow/green: 0.25 m maximum displacement in point reduction. Orange/cyan: 1.0 m maximum displacement. (Color figure online)

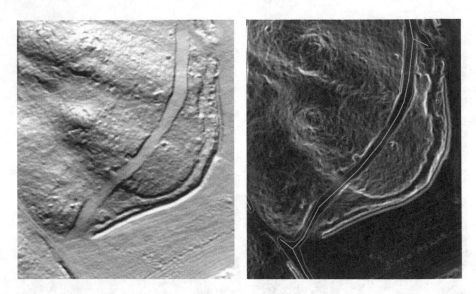

Fig. 4. A road that was mapped by the automatic method (right), but missing in the existing vector data (left).

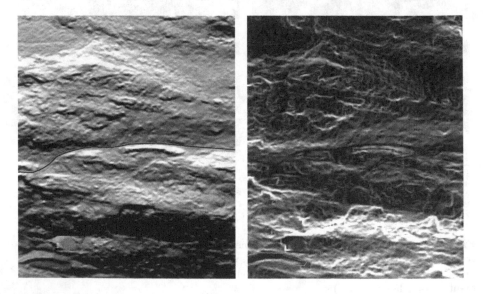

Fig. 5. Tractor road that was missed by the automatic method. Left: existing road centre line. Right: the road is difficult to see in the gradient image.

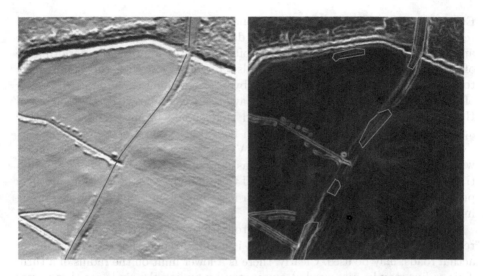

Fig. 6. Tractor road crossing a field. Left: existing centre line. Right: result of automatic detection.

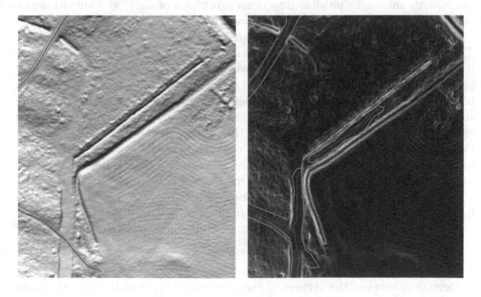

Fig. 7. False automatic detection of road.

5   Discussion

Even though the deep fully convolutional neural network provides very good results for mapping forest roads in lidar data, there is a potential to improve the approach. Adding more training data often helps to improve the performance of deep neural networks. More data also provides you with the opportunity to increase the network size and thereby it's modelling capabilities. In terms of network architectures, the topology aware FCN proposed by BenTaieb and Hamarneh [2] is one promising method that should be investigated. Another approach is to use a conditional random field (CRF) based post-processing.

Pre-trained networks, e.g. Alexnet, VGGnet or GoogleNet, in combination with fine-tuning, could also be applied as part of the FCN [14]. The use of pre-trained networks has become a standard technique in computer vision and may provide a performance gain, in particular if we have a limited number of training images.

As a post-processing step for the automatic road detection method, a point reduction method [4], or a method that is good in replicating the curvatures of actual roads, may be used. Clearly, there is a lower limit on the radius of a turn. This radius may be measured at any vertex by finding the circle arc that passes through that vertex, the preceding vertex and the succeeding vertex.

Another alternative could be to grow the skeleton image (by creating a distance map with a maximum distance limit) and then to re-create the skeleton image by thinning a thresholded distance map. This may produce a smoother skeleton image. However, the skeleton image will always result in a vectorised result with only eight possible directions (multiples of 45°), so a smoothing or point reduction of the vector data is always needed.

Training of the detection method was done on a subset of the Etnedal dataset. There is always a trade-off between training and testing. If the training data set is too small, then the method may be over-fitted on the training data and may produce bad results on other areas. E.g., if the training data only includes steep hillsides with roads with many turns, then the method may be bad at detecting straight roads in flat terrain, and vice versa. However, if the method is trained on representative parts of all of Norway, then the method may be bad at making local adaptions. So, a solution may be to run training or fine-tuning with existing road centre line data for each dataset, and then run automatic detection on the same dataset, or combine the results of a model for whole Norway with the results of from a local model. In both cases, the result may be improved centre lines in those parts of the terrain where the original centre lines were inaccurate or missing. Further, it could be interesting to compute quality metrics by comparing the new centre lines with the existing:

1. For all roads where there is an old and a new centre line, what is the average deviation between the vertices of the new centre line and the corresponding closest locations on the old centre line?
2. How many metres of new road centre lines do not match an existing road centre line?

3. How many metres of existing road centre line do not have a corresponding new road centre line?

6 Conclusions

In this paper, we have demonstrated that end-to-end segmentation using a fully convolutional neural networks provides very good results in terms of mapping forests roads in lidar data. Do to these promising results, we conclude that deep neural network methods provides a good basis for designing algorithms for large scale mapping of roads, but also other objects like e.g. cultural heritages, in lidar data.

Acknowledgments. Part of this research was financed by the Norwegian Mapping Authority, Hamar regional office, which also provided vector data. Airborne laser scanning data was provided by Oppland County Administration.

References

1. Badrinarayanan, V., Kendall, A., Cipolla, R.: Segnet: a deep convolutional encoder-decoder architecture for image segmentation (2015). arXiv preprint arXiv:1511.00561
2. BenTaieb, A., Hamarneh, G.: Topology aware fully convolutional networks for histology gland segmentation. In: Ourselin, S., Joskowicz, L., Sabuncu, M.R., Unal, G., Wells, W. (eds.) MICCAI 2016. LNCS, vol. 9901, pp. 460–468. Springer, Cham (2016). doi:10.1007/978-3-319-46723-8_53
3. Camp-Valls, G., Bruzzone, L.: Kernel Methods for Remote Sensing Data Analysis/Edited by Gustavo Camps-Valls Lorenzo Bruzzone. Wiley, Chichester (2009)
4. Douglas, D.H., Peucker, T.K.: Algorithms for the reduction of the number of points required to represent a digitized line or its caricature. Cartographica: Int. J. Geograph. Inf. Geovisualization **10**(2), 112–122 (1973)
5. Eigen, D., Fergus, R.: Predicting depth, surface normals and semantic labels with a common multi-scale convolutional architecture. In: Proceedings of IEEE International Conference on Computer Vision, pp. 2650–2658 (2015)
6. Farabet, C., Couprie, C., Najman, L., LeCun, Y.: Learning hierarchical features for scene labeling. IEEE Trans. Pattern Anal. Machine Intell. **35**(8), 1915–1929 (2013)
7. Ferraz, A., Mallet, C., Chehata, N.: Large-scale road detection in forested mountainous areas using airborne topographic lidar data. ISPRS J. Photogramm. Remote Sens. **112**, 23–36 (2016)
8. Hariharan, B., Arbeláez, P., Girshick, R., Malik, J.: Hypercolumns for object segmentation and fine-grained localization. In: Proceedings of IEEE Conference Computer Vision Pattern Recognition, pp. 447–456 (2015)
9. He, K., Zhang, X., Ren, S., Sun, J.: Delving deep into rectifiers: surpassing human-level performance on imagenet classification. In: Proceedings of IEEE International Conference on Computer Vision, pp. 1026–1034 (2015)
10. Ioffe, S., Szegedy, C.: Batch normalization: accelerating deep network training by reducing internal covariate shift (2015). arXiv preprint arXiv:1502.03167

11. ISPRS: ISPRS 2D Semantic Labeling Contest (2015). http://www2.isprs.org/commissions/comm3/wg4/semantic-labeling.html
12. Kampffmeyer, M., Salberg, A.B., Jenssen, R.: Semantic segmentation of small objects and modeling of uncertainty in Urban remote sensing images using deep convolutional neural networks. In: Proceedings of IEEE Conference on Computer Vision Pattern Recognition Workshops, pp. 1–9 (2016)
13. Lagrange, A., Saux, B.L., Beaupère, A., Boulch, A., Chan-Hon-Tong, A., Herbin, S., Randrianarivo, H., Ferecatu, M.: Benchmarking classification of earth-observation data: from learning explicit features to convolutional networks. In: 2015 IEEE International Geoscience Remote Sensing Symposium (IGARSS), pp. 4173–4176 (2015)
14. Long, J., Shelhamer, E., Darrell, T.: Fully convolutional networks for semantic segmentation. In: Proceedings of IEEE Conference on Computer Vision Pattern Recognition, pp. 3431–3440 (2015)
15. Mnih, V., Hinton, G.E.: Learning to detect roads in high-resolution aerial images. In: Daniilidis, K., Maragos, P., Paragios, N. (eds.) ECCV 2010. LNCS, vol. 6316, pp. 210–223. Springer, Heidelberg (2010). doi:10.1007/978-3-642-15567-3_16
16. Paisitkriangkrai, S., Sherrah, J., Janney, P., Hengel, A.: Effective semantic pixel labelling with convolutional networks and conditional random fields. In: Proceedings of IEEE Conference on Computer Vision and Pattern Recognition Workshops, pp. 36–43 (2015)
17. Penatti, O.A.B., Nogueira, K., dos Santos, J.A.: Do deep features generalize from everyday objects to remote sensing and aerial scenes domains? In: Proceedings of IEEE Conference on Computer Vision Pattern Recognition Workshops, pp. 44–51 (2015)
18. Pinheiro, P., Collobert, R.: Recurrent convolutional neural networks for scene parsing (2013). arXiv preprint arXiv:1306.2795
19. Salberg, A.B.: Detection of seals in remote sensing images using features extracted from deep convolutional neural networks. In: 2015 IEEE International Geoscience Remote Sensing Symposium (IGARSS), pp. 1893–1896 (2015)
20. Sermanet, P., Eigen, D., Zhang, X., Mathieu, M., Fergus, R., LeCun, Y.: OverFeat: integrated recognition, localization and detection using convolutional neural networks. In: International Conference on Learning Representations (ICLR), CBLS, Banff, Canada, April 2014
21. Trier, Ø.D.: Evaluation of methods for detection of roads in laser data - preliminary results. LasTrak pilot project (in Norwegian). NR-Note SAMBA/09/15, Norwegian Computing Center, Oslo (2015)
22. Wang, W., Yang, N., Zhang, Y., Wang, F., Cao, T., Eklund, P.: A review of road extraction from remote sensing images. J. Traffic Transp. Eng. (Engl. Ed.) 3(3), 271–282 (2016)

Physics-Aware Gaussian Processes for Earth Observation

Gustau Camps-Valls[1]([X]), Daniel H. Svendsen[1], Luca Martino[1],
Jordi Muñoz-Marí[1], Valero Laparra[1], Manuel Campos-Taberner[2],
and David Luengo[3]

[1] Image Processing Laboratory (IPL), Universitat de València, València, Spain
gustau.camps@uv.es
[2] Faculty of Physics, Universitat de València, València, Spain
[3] Signal Processing and Comunications Department,
University of Politécnica de Madrid, Madrid, Spain
http://isp.uv.es

Abstract. Earth observation from satellite sensory data pose challenging problems, where machine learning is currently a key player. In recent years, Gaussian Process (GP) regression and other kernel methods have excelled in biophysical parameter estimation tasks from space. GP regression is based on solid Bayesian statistics, and generally yield efficient and accurate parameter estimates. However, GPs are typically used for inverse modeling based on concurrent observations and *in situ* measurements only. Very often a *forward model* encoding the well-understood physical relations is available though. In this work, we review three GP models that respect and learn the physics of the underlying processes in the context of *inverse modeling*. First, we will introduce a Joint GP (JGP) model that combines *in situ* measurements and simulated data in a single GP model. Second, we present a latent force model (LFM) for GP modeling that encodes ordinary differential equations to blend data-driven modeling and physical models of the system. The LFM performs multi-output regression, adapts to the signal characteristics, is able to cope with missing data in the time series, and provides explicit latent functions that allow system analysis and evaluation. Finally, we present an Automatic Gaussian Process Emulator (AGAPE) that approximates the forward physical model via interpolation, reducing the number of necessary nodes. Empirical evidence of the performance of these models will be presented through illustrative examples of vegetation monitoring and atmospheric modeling.

Keywords: Earth observation · Remote sensing · Vegetation · Kernel methods · Gaussian processes · Inverse modeling · Geosciences · Transfer models

---

G. Camps-Valls—The research was funded by the European Research Council (ERC) under the ERC-CoG-2014 SEDAL project (grant agreement 647423), and the Spanish Ministry of Economy and Competitiveness (MINECO) through the project TIN2015-64210-R.

P. Sharma and F.M. Bianchi (Eds.): SCIA 2017, Part II, LNCS 10270, pp. 205–217, 2017.
DOI: 10.1007/978-3-319-59129-2_18

1   Introduction

Solving inverse problems is a recurrent topic of research in Physics in general, and in Earth Observation (EO) in particular. Earth Observation encompasses geosciences, climate science and remote sensing. After all, Science is about making inferences about physical parameters from sensory data. A very relevant inverse problem is that of estimating vegetation properties from remotely sensed images. Accurate inverse models help to determine the phenological stage and health status (e.g., development, productivity, stress) of crops and forests [12], which has important societal, environmental and economical implications. Leaf chlorophyll content (Chl), leaf area index (LAI), and fractional vegetation cover (FVC) are among the most important vegetation parameters to retrieve from space observations [15, 24].

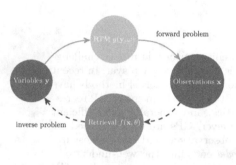

Fig. 1. Forward (solid lines) and inverse (dashed lines) problems in remote sensing.

In general, mechanistic models implement the laws of Physics and allow us to compute the data values given a model [21]. This is known as the *forward* problem. In the *inverse* problem, the aim is to reconstruct the model from a set of measurements, see Fig. 1. Notationally, a forward model describing the system is expressed as $x = g(y, \omega)$, where x is a measurement obtained by the satellite (e.g. radiance); the vector y represents the state of the biophysical variables on the Earth (which we desire to infer or predict and is often referred to as *outputs* in the inverse modeling approach); ω contains a set of controllable conditions (e.g. wavelengths, viewing direction, time, Sun position, and polarization); and $g(\cdot)$ is a function which relates y with x. Such a function g is typically considered to be nonlinear, smooth and continuous. Our goal is to obtain an inverse model, $f(\cdot) \approx g^{-1}(\cdot)$, parametrized by θ, which approximates the biophysical variables y given the data x received by the satellite, i.e. $\hat{y} = f(x, \theta)$. Radiative transfer models (RTMs) are typically used to implement the forward direction [13, 22]. However, inverting RTMs directly is very complex because the number of unknowns is generally larger than the number of independent radiometric information [14]. Also, estimating physical parameters from RTMs is hampered by the presence of high levels of uncertainty and noise, primarily associated to atmospheric conditions and sensor calibration. This translates into inverse problems where deemed similar spectra may correspond to very diverse solutions. This gives raise to undetermination and ill-posed problems.

Methods for model inversion and parameter retrieval can be roughly separated in three main families: statistical, physical and hybrid methods [10]. *Statistical inversion* predicts a biogeophysical parameter of interest using a training dataset of input-output data pairs coming from concurrent measurements of the parameter of interest (e.g. leaf area index -LAI-) and the corresponding satellite observations (e.g. reflectances). Statistical methods typically outperform other approaches, but ground truth measurements involving a terrestrial campaign are necessary. *Physical inversion* reverses RTMs by searching for similar spectra in look-up-tables (LUTs) and assigning the parameter corresponding to the most similar observed spectrum. This requires selecting an appropriate cost function, and generating a rich, representative LUT from the RTM. The use of RTMs to generate data sets is a common practice, and especially convenient because acquisition campaigns are very costly in terms of time, money, and human resources, and usually limited in terms of parameter combinations. Finally, *hybrid inversion* exploits the input-output data generated by RTM simulations and train statistical regression models to invert the RTM model. Hybrid models combine the flexibility and scalability of machine learning while respecting the physics encoded in the RTMs. Currently, kernel machines in general [8], and Bayesian non-parametric approaches such as Gaussian Process (GP) regression [19] in particular, are among the preferred regression models [9, 23].

While hybrid inversion is practical when no *in situ* data is available, intuitively it makes sense to let predictions be guided by actual measurements whenever they are present. Likewise, when only very few real in situ measurements are available, it is sensible to incorporate simulated data from RTMs to properly ground the models. This is the first pathway considered in this paper, which extends the hybrid inversion by proposing a statistical method that performs nonlinear and nonparametric inversion blending both real and simulated data. The so-called joint GP (JGP) essentially learns how to trade off noise variance in the real and simulated data.

A second topic covered in this paper follows an alternative pathway to *learn* latent functions that generated the observations using GP models. We introduce a *latent force model* (LFM) for GP modelling [1]. The proposed LFM-GP combines the ordinary differential equations of the forward model (through smoothing kernels) and empirical data (from *in situ* campaigns). The LFM presented here performs multi-output structured regression, adapts to the signal characteristics, is able to cope with missing data in the time series, and provides explicit latent functions that allow system analysis and evaluation.

Finally, we deal with the important issue of *emulation*, that is *learning* surrogate GP models to approximate costly RTMs. The proposed Automatic Gaussian Process Emulator (AGAPE) methodology combines the interpolation capabilities of Gaussian processes (GPs) with the accurate design of an acquisition function that favours sampling in low density regions and flatness of the interpolation function.

2   Gaussian Process Models for Inverse Modeling

GPs are state-of-the-art tools for regression and function approximation, and have been recently shown to excel in biophysical variable retrieval by following both statistical [9, 23] and hybrid approaches [6, 7]. Let us consider a set of n pairs of observations or measurements, $\mathcal{D}_n := \{\mathbf{x}_i, y_i\}_{i=1}^{n}$, perturbed by an additive independent noise. The input data pairs $(\mathbf{X}, \mathbf{y})$ used to fit the inverse machine learning model $f(\cdot)$ come from either *in situ* field campaign data (statistical approach) or simulations by means of an RTM (hybrid approach). We assume the following model,

$$y_i = f(\mathbf{x}_i) + e_i, \ e_i \sim \mathcal{N}(0, \sigma_n^2),$$ (1)

where $f(\mathbf{x})$ is an unknown latent function, $\mathbf{x} \in \mathbb{R}^d$, and σ_n^2 stands for the noise variance. Defining $\mathbf{y} = [y_1, \ldots, y_n]^\mathsf{T}$ and $\mathbf{f} = [f(\mathbf{x}_1), \ldots, f(\mathbf{x}_n)]^\mathsf{T}$, the conditional distribution of $\mathbf{y}$ given $\mathbf{f}$ becomes $p(\mathbf{y}|\mathbf{f}) = \mathcal{N}(\mathbf{f}, \sigma_n^2 \mathbf{I})$, where $\mathbf{I}$ is the $n \times n$ identity matrix. Now, in the GP approach, we assume that $\mathbf{f}$ follows a n-dimensional Gaussian distribution $\mathbf{f} \sim \mathcal{N}(\mathbf{0}, \mathbf{K})$ [3].

Fig. 2. Statistical inverse modelling.

The covariance matrix $\mathbf{K}$ of this distribution is determined by a kernel function with entries $\mathbf{K}_{ij} = k(\mathbf{x}_i, \mathbf{x}_j) = \exp(-\|\mathbf{x}_i - \mathbf{x}_j\|^2/(2\sigma^2))$, encoding similarity between the input points [19]. The intuition here is the following: the more similar input i and j are, according to some metric, the more correlated output i and j ought to be. Thus, the marginal distribution of $\mathbf{y}$ can be written as

$$p(\mathbf{y}) = \int p(\mathbf{y}|\mathbf{f}) p(\mathbf{f}) d\mathbf{f} = \mathcal{N}(\mathbf{0}, C_n),$$

where $C_n = \mathbf{K} + \sigma_n^2 \mathbf{I}$. Now, what we are really interested in is predicting a new output y_*, given an input x_* (Fig. 2). The GP framework handles this by constructing a joint distribution over the training and test points,

$$\begin{bmatrix} \mathbf{y} \\ y_* \end{bmatrix} \sim \mathcal{N}\left(\mathbf{0}, \begin{bmatrix} C_n & \mathbf{k}_*^\mathsf{T} \\ \mathbf{k}_* & c_* \end{bmatrix}\right),$$

where $\mathbf{k}_* = [k(\mathbf{x}_*, \mathbf{x}_1), \ldots, k(\mathbf{x}_*, \mathbf{x}_n)]^\mathsf{T}$ is an $n \times 1$ vector and $c_* = k(\mathbf{x}_*, \mathbf{x}_*) + \sigma_n^2$. Then, using standard GP manipulations, we can find the distribution over y_* conditioned on the training data, which is a normal distribution with predictive mean and variance given by

$$\mu_{\mathrm{GP}}(\mathbf{x}_*) = \mathbf{k}_*^\mathsf{T}(\mathbf{K} + \sigma_n^2 \mathbf{I}_n)^{-1}\mathbf{y},$$
$$\sigma_{\mathrm{GP}}^2(\mathbf{x}_*) = c_* - \mathbf{k}_*^\mathsf{T}(\mathbf{K} + \sigma_n^2 \mathbf{I}_n)^{-1}\mathbf{k}_*.$$ (2)

Thus, GPs yield not only predictions $\mu_{\mathrm{GP}*}$ for test data, but also the so-called "error-bars", $\sigma_{\mathrm{GP}*}$, assessing the uncertainty of the mean prediction. The hyperparameters $\boldsymbol{\theta} = [\sigma, \sigma_n]$ to be tuned in the GP determine the width of the squared exponential kernel function and the noise on the observations. This can be done by marginal likelihood maximization or simple grid search, attempting to minimize the squared prediction errors.

3   Forward and Inverse Joint GP Models

Let us assume that the previous dataset $\mathcal{D}_n$ is formed by two disjoint sets: one set of r real data pairs, $\mathcal{D}_r = \{(\mathbf{x}_i, y_i)\}_{i=1}^r$, and one set of s RTM-simulated

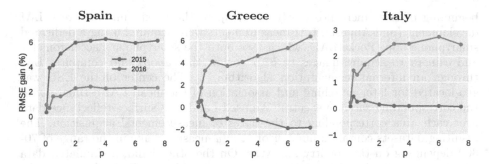

Fig. 3. Obtained accuracy gains in RMSE of JGP over GP for the different sites, campaign dates and simulated-to-real data ratios.

pairs $\mathcal{D}_s = \{(y_j, \mathbf{x}_j)\}_{j=r+1}^n$, so that $n = r + s$ and $\mathcal{D}_n = \mathcal{D}_r \cup \mathcal{D}_s$. In matrix form, we have $\mathbf{X}_r \in \mathbb{R}^{r \times d}$, $\mathbf{y}_r \in \mathbb{R}^{r \times 1}$, $\mathbf{X}_s \in \mathbb{R}^{s \times d}$ and $\mathbf{y}_s \in \mathbb{R}^{s \times 1}$, containing all the inputs and outputs of $\mathcal{D}_r$ and $\mathcal{D}_s$, respectively. Finally, the $n \times 1$ vector $\mathbf{y}$ contains all the n outputs, sorted with the real data first, followed by the simulated data. Now, we define a different model, where the observation noise depends on the origin of the data: σ_n^2 for real observations ($\mathbf{x}_i \in \mathcal{D}_r$) or σ_n^2/γ for RTM simulations ($\mathbf{x}_i \in \mathcal{D}_s$), where the parameter $\gamma > 0$ accounts for the importance of the two sources of information relative to each other.

The resulting distribution of $\mathbf{y}$ given $\mathbf{f}$ is only slightly different from that of the regular GP, namely $p(\mathbf{y}|\mathbf{f}) = \mathcal{N}(\mathbf{f}, \sigma_n^2 \mathbf{V})$ where $\mathbf{V}$ is an $n \times n$ diagonal matrix in which the first r diagonal elements are equal to 1 and the remaining s are equal to γ^{-1}: $\mathbf{V} = \text{diag}(1, \ldots, 1, \gamma^{-1}, \ldots, \gamma^{-1})$. The predictive mean and variance of a test output y_*, conditioned on the training data, then becomes

$$\mu_{\text{JGP}}(\mathbf{x}_*) = \mathbf{k}_*^{\mathsf{T}}(\mathbf{K} + \sigma_n^2 \mathbf{V})^{-1}\mathbf{y},$$
$$\sigma_{\text{JGP}}^2(\mathbf{x}_*) = c_* - \mathbf{k}_*^{\mathsf{T}}(\mathbf{K} + \sigma_n^2 \mathbf{V})^{-1}\mathbf{k}_*. \tag{3}$$

Note that when $\gamma = 1$ the standard GP formulation is obtained. Otherwise γ acts as an extra regularization term accounting for the relative importance of the real and the simulated data points. The hyperparameters of the JGP are $\boldsymbol{\theta} = [\sigma, \sigma_n, \gamma]$, which can be selected by maximizing the marginal likelihood of the observations as usual in the GP framework [19]. It is important to note that hyperparameter fitting should be performed with respect to real data, so that the method learns the mapping from *real* input to output.

3.1   Experimental Results

We are concerned with the prediction of leaf area index (LAI) parameter from space, a parameter that characterizes plant canopies and is roughly defined as the total needle surface area per unit ground area. Non-destructive real LAI data were acquired over Elementary Sampling Units (ESUs) within rice fields in Spain, Italy and Greece during field campaigns in 2015 and 2016. The temporal frequency of the campaigns was approximately 10 days starting from the very beginning of rice emergence (early-June) up to the maximum rice green LAI development (mid-August). LAI measurements were acquired using a dedicated smartphone app (PocketLAI), which uses both the smartphone's accelerometer and camera to acquire images at $57.5°$ below the canopy and computes LAI through an internal segmentation algorithm [7]. The center of the ESU was geo-located for later matching and association of the mean LAI estimate with the corresponding satellite spectra. We used Landsat 8 surface reflectance data over each area corresponding to the dates of measurements' acquisition. The resulting datasets contain a number of *in situ* measurements in the range of 70-300 depending on the country and year. On the other hand, a simulated data set of $s = 2000$ pairs of Landsat 8 spectra and LAI was obtained running the PROSAIL RTM in forward mode.[1] The leaf and canopy variables as well as the soil brightness parameter, were generated following a PROSAIL site-specific parameterization to constrain the model to Mediterranean rice areas [6].

We assessed the performance of JGP for different amounts of real and simulated data. The gain in accuracy was measured as the reduction in root mean square error (RMSE gain [%] $= 100 \times (\text{RMSE}_{\text{GP}}\text{-RMSE}_{\text{JGP}})/\text{RMSE}_{\text{GP}})$. We evaluated performance in the 6 datasets generated for different countries (SP, GR, IT) and years (2015, 2016). Figure 3 shows the effect of the ratio between simulated and real data points $p = s/r$ on the RMSE gain evaluated using 10-fold crossvalidation. When no simulated data is used, the JGP model reduces to the standard GP model, but when introducing an amount of PROSAIL-datapoints similar to the amount of real datapoints, i.e. $p \sim 1$, a noticeable gain is for datasets gathered in 2016. In the case of the data from Spain, the gain appears rather stable (between 6 and 2% in 2015 and 2016 respectively) after reaching a ratio of $p = 2$, indicating what size of the simulated dataset is needed for an increase in accuracy. The results for Greece and Italy, however, show that the use of simulated data attempting to fill in the under-represented domain of the real data, is not always useful.

4   Inverse Modelling with Latent Force Models

We are interested in inverse modelling from real *in situ* data, and to *learn* not only an accurate retrieval model but also about the physical mechanism that generated the input-output observed relations without even accessing any RTM, see Fig. 4. Here, we assume that our observations correspond simply to the temporal variable, $\mathbf{x} \sim t$, so the latent functions are defined in the time domain, $f_r(t)$. Nevertheless, extension to multidimensional objects such as radiances is straightforward by using kernels. Notationally, let us consider a multioutput scenario with Q correlated observed time series, $y_q(t)$ for $1 \leq q \leq Q$, and let us assume that we have n samples available for each of these signals, taken at sampling points t_i, s.t. $y_q[i] = y_q(t_i)$ for $1 \leq i \leq n$. This is the *training set*, which is composed of an input vector, $\mathbf{t} = [t_1, \ldots, t_n]^\top$, and an output matrix, $\mathbf{Y} = [\mathbf{y}_1, \ldots, \mathbf{y}_Q]$ with $\mathbf{y}_q = [y_q[1], \ldots, y_q[n]]^\top$. We aim to build a GP

---

[1] PROSAIL simulates leaf reflectance for the optical spectrum, from 400 to 2500 nm with a 1 nm spectral resolution, as a function of biochemistry and structure of the canopy, its leaves, the background soil reflectance and the sun-view geometry.

model for the Q outputs that can be used to perform inference on the *test set*:
$\widetilde{\mathbf{t}} = [\widetilde{t}_1, \ldots, \widetilde{t}_m]^\top$ and $\widetilde{\mathbf{Y}} = [\widetilde{\mathbf{y}}_1, \ldots, \widetilde{\mathbf{y}}_Q]$ with $\widetilde{\mathbf{y}}_q = [\widetilde{y}_q[1], \ldots, \widetilde{y}_q[m]]^\top$ and
$\widetilde{y}_q[m'] = y_q(\widetilde{t}_{m'})$ for test inputs at $t_{m'}$.

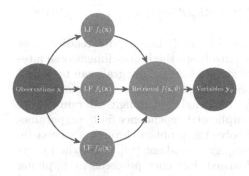

Fig. 4. Inverse modeling with latent forces.

Formulation. Let us assume that a set of R independent latent functions (LFs), $f_r(t)$ with $1 \leq r \leq R$, are responsible for the observed correlation between the outputs. Then, the cross-correlation between the outputs arises naturally as a result of the coupling between the set of independent LFs, instead of being imposed directly on the set of outputs. Let us define the form of these latent functions and the coupling mechanism between them. In this work, we model the LFs as zero-mean Gaussian processes (GPs), and the coupling system emerges through a linear convolution operator described by an *impulse response*, $h_q(t)$, as follows:

$$y_{r,q}(t) = L_q[t]\{f_r(t)\} = f_r(t) * h_q(t) = \int_0^t f_r(\tau)h_q(t-\tau)\mathrm{d}\tau. \qquad (4)$$

where $L_q[t]\{f_r(t)\}$ indicates the linear operator associated to the linear convolution. The resulting outputs are finally obtained as a linear weighted combination of these pseudo-outputs plus an additive white Gaussian noise (AWGN) term:

$$y_q(t) = \sum_{r=1}^R S_{r,q}y_{r,q}(t) + w_q(t), \qquad (5)$$

where $S_{r,q}$ represents the coupling strength between the r-th LF and the q-th output, and $w_q(t) \sim \mathcal{N}(0, \eta_q^2)$ is the AWGN term. In practice, we consider only the squared exponential auto-covariance function for the LFs, $k_{f_r f_r}(t'-t)$ $\propto \exp(-\frac{(t'-t)^2}{2\ell_r^2})$, where the hyperparameter ℓ_r controls the length-scale of the process. The smoothing kernel encodes our knowledge about the linear system (that relates the unobserved LFs and the outputs), and can be based on basic physical principles of the system at hand (as in [1]) or selected arbitrarily (as in [4,11]). In this paper, we consider also the Gaussian smoothing kernel, $h_q(t) \propto \exp(-\frac{t^2}{2\nu_q^2})$. Now, since the LFs are zero-mean GPs, the noise is also zero-mean and Gaussian, and all the operators involved are linear, the joint LFs-outputs process is also a GP. Therefore, the mean function of the q-th output is $\mu_{y_q}(t) = 0$, whereas the cross-covariance function between two outputs is

$$k_{y_p y_q}(t, t') = \sum_{r=1}^R S_{r,p}S_{r,q}L_p[t]\{L_q[t']\{k_{f_r f_r}(t,t')\}\} + \eta_q^2\delta[p-q]\delta[t'-t], \qquad (6)$$

where the term $\mathrm{L}_p[t]\,\{\mathrm{L}_q[t']\,\{k_{f_r f_r}(t,t')\}\}$ denotes the application of the convolutional operator twice to the autocorrelation function of the LFs, which results in the following double integral:

$$\mathrm{L}_p[t]\,\{\mathrm{L}_q[t']\,\{k_{f_r f_r}(t,t')\}\} = \int_0^t \int_0^{t'} h_p(t-\tau)h_q(t'-\tau') \times k_{f_r f_r}(\tau,\tau')\mathrm{d}\tau'\mathrm{d}\tau.$$

Finally, the cross-correlation between the LFs and the outputs readily gives $k_{f_r y_q}(t,t') = S_{r,q}\mathrm{L}_q[t']\,\{k_{f_r f_r}(t,t')\}$, which involves a single one-dimensional integral already computed in an intermediate step before. All integrals can be solved analytically when both the LFs and the smoothing kernel have a Gaussian shape.

Learning hyperparameters is very challenging through marginal log-likelihood maximization because of its complicated dependence on hyperparameters $\boldsymbol{\theta} = [\nu_q, l_r, \sigma, \sigma_n, \eta_q]$. We propose to solve the problem through a stochastic gradient descent technique, the scaled conjugate gradient [18]. Once the hyperparameters $\boldsymbol{\theta}$ of the model have been learned, inference proceeds by applying standard GP regression formulas [19] (cf. Sect. 2). Now, since the conditional PDF is Gaussian, the minimum mean squared error (MMSE) prediction is simply given by the conditional mean:

$$\hat{\mathbf{y}} = \boldsymbol{\mu}_{\tilde{\mathbf{y}}|\mathbf{y}} = \mathbf{K}_{\tilde{\mathbf{y}}\mathbf{y}}\mathbf{K}_{\mathbf{y}\mathbf{y}}^{-1}\mathbf{y}, \qquad (7)$$

where $\hat{\mathbf{y}} = [\hat{\mathbf{y}}_1^{\top}, \ldots, \hat{\mathbf{y}}_Q^{\top}]$ is the vectorized version of the inferred outputs, which can be expressed in matrix form as $\hat{\mathbf{Y}} = [\hat{\mathbf{y}}_1, \ldots, \hat{\mathbf{y}}_Q]$ with $\hat{\mathbf{y}}_q = [\hat{y}_q[1], \ldots, \hat{y}_q[m]]^{\top}$ and $\hat{y}_q[m'] = \hat{y}_q(\tilde{t}'_m)$.

4.1   Experimental Results

We are concerned about multiple time series of two (related) biophysical parameters, LAI and fAPAR (Fraction of Absorbed Photosynthetically Active Radiation), in the locations of the experiments in Sect. 3.1. We focus on a set of representative rice pixels of each area, thus allowing us to observe the inter-annual variability of rice from 2003 to 2014 at a coarse spatial resolution (2 Km), which is useful for regional vegetation modelling. We focus on learning the latent forces for the multi-output time series composed of the LAI and fAPAR data for Spain and Italy (i.e., the number of outputs is $Q = 4$) from the beginning of 2003 until the end of 2013. We use all the LAI data available from the MODIS sensor for Spain ($N = 506$ samples), and the first half (years 2003–2009) of the other three time series. The recovered LF ($R = 1$) and two examples of the modelled time series are displayed in Fig. 5. Note that the model has succeeded in capturing the dynamics of the data by using a single LF. Good numerical results were obtained: for Spain, we have MSE = 0.1139 and MSE = 0.0080 for LAI and fAPAR respectively, whereas for Italy we have MSE = 0.2422 and MSE = 0.0046, respectively.

5   Automatic GP Emulation

Emulation deals with the challenging problem of building statistical models for complex physical RTMs. The emulators are also called *surrogate* or *proxy* models, and they try to learn the equations encoded from data. Namely, an emulator

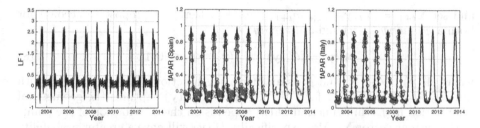

Fig. 5. Gap filling example using a single LF (i.e., $R = 1$). Training used all the LAI data from Spain (years 2003–2013) and the first half (years 2003–2009) of the other three time series: fAPAR (ES), LAI (IT) and fAPAR (IT). The second half constitutes the test set of such time series. Training data (red circles), test data (red dashed line), predicted time series (black line) and uncertainty measured by ±2 standard deviations about the mean predicted value (gray shaded area). (Color figure online)

is a statistical model which tries to reproduce the behavior of a deterministic and very costly physical model. Emulators built with GPs are gaining popularity in remote sensing and geosciences, since they allow efficient data processing and sensitivity analysis [5,9,20]. Here, we are interested in optimizing emulators such that a minimal number of simulations is run. The technique is called AGAPE (automatic Gaussian Process emulator), and is related to some Bayesian optimization and active learning techniques.

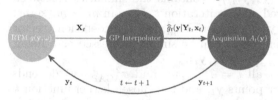

Fig. 6. Scheme of an automatic emulator.

The goal is to interpolate a costly function $g(\mathbf{y})$ choosing adequately the nodes, in order to reduce the error in the interpolation with the smallest possible number of evaluation of $g(\mathbf{y})$. Given an input matrix of nodes (used for the interpolation) at the t-th iterations, $\mathbf{Y}_t = [\mathbf{y}_1 \cdots \mathbf{y}_{m_t}]$, of dimension $d \times m_t$ (where d is the dimension of each $\mathbf{y}_i$ and m_t is the number of points), we have a vector of outputs, $\mathbf{x}_t = [x_1, \ldots, x_{m_t}]^{\mathsf{T}}$, where $x_t = g(\mathbf{y}_t)$ is the estimation of the observations (e.g., radiances) at iteration $t \in \mathbb{N}^+$ of the algorithm. Figure 6 shows a graphical representation of a generic automatic emulator. At each iteration t one performs an *interpolation*, $\widehat{g}_t(\mathbf{y}|\mathbf{Y}_t, \mathbf{x}_t)$, followed by an *optimization* step that updates the acquisition function, $A_t(\mathbf{y})$, updates the set $\mathbf{Y}_{t+1} = [\mathbf{Y}_t, \mathbf{y}_{m_t+1}]$ adding a new node, set $m_t \leftarrow m_t + 1$ and $t \leftarrow t + 1$. The procedure is repeated until a suitable stopping condition is met, such as a certain maximum number of points is included or a desired precision error ϵ is achieved, $\|\widehat{g}_t(\mathbf{y}) - \widehat{g}_{t-1}(\mathbf{y})\| \leq \epsilon$.

Formulation. the acquisition function, $A_t(\mathbf{y})$, encodes useful information for proposing new points to build the emulator. At each iteration, a new node is added maximizing $A_t(\mathbf{y})$, i.e.,

$$\mathbf{y}_{m_t+1} = \arg\max A_t(\mathbf{y}),$$

and set $\mathbf{Y}_{t+1} = [\mathbf{Y}_t, \mathbf{y}_{m_t+1}]$, $m_{t+1} = m_t + 1$. Here, we propose to account for both a *geometry* $G_t(\mathbf{y})$ and a *diversity* $D_t(\mathbf{y})$,

$$A_t(\mathbf{y}) = [G_t(\mathbf{y})]^{\beta_t} D_t(\mathbf{y}), \quad \beta_t \in [0, 1], \tag{8}$$

where $A_t(\mathbf{y}) : \mathcal{Y} \mapsto \mathbb{R}$, and β_t is an increasing function with respect to t, with $\lim_{t\to\infty} \beta_t = 1$ (or $\beta_t = 1$ for $t > t'$). Function $G_t(\mathbf{y})$ captures the geometrical information in g, while function $D_t(\mathbf{x})$ depends on the distribution of the points in the current vector $\mathbf{Y}_t$. More specifically, $D_t(\mathbf{y})$ will have a greater probability mass around empty areas within $\mathcal{Y}$, whereas $D_t(\mathbf{y})$ will be approximately zero close to the support points and exactly zero at the support points, i.e., $D_t(\mathbf{y}_i) = 0$, for $i = 1, \ldots, m_t$ and $\forall t \in \mathbb{N}$. Since g is unknown, the function $G_t(\mathbf{y})$ can be only derived from information acquired in advance or by considering the approximation $\widehat{g}$. The tempering value, β_t, helps to downweight the likely less informative estimates in the very first iterations. If $\beta_t = 0$, we disregard $G_t(\mathbf{y})$ and $A_t(\mathbf{y}) = D_t(\mathbf{y})$, whereas, if $\beta_t = 1$, we have $A_t(\mathbf{y}) = G_t(\mathbf{y})D_t(\mathbf{y})$.

We consider a GP for emulation, so the inputs and outputs are now reversed. In addition, note that interpolation fixes $\sigma_n = 0$. Therefore, the AGAPE predictive mean and variance at iteration t for a new point $\mathbf{y}_*$ become simply

$$\mathbb{E}[\widehat{g}_t(\mathbf{y}_*)] = \mu_{\text{AGAPE}}(\mathbf{y}_*) = \mathbf{k}_*^\mathsf{T} \mathbf{K}^{-1} \mathbf{x} = \mathbf{k}_*^\mathsf{T} \boldsymbol{\alpha},$$
$$\mathbb{V}[\widehat{g}_t(\mathbf{y}_*)] = \sigma^2_{\text{AGAPE}}(\mathbf{y}_*) = k(\mathbf{y}_*, \mathbf{y}_*) - \mathbf{k}_*^\mathsf{T} \mathbf{K}^{-1} \mathbf{k}_*,$$

where now $\mathbf{k}_* = [k(\mathbf{y}_*, \mathbf{y}_1), \ldots, k(\mathbf{y}_*, \mathbf{y}_{m_t})]^\mathsf{T}$ contains the similarities between the input point $\mathbf{x}_*$ and the observed ones at iteration t, $\mathbf{K}$ is an $m_t \times m_t$ kernel matrix with entries $\mathbf{K}_{i,j} := k(\mathbf{y}_i, \mathbf{y}_j)$, and $\boldsymbol{\alpha} = \mathbf{K}_{nn}^{-1}\mathbf{x}_t$ is the coefficient vector for interpolation. The interpolation for $\mathbf{y}_*$ can be simply expressed as a linear combination of $\widehat{g}_t(\mathbf{y}_*) = \mathbf{k}_*^\mathsf{T} \boldsymbol{\alpha} = \sum_{i=1}^{m_t} \alpha_i k(\mathbf{y}_*, \mathbf{y}_i)$.

Note that $\sigma^2_{\text{AGAPE}}(\mathbf{y}_i) = 0$ for all $i = 1, \ldots, m_t$ and $\sigma^2_{\text{AGAPE}}(\mathbf{y})$ depends on the distance among the support points $\mathbf{y}_t$, and the chosen kernel function k and associated hyper-parameter σ. For this reason, the function $\sigma^2_{\text{AGAPE}}(\mathbf{y})$ is a good candidate to represent the distribution of the $\mathbf{y}_t$'s since it is zero at each $\mathbf{y}_i$, and higher far from the points $\mathbf{y}_i$'s. Moreover, $\sigma^2_{\text{AGAPE}}(\mathbf{y})$ takes into account the information of the GP interpolator. Therefore, we consider as the diversity term $D(\mathbf{y}) := \sigma^2_{\text{AGAPE}}(\mathbf{y})$, i.e., $D(\mathbf{y})$ is induced by the GP interpolator.

As geometric information, we consider enforcing flatness on the interpolation function, and thus aim to minimize the norm of the the gradient of the interpolating function $\widehat{g}_t$ w.r.t. the input data $\mathbf{y}$, i.e., $G(\mathbf{y}) = \|\nabla_y \widehat{g}_t(\mathbf{y}|\mathbf{Y}_t, \mathbf{x}_t)\| = \|\sum_{i=1}^{m_t} \alpha_i \nabla_y k(\mathbf{y}, \mathbf{y}_i)\|$. This intuitively makes wavy regions of g require more support points than flat regions. The gradient vector for the squared exponential kernelm $k(\mathbf{y}, \mathbf{y}') = \exp(-\|\mathbf{y} - \mathbf{y}'\|^2/(2\sigma^2))$ with $\mathbf{y} = [y_1, \ldots, y_d]^\mathsf{T}$, can be computed in closed-form, $\nabla_y k(\mathbf{y}, \mathbf{y}') = -\frac{k(\mathbf{y}, \mathbf{y}')}{\sigma^2}[(y_1 - y_1'), \ldots, (y_d - y_d')]^\mathsf{T}$, so the geometry term $G(\mathbf{y})$ can be defined, for instance, as follows:

$$G(\mathbf{y}) = \|\nabla_y \widehat{g}_t(\mathbf{y}|\mathbf{Y}_t, \mathbf{x}_t)\| = \left\| \frac{1}{m_t} \sum_{i=1}^{m_t} \nabla_y [k(\mathbf{y}, \mathbf{y}_i)] \right\|, \tag{9}$$

which reduces the dependence to the current approximation $\widehat{g}_t$. Therefore, the acquisition function can be readily obtained by defining $\beta_t = 1 - \exp(-\gamma t)$, where $\gamma \geq 0$ is a positive scalar and plugging Eq. (9) into Eq. (8). We optimized $A(\mathbf{x})$ using interacting parallel simulated annealing methods [16,17].

5.1 Experimental Results

We show empirical evidence of performance on the optimization of selected points for a complex and computationally expensive RTM: the MODTRAN5-based LUT. MODTRAN5 is considered as the *de facto* standard atmospheric RTM for atmospheric correction applications [2]. In our test application, and for the sake of simplicity, we have considered $d = 2$ with the Aerosol Optical Thickness at 550 nm (τ) and ground elevation (h) as key input parameters. The underlying function $g(\mathbf{y})$ consists therefore on the execution of MODTRAN5 at given values of τ and h and wavelength of 760 nm. The input parameter space is bounded to 0.05–0.4 for τ and 0–3 km for h. In order to test the accuracy of the different schemes, we have evaluated $g(\mathbf{y})$ at all the possible 1750 combinations of 35 values of τ and 50 values of h. Namely, this thin grid represents the ground-truth in this example.

We tested (a) a standard, yet suboptimal, random approach choosing points uni-
formly within $\mathcal{Y} = [0.05, 0.4] \times [0, 3]$, (b) the
Latin Hypercube sampling [5], and (c) the
proposed AGAPE. We start with $m_0 = 5$

Table 1. Averaged number of nodes m_t.

Random	Latin Hypercube	AGAPE
28.43	16.69	9.16

points $\mathbf{y}_1 = [0.05, 0]^\top$, $\mathbf{y}_2 = [0.05, 3]^\top$, $\mathbf{y}_3 = [0.4, 0]^\top$, $\mathbf{y}_4 = [0.4, 3]^\top$ and $\mathbf{y}_5 = [0.2, 1.5]^\top$ for all the techniques. We compute the final number of nodes m_t required to obtain an ℓ_2 distance between g and $\widehat{g}$ smaller than $\epsilon = 0.03$, with the different methods. The results, averaged over 10^3 runs, are shown in Table 1. AGAPE requires the addition of ≈ 4 new points to obtain a distance smaller than 0.03.

6 Conclusions

We introduced three different schemes based on GP modeling in the interplay between Physics and Machine Learning, with the focus on the Earth system modeling. Canonical machine learning for EO problems rely on in situ observational data, and often disregard the physical knowledge and models available. We argue that the equations encoded in forward physical models may be very useful in inverse GP modeling, such that models may give consistent, physically meaningful estimates. Three types of physics-aware GP models were introduced: a simple approach to combine *in situ* measurements and simulated data in a single GP model, a latent force model that incorporates ordinary differential equations, and an automatic compact emulator of physical models through GPs. The developed models demonstrated good performance, adaptation to the signal characteristics and transportability to unseen situations.

References

1. Álvarez, M.A., Luengo, D., Lawrence, N.D.: Linear latent force models using gaussian processes. IEEE Trans. Patt. Anal. Mach. Intell. **35**(11), 2693–2705 (2013)
2. Berk, A., Anderson, G., Acharya, P., Bernstein, L., Muratov, L., Lee, J., Fox, M., Adler-Golden, S., Chetwynd, J., Hoke, M., Lockwood, R., Gardner, J., Cooley, T., Borel, C., Lewis, P., Shettle, E.: MODTRAN5: 2006. In: International Society for Optics and Photonics (2006)
3. Bishop, C.M.: Pattern recognition. Mach. Learn. **128**, 1–58 (2006)
4. Boyle, P., Frean, M.: Dependent gaussian processes. In: NIPS, pp. 217–224 (2004)
5. Busby, D.: Hierarchical adaptive experimental design for gaussian process emulators. Reliab. Eng. Syst. Saf. **94**, 1183–1193 (2009)
6. Campos-Taberner, M., García-Haro, F., Camps-Valls, G., Grau-Muedra, G., Nutini, F., Crema, A., Boschetti, M.: Multitemporal and multiresolution leaf area index retrieval for operational local rice crop monitoring. Rem. Sens. Environ. **187**, 102–118 (2016)
7. Campos-Taberner, M., Garcia-Haro, F., Moreno, A., Gilabert, M., Sanchez-Ruiz, S., Martinez, B., Camps-Valls, G.: Mapping leaf area index with a smartphone and gaussian processes. IEEE Geosci. Remote Sens. Lett. **12**(12), 2501–2505 (2015)
8. Camps-Valls, G., Bruzzone, L. (eds.): Kernel Methods for Remote Sensing Data Analysis. Wiley, UK (2009)
9. Camps-Valls, G., Muñoz-Marí, J., Verrelst, J., Mateo, F., Gomez-Dans, J.: A survey on gaussian processes for earth observation data analysis. IEEE Geosci. Remote Sens. Mag. **3**(2), 1–20 (2016)
10. Camps-Valls, G., Tuia, D., Gómez-Chova, L., Jiménez, S., Malo, J. (eds.): Remote Sens. Image Process. Morgan & Claypool Publishers, USA (2011)
11. Higdon, D.: Space and space-time modeling using process convolutions. In: Anderson, C.W., Barnett, V., Chatwin, P.C., El-Shaarawi, A.H. (eds.) Quantitative Methods for Current Environmental Issues, pp. 37–54. Springer, London (2002)
12. Hilker, T., Coops, N.C., Wulder, M.A., Black, T.A., Guy, R.D.: The use of remote sensing in light use efficiency based models of gross primary production: a review of current status and future requirements. Sci. Total. Environ. **404**(2–3), 411–423 (2008)
13. Jacquemoud, S., Bacour, C., Poilvé, H., Frangi, J.P.: Comparison of four radiative transfer models to simulate plant canopies reflectance: direct and inverse mode. Remote Sens. Environ. **74**(3), 471–481 (2000)
14. Liang, S.: Advances in Land Remote Sensing: System, Modeling, Inversion and Applications. Springer, Germany (2008)
15. Lichtenthaler, H.K.: Chlorophylls and carotenoids: pigments of photosynthetic biomembranes. Methods Enzymol. **148**, 350–382 (1987)
16. Martino, L., Elvira, V., Luengo, D., Corander, J., Louzada, F.: Orthogonal parallel MCMC methods for sampling and optimization. Dig. Sign Proc. **58**, 64–84 (2016)
17. Read, J., Martino, L., Luengo, D.: Efficient monte carlo optimization for multi-label classifier chain. In: IEEE International Conference on Acoustics, Speech, and Signal Processing, ICASSP, pp. 1–5 (2013)
18. Nabney, I.: NETLAB: Algorithms for Pattern Recognition. Springer, London (2002)
19. Rasmussen, C.E., Williams, C.K.I.: Gaussian Processes for Machine Learning. The MIT Press, New York (2006)
20. Rivera, J., Verrelst, J., Gómez-Dans, J., Muñoz-Marí, J., Moreno, J., Camps-Valls, G.: An emulator toolbox to approximate radiative transfer models with statistical learning. Remote Sens. **7**(7), 9347–9370 (2015)

21. Snieder, R., Trampert, J.: Inverse Problems in Geophysics. Springer, Vienna (1999)
22. Verhoef, W., Bach, H.: Simulation of hyperspectral and directional radiance images using coupled biophysical and atmospheric radiative transfer models. Remote Sens. Environ. **87**, 23–41 (2003)
23. Verrelst, J., Alonso, L., Camps-Valls, G., Delegido, J., Moreno, J.: Retrieval of vegetation biophysical parameters using gaussian process techniques. IEEE Trans. Geosci. Remote Sens. **50**(5), 1832–1843 (2012)
24. Whittaker, R.H., Marks, P.L.: Methods of assessing terrestrial productivity. In: Lieth, H., Whittaker, R.H. (eds.) Primary Productivity of the Biosphere, vol. 14, pp. 55–118. Springer, Heidelberg (1975)

Medical and Biomedical Image Analysis

Automatic Segmentation of Bone Tissue from Computed Tomography Using a Volumetric Local Binary Patterns Based Method

Jukka Kaipala[(⊠)], Miguel Bordallo López, Simo Saarakkala,
and Jérôme Thevenot

University of Oulu, Oulu, Finland
jukka.kaipala@icloud.com, miguelbl@ee.oulu.fi,
{simo.saarakkala,jerome.thevenot}@oulu.fi

Abstract. Segmentation of scanned tissue volumes of three-dimensional (3D) images often involves - at least partially - some manual process, as there is no standardized automatic method. A well-performing automatic segmentation would be preferable, not only because it would improve segmentation speed, but also because it would be user-independent and provide more objectivity to the task. Here we extend a 3D local binary patterns (LBP) based trabecular bone segmentation method with adaptive local thresholding and additional segmentation parameters to make it more robust yet still perform adequately when compared to traditional user-assisted segmentation. We estimate parameters for the new segmentation method (AMLM) in our experimental setting, and have two micro-computed tomography (μCT) scanned bovine trabecular bone tissue volumes segmented by both the AMLM and two experienced users. Comparison of the results shows superior performance of the AMLM.

Keywords: Segmentation · 3D · LBP · Micro-CT

1 Introduction

Image segmentation is a process that classifies image constituents into two or more groups that represent some distinct aspects of the data to highlight relevant image features for further analysis. The simplest segmentation task is binary thresholding, which defines imaged object boundaries by dividing constituents into background and foreground depending on whether their grayscale value meets a selected gray-value threshold. There are several more complex thresholding methods that are classifiable by the locality of their thresholds, by the assumptions made about voxel connectivity, and by the extent that they can be automated. Examples of common methods are Otsu's method [1], region growing methods [2], and shape-based methods like geodesic active contours [3].

Segmentation is one of the most difficult problems in image processing and remains an active area of development [4]. Despite decades of research, there exists no universal segmentation method that would produce best results in all cases. One reason is that there is no absolute ground truth (GT), and therefore, a single right answer.

© Springer International Publishing AG 2017
P. Sharma and F.M. Bianchi (Eds.): SCIA 2017, Part II, LNCS 10270, pp. 221–232, 2017.
DOI: 10.1007/978-3-319-59129-2_19

Another reason is that segmentation methods vary in their sensitivity to artifacts and image quality. Most segmentation methods are specialized and perform best on the object and modality they were designed for. In practice, segmentation methods are considered and selected separately for each task. Three-dimensional (3D) segmentation faces additional challenges due to large datasets, and because computationally intensive segmentation algorithms, like shape-based algorithms, require much more memory and calculation power than simple thresholding.

Nowadays, volumetric imaging is often used in the medical field, from devices such as magnetic resonance imaging or computed tomography (CT), and several applications require volumetric segmentation of tissues [5]. One of the biological structures that is relevant to assess in 3D is the metabolically active trabecular bone, this one being at upmost interest in studies of musculoskeletal disorders. Pathological bone conditions such as osteoarthritis and osteoporosis are linked to small changes in trabecular bone microstructure that can be assessed from 3D micro-computed tomography (μCT) images [6].

However, while high-resolution imaging can be used for *in vitro* CT studies, the clinical resolution is limited due to radiation levels which must be kept low [7]. Low resolution imaging increases the impact of artifacts such as the partial volume effect (PVE), and leads to challenging segmentations. PVE comprises a class of imaging resolution related artifacts that limit how well a reconstructed voxel can represent its object location [8].

The lack of generic automated algorithms increases the effect of human factors like time consumed, interpersonal variance and systematic error. Together these factors have negative effect on 3D scanned image segmentation and subsequent structural analysis as well as clinical diagnosis. Local binary patterns (LBP) based methods represent a promising but still under-investigated alternative solution to conventional manual and semi-automatic 3D segmentation methods [5, 9].

This paper presents a novel automated adaptive multiscale local binary patterns (LBP)-based 3D segmentation method (AMLM), which extends an existing trabecular bone tissue specific segmentation method with customized adaptive local thresholding to improve its robustness. We estimate scanner and resolution specific thresholding parameters using bone phantom scans at three resolutions, and then evaluate the adaptive thresholding by comparing automated and traditional binarization results of two bovine subchondral bone samples scanned with the same equipment. Section 2 introduces the used LBP-based method. Section 3 describes the measurements in detail. Section 4 presents the results and Sect. 5 concludes the paper.

2   Multiscale LBP-Based 3D Segmentation Method Using Adaptive Local Thresholding

2.1   Introduction to LBP

LBP is an image operator that assigns each pixel a descriptor value, which is obtained by thresholding gray-values of a pixel neighborhood point pattern and interpreting the result as a binary number. This LBP code can be used to classify different neighborhood patterns. The original LBP method was introduced for image texture analysis, where it has been shown to be both computationally efficient and insensitive to global variations of grayscale values [10–13].

The original LBP neighborhood consisted of the eight pixels adjacent to a center pixel, but the model has since been extended to larger and circularly shaped neighborhoods with bilinear interpolation [11]. Figure 1 (left) illustrates the principle of the 8-bit LBP code calculation using a circular neighborhood with eight ordered members.

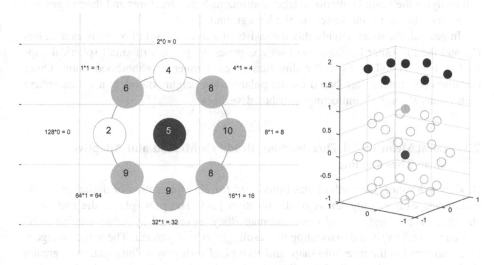

Fig. 1. Left: 8-bit LBP code calculation of a pixel (solid black) in 2D with eight LBP neighbors (gray). The code value is the sum of the bit values of the neighborhood elements (solid gray) whose interpolated gray-value (number inside the element) equals or exceeds the center pixel gray-value. In this example, the LBP code is $1 + 4 + 8 + 16 + 32 + 64 = 125$. Right: Partial multiscale neighborhood of a voxel (solid black, at the origo of this frame) in the default neighborhood configuration. The spherical inner neighborhood at radius $R_1 = 1$ consists of 26 points (gray). The remaining points (solid black) at radius 2 at the top represent the outer neighborhood cluster of the topmost inner neighborhood point (solid gray).

The circular LBP operator takes or interpolates the P neighboring point gray-values g_p at radius R from the center pixel, thresholds them using the center pixel gray-value g_c, then encodes non-negative results at (zero-based) neighbor positions p into binary values 2^p and calculates their sum [11].

2.2   Multiscale LBP-Based 3D Segmentation

Multiscale LBP-based 3D segmentation method (MLM) is a new automated segmentation method that has been suggested as an alternative to the common binary thresholding for analyzing bone microstructures in CT images [5]. It has recently been validated for the segmentation of μCT scans of osteoarthritic trabecular bone and the subsequent statistical analysis [9].

MLM is based on LBP but differs from it in certain ways. Major difference from traditional LBP is that patterns are evaluated using a global threshold instead of the voxel gray-value or other local threshold. In other words, pattern elements are defined in effect by their tissue membership instead of their relative local activity.

In addition, MLM examines neighborhood patterns on two nested levels in a 3D volume, firstly for the voxel itself and secondly for its inner neighborhood points. Figure 1 (right) illustrates the geometry of the MLM neighborhood. The MLM groups and analyzes the nested patterns to label continuous bone structures and their edges and to set disconnected bone voxels to the background.

In general, the inner neighborhood consists of a spherical set of N_1 vertices at radius R_1, and the full outer neighborhood set comprises N_1 patches of small spherical caps further away at radius R_2 in the direction of each inner neighborhood point. Outer neighborhood patches are based on the polar vertex neighborhood of a vertex sphere with radius R_2 and N_2 uniformly distributed vertices.

2.3   Local Mean Based Thresholding: Bradley's Method and Adaptive Mean Thresholding

Bradley's thresholding method was introduced for 2D binarization of digital grayscale documents with varying levels of illumination [14]. The principle of the method is straightforward. First, a local threshold map $B(x,y)$ is created by mean filtering a 2D source image $f(x,y)$ and downscaling the result $f_\mu(x,y)$ by T percent. The source image is then compared to the threshold map, and each pixel with gray-value equal to or greater than its corresponding local threshold value is assigned to the foreground. The method requires two external parameters, (isotropic) mean filter kernel size W and downscaling adjustment percentage T. The actual threshold map $B(x,y)$ is

$$B(x,y) = \left(1 - \frac{T}{100}\right) f_\mu(x,y).$$ (1)

The calculation of the mean filtered image f_μ can be expressed as

$$f_\mu(x,y) = \frac{1}{W^2} \sum_{i=x-r}^{x+r} \sum_{j=y-r}^{y+r} f(x,y).$$ (2)

where r is the axial extent of the filter mask from its center, and W is short for $2r + 1$, the actual width of the mean filter kernel.

The other 2D method, adaptive mean thresholding (also known as mean–C) [15], is very much like Bradley's method. The difference is that it subtracts an arbitrary constant value C from the mean filtered image instead of downscaling it with a weight factor. The threshold calculation can be done using the following:

$$B(x,y) = f_\mu(x,y) - C.$$ (3)

An important feature of both mean thresholding methods is that the calculation of neighborhood mean values can be conducted very efficiently by using a pre-calculated integral image, which eliminates explicit sum calculation for local means. The revised calculations comprise simple addition and subtraction of indexed values, and the number of operations is fixed regardless of the kernel size W. This makes mean filtering based adaptive thresholding methods fast compared to e.g. median filtering. The performance gain is advantageous in 3D image analysis, where datasets can be very large. Hence, for this application, these computationally simple and relatively fast mean adaptive thresholding algorithms present therefore a promising local alternative to global thresholding.

2.4   MLM with Adaptive Local Thresholding (AMLM)

The AMLM is an extension of the MLM with the same basic principles. It integrates the adaptive local methods in 3D to threshold and evaluate neighborhood patterns. This and other distinguishing features of AMLM are highlighted in Fig. 2.

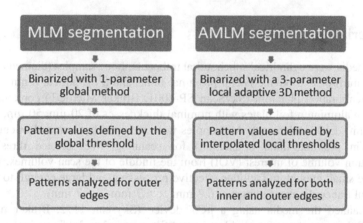

Fig. 2. Major features that make the AMLM segmentation (right) different from MLM.

The MLM uses a global value for its initial thresholding and LBP calculations. We replaced it with a map of local threshold values for this experiment to make the method more robust, especially if the gray-value intensity varies over the imaged region.

Our thresholding method is a generalized 3D adaptation of two existing 2D binary thresholding methods, the adaptive mean thresholding and Bradley's method. The volumetric extension of the mean formula can be expressed as

$$f_\mu(x,y,z) = \frac{1}{W^3} \sum_{i=x-r}^{x+r} \sum_{j=y-r}^{y+r} \sum_{k=z-r}^{z+r} f(x,y,z). \tag{4}$$

and the corresponding hybrid formula that incorporates both previously introduced 2D methods becomes

$$B(x,y,z) = \left(1 - \frac{T}{100}\right) f_\mu(x,y,z) + C. \tag{5}$$

The hybrid method requires three external parameters: mean filter kernel size W, local mean volume scaling percent adjustment T and constant adjustment C.

Inner border labeling was added to AMLM based on the hypothesis that its inclusion would complement the set of likely partial volume voxels in the tissue border area and give more calculation options to improve the accuracy of subsequent volumetric LBP analysis.

AMLM segmentation is affected by several parameters. One set of parameters configures the initial volumetric thresholding, which in effect determines the segmented tissue volume. Another parameter set defines the neighborhood parameters for the LBP-based multiscale segmentation algorithm, which further segments the inner and outer object borders in the initial binary volume and relabels apparent disconnected voxels to the background. We estimate these two parameter sets separately, firstly because they do not have a large effect on each other, and secondly because coupling would be impractical due to large number of possible parameter combinations.

3 Materials and Methods

Scanned media were cylindrical subchondral trabecular bone samples (diameter 10 mm, length 20 mm) of two bovine lateral proximal tibias and a cylindrical 8 mm diameter thickness calibration phantom (SkyScan SP-4001, Bruker MicroCT) containing four 2 mm wide aluminum foil plates with nominal thicknesses of 20 µm, 50 µm, 125 µm and 250 µm ($\pm 10\%$ tolerance). All samples were wrapped in foam and oriented horizontally (in proximal-tibial orientation) for scanning. We selected thresholding/ segmentation volume of interest (VOI) from the middle of the scan volumes. The volumes were small enough to facilitate effective processing and large enough to produce meaningful trabecular measurements (6.7 mm $\times$ 6.7 mm $\times$ 6.7 mm).

We scanned the media using a µCT device (Skyscan 1172, Bruker microCT, Kontich, Belgium) set up with an Al (0.5 mm) filter, complete 360° rotation, a step size of 0.5°, and averaging of 3 frames. We used three camera settings: 4000 $\times$ 2672 pixels (1 $\times$ 1 binning), 2000 $\times$ 1336 pixels (2 $\times$ 2 binning), and 1000 $\times$ 668 pixels (4 $\times$ 4 binning), enabling resolutions of 8.71 µm, 17.42 µm, and 34.84 µm. These resolutions are referred to by the next highest integer sizes (9 µm, 18 µm, and 35 µm). Respective exposure time, acceleration voltage, and current settings were: 1300 ms, 50 kV, 500 µA; 350 ms, 50 kV, 500 µA; 90 ms, 40 kV, 476 µA (phantom) / 500 µA (samples).

We developed the automatic segmentation scripts with MATLAB (version R2016a 64-bit, MathWorks) and performed the automatic segmentation of the VOIs on a laptop (Fujitsu Lifebook NH532, Intel® Core™ i5-3210M CPU @ 2.50 GHz with 2 cores, 4 logical processors, 16 GB RAM) using the previously selected thresholding and neighborhood parameters.

3.1 Estimation of Segmentation Parameters

We calculated the volumetric trabecular bone thickness (Tb.Th) of segmented µCT thickness phantom scans for all three resolutions using a number of different thresholding parameter sets. We ignored the least reliable measurements of plates with nominal thickness less than double the voxel size (20 μm plate at the 18 μm resolution, 20 μm and 50 μm plates at the 35 μm resolution). We calculated parameter combinations that best matched the nominal and measured phantom plate thicknesses for each resolution. We used the mean of relative (percent) plate thickness errors between the nominal and measured values as a goodness indicator. The results yielded resolution specific thresholding parameters (W, T, C) for configuring the automatic segmentation.

We segmented all trabecular bone sample VOIs using AMLM with the default (global) MLM thresholding for different neighborhood parameter sets $\{R_1 \times R_2 \times N_1 \times N_2\}$. The set of all test neighborhood parameter sets was $R_1 \times R_2 \times N_1 \times N_2$, where $R_1 \in \{0.69, 1.00, 1.46\}$, $R_2 \in \{1.69, 2.00, 2.46\}$, $N_1 \in \{18, 26, 38\}$, and $N_2 \in \{54, 78, 114\}$. Subsequently, we measured the thresholding and processing times and calculated the structural similarity index (SSIM) between reconstructed and segmented CT volumes and its linear correlation with different parameters. After that, we analyzed the results and selected the best neighborhood parameter set for the automatic segmentation.

We measured segmentation times and calculated the bone volume fraction (BV/TV) and average Tb.Th of each segmented volume with the CT Analyzer (CTAn) application (version 1.16.4.1+ 64-bit, Bruker microCT). We configured the 35 μm AMLM thresholding also visually with an experimental segmentation preview tool that we developed with MATLAB for this purpose. With its help, we previewed the sample 1 to select a set of thresholding parameters and used them to segment both samples with AMLM. In addition, we calculated the mean BV/TV and Tb.Th measures from the traditionally segmented VOIs.

3.2 Evaluation of Automated Adaptive Segmentation

The adaptive 3D thresholding method was tested against traditional user-dependent segmentation. The same segmentation scans, of the two trabecular bone samples scanned at three different resolutions with a μCT device, were binarized by two experienced users separately with CTAn (version 1.14.4.1). Both people experimented with their preferred image operations and parameters (filters, morphological, segmentation) until the results were visually acceptable, and measured the time spent on each VOI.

Subsequently, we took the GT reference measures from the highest resolution VOIs, which we thresholded (Otsu's method) and analyzed with CTAn (version 1.16.4.1+ 64-bit). Finally, we compared the measurements of the traditional computer-assisted segmentation and AMLM segmentation to each other and to the GT.

4 Results

4.1 Estimation of Segmentation Parameters

We selected the thresholding parameter values in Table 1 for each resolution based on the measurements of phantom scans segmented with different parameters. Parameter estimates were expected to be less reliable at lower resolutions due to lower image

Table 1. Selected thresholding parameters for each resolution based on segmented phantom scan measurements; W (px) = mean filter kernel size; T (%) = mean image downscaling adjustment; $C * 256$ = downscaled mean image absolute adjustment as normalized fraction value multiplied by 256; μ_δ (%) = mean percent error from the nominal phantom plate thickness

Resolution	W (px)	T (%)	$C * 256$	μ_δ (%)
9 μm (4 plates)	7	17	18	2.951
18 μm (3 plates)	5	30	30	2.492
35 μm (2 plates)	3	40	60	7.072

quality and smaller number of measurable plates. The large mean percent error at the lowest resolution is therefore not surprising.

We calculated sample correlation coefficients μ_r of the values of tested neighborhood parameters with corresponding measurements of the SSIM index and thresholding time and averaged them over the VOIs in Table 2. The results show that both SSIM and segmentation time express the strongest linear dependency on the inner neighborhood radius parameter R_1.

Table 2. Average sample correlation coefficients μ_r of the tested neighborhood parameter values with similarity and segmentation time; constant values are $R_1 = 1.46$ and $N_1 = 38$; SD = standard deviation

Constant parameters	Correlated parameter	SSIM		Segmentation time	
		μ_r	SD	μ_r	SD
–	R_1	0.9967423	0.000789079	0.870272193	0.012790265
	N_1	0.055986465	0.002272591	0.170243243	0.063093335
	R_2	−0.003343024	0.000537735	0.007627258	0.038023432
	N_2	0.000336614	0.001993851	−0.030461281	0.005478575
R_1	N_1	0.912460769	0.011077648	0.826272783	0.041962928
	R_2	−0.031314714	0.045380142	0.053173526	0.121071452
	N_2	0.047020701	0.053080101	−0.10590968	0.013584591
R_1, N_1	R_2	−0.003343024	0.000537735	0.007627258	0.038023432
	N_2	0.000336614	0.001993851	−0.030461281	0.005478575

Keeping the inner radius R_1 constant with the parameter value that gives the greatest similarity, the next set of averaged correlations show that N_1 is the next relevant parameter whose value correlates with SSIM. The measurements of the remaining parameters R_2 and N_2 do not significantly correlate with SSIM when the more significant parameters R_1 and N_1 were set constant with values that maximized their contribution to SSIM. Based on these results, we selected the neighborhood parameter values for automatic segmentation as $R_1 = 1.46$, $N_1 = 38$, $R_2 = 2$, and $N_2 = 78$.

4.2   Evaluation of Automated Adaptive Segmentation

The AMLM segmentation was performed using the previously selected thresholding and neighborhood parameters. We measured the average BV/TV and Tb.Th as well as binary thresholding and complete trabecular sub-label segmentation times as shown in Table 3. The data in the table include measurements using the alternate visual setup for the 35 μm volumes, as well as the GT reference measurements. The 35 μm resolution BV/TV and Tb.Th measurements of the original AMLM configuration are markedly different from the other measurements. Binarization times are very small compared to trabecular segmentation times.

Table 3. Measurements from AMLM segmentation results except where indicated; * = Automatic Otsu's thresholding of the best resolution (GT); ** = Binarized using the alternative thresholding parameters ($W = 3$, $T = 53$, $C = 60$)

Sample, resolution	BV/TV (%)	Tb.Th (μm)	Binarization time (s)	Segmentation total time (s)
1, 9 μm	23.22052	184.70	2	404
1, 9 μm*	21.42417	180.18	n/a	n/a
1, 18 μm	24.30018	193.43	1	63
1, 35 μm	18.09582	156.93	0	8
1, 35 μm**	24.34506	190.31	0	n/a
2, 9 μm	15.78273	205.81	1	397
2, 9 μm*	14.39293	201.80	n/a	n/a
2, 18 μm	16.25249	210.43	0	60
2, 35 μm	11.60439	176.42	0	7
2, 35 μm**	12.53121	208.24	0	n/a

We measured the BV/TV and Tb.Th of the traditionally binarized volumes as reference for comparison with the automatic segmentation as shown in Table 4. Segmentation takes several minutes, and time deviation is large. Measured BV/TV and Tb.Th values tend to be greater at lower resolutions.

Comparison of automatic AMLM segmentation results to traditional segmentation in Fig. 3 shows that the AMLM structural measurements are slightly lower but comparable, except for the measures taken from the lowest resolution VOIs, which are

Table 4. Average measures from traditional binary segmentation; SD = standard deviation

Sample, resolution	BV/TV (%)	SD	Tb.Th (μm)	SD	Binarization time (s)	SD
1, 9 μm	24.6	1.62	201	10.7	573	216
1, 18 μm	24.7	3.52	204	22.2	227	66
1, 35 μm	27.1	2.26	237	1.98	1069	694
2, 9 μm	16.8	1.06	222	9.25	562	308
2, 18 μm	17.1	1.41	225	13.0	355	177
2, 35 μm	18.3	1.26	243	9.41	366	93

significantly lower. Binarization time with the parameterized 3D hybrid method is negligible compared to traditional thresholding even at the highest resolution, where the VOI consists of 2563 voxels.

Figure 4 compares measured structural parameters of AMLM and traditional segmentation to those of the best resolution VOIs (GT). All measurements from AMLM segmented VOIs are closer to our GT than traditionally measured values. This is true especially for the measurements of Tb.Th, which is not surprising, because AMLM was parameterized with thickness measurements. Both the traditional and automatic AMLM segmentation tend to overestimate the bone volume fraction and trabecular thickness, except for the AMLM measures at the lowest resolution.

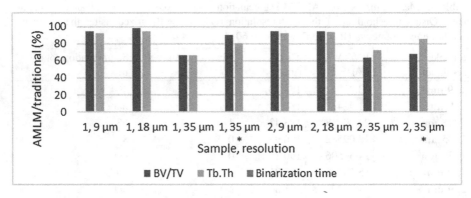

Fig. 3. Percent ratio of the automatic and traditional segmentation measurements for different VOIs (sample, resolution). * = Segmented using the alternative AMLM thresholding parameters.

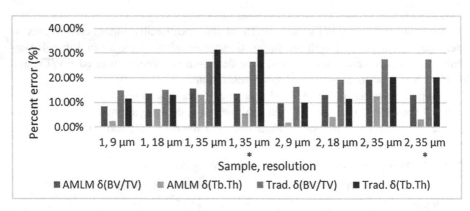

Fig. 4. Percent error of the automatic and traditional segmentation measurements compared to the best resolution raw image measurements. * = Segmented using the alternative AMLM thresholding parameters.

5 Conclusion

In this study, we segmented trabecular bone µCT scans with the LBP-based AMLM segmentation. We have demonstrated a successful application of the generalization of the adaptive mean thresholding and Bradley's method in 3D as part of the segmentation. To our knowledge, neither of these thresholding methods has been adapted into 3D and applied to analysis of medical images before.

We have shown how neighborhood parameter adjustment affects the perceived similarity of the segmentation result to the source volume and suggested more optimal values based on the results.

We compared the automatic method to the traditional segmentation performed by two experienced users. The automatic method outperformed traditional segmentation in binarization speed by about 20000% in the worst case, and its relative BV/TV and Tb. Th measurement errors were respectively 33% and 57% smaller on the average. This indicates that the method was successful for the used equipment and samples. However, the varied Tb.Th measurement results indicate that the automatic segmentation method configuration is not reliable at low resolutions. Corresponding measurements of low resolution VOI segmentation configured visually with adaptive thresholding tool are more in line with the other measurements, and further suggest that thresholding parameter configuration is a problem.

The hybrid adaptive mean thresholding method was selected, because neither Bradley's method nor mean–C could produce generally satisfactory results by themselves, and mean filter was preferred over median and Gaussian filters because of its superior speed performance in this setting. This was also the reason why more sophisticated segmentation methods were rejected. However, these considerations will change if more efficient methods become available, or if computation power and memory become significantly less of a factor in the future.

A drawback of the new adaptive local thresholding is that it requires three external parameters. Optimal parameters cannot be based on simple rules, and they must be determined experimentally for each tissue, imaging modality, and resolution, for example with a thickness phantom like here. Even then, parameter configuration takes time and the manual process is prone to inaccuracies, especially at low resolutions, which undermines the repeatability of automated segmentation. Configuration can be made easier and more reliable with an interactive thresholding preview tool, which facilitates experimenting with different parameters starting with reasonable default values. We developed such tool and used it with improved results.

As future work, the segmentation method could be tested with more varied data to determine how our results are able to generalize and to which extent they depend on the characteristics of the data. Also, the performance of the AMLM compared to the original method remains to be evaluated in statistical bone microstructure analysis. The 3D thresholding method might find use in other applications where speed or computational simplicity is important.

References

1. Otsu, N.: A threshold selection method from grey-level histograms. IEEE Trans. Syst. Man Cybern. **9**, 62–66 (1979)
2. Modayur, B., Prothero, J., Ojemann, G., Maravilla, K., Brinkley, J.: Visualization-based mapping of language function in the brain. Neuroimage **6**, 245–258 (1997)
3. Caselles, V., Kimmel, R., Sapiro, G.: Geodesic active contours. Int. J. Comput. Vis. **22**, 61–79 (1997)
4. Uchida, S.: Image processing and recognition for biological images. Dev. Growth Differ. **55**, 523–549 (2013)
5. Thevenot, J., Chen, J., Finnilä, M., Nieminen, M., Lehenkari, P., Saarakkala, S., Pietikäinen, M.: Local binary patterns to evaluate trabecular bone structure from Micro-CT data: application to studies of human osteoarthritis. In: Agapito, L., Bronstein, M.M., Rother, C. (eds.) ECCV 2014. LNCS, vol. 8926, pp. 63–79. Springer, Cham (2015). doi:10.1007/978-3-319-16181-5_5
6. Zhang, Z.M., Li, Z.C., Jiang, L.S., Jiang, S.D., Dai, L.Y.: Micro-CT and mechanical evaluation of subchondral trabecular bone structure between postmenopausal women with osteoarthritis and osteoporosis. Osteoporis Int. **21**, 1383–1390 (2010)
7. Tabor, Z., Latała, Z.: 3D gray-level histomorphometry of trabecular bone - a methodological review. Image Anal. Stereol. **33**, 1–12 (2014)
8. Erlandsson, K., Buvat, I., Pretorius, P.H., Thomas, B.A., Hutton, B.F.: A review of partial volume correction techniques for emission tomography and their applications in neurology. Cardiol. Oncol. Phys. Med. Biol. **57**, R119–R159 (2012)
9. Finnilä, M.A., Thevenot, J., Aho, O.M., Tiitu, V., Rautiainen, J., Kauppinen, S., Nieminen, M.T., Pritzker, K., Valkealahti, M., Lehenkari, P., Saarakkala, S.: Association between subchondral bone structure and osteoarthritis histopathological grade. J. Orthop. Res. **35**(4), 785–792 (2016)
10. Ojala, T., Pietikäinen, M., Harwood, D.: A comparative study of texture measures with classification based on featured distributions. Pattern Recogn. **29**, 51–59 (1996)
11. Ojala, T., Pietikäinen, M., Mäenpää, T.: Gray scale and rotation invariant texture classification with local binary patterns. In: Vernon, D. (ed.) ECCV 2000. LNCS, vol. 1842, pp. 404–420. Springer, Heidelberg (2000). doi:10.1007/3-540-45054-8_27
12. Huang, D., Shan, C., Ardabilian, M., Wang, Y., Chen, L.: Local binary patterns and its application to facial image analysis: a survey. IEEE Trans. Syst. Man. Cybern. C **41**, 765–781 (2011)
13. Bordallo López, M., Nieto, A., Boutellier, J., Hannuksela, J., Silvén, O.: Evaluation of real-time LBP computing in multiple architectures. J. Real-Time Image Proc. (2014). doi:10.1007/s11554-014-0410-5
14. Bradley, D., Roth, G.: Adaptive thresholding using the integral image. J. Graph. Tools **12**, 13–21 (2007)
15. Department of Artificial Intelligence in the University of Edinburgh. http://homepages.inf.ed.ac.uk/rbf/HIPR2/adpthrsh.htm

Local Adaptive Wiener Filtering for Class Averaging in Single Particle Reconstruction

Ali Abdollahzadeh[1,2(✉)], Erman Acar[1], Sari Peltonen[1],
and Ulla Ruotsalainen[1]

[1] Laboratory of Signal Processing, Tampere University of Technology,
P.O.Box 553, 33101 Tampere, Finland
ali.abdollahzadeh@uef.fi
[2] Biomedical Imaging Unit, A.I.Virtanen Institute for Molecular Sciences,
University of Eastern Finland, P.O.Box 1627, 70211 Kuopio, Finland

Abstract. In cryo-electron microscopy (cryo-EM), the Wiener filter is the optimal operation – in the least-squares sense – of merging a set of aligned low signal-to-noise ratio (SNR) micrographs to obtain a class average image with higher SNR. However, the condition for the optimal behavior of the Wiener filter is that the signal of interest shows stationary characteristic thoroughly, which cannot always be satisfied. In this paper, we propose substituting the conventional Wiener filter, which encompasses the whole image for denoising, with its local adaptive implementation, which denoises the signal locally. We compare our proposed local adaptive Wiener filter (LA-Wiener filter) with the conventional class averaging method using a simulated dataset and an experimental cryo-EM dataset. The visual and numerical analyses of the results indicate that LA-Wiener filter is superior to the conventional approach in single particle reconstruction (SPR) applications.

Keywords: Electron microscopy · Local adaptive Wiener filter · Class averaging · Single particle reconstruction · Spectral signal-to-noise ratio

1 Introduction

Revealing 3D structures of biomolecular assemblies in their native environment, with sub-nanometer to more recently near-atomic resolution, single particle reconstruction (SPR) is a highly demanded technique in the field of molecular and cellular biology. The technique is especially desired for the structural studies of biomolecules when their crystals are difficult to obtain with sufficient quality for crystallography [1]. SPR uses a collection of cryo-electron microscopy (cryo-EM) images representing the projections of the same particle from different spatial views. The standard procedure is comprised of detecting the particles inside the cryo-EM images, classifying them according to their orientations, averaging the images in each class and reconstructing the particle in 3D. Due to very high radiation sensitivity of the biological specimens and thus low

A. Abdollahzadeh and E. Acar – These authors contributed equally to this work.

© Springer International Publishing AG 2017
P. Sharma and F.M. Bianchi (Eds.): SCIA 2017, Part II, LNCS 10270, pp. 233–244, 2017.
DOI: 10.1007/978-3-319-59129-2_20

maximal allowed electron exposure [2], individual particle images suffer from extremely low signal-to-noise ratio (SNR). In addition, optical characteristic of the microscope that can be summarized in the contrast transfer function (CTF) causes loss of information in the Fourier domain at regular intervals. In order to overcome these problems, large number of images (typically more than 5000) is used in SPR. Therefore, the image processing techniques to classify and merge these images play an important role to reconstruct the particle in 3D with high resolution and accuracy.

The Wiener filter is the optimal estimator of the desired signal from noisy measurements in a stationary Gaussian process by minimizing the mean-squared error [3, 4]. It is used in SPR to obtain class averages with enhanced SNR by merging the images corresponding to the same spatial orientation [5]. The filter utilizes the spectral-SNR (SSNR) of the images in the class averaging process. The ability of the Wiener filter to restore the signal depends on the accuracy of the SSNR estimation. Several studies suggest to replace the SSNR with a constant value in the absence of an accurate SSNR estimate [6, 7]. Alternatively, using other sources of information such as pre-existing low resolution models is another approach to calculate the SSNR [8]. These methods may suffer from bias to the model. Other earlier works for tackling the SSNR problem suggest using the average image as reference to determine the SSNR of each individual image: averaging N micrographs reduces the variance of the noise by a factor of N relative to an individual micrograph, provided that the noise is independent and identically distributed in all images in the set [5]. However, a more recent study demonstrates that the conventional Wiener filter (conv-Wiener filter) fails in the optimal estimation of the particle density itself, as the particle is substantially surrounded by signal-free solvent region in the raw data images. Therefore, the output of conv-Wiener filter is optimal considering the whole image, but not when the particle, i.e. the region of interest, is considered only [9]. Their modified version of the Wiener filter is designed to optimize the signal estimate within a predefined mask region.

In this paper, we present a modified Wiener filter to merge the micrographs and determine the class average images in SPR. Since the optimality of the Wiener filter is based on the assumption that the signal is stationary and the additive noise is Gaussian distributed, violation of this condition reduces the denoising accuracy of the filter. It is easier to satisfy these conditions locally rather than globally as different locations of the experimental EM micrographs may contain diverse distributions. This insight motivated us to alter conv-Wiener filter into a local adaptive one. This filter is extensively used in image/video denoising and it is successful in the sense that it effectively removes noise while preserving important image features such as edges [10]. This definition of the Wiener filter advantages from a sliding window applied throughout the aligned micrograph images to establish a least-squares estimate of the class average. The extraction of the SSNR is implemented based on the average of the projection images. In the next sections, we will formulate the problem and present the qualitative and quantitative evaluation of the proposed method in comparison to the conventional implementation.

2 Materials and Methods

2.1 Datasets

2.1.1 Synthetic Data

Using UCFS Chimera [11], a 3D phantom model was generated from Protein Data Bank (PDB) atomic coordinate data of the 50S ribosomal subunit. The simulated model was a $256 \times 256 \times 256$ voxel array with isotropic voxel size of 1.7 Å/voxel. A 2D noise-free projection was obtained by taking the Radon transform of the phantom in a certain spatial angle. To construct a synthetic dataset, the noise-free projection was firstly contaminated with structural noise. The structural noise represents the irreproducible component of the specimen resulted from ice matrix, thin carbon film or incomplete particles [12]. Then, a mixture of Poisson and Gaussian noise was added to the result to represent the electron scattering statistics and CCD detector noise (see Fig. 1a–d). In each step, noise variance was determined in such a way that the final image shows some extent of similarity to the experimental EM projections. The SNR of the final synthetic image was ~ 0.39. We also introduced random rotation in the range of $[-1, 1]$ degree, and random translation of $[-1, 1]$ pixel to the images to represent the alignment errors. Three random synthetic projection images are depicted in Fig. 1d–f. We generated three sets of projection images containing $N = 5$, 30 and 100 micrographs.

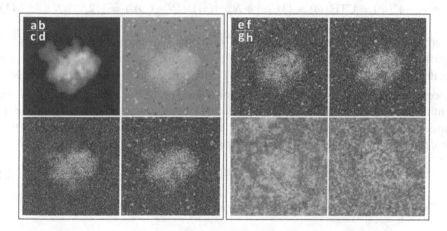

Fig. 1. (a)–(d) Steps of generating a synthetic particle image: (a) 256×256 noise-free projection image. (b) Structural noise represents the irreproducible component of the specimen; we added the structural noise to the noise-free projection image before image formation. (c) Noise-free projection image is contaminated with mixture of Poisson and Gaussian noise. (d) Synthetic projection image resulted from contaminating (b) with mixture of Poisson and Gaussian noise; SNR of the synthetic particle images is ~ 0.39. (e), (f) 2 random particle images of the synthetic dataset. (g), (h) 2 random particle images of the experimental dataset.

2.1.2   Experimental Data

In order to evaluate the effect of the proposed merging approach on real data, we examined the method on the previously published cryo-EM dataset. The dataset is comprised of 82,575 images of the 50S ribosomal subunit [13]. The images are 100×100 pixels with isotropic pixel size of 3.26 Å/pixel. They were CTF-corrected with ACE2, a variation of the ACE1 [14] algorithm. A subset of 10,000 images was randomly selected from the original dataset. Two random projection images are shown in Fig. 1g, h. The 2D translations and the 3D projection angles of the selected particle images were estimated by a reference-based method using rotation-free cross-correlation. The reconstruction result in the cryo-EM dataset was low-pass filtered to 70 Å and used as the reference model. The particle images were classified into 196 classes. In each class, projections are sorted in a descending order based on their consistency with the reference image. We used classes with large number of particle images (over 200 projections) for our experiments.

2.2   Class Averaging Methods

2.2.1   Class Averaging with Conventional Wiener Filter in SPR

Defining a set of N images representing different realizations of the particle images in the same spatial orientation, and considering the additive noise model, the image formation in the Fourier domain can be expressed as:

$$Y^i(\omega) = \text{CTF}^i(\omega) \times \left(M(\omega) + N^i_{strc}(\omega)\right) + N^i_{img}(\omega), i = 1, 2, \ldots, N \qquad (1)$$

where ω is a 2D Fourier coordinate and $Y^i(\omega)$ is the ith image of the set. $M(\omega)$ is the signal of interest common to all images, $N^i_{strc}(\omega)$ is the structural noise related to the fluctuations from the sample itself, $N^i_{img}(\omega)$ is the image noise from the measurement process and $\text{CTF}^i(\omega)$ is the contrast transfer function of the ith micrograph. In the continuation, we consider two assumptions: (1) both structural and image noise is considered with an overall term $N^i(\omega)$ [9]; and (2) images are CTF corrected, so that we define CTF = 1. Then, Eq. (1) can be simplified as:

$$Y^i(\omega) = M(\omega) + N^i(\omega). \qquad (2)$$

Defining the estimated signal by the Wiener filter as $M(\omega)$, we have:

$$\widehat{M}(\omega) = \sum_i W(\omega) Y^i(\omega), \qquad (3)$$

$$W(\omega) = \frac{1}{N + 1/\text{SSNR(R)}}. \qquad (4)$$

Here, $W(\omega)$ is the Wiener filter coefficients and R is the annulus corresponding to the radial frequency in the Fourier domain [15]. Averaging all the projection images to provide a good estimation for SSNR in Eq. (4), we write:

$$SSNR(R) = \frac{\sum_R \sum_{i=1}^{N} |Y^i(\omega)|^2}{\sum_R \sum_{i=1}^{N} \left|Y^i(\omega) - \frac{1}{N}\sum_{i=1}^{N} Y^i(\omega)\right|^2}. \tag{5}$$

In a previous study, SSNR was divided by a coefficient $f_{particle}$ to optimize the Wiener filter considering only the particle and not the whole image. $f_{particle}$ is defined as the fraction of the image containing the particle [9]. In our implementation of conv-Wiener filter, we also employed $f_{particle}$ in the SSNR calculations.

2.2.2   Class Averaging with Local Adaptive Wiener Filter in SPR

In order to focus on the local regions in the image plane, we extract overlapping blocks with size $s \times s$ in one-pixel-sliding manner in the spatial domain. We treat the set of extracted sub-images independently [16]. To implement the local adaptive Wiener filter (LA-Wiener filter), we can reformulate Eqs. (3)–(5) as:

$$SSNR_{s,x}(R) = \frac{\sum_R \sum_{i=1}^{N} \left|FT(y_{s,x}^i)\right|^2}{\sum_R \sum_{i=1}^{N} \left|FT(y_{s,x}^i) - \frac{1}{N}\sum_{i=1}^{N} FT(y_{s,x}^i)\right|^2}, \tag{6}$$

$$\hat{\mu}_{s,x} = FT^{-1}\left(\frac{\sum_{i=1}^{N} FT(y_{s,x}^i)}{N + \frac{1}{SSNR_{s,x}(R)}}\right), \tag{7}$$

where, FT stands for Fourier transform, $y_{s,x}^i$ denotes an $s \times s$ block, centered at spatial location x of the image y^i in the spatial domain. $SSNR_{s,x}$ and $\hat{\mu}_{s,x}$ are the SSNR and signal estimate of each block, respectively. After denoising all the blocks separately, they are concatenated to reconstruct the whole image, in which overlapping regions of the blocks are averaged.

2.3   Numerical Analysis Method

Numerically evaluating the quality of the synthetic class averages, we calculated reference-based Fourier ring correlation (FRC_{ref}): the noise-free particle image (Fig. 1a) is the reference image. For the numerical analysis of the experimental data, we used even/odd FRC ($FRC_{e/o}$) analysis. Therefore, as the particle images of each class are sorted based on their consistency with the reference model, they are divided into even/odd half-sets. The class average of each half-set is calculated and the consistency between the results is calculated by the $FRC_{e/o}$ curves [17].

3 Results and Discussion

In order to compare LA-Wiener filter with the conventional method of class averaging, we analyzed the result of merging different number of synthetic projection images by means of visual and numerical analysis. The study is extended into the experimental dataset for different number of projections, classes and window sizes.

3.1 Synthetic Dataset

Results of merging 5, 30, and 100 particle images with conv- and LA-Wiener filter are shown in Fig. 2. The window size of 3×3 is selected for LA-Wiener filter. Visual impression of the resultant images indicate that the proposed method is better than conv-Wiener filter in all cases. Increasing the number of images, results of LA-Wiener filter are smoother and at the same time contain sharper edges. For instance, the boundary of the L1 stalk of the 50S ribosome is distinguishable from the solvent only with 30 particle images, when LA-Wiener filter is used. The numerical analysis of the results is also in line with the visual impression. The FRC_{ref} curves are measured compared to the noise free projection shown (i.e. ground truth) in Fig. 2. Considering the pixel size as 1.7 Å/pixel and 0.5 cut-off frequency, our proposed method gains at least 2 Å higher resolution compared to conv-Wiener filter for different number of projection images.

3.2 Experimental Dataset

We compared the result of merging different number of particle images ($N_{half-set}$ = 3, 10, 30, 50 and 100) with conv- and LA-Wiener filter for one class of the experimental dataset. The results are presented in Fig. 3. The window size of 3×3 is selected for LA-Wiener filter. Visual evaluation of the results indicate that LA-Wiener filter out-performs the conventional method for all cases, as it does in the synthetic experiment. The performance of LA-Wiener filter enhances considerably by increasing the number of projections. The images merged with LA-Wiener filter are less noisy and the edges were preserved better compared to the conventional merging method. For the numerical analysis, projection images of this class are divided into half-sets and $FRC_{e/o}$ is calculated for each case. Considering 0.5 cut-off frequency and the pixel size of 3.26 Å/pixel, the proposed method gains ~ 0 Å, ~ 1.5 Å, ~ 2 Å, ~ 3–4 Å and ~ 4–5 Å better resolution compared to conv-Wiener filter for $N_{half-set}$ = 3, 10, 30, 50 and 100, respectively. In Fig. 4, the superiority of LA-Wiener (window size of 3×3) compared to the conventional method is extended to 5 different classes for a fixed number of projection images ($N_{half-set}$ = 30). Visual impression and $FRC_{e/o}$ measurements of the results indicate that LA-Wiener filter is better than conv-Wiener filter at least by 2 Å for all classes. To evaluate the performance of LA-wiener more thoroughly, all the classes containing over 200 projections (19 classes) were analyzed. From each class, the first 6, 30, and 100 particle images were divided into half-sets ($N_{half-set}$ = 3, 15, 50). The consistency between resultant merged images of the half-sets were analyzed by $FRC_{e/o}$. Considering the 0.5 cut-off frequency and the pixel size of 3.26 Å/pixel, Fig. 5 plots the behavior of conv- and LA-Wiener filter in a compact representation. The results indicate that the LA-Wiener filter performs better for all classes: for lower number of images, LA-Wiener filter performs slightly better than conv-Wiener filter, however, increasing the number of projections, considerably enhances the performance of LA-Wiener filter almost for all classes.

The effect of the window size on LA-Wiener filter, for a fixed number of projection images ($N_{half-set} = 30$) in a single class is shown in Fig. 6. We examined window size of $3 \times 3, 5 \times 5, 7 \times 7, 9 \times 9$, and 11×11. The results indicate that enlarging the window

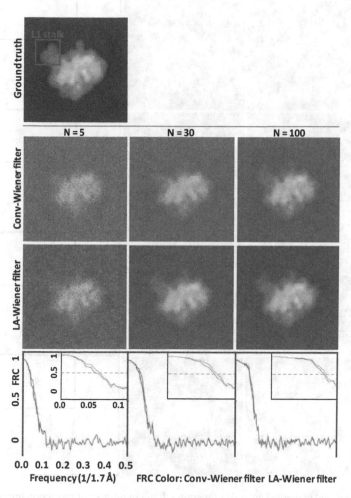

Fig. 2. Visual and numerical analysis of merging different number of projection images from the synthetic dataset. The visual impression of merging N = 5, 30 and 100 particle images indicates that LA-Wiener filter performs better than conv-Wiener filter in denoising and preserving the edges. LA-Wiener filter defines the boundary of the L1 stalk when only 30 projection images are used. Considering 0.5 cut-off frequency and 1.7 Å/pixel, LA-Wiener filter gains at least ~ 2 Å better resolution compared to conv-Wiener filter.

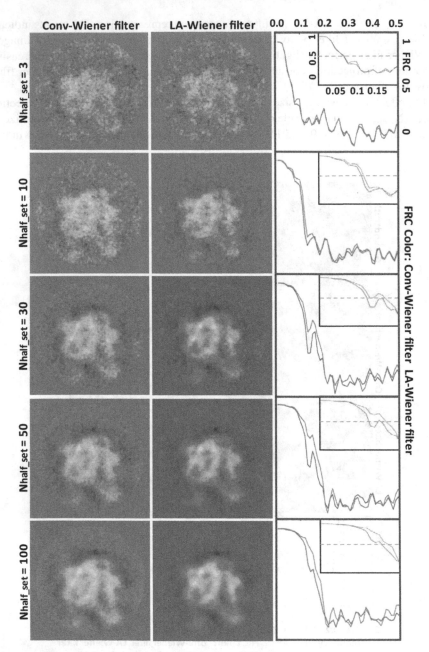

Fig. 3. Visual and numerical analysis of merging different number of projection images from the experimental dataset. The visual impression of merging $N_{half\text{-}set}$ = 3, 10, 30, 50 and 100 projection images of one class indicates that LA-Wiener filter performs better than conv-Wiener filter. Considering the pixel size of 3.26 Å/pixel, the $FRC_{e/o}$ curves imply that LA-Wiener filter gains at least $\sim$0 Å, $\sim$1.5 Å, $\sim$2 Å, $\sim$3–4 Å and $\sim$4–5 Å better resolution compared to conv-Wiener filter for $N_{half\text{-}set}$ = 3, 10, 30, 50 and 100, respectively.

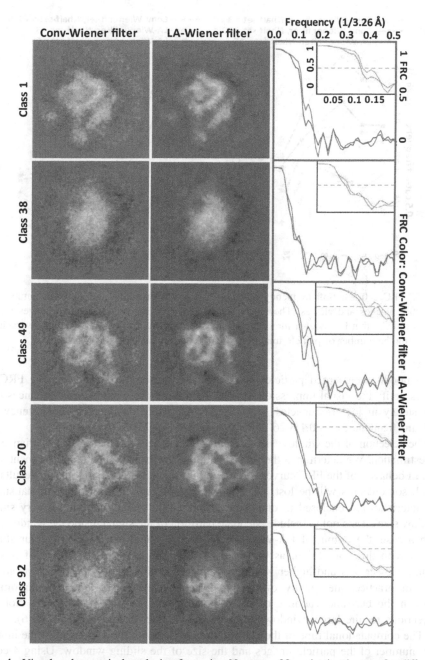

Fig. 4. Visual and numerical analysis of merging $N_{half-set} = 30$ projection images for different classes of the experimental dataset. The results of merging 30 projection images by conv- and LA-Wiener filter for different classes indicate that LA-Wiener filter performs better than the conventional implementation of the Wiener filter. Considering 0.5 cut-off frequency and pixel size of 3.2 Å/pixel, LA-Wiener filter gains ~ 2 Å better resolution compared to conv-Wiener filter in all classes.

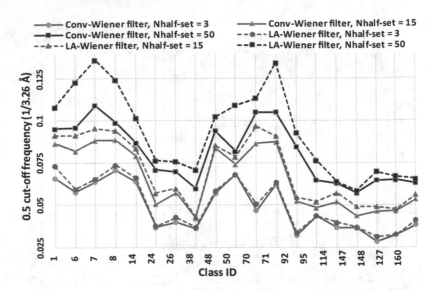

Fig. 5. FRC$_{e/o}$ 0.5 cut-off frequency of conv- and LA-Wiener filter for different number of projection images and classes. The results for N$_{half-set}$ = 3, 15, and 50 particle images in 19 classes indicate that LA-Wiener filter outperforms the conventional approach in class averaging. Increasing the number of particle images boosts the performance of LA-Wiener filter.

size smooths the merged particle images without preserving the edges. The FRC$_{e/o}$ curves verify the visual impression as well. Increasing the window size, enhances the consistency in low frequencies ($< {\sim} 0.1/3.26$ Å), while declines the consistency in medium frequencies ($\sim 0.1/3.26$ Å $<$ frequency $< \sim 0.2/3.26$ Å).

The selection of the window size is a compromise between the bias and variance of the estimation. We determined the proper window size based on visual assessment and overall behavior of the FRC curves. A large moving window oversmooths the results in which some details will be lost, i.e. high bias, since regions with different statistical characteristics are enclosed in one frame. On the other hand, applying a very small window size means not to include adequate partition of the signal inside the window. In such a case, the output of LA-Wiener filter contains high variance [16]. Our study agrees with this conclusion, as the results of merging with window size of 3×3 contains lower bias and higher variance compared to the results of 11×11 window size. In practice, the quality of the micrographs determines how to compromise between the bias and variance of the estimation, and so the window size. Noisier projections require larger windows to generate smoother class averages (Fig. 6).

The computational time of the algorithm increases with the increment of the image size, number of the particle images and the size of the sliding window. Using 4-core Intel CPU 3.41 GHz machine with 64 GB RAM, processing time of LA-Wiener filter in MATLAB for 30 synthetic particle images of size 256×256 with window size of 3×3 and 5×5 is 7.5 and 8.3 s, respectively.

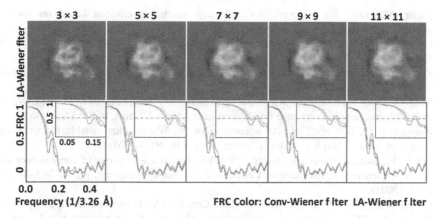

Fig. 6. Effect of window size on LA-Wiener filter. Window size of 3×3, 5×5, 7×7, 9×9 and 11×11 is assessed for a fixed number of projection images ($N_{half\text{-}set} = 30$) in a single class. The results indicate that increasing the window size of LA-Wiener filter blurs the merged particle images. The $FRC_{e/o}$ curves are in line with the visual impression: enlarging the window size enhances the consistency in low frequencies ($< \sim 0.1/3.26$ Å), while declines the consistency in medium frequencies ($\sim 0.1/3.26$ Å $<$ frequency $< \sim 0.2/3.26$ Å).

4 Conclusion

In this paper, we introduced LA-Wiener filter for merging a set of cryo-EM images in the class averaging step of SPR. To reformulate conv-Wiener filter, we considered local minimization of the mean-squared error by employing a sliding window throughout the image plane. As statistical properties exhibit more homogeneity in smaller blocks compared to the whole image, LA-Wiener filter performs more accurately compared to the conventional implementation of the Wiener filter. We verified the superiority of our proposed method for both synthetic and experimental datasets, when different number of projection images are included and for different classes to show its application for variety of cases in cryo-EM.

References

1. Frank, J.: Three-Dimensional Electron Microscopy of Macro-molecular Assemblies: Visualization of Biological Molecules in Their Native State. Oxford University Press, New York (2006)
2. Reimer, L., Kohl, H.: Transmission Electron Microscopy Physics of Image Formation. Springer, New York (2008)
3. Kuan, D.T., Sawchuk, A.A., Strand, T.C., Chavel, P.: Adaptive noise smoothing filter for images with signal-dependent noise. IEEE Trans. Pattern Anal. Mach. Intell. **7**, 165–177 (1985)
4. Lee, J.S.: Digital image enhancement and noise filtering by use of local statistics. IEEE Trans. Pattern Anal. Mach. Intell. **2**, 165–168 (1980)

5. Unser, M., Trus, B.L., Steven, A.C.: A new resolution criterion based on spectral signal-to-noise ratios. Ultramicroscopy **23**, 39–51 (1987)
6. Grigorieff, N.: FREALIGN: high-resolution refinement of single particle structures. J. Struct. Biol. **157**, 117–125 (2007)
7. Zeng, X., Stahlberg, H., Grigorieff, N.: A maximum likelihood approach to two-dimensional crystals. J. Struct. Biol. **160**, 362–374 (2007)
8. Tang, G., Peng, L., Baldwin, P.R., Mann, D.S., Jiang, W., Rees, I., Ludtke, S.J.: EMAN2: an extensible image processing suite for electron microscopy. J. Struct. Biol. **157**, 38–46 (2007)
9. Sindelar, C.V., Grigorieff, N.: An adaptation of the Wiener filter suitable for analyzing images of isolated single particles. J. Struct. Biol. **176**, 60–74 (2011)
10. Jin, F., Fieguth, P., Winger, L., Jernigan, E.: Adaptive Wiener filtering of noisy images and image sequences. In: International Conference on Image Processing, pp. III-349–III-352. IEEE (2003)
11. Pettersen, E.F., Goddard, T.D., Huang, C.C., Couch, G.S., Greenblatt, D.M., Meng, E.C., Ferrin, T.E.: UCSF Chimera—a visualization system for exploratory research and analysis. J. Comput. Chem. **25**, 1605–1612 (2004)
12. Baxter, W.T., Grassucci, R.A., Gao, H., Frank, J.: Determination of signal-to-noise ratios and spectral SNRs in cryo-EM low-dose imaging of molecules. J. Struct. Biol. **166**, 126–132 (2009)
13. Voss, N.R., Lyumkis, D., Cheng, A., Lau, P.W., Mulder, A., Lander, G.C., Brignole, E.J., Fellmann, D., Irving, C., Jacovetty, E.L., Leung, A., Pulokas, J., Quispe, J.D., Winkler, H., Yoshioka, C., Carragher, B., Potter, C.S.: A toolbox for ab initio 3-D reconstructions in single-particle electron microscopy. J. Struct. Biol. **169**, 389–398 (2010)
14. Mallick, S.P., Carragher, B., Potter, C.S., Kriegman, D.J.: ACE: automated CTF estimation. Ultramicroscopy **104**, 8–29 (2005)
15. Saxton, W.O.: Computer Techniques for Image Processing in Electron Microscopy. Academic Press Inc., New York (1978)
16. Lim, J.S.: Image restoration by short space spectral subtraction. In: IEEE International Conference on Acoustics, Speech, and Signal Processing, pp. 191–197. IEEE (1980)
17. Cardone, G., Grünewald, K., Steven, A.C.: A resolution criterion for electron tomography based on cross-validation. J. Struct. Biol. **151**, 117–129 (2005)

Comparison of Concave Point Detection Methods for Overlapping Convex Objects Segmentation

Sahar Zafari[1]([✉]), Tuomas Eerola[1], Jouni Sampo[2], Heikki Kälviäinen[1], and Heikki Haario[2]

[1] Machine Vision and Pattern Recognition Laboratory,
School of Engineering Science, Lappeenranta University of Technology,
Lappeenranta, Finland
{sahar.zafari,tuomas.eerola,heikki.kalviainen}@lut.fi
[2] Mathematics Laboratory, School of Engineering Science, Lappeenranta University
of Technology, Lappeenranta, Finland
{jouni.sampo,heikki.haario}@lut.fi

Abstract. Segmentation of overlapping convex objects has gained a lot of attention in numerous biomedical and industrial applications. A partial overlap between two or more convex shape objects leads to a shape with concave edge points that correspond to the intersections of the object boundaries. Therefore, it is a common practice to utilize these concave points to segment the contours of overlapping objects. Although a concave point has a clear mathematical definition, the task of concave point detection (CPD) from noisy digital images with limited resolution is far from trivial. This work provides the first comprehensive comparison of CPD methods with both synthetic and real world data. We further propose a modification to an earlier CPD method and show that it outperforms the other methods. Finally, we demonstrate that by using the enhanced concave points we obtain segmentation results that outperform the state-of-the-art in the task of partially overlapping convex object segmentation.

Keywords: Segmentation · Concavity · Concave points · Overlapping objects · Cell segmentation · Convex objects

1 Introduction

Segmentation of overlapping objects aims to address the issue of representation of multiple objects with partial views. Overlapping or occluded objects occur in various applications, such as morphology analysis of molecular or cellular objects in biomedical and industrial imagery where the quantitative analysis of individual objects by their size and shape is desired [1–3]. In many such applications, the objects can often be assumed to contain approximately convex shapes. Even with rather strong shape prior, segmentation of overlapping objects remains a challenging task. Deficient information from the objects with occluded or overlapping parts introduces considerable complexity into the segmentation process.

© Springer International Publishing AG 2017
P. Sharma and F.M. Bianchi (Eds.): SCIA 2017, Part II, LNCS 10270, pp. 245–256, 2017.
DOI: 10.1007/978-3-319-59129-2_21

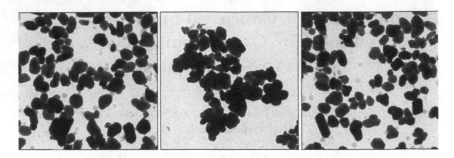

Fig. 1. Overlapping nanoparticles.

For example, in the context of contour estimation, the contours of objects intersecting with each other do not usually contain enough visible geometrical evidence, which can make contour estimation problematic and challenging. Frequently, the segmentation method has to rely purely on edges between the background and the foreground, which makes the processed image essentially a silhouette image (see Fig. 1). Furthermore, the task involves simultaneous segmentation of multiple objects. A large number of objects in the image causes a large number of variations in pose, size, and shape of the objects, and leads to a more complex segmentation problem.

A robust family of overlapping convex object segmentation approaches is based on concave point detection (CPD). These approaches consist of finding the concavity locations on the objects boundaries and utilize these concave points to segment the contours of overlapping objects in such way that each contour segment contains edge points from one object only. This provides an important cue for further object segmentation (contour estimation).

Several methods utilizing CPD have been proposed for overlapping object segmentation. In [2,4–6], the problem of overlapping objects segmentation are addressed using concave points detection and ellipse fitting. These approaches involve two main steps: contour evidence extraction and contour estimation. Contour evidence extraction starts by recovering contour segments from a binarized image using concave contour point detection. The contour segments which belong to the same objects are grouped using various heuristics. Finally, the contour estimation is implemented through a non-linear ellipse fitting problem in which partially observed objects are modeled in the form of ellipse-shape objects. In [7] the task of overlapping fibers segmentation is approached using the concave point detection and fiber skeleton information.

While CPD has already been shown to perform rather efficiently in the task of overlapping convex objects segmentation, no complete comparative analysis of existing CPD methods has been done. The key contribution of this work is to perform comprehensive comparison to ascertain the validity of CPD for overlapping convex objects segmentation using synthetic and real world datasets. Another contribution is the integration of the best CPD method into the segmentation of overlapping objects, enabling improvements compared to five existing segmentation methods with higher detection rate and segmentation accuracy.

The paper is organized as follows. Sect. 2 introduces the methodology used for the CPD methods. The CPD methods are applied to synthetic and real datasets and the results are presented in Sect. 3. Conclusions are drawn in Sect. 4.

2   Concave Point Detection

There exists a wide range of the CPD methods from which we selected those that have been applied for the task of overlapping objects segmentation. These methods can be categorized into four groups: curvature, skeleton, chord, and polygon approximation based approaches.

2.1   Curvature

In the curvature based methods, given the sequence of extracted contour points $C = \{c_1, c_2, ...\}$, the curvature value k is first computed for every contour points $c_i = (x_i, y_i)$ as

$$k = (x_i' y_i'' - y_i' x_i'')/(x_i'^2 + y_i'^2)^{3/2}. \tag{1}$$

After this, contour point c_i is classified as dominant point if i is local extrema point for curvature value k.

The detected dominant points $c_{d,i} \in C_{dom}$ may locate in both concave and convex regions of the objects contours. To detect the dominant points that are concave, various approaches exists. In the method proposed by Wen *et al.* [8] the detected dominant points are classified as concave if the value of maximum curvature is larger than a preset threshold value. In the method proposed by Zafari *et al.* [5], the dominant points are concave points if the line connecting $c_{d,i+1}$ to $c_{d,i-1}$ does not reside inside the object. In the method proposed by Dai *et al.* [7], the dominant points are qualified as concave points if the value of triangle area S for the points $c_{d,i+1}, c_{d,i}, c_{d,i-1}$ is positive and the points are arranged in order of counter-clockwise.

2.2   Skeleton

In skeleton based methods, the skeleton and boundary information of the objects are used to detect the concave points. In a method proposed by Samma *et al.* [9], the original image I is first eroded using a 3×3 structuring element SE:

$$E = I \ominus SE. \tag{2}$$

Next, the original image I is subtracted from the eroded image E in order to extract the contour points

$$C = I - E \tag{3}$$

Then, the complement of the original image $\bar{I}$ is dilated

$$D = \bar{I} \oplus SE \tag{4}$$

and the skeleton SK of the dilated image is determined. Finally the concave points are identified as the intersections between the skeleton and contour points as

$$C_{con} = C \cap SK. \tag{5}$$

In the method proposed by Wang et al. [10], the objects are first skeletonized and contours are extracted. Next, for every contour points c_i the shortest distance from the skeleton is computed. The concave points are detected as the local minimum on the distance histogram whose value is higher than a certain threshold.

2.3   Chord

In chord based methods the concave points are identified as points in concave regions of contours that have the maximum distance to the concave area chord. In the method proposed by Farhan et al. [11], the concave region is obtained by evaluating the line fitted to the contour points. The concavity criterion of contour segments along two contour points is satisfied if the line that joins the two contour points does not reside inside an object. The local chord corresponding to that concave region is a line connecting the sixth adjacent contour point on the either side of the current point on the concave region of the object contours.

In the method proposed by Kumar et al. [12], the concave points are extracted using the boundaries of concave regions and their corresponding convex hull chords. In this way, the concave points are defined as points on the boundaries of concave regions that have maximum perpendicular distance from the convex hull chord. The boundaries and convex hull are obtained using a method proposed in [13].

2.4   Polygon Approximation

Polygonal approximation is a well-known method to represent the objects contours by a sequence of dominant points. It can be used to reduce the complexity, smooth the objects contours, and avoid detection of false concave points.

Given the sequence of extracted contour points $C = \{c_1, c_2, ...\}$, the dominant points are determined by co-linear suppression. To be specific, every contour point c_i is examined for co-linearity, while it is compared to the previous and the next successive contour points. The point c_i is considered as the dominant point if it is not located on the line connecting $c_{i-1}(x_{i-1}, y_{i-1})$ and $c_{i+1}(x_{i+1}, y_{i+1})$ and the distance d_i from c_i to the line connecting $c_{i-1}(x_{i-1}, y_{i-1})$ to $c_{i+1}(x_{i+1}, y_{i+1})$ is bigger than a pre-set threshold $d_i > d_{th}$.

The distance d_i is obtained by:

$$d_i = \sqrt{\frac{((x_i - x_{i-1})(y_{i+1} - y_{i-1}) - (y_i - y_{i-1})(x_{i+1} - x_{i-1}))^2}{(x_{i-1} - x_{i+1})^2 + (y_{i-1} - y_{i+1})^2}}. \qquad (6)$$

After polygon approximation and dominant points detection, the concave points can be obtained using methods proposed by Bai et al. [4] and Zhang et al. [2].

In the method proposed by Bai et al. [4], the dominant point $c_{d,i} \in C_{dom}$ is defined as concave if the concavity value of $c_{d,i}$ is within the range of a_1 to a_2 and the line connecting $\overline{c_{d,i+1}c_{d,i-1}}$ does not pass through the inside the objects.

The concavity value of $c_{d,i}$ is defined as the angle between lines $(c_{d,i-1}, c_{d,i})$ and $(c_{d,i+1}, c_{d,i})$ as follows:

$$\begin{cases} |\, a(c_{d,i-1}, c_{d,i}) - a(c_{d,i+1}, c_{d,i})\,| & \text{if } |a(c_{d,i-1}, c_{d,i}) - a(c_{d,i+1}, c_{d,i})| < \pi \\ \pi - |a(c_{d,i-1}, c_{d,i}) - a(c_{d,i+1}, c_{d,i})| & \text{otherwise} \end{cases} \qquad (7)$$

where

$$a(c_{d,i-1}, c_{d,i}) = \tan^{-1}((y_{d,i-1} - y_{d,i})/(x_{d,i-1} - x_{d,i})) \qquad (8)$$

and

$$a(c_{d,i+1}, c_{d,i}) = \tan^{-1}((y_{d,i+1} - y_{d,i})/(x_{d,i+1} - x_{d,i})). \qquad (9)$$

In the method proposed by Zhang et al. [2], the dominant point $c_{d,i} \in C_{dom}$ is considered to be a concave point if $\overrightarrow{c_{d,i-1}c_{d,i}} \times \overrightarrow{c_{d,i}c_{d,i+1}}$ is positive:

$$C_{con} = \{c_{d,i} \in C_{dom} \; : \; \overrightarrow{c_{d,i-1}c_{d,i}} \times \overrightarrow{c_{d,i}c_{d,i+1}} > 0\}. \qquad (10)$$

Both Bai et al. [4] and Zhang et al. [2] define the threshold value d_{th} manually. Our proposal is to use the method proposed by Prasad et al. [14] for the polygonal approximation and detection of the dominant points. In this method first the line connecting $c_{i-1}(x_{i-1}, y_{i-1})$ and $c_{i+1}(x_{i+1}, y_{i+1})$ is digitized and then the threshold value d_{th} is selected automatically based on the angular distance between the slope of the actual line and the digitized line. When the Prasad method combined with the concave point criterion proposed by Zhang et al. [2] the concave point detection becomes fully parameter free. We refer this method as Prasad+Zhang.

3   Experiments

This section presents the data, the performance measures, and the results for the concave point detection and segmentation.

3.1 Data

The experiments were carried out using one synthetically generated dataset and one dataset from a real-world application.

The synthetic dataset (see Fig. 2(a)) consists of images with overlapping ellipse-shape objects that are uniformly randomly scaled, rotated, and translated. Three subsets of images were generated to represent different degrees of overlap between objects. The dataset consists of 150 sample images divided into three classes of overlap degree. The maximum rates of overlapping area allowed between two objects are 40%, 50%, and 60%, respectively, for the first, second, and third subset. Each subset of images in the dataset contains 50 images of 40 objects. The minimum and maximum width and length of the ellipses are 30, and 45 pixels. The image size is 300 × 400 pixels.

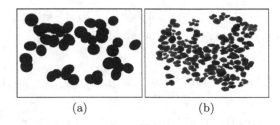

(a)   (b)

Fig. 2. Example images from the datasets studied: (a) Synthetic dataset with the maximum overlap of 40%; (b) Nanoparticles dataset.

The real dataset (nanoparticles dataset) contains nanoparticles images captured using transmission electron microscopy (see Fig. 2(b)). In total, the dataset contains 11 images of 4008 × 2672 pixels. Around 200 particles were marked manually in each image by an expert. The annotations consist of manually drawn contours of the objects. This information was also used to determine the concave points. Since not all the objects are marked, a pre-processing step is applied to eliminate the unmarked objects from the images. It should be noted that the images consist of dark objects on a white background and, therefore, pixels outside the marked objects could be colored white without making the images considerably easier to analyze.

3.2 Performance Measures

To evaluate the method performance and to compare the methods, two specific performance measures, True Positive Rate (TPR), $TPR = \frac{TP}{TP+FN}$, and Positive Predictive Value (PPV), $PPV = \frac{TP}{TP+FP}$, were used. True Positive (TP) is the number of correctly detected concave points or segmented objects, False Positive (FP) is the number of incorrectly detected concave points or segmentation results, and False Negative (FN) is the number of missed concave points or objects.

To determine whether a concave point was correctly detected (TP), the distance to the ground truth concave points was computed and the decision was made using a predefined threshold value. The threshold value was set to 10 pixels. The average distance (AD) from detected concave points to the ground truth concave points was used as the third performance measure for the concave points detection.

To decide whether the segmentation result was correct or incorrect, Jaccard Similarity coefficient (JSC) [15] was used. Given a binary map of the segmented object O_s and the ground truth particle O_g, JSC is computed as

$$JSC = \frac{O_s \cap O_g}{O_s \cup O_g}. \tag{11}$$

The threshold value for the ratio of overlap (JSC threshold) was set to 0.5. The average JSC (AJSC) value was also used as the third measure to evaluate the segmentation performance, besides TPR and PPV.

The parameters of the CPD methods were set experimentally to obtain the best possible result for each method. All the methods have been implemented by ourselves based on the original papers. The implementation made by the corresponding authors [14] was used for polygonal approximation in the Prasad+Zhang method.

3.3   Results

Concave Point Detection. The results of the concave points detection methods applied to the synthetic and nanoparticles datasets are presented in Tables 1 and 2 respectively.

From the result with the synthetic dataset (Table 1) it can be seen that Zhang [2] scored the highest value of TPR and AD, but it suffers from the low value of PPV. This is because the method produces a large number of false concave points. In terms of PPV, Zafari [5] and Kumar [6] ranked the highest. However these approaches tend to under estimate the number of concave points as they often fail to detect the concave points at low depth of concave regions. Considering all the scores together, the results obtained from the synthetic dataset showed that the proposed Prasad+Zhang method achieved the best performance.

The result on the nanoparticles dataset (Table 2) show similar results and Prasad [14]+Zhang [2] outperforms the others with the highest TPR value and competitive PPV and AD values. It should be noted that Prasad [14]+Zhang [2] performs better than Zhang [2] in the nanoparticles dataset since it is more robust in presence of objects with noisy boundaries.

Figure 3 shows example results of concave point detection methods applied to a patch of a nanoparticles image. It can be seen that while Zafari [5], Dai [7], Kumar [6] and Samma [9] suffer from false negatives and Zhang [2], Bai [4],

Table 1. Comparison of the performance of concave point detection methods on the synthetic dataset.

Methods	Type	Overlapping rate[%]	TPR [%]	PPV [%]	AD [%]
Prasad [14]+Zhang [2]	Polygon approximation	40	97	97	1.47
Zhang [2]	Polygon approximation	40	**98**	37	**1.25**
Bai [4]	Polygon approximation	40	92	46	1.34
Zafari [5]	Curvature	40	81	**98**	2.04
Wen [8]	Curvature	40	97	61	1.75
Dai [7]	Curvature	40	71	97	2.58
Samma [9]	Skeleton	40	75	83	2.20
Wang [10]	Skeleton	40	87	86	1.87
Kumar [6]	Chord	40	64	**98**	2.64
Farhan [11]	Chord	40	91	18	1.77
Prasad [14]+Zhang [2]	Polygon approximation	50	**97**	96	1.50
Zhang [2]	Polygon approximation	50	96	38	**1.25**
Bai [4]	Polygon approximation	50	86	47	1.34
Zafari [5]	Curvature	50	77	**98**	2.08
Wen [8]	Curvature	50	94	62	1.81
Dai [7]	Curvature	50	68	97	2.57
Samma [9]	Skeleton	50	73	83	2.24
Wang [10]	Skeleton	50	83	86	1.95
Kumar [6]	Chord	50	60	**98**	2.62
Farhan [11]	Chord	50	87	19	1.81
Prasad [14]+Zhang [2]	Polygon approximation	60	94	96	1.51
Zhang [2]	Polygon approximation	60	**96**	40	**1.31**
Bai [4]	Polygon approximation	60	82	37	1.47
Zafari [5]	Curvature	60	73	**98**	2.13
Wen [8]	Curvature	60	91	64	1.79
Dai [7]	Curvature	60	63	98	2.61
Samma [9]	Skeleton	60	67	82	2.26
Wang [10]	Skeleton	60	80	88	1.96
Kumar [6]	Chord	60	58	**98**	2.71
Farhan [11]	Chord	60	82	19	1.91

Wen [8] and Farhan [11] suffer from false positives, Prasad [14]+Zhang [2] suffers from neither false positives or false negatives.

To provide an assessment of the usability of CPD methods for the task of overlapping object segmentation, comparison of processing times is presented in Table 2. The processing times were computed for a single nanoparticles image using a PC with a 3.20 GHz CPU and 8 GB of RAM, running the MATLAB implementation of all the methods. Prasad [14]+Zhang [2] achieved the lowest computation time.

Segmentation. To demonstrate the effect of the more accurate concave points on the segmentation of partially overlapping objects we replaced the concave

Table 2. Comparison of the performance of concave point detection methods on the nanoparticles dataset.

Methods	Type	TPR [%]	PPV [%]	AD [%]	Computation Time (s)
Prasad [14]+Zhang [2]	Polygon approximation	**96**	90	1.47	**0.50**
Zhang [2]	Polygon approximation	84	38	**1.42**	2.28
Bai [4]	Polygon approximation	79	40	1.55	2.52
Zafari [5]	Curvature	65	**95**	2.30	2.48
Wen [8]	Curvature	94	51	1.97	2.13
Dai [7]	Curvature	54	87	2.84	2.50
Samma [9]	Skeleton	80	84	2.28	2.95
Wang [10]	Skeleton	65	58	2.37	4.95
Kumar [6]	Chord	33	94	3.09	8.17
Farhan [11]	Chord	77	15	2.09	8.82

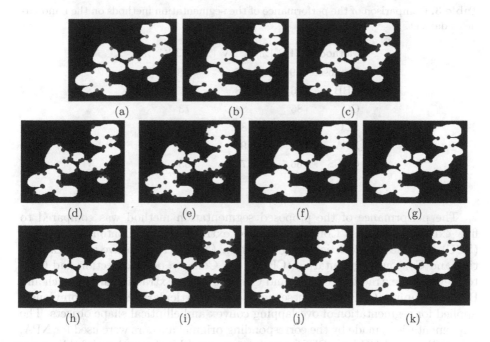

Fig. 3. Comparison of the performance of the concave point detection methods on the nanoparticles dataset: (a) Ground Truth; (b) Prasad [14]+Zhang [2]; (c) Zhang [2]; (d) Bai [4]; (e) Wen [8]; (f) Zafari [5]; (g) Dai [7]; (h) Kumar [6]; (i) Farhan [11]; (j) Wang [10]; (k) Samma [9].

points detection part of the segmentation framework proposed in [16] with the proposed Prasad [14]+Zhang [2] method. The original implementation used the Zafari method [5] for concave points detection.

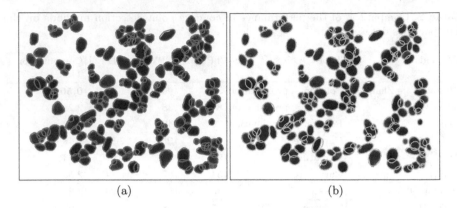

(a)   (b)

Fig. 4. An example of the proposed method segmentation result on nanoparticles dataset: (a) Ground truth; (b) Proposed method.

Table 3. Comparison of the performance of the segmentation methods on the nanoparticles dataset.

Methods	TPR [%]	PPV [%]	AJSC [%]
Proposed	**89**	87	**78**
CBB	86	88	77
CC	85	84	73
SCC	79	89	72
NPA	62	**90**	58
CECS	66	73	53

The performance of the proposed segmentation method was compared to five existing methods: Contour evidence extraction and Contour estimation (SCC) [17] and Nanoparticles Segmentation (NPA) [1], Concave-point Extraction and Contour Segmentation (CECS) [2], Concave point detection and Contour evidence extraction (CC) [5] and Concave-point Extraction and Branch and Bound grouping (CBB) [16]. These methods are particularly chosen as previously applied for segmentation of overlapping convex and elliptical shape objects. The implementations made by the corresponding original authors were used for NPA, CC, CBB, and SCC [18]. CECS was implemented by the authors of this paper based on [2]. Figure 4 shows an example of segmentation result.

The corresponding performance statistics of the competing methods applied to the nanoparticles dataset are shown in Table 3. As it can be seen, the proposed method outperformed the other five methods with respect to TPR and JSC, and achieved comparable performance with NPA in terms of PPV. The high JSC value of the proposed method indicates its superiority with respect to the resolved overlap ratio.

4 Conclusions

In this paper, we reviewed existing the concave point detection methods for the problem of partially overlapping convex objects segmentation and evaluated the methods using synthetic and real world datasets. We further proposed a novel parameter-free modification of an earlier method and showed that it outperforms the earlier methods. Finally we utilized the best concave point detection method with partially overlapping convex objects segmentation framework [16]. The experiments showed that the enhanced segmentation method achieved high detection and segmentation accuracies and outperformed five competing methods on real world dataset. The proposed method relies only on edge information and can be applied also to other segmentation problems where the objects are partially overlapping and have an approximately convex shape.

References

1. Park, C., Huang, J.Z., Ji, J.X., Ding, Y.: Segmentation, inference and classification of partially overlapping nanoparticles. IEEE Trans. Pattern Anal. Mach. Intell. **35**, 669–681 (2013)
2. Zhang, W.H., Jiang, X., Liu, Y.M.: A method for recognizing overlapping elliptical bubbles in bubble image. Pattern Recogn. Lett. **33**, 1543–1548 (2012)
3. Kothari, S., Chaudry, Q., Wang, M.: Automated cell counting and cluster segmentation using concavity detection and ellipse fitting techniques. In: IEEE International Symposium on Biomedical Imaging, pp. 795–798 (2009)
4. Bai, X., Sun, C., Zhou, F.: Splitting touching cells based on concave points and ellipse fitting. Pattern Recogn. **42**, 2434–2446 (2009)
5. Zafari, S., Eerola, T., Sampo, J., Kälviäinen, H., Haario, H.: Segmentation of partially overlapping nanoparticles using concave points. In: Bebis, G., Boyle, R., Parvin, B., Koracin, D., Pavlidis, I., Feris, R., McGraw, T., Elendt, M., Kopper, R., Ragan, E., Ye, Z., Weber, G. (eds.) ISVC 2015. LNCS, vol. 9474, pp. 187–197. Springer, Cham (2015). doi:10.1007/978-3-319-27857-5_17
6. Yeo, T., Jin, X., Ong, S., Sinniah, R., et al.: Clump splitting through concavity analysis. Pattern Recogn. Lett. **15**, 1013–1018 (1994)
7. Dai, J., Chen, X., Chu, N.: Research on the extraction and classification of the concave point from fiber image. In: IEEE 12th International Conference on Signal Processing (ICSP), pp. 709–712 (2014)
8. Wen, Q., Chang, H., Parvin, B.: A delaunay triangulation approach for segmenting clumps of nuclei. In: IEEE Sixth International Conference on Symposium on Biomedical Imaging: From Nano to Macro, ISBI 2009, pp. 9–12. IEEE Press, Piscataway (2009)
9. Samma, A.S.B., Talib, A.Z., Salam, R.A.: Combining boundary and skeleton information for convex and concave points detection. In: IEEE Seventh International Conference on Computer Graphics, Imaging and Visualization (CGIV), pp. 113–117 (2010)
10. Wang, W., Song, H.: Cell cluster image segmentation on form analysis. In: IEEE Third International Conference on Natural Computation (ICNC), vol. 4, pp. 833–836 (2007)

11. Farhan, M., Yli-Harja, O., Niemistö, A.: A novel method for splitting clumps of convex objects incorporating image intensity and using rectangular window-based concavity point-pair search. Pattern Recogn. **46**, 741–751 (2013)
12. Kumar, S., Ong, S.H., Ranganath, S., Ong, T.C., Chew, F.T.: A rule-based approach for robust clump splitting. Pattern Recogn. **39**, 1088–1098 (2006)
13. Rosenfeld, A.: Measuring the sizes of concavities. Pattern Recogn. Lett. **3**, 71–75 (1985)
14. Prasad, D.K., Leung, M.K.H.: Polygonal representation of digital curves. In: Stanciu, S.G. (ed.) Digital Image Processing, pp. 71–90. InTech (2012)
15. Choi, S.S., Cha, S.H., Tappert, C.C.: A survey of binary similarity and distance measures. J. Systemics Cybern. Inform. **8**, 43–48 (2010)
16. Zafari, S., Eerola, T., Sampo, J., Kälviäinen, H., Haario, H.: Segmentation of partially overlapping convex objects using branch and bound algorithm. In: Chen, C.-S., Lu, J., Ma, K.-K. (eds.) ACCV 2016. LNCS, vol. 10118, pp. 76–90. Springer, Cham (2017). doi:10.1007/978-3-319-54526-4_6
17. Zafari, S., Eerola, T., Sampo, J., Kälviäinen, H., Haario, H.: Segmentation of overlapping elliptical objects in silhouette images. IEEE Trans. Image Process. **24**, 5942–5952 (2015)
18. Zafari, S., Eerola, T., Sampo, J., Kälviäinen, H., Haario, H.: Segmentation of overlapping objects (2016). http://www2.it.lut.fi/project/comphi1/index.shtml. Accessed Aug 2016

Decoding Gene Expression in 2D and 3D

Maxime Bombrun[1(✉)], Petter Ranefall[1], Joakim Lindblad[1], Amin Allalou[1],
Gabriele Partel[1], Leslie Solorzano[1], Xiaoyan Qian[2], Mats Nilsson[2],
and Carolina Wählby[1]

[1] Division of Visual Information and Interaction, Science for Life Laboratory,
Department of Information Technology, Uppsala University,
Lägerhyddsvägen 2, 751 37 Uppsala, Sweden
{maxime.bombrun,petter.ranefall,joakim.lindblad,amin.allalou,
gabriele.partel,leslie.solorzano,carolina.wahlby}@it.uu.se
[2] Science for Life Laboratory, Department of Biochemistry and Biophysics,
Stockholm University, Tomtebodavägen 23, 171 65 Solna, Sweden
{xiaoyan.qian,mats.nilsson}@scilifelab.se

Abstract. Image-based sequencing of RNA molecules directly in tissue samples provides a unique way of relating spatially varying gene
expression to tissue morphology. Despite the fact that tissue samples
are typically cut in micrometer thin sections, modern molecular detection methods result in signals so densely packed that optical "slicing"
by imaging at multiple focal planes becomes necessary to image all signals. Chromatic aberration, signal crosstalk and low signal to noise ratio
further complicates the analysis of multiple sequences in parallel. Here a
previous 2D analysis approach for image-based gene decoding was used
to show how signal count as well as signal precision is increased when
analyzing the data in 3D instead. We corrected the extracted signal measurements for signal crosstalk, and improved the results of both 2D and
3D analysis. We applied our methodologies on a tissue sample imaged in
six fluorescent channels during five cycles and seven focal planes, resulting in 210 images. Our methods are able to detect more than 5000 signals representing 140 different expressed genes analyzed and decoded in
parallel.

Keywords: 2D and 3D signal detection · Microscopy based in situ
sequencing · Image processing & analysis · Crosstalk compensation

1 Introduction

Digital pathology is making its way into modern clinical diagnosis, increasing
the need for automated digital image analysis methods for fast and reproducible
quantification of tissue morphology [1]. In multi-cellular organisms, all cells have
the same genes, while at the same time different cell types have different functions. The identity and function of a cell is defined by the gene expression (i.e.,
transcription). Thus, analysis of gene expression provides valuable information
on health and disease, e.g. by identifying different types of immune cells or

© Springer International Publishing AG 2017
P. Sharma and F.M. Bianchi (Eds.): SCIA 2017, Part II, LNCS 10270, pp. 257–268, 2017.
DOI: 10.1007/978-3-319-59129-2_22

metastatic tumor cells. Most existing approaches for analysis of gene expression are based on bulk analysis of larger tissue samples, making it impossible to correlate gene expression with individual cells. More recently, analysis of individual cells has been made possible [2], but requires cells to be removed from the tissue architecture, resulting in loss of spatial information.

Our previously published methods for image-based *in situ* sequencing of expressed genes allow multiplexed gene expression profiling at cellular resolution in intact tissue samples, and thus opens up for detailed large-scale comparison of genotype and phenotype [3]; similar approaches have later been developed by others [4,5]. Expressed genes are detected by molecular probes, locally amplified by rolling circle amplification, and decoded by sequential staining and imaging cycles. Each cycle targets the four letters of the genetic code with different fluorescent colors (see Fig. 1). By controlled design of probes, such that each probe contains a known "barcode" (i.e., sequence of nucleotide bases), it is known a priori what sequences of signals to expect across fluorescent colors and sequencing cycles, and only the number of signals as well as their location are unknown. Multiple molecular probes, targeting genes are typically used in parallel, and as little as five cycles of decoding can detect as many as $4^5 = 1024$ distinct barcodes in the same tissue sample. By comparing the number of expected barcodes to

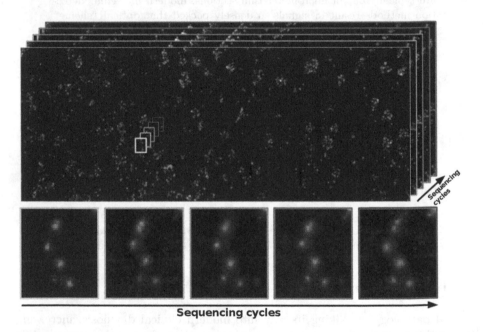

Sequencing cycles

Fig. 1. Amplified expressed genes (here enhanced for visualization purposes) in a tissue sample imaged in five sequencing cycles. In each cycle, four different fluorescent probes target each of the four letters of the genetic code. In this illustration, cyan=A, orange=C, magenta=G, green=T. The sequence of colors in a given position reveals the barcode of a unique expressed gene. (Color figure online)

the number of unexpected barcodes (most likely originating from random noise and autofluorescence), it is possible to evaluate precision as well as efficiency (number of detected signals) of an image analysis approach.

Considering image size and richness of information, computerized image processing provides tools for enabling spatially resolved information and quantitative measurements. Tissue samples are typically cut in slices of a few micrometers prior to analysis, yet the data is typically collected by imaging the sample at multiple focal planes, acquiring a stack of images representing a 3D volume. An argument for this is that the different micrometer-sized signals often lie in different focal planes, making it impossible to collect an image where all are focused at once. Despite the data being 3D, all analysis approaches previously described were based on 2D projections of such 3D volumes. This is true also for our own previous approach [3,6] implemented in TissueMaps [7], a platform for 2D giga-pixel image analysis and visualization built on free and open-source software.

As more genes are targeted in parallel, and the efficiency of the molecular methods increases, the signals in the tissue samples become denser. This means that a lot of information will be lost when relying on 2D projections for signal decoding. To avoid over-crowding, one has to limit the amplification step, meaning that a more complete analysis of gene expression comes at the cost of lower signal-to-noise ratios and signals close to the resolution limits of the microscope. Images are shifted between imaging cycles due to the manual staining/washing procedures, and signals from different fluorophores may be shifted due to chromatic aberration which further complicates the data analysis. Correction for chromatic aberrations has been suggested for similar methods by others: Briefly, the methods of [4,5] first correct the effects of chromatic aberrations, respectively, through deconvolution and morphological opening followed by background subtraction. Then, the alignment is done on the maximum-intensity projection (MIP) along the z-dimension and using brick-based algorithm and cross-correlation of MIP along the c-dimension. Finally, the signal detection is completed by a per-pixel base calling and barcoding evaluation for maxima above a specified threshold value in a log-filtered version of the aligned images.

In this study, we approached the challenge of analyzing a full 3D data set with four color channels and five sequencing cycles. We compared the output to our previously published 2D approach [3,6] applied to the same dataset projected to 2D. Finally, these results were improved using a post-processing crosstalk compensation to better separate the different color channels, and thus, correct some unexpected transcripts to expected barcodes. The methods were evaluated by comparing the number of detected signals by each method as well as the ratio of expected versus unexpected barcodes of targeted genes.

2   Image Acquisition

A total of 140 gene transcripts were targeted within a 10 μm thick tissue sample, and subjected to five cycles of sequencing by ligation as previously described [3].

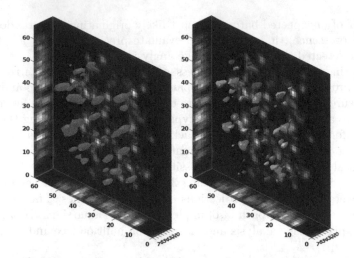

Fig. 2. 3D visualizations of the image showing that more than one signal can appear along the z-direction (*blue axis*). The background images are the maximum-intensity projection of the slices in the general stain channel. The left image shows an example of the spatial distribution of the signals in the general stain. The right image shows the spatial distribution of the individual signal detection separated channel-wise (one per color). This illustrates the need for a detection in 3D since some signals are merged during the projection. (Color figure online)

Images were acquired at seven different focal depths, 1.4 μm apart, to create a 3D image volume using an Axioplan II epifluorescence microscope with a numerical aperture of 0.8 and a nominal magnification of 20.0 at 610 μm distance. In each sequencing cycle, the four letters of the genetic code, A, C, G, and T, were fluorescence stained with Cy5, Texas Red, Cy3, and Fluorescein respectively. Furthermore, a general stain (AF750) marking all targets and a nuclear stain (DAPI) were also added to visualize signal distribution and tissue morphology, resulting in a total of six color channels. The resulting image volume is 2048 × 2048 pixels, with a z-dimension of seven, a color dimension of six and a time dimension (=sequencing cycles or t) of five, for a total of 210 images to process. A cut out volume of 63 × 66 × 7 voxels from one color channel at one time point is shown in Fig. 2, illustrating signal size, noise and resolution in different spatial dimensions. Note that signals located in the same (x,y) position, but different z-positions will be merged when working with projected data in 2D.

3   Image Analysis

The global workflow aims to align and normalize the data prior to sequence decoding, as illustrated in Fig. 3. The challenges lie in (i) image registration, compensating for alignment shifts along the sequencing cycles and chromatic shift,

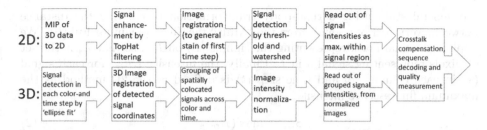

Fig. 3. Workflow of the original 2D method and the proposed 3D approach. The signal decoding relies on measuring the signal intensity at the same position for all fluorescent channels at each sequencing step. Therefore, registration is needed, in the 2D case it is a registration of the image data, while in 3D a registration of detected signals. Post-processing including crosstalk compensation increase signal confidence, as defined by the quality measurement.

(ii) signal detection and normalization, and (iii) signal decoding. The signal-to-noise ratio (SNR) in the images is limited by the trade off between exposure time during image acquisition and bleaching of the stains. The longer the exposure time, the higher the SNR, but at the same time there is an increased risk of bleaching signals in neighboring focal planes. In order to detect as many true signals as possible, we have decided to have a more inclusive approach for signal detection. Following signal decoding noise and true signals were discriminated using a quality measurement as described in Sect. 3.3.

3.1 2D Approach

For the 2D approach, we used our previously published method [3,6], implemented in the TissueMaps workflow [7]. TissueMaps is built on free and open source tools, and the analysis workflow makes use of the CellProfiler software [8]. The 2D analysis started with the MIP of the image stack (reducing the z-dimension to one). Following the MIP, for each cycle (t), each image channel (I_{tc}, c representing either the general stain or one of the four letters of the genetic code) was first enhanced by a top-hat transformation (F_{tc}) with a structuring element (B) consisting of a disc with radius 10 pixels:

$$F_{tc} = (I_{tc} - I_{tc} \circ B), \tag{1}$$

where $\circ$ is a morphological opening. Individual signals were then defined by a labeled mask (L) in the general stain channel (D) of the first sequencing cycle ($t = 1$), by a fixed intensity threshold, low enough to detect the signals after the top-hat (here, equal to 0.5). Finally, clustered signals were separated by shape-based watershed segmentation, i.e., a watershed applied on the negated distance transform, since signals are relatively circular [9]. Filtered images (F_{tc}) from the same sequencing cycle were thereafter registered ($R_{tc} = registered(F_{tc})$) towards the general stain using a rigid-body transformation (preserving the distance between every pair of points), from the "MultiStackReg" plugin for Fiji [10]. We applied the final mask representing the signals (L) on R_{tc}, so that L_s is the set of pixels representing signal s in L. Finally, the intensity (S_{stc}) for each signal (s) in each channel (c) and time step (t) is defined in the 2D method as the maximum fluorescence intensity:

$$S_{stc} = \max_{p \in L_s} (R_{tc})_p \tag{2}$$

We specifically extracted its (x, y) location as well as the intensity of this location in each of the five color channels (general stain and four letters of the genetic code), and five time steps in order to later decode and evaluate the signal as described in Sect. 3.3.

3.2   3D Approach

In the 3D approach, signals were separately detected in all color channels at all time steps using a local thresholding approach referred to as Per Object Ellipsefit (POE) [11]. The POE method computes local adaptive thresholds for each individual object (signal) where the threshold values are set to optimize the ellipse (ellipsoid in 3D) fit. This is done by creating a component tree [12] and traversing the pixels in order of decreasing intensities. Ellipsoid fit is defined by computing the moment matrix M for each object, extracting the axes from the eigenvalues of M, and computing the ratio between the actual object volume and an ideal ellipsoid with the dimensions given by these axes. The search for the best ellipsoid fit is done within given ranges for object volume (36–96 voxels), major and minor axis length (3–8 pixels), and value of the ellipsoid fit (≥ 0.5).

Following signal detection, 3D spatial coordinates of detected signals were aligned and grouped. Within each time step (sequencing cycle) the color channels representing A, C, G, and T were affinely registered to the general stain of that same time step, using Iterative Closest Point (ICP) [13], followed by a spline based ICP version [14,15], with a grid of $6 \times 6 \times 5$ control points, that further corrects any chromatic aberration. Once the channels were aligned within each cycle, the general stain of each cycle was aligned with the general stain of the first cycle, used as a reference, utilizing rigid ICP registration. The associated channels of the time step were aligned using the same transformation.

Due to digitization effects and noise, slight shifts in the detected signals, for different color channels and time steps, remain also after the registration. Detected signals closer to each other than 3.4 pixels were merged together as one spot. As (x, y) location of a merged signal s we use the centroid of the corresponding cluster of (registered) signals. The intensity values of all merged signals were extracted from the smoothed (gaussian filter with $\sigma = 0.5$) and dilated (ball-shaped structural element with five pixels diameter) original images, utilizing the inverse of the respective registration transformations.

Intensity measures were normalized separately for each channel (c) and time (t), such that signals with a brightness equal to the mode of the respective image volume gets the value zero, and the mean detected signal intensity is mapped to the value one:

$$S_{stc} = \frac{R_{stc} - mode(R_{tc})}{\frac{1}{N_s}(\sum_{l=1}^{N_s} R_{ltc}) - mode(R_{tc})},$$

(3)

where S_{stc} is the intensity of signal s in channel c and time t for the 3D method, and N_s is the total number of detected signals.

Due to the inclusive intensity threshold used for the signal detection, artifacts from random background noise may have been detected as well. After normalization, a quality check based on the general stain was applied to reduce such noise. We require that the general stain channel (D), for each cycle (c), presents each signal detected (s), so that the following condition holds for all cycles:

$$\frac{S_{stc|c=D}}{\max_{c\in\{A,C,G,T\}} S_{stc}} \geq 0.1$$

(4)

This step reduces the number of signals by approximately 2%.

3.3   Sequence Decoding and Quality Measurement

We measured the respective quality value for the 2D and 3D methods to evaluate the consistency of the signals detected. For each sequencing cycle, every location containing a signal is assigned the base, A, C, G or T, decided on the highest image intensity (following top-hat (2D) or normalization according to Eq. (3) (3D)). Autofluorescence may result in false signals that have a high intensity across all sequencing cycles, but always display the same color (that is, always appear in the same color channel). Such signals will appear as "homopolymers", e.g. barcodes consisting of a single letter, such as 'AAAAA' or 'GGGGG'. No such signals were included in the expected barcodes, and they are removed from our set of detected signals, reducing the number of signals by 0.6%.

To evaluate the signals detected, a quality Q_{st} of a signal s, in the cycle t was defined as:

$$Q_{st} = \frac{\max_{c\in\{A,C,G,T\}} S_{stc}}{\sum_{c\in\{A,C,G,T\}} S_{stc}}$$

(5)

The quality score of the full sequence Q_s of signal s is further defined by the quality of its "weakest" cycle:

$$Q_s = \min_{t\in\{1,2,...,N_t\}} Q_{st}$$

(6)

The quality score ranges from $\frac{1}{N_t}$ (i.e., all signals equal) to 1 (all non-max signals equal to 0).

3.4 Crosstalk Compensation

Intensity values detected from each of the five sequencing cycles were crosstalk compensated in order to color-correct the intensities and determine the real dye concentration present in each signal. The sequencing cycles can not be assumed to be independent from each other, but the sequencing process and the image acquisition is affected by several kinds of cycle-dependent noise (e.g., focus, imperfect image registration, chromatic aberration, photobleaching, and other experimental conditions), meaning that the crosstalk between channels may vary cycle to cycle. Therefore a separate crosstalk compensation matrix for each of the sequencing cycles was estimated. Each crosstalk matrix X_t was estimated as in Sect. 2.2.6 of Li and Speed [16], inverted and multiplied by the matrix of the intensities of all signals s of cycle t, producing crosstalk compensated intensity values:

$$\begin{bmatrix} X_{stA} \\ X_{stC} \\ X_{stG} \\ X_{stT} \end{bmatrix} = X_t^{-1} \begin{bmatrix} S_{stA} \\ S_{stC} \\ S_{stG} \\ S_{stT} \end{bmatrix} \tag{7}$$

We measured a new quality value for the methods by replacing the intensity S_{stc} in Eq. 5 by the compensated intensity value X_{stc}.

3.5 Validation Approach

The only "ground truth" available for this type of image data is the a priori knowledge of the barcodes of the probes applied to the tissue section. In this particular experiment, 140 different probes were applied. The barcode length is five letters, meaning that our decoding approach may find $4^5 = 1024$ different codes, but only 140 out of these codes are to be expected (TP), and it can be assumed that any other code found is noise due to poor signal detection/decoding and is considered as unexpected (FP). There are of course also other sources of error, such as actual errors in the probes, but these will affect the 2D and 3D approach equally. Using the quality measure described in Sect. 3.3 an acceptance threshold can be set to balance the signal count vs. the signal precision (TP/(TP+FP)).

4 Results

4.1 Validation

Out of the 1024 possible barcodes, only 140 correspond to the barcodes of our targeted gene transcripts. If decoded signals were completely random (and the four homopolymers removed), a precision of $140/(1024-4) = 0.14$ would be expected. From Fig. 4, showing number of detected TP versus number of detected FP depending on the quality threshold value, we can see that the new 3D approach detects more signals than the 2D method with a higher ratio of TP over

FP (respectively red curve and blue curve). The alignment method in the 2D workflow produces part of the FP signals due to its difficulty to find control points to define the transformation, especially in this noisy dataset. Moreover, the MIP tends to overcrowd the working plane so that two signals may overlap and corrupt the decoding process. The crosstalk compensation improves the 2D workflow by correcting some of the unexpected barcodes and thus, improving these results (blue dashed curve). On the other hand, the 3D approach is able to extract more robust information through the z-dimension which helps for both the registration process and for the spatial localization of the signal as they are better separated. These better results are also improved through the crosstalk compensation (red dashed curve). Consequently, assuming an acceptable ratio of one FP for four TP, i.e. a precision of 0.8, then we obtained respectively 2641 and 2968 TP for the 2D and 3D method, which increase to 3622 and 4742 TP with the crosstalk compensation (black square markers).

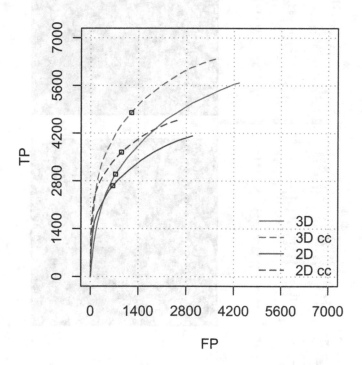

Fig. 4. Comparison of the original 2D method and the proposed 3D approach by plotting true positive signals (TP) against the false positive signals (FP) at various quality threshold settings. *The red and blue curves* show the signals detected by the 3D and 2D approaches respectively, before compensation for crosstalk. *The dashed curves* show the results after the application of crosstalk compensation. Precision, i.e., TP/(TP+FP), increases for both the 2D and the 3D approach when crosstalk compensation is applied as shown by the *black square markers* for a precision of 0.8. (Color figure online)

4.2 Visualisation

We confirmed the spatial localization of the transcripts detected by using the TissueMaps platform. Currently, this platform allows the display of 54 different symbols to localize the genes on a 2D image at different resolutions. We chose the projected general stain image as background and displayed the 54 most common barcodes (sum of the two methods) among the total 140 genes detected by our methods (Fig. 5).

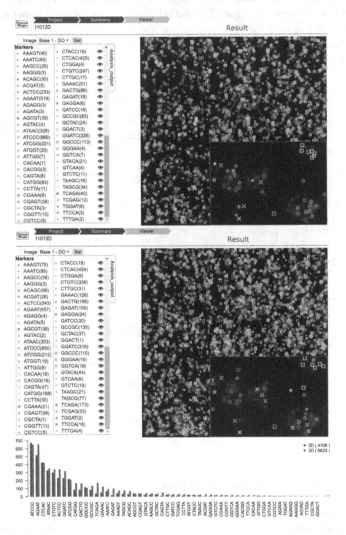

Fig. 5. Visualization of the 54 most common transcripts among the total 140 expected barcodes using the TissueMaps platform. The top image is the result of the 2D approach while the middle visualization corresponds to the 3D method. The bottom bar plot represents these 54 transcript counts for the 2D (in blue) and 3D (in red) methods. (Color figure online)

5   Discussions

Digital microscopes can capture 3D images of signals emitted by molecular detection probes by recording data at multiple focal planes of the imaged tissue samples. While the current 2D method of gene decoding, by applying a MIP, provides an overview of the stack with a good SNR, it tends to overcrowd the signals and lose some of the individuality. The 3D approach, presented in this study, analyzes the different slices of the tissue volume to detect more of separated signals. The individual transcripts are in the same proportion (in respect to the total number detected) and present the same global pattern in the tissue (Fig. 5).

The advantages of the 3D method also come from the improvement and use of new steps. The images were normalized based on their mode and mean, and the segmentation was applied to each 3D volume (four channels × five sequencing cycles) individually rather than on the general stain. This allows the 3D method to compensate the SNR channel-wise, similarly to the top-hat in the 2D approach, but also to have a better definition of the individuality in each channel where signal overlap could occur in the general stain.

We also improved the general quality measurement and gene decoding by incorporating crosstalk compensation. This allows us to correct some of the unexpected barcodes based on the signal intensities (in each channel) and the general tendency of the signals to switch from one base to another. For both methods, the crosstalk compensation as a post-process converts around a thousand of false positive signals into true positive signals (Fig. 4) and increases the precision of our results.

Acknowledgments. The authors would like to thank Johan Oelrich for contributions to the development of TissueMaps, and the European Research council for funding via ERC Consolidator grant 682810 to C. Wählby.

References

1. Gurcan, M.N., Boucheron, L.E., Can, A., Madabhushi, A., Rajpoot, N.M., Yener, B.: Histopathological image analysis: a review. IEEE Rev. Biomed. Eng. **2**, 147–171 (2009)
2. Darmanis, S., Sloan, S.A., Zhang, Y., Enge, M., Caneda, C., Shuer, L.M., Gephart, M.G.H., Barres, B.A., Quake, S.R.: A survey of human brain transcriptome diversity at the single cell level. Proc. Nat. Acad. Sci. **112**(23), 7285–7290 (2015)
3. Ke, R., Mignardi, M., Pacureanu, A., Svedlund, J., Botling, J., Wählby, C., Nilsson, M.: In situ sequencing for RNA analysis in preserved tissue and cells. Nat. Methods **10**(9), 857–860 (2013)
4. Lee, J.H., Daugharthy, E.R., Scheiman, J., Kalhor, R., Yang, J.L., Ferrante, T.C., Terry, R., Jeanty, S.S., Li, C., Amamoto, R., Peters, D.T., Turczyk, B.M., Marblestone, A.H., Inverso, S.A., Bernard, A., Mali, P., Rios, X., Aach, J., Church, G.M.: Highly multiplexed subcellular RNA sequencing in situ. Science **343**(6177), 1360–1363 (2014)
5. Shah, S., Lubeck, E., Zhou, W., Cai, L.: In situ transcription profiling of single cells reveals spatial organization of cells in the mouse hippocampus. Neuron **92**(2), 342–357 (2016)

6. Pacureanu, A., Ke, R., Mignardi, M., Nilsson, M., Wählby, C.: Image based in situ sequencing for RNA analysis in tissue. In: 2014 IEEE 11th International Symposium on Biomedical Imaging (ISBI), pp. 286–289. IEEE (2014)

7. Ranefall, P., Pacureanu, A., Avenel, C., Carpenter, A.E., Wählby, C.: The gigapixel challenge: full resolution image analysis without losing the big picture: an open-source approach for multi-scale analysis and visualization of slide-scanner data. In: SSBA 2014, Symposium of the Swedish Society for Automated Image Analysis, Luleå, Sweden (2014)

8. Carpenter, A.E., Jones, T.R., Lamprecht, M.R., Clarke, C., Kang, I.H., Friman, O., Guertin, D.A., Chang, J.H., Lindquist, R.A., Moffat, J., Golland, P., Sabatini, M.: Cell profiler: image analysis software for identifying and quantifying cell phenotypes. Genome Biol. 7(10), R100 (2006)

9. Malpica, N., Ortiz de Solorzano, C., Vaquero, J.J., Santos, A., Vallcorba, I., Garcia-Sagredo, J.M., Pozo, F.D.: Applying watershed algorithms to the segmentation of clustered nuclei. Cytometry 28, 289–297 (1997)

10. Thevenaz, P., Ruttimann, U.E., Unser, M.: A pyramid approach to subpixel registration based on intensity. IEEE Trans. Image Process. 7(1), 27–41 (1998)

11. Ranefall, P., Sadanandan, S.K., Wählby, C.: Fast adaptive local thresholding based on ellipse fit. In: International Symposium on Biomedical Imaging (ISBI 2016), Prague, Czech Republic (2016)

12. Najman, L., Couprie, M.: Building the component tree in quasi-linear time. IEEE Trans. Image Process. 15(11), 3531–3539 (2006)

13. Amberg, B., Romdhani, S., Vetter, T.: Optimal step nonrigid ICP algorithms for surface registration. In: 2007 IEEE Conference on Computer Vision and Pattern Recognition, pp. 1–8. IEEE (2007)

14. Rueckert, D., Sonoda, L.I., Hayes, C., Hill, D.L., Leach, M.O., Hawkes, D.J.: Nonrigid registration using free-form deformations: application to breast MR images. IEEE Trans. Med. Imaging 18(8), 712–721 (1999)

15. Lee, S., Wolberg, G., Shin, S.Y.: Scattered data interpolation with multilevel B-splines. IEEE Trans. Vis. Comput. Graph. 3(3), 228–244 (1997)

16. Li, L., Speed, T.P.: An estimate of the crosstalk matrix in four-dye fluorescence-based DNA sequencing. Electrophoresis 20(7), 1433–1442 (1999)

Estimation of Heartbeat Peak Locations
and Heartbeat Rate from Facial Video

Mohammad A. Haque[✉], Kamal Nasrollahi,
and Thomas B. Moeslund

Visual Analysis of People (VAP) Laboratory,
Aalborg University, Aalborg, Denmark
{mah, kn, tbm}@create.aau.dk

Abstract. Available systems for heartbeat signal estimations from facial video only provide an average of Heartbeat Rate (HR) over a period of time. However, physicians require Heartbeat Peak Locations (HPL) to assess a patient's heart condition by detecting cardiac events and measuring different physiological parameters including HR and its variability. This paper proposes a new method of HPL estimation from facial video using Empirical Mode Decomposition (EMD), which provides clearly visible heartbeat peaks in a decomposed signal. The method also provides the notion of both color- and motion-based HR estimation by using HPLs. Moreover, it introduces a decision level fusion of color and motion information for better accuracy of multi-modal HR estimation. We have reported our results on the publicly available challenging database MAHNOB-HCI to demonstrate the success of our system in estimating HPL and HR from facial videos, even when there are voluntary internal and external head motions in the videos. The employed signal processing technique has resulted in a system that could significantly advance, among others, health-monitoring technologies.

Keywords: Heartbeat Rate · Facial video · Head motion · Facial skin color · Empirical Mode Decomposition · Multimodal fusion

1 Introduction

Heartbeat signals represent Heartbeat Peak Locations (HPLs) in a temporal domain and help physicians assess the condition of a human cardiovascular system by detecting cardiac events and measuring different important physiological parameters such as Heartbeat Rate (HR) and its variability [1]. When the human heart pumps blood, subtle chromatic changes in the facial skin and slight head motion occur periodically. These changes and motion are associated with the periodic heartbeat signal and can be detected in a facial video [2].

Takano et al. first utilized the trace of skin color changes in facial video to extract heartbeat signal and estimate HR [3]. They recorded the variations in the average brightness of the Region of Interest (ROI) – a rectangular area on the subject's cheek – to estimate HR. About two years later, Poh et al. proposed a method that used ROI mean color values as color traces from facial video acquired by a simple webcam, and employed Independent Component Analysis (ICA) to separate the periodic signal

© Springer International Publishing AG 2017
P. Sharma and F.M. Bianchi (Eds.): SCIA 2017, Part II, LNCS 10270, pp. 269–281, 2017.
DOI: 10.1007/978-3-319-59129-2_23

sources and a frequency domain analysis of an ICA component to measure HR [4]. Kwon et al. improved Poh's method in [4] by using merely green color channel instead of all three Red-Green-Blue (RGB) color channels [5]. Wei et al. employed a Laplacian Eigenmap (LE) and Cheng et al. employed an Empirical Mode Decomposition (EMD), rather than ICA, to obtain the uncontaminated heartbeat signal and demonstrated better performance than the ICA-based method [6, 7]. Other articles contributed a peripheral improvement of the color-based HR measurement by using a better estimation of ROI [8], adding a supervised machine learning component to the system [9], analyzing the distance between the camera and the face during data capture [10], and selecting best facial patches using majority voting [11].

Color-based methods suffer from tracking sensitivity to color noise and illumination variation. Thus, Balakrishnan et al. proposed a method for HR estimation which was based on invisible motion in the head due to pulsation of the heart muscles, which can be obtained by a Ballistocardiogram [12]. In this approach, some feature points were automatically selected on the ROI of the subject's facial video frames. These feature points were tracked by the Kanade–Lucas–Tomasi (KLT) feature tracker [13] to generate some trajectories, and then a Principle Component Analysis (PCA) was applied to decompose trajectories into a set of orthogonal signals based on Eigen values. Selection of the heartbeat rate was accomplished by using the percentage of the total spectral power of the signal, which accounted for the frequency with the maximum power and its first harmonic. A semi-supervised method in [14] was proposed to improve the results of Balakrishnan's method by using the Discrete Cosine Transform (DCT) along with a moving average filter rather than the Fast Fourier Transform (FFT) employed in Balakrishnan's work. The method in [15] also utilized motion information; however, unlike [12] it used ICA (previously used in color-based methods) to decompose the signal.

Though estimation of heartbeat signal from facial video was investigated a lot with different applications in the literature [2, 16–18], in view of the previous methods described above we summarized the following demands/challenges to investigate in this paper:

i. Previous methods provide an average HR over a certain time period, e.g. 30–60 s. Average HR alone is not sufficient to reveal some conditions of the cardiovascular system [12]. Health monitoring personnel often ask for a more detailed view of heartbeat signals with visible peaks that indicate the beats. Moreover, an important vital sign, heartbeat rate variability, cannot be obtained without locating heartbeat peaks in the signal. However, employing frequency domain decompositions along with some filters on the extracted color or motion traces from the facial video, as used in state-of-the-art methods of [4, 5, 8, 12, 14, 18, 19], does not provide visible HPLs for further analysis.

ii. The accuracy of HR estimation from facial video has yet to reach the level of ECG-based HR estimation. This compels investigations of a more effective signal processing method than the methods used in the literature to estimate HR.

iii. When a facial video is available, the beating of a heart typically shows in the face through changing skin color and head movement. Thus, merely estimating HR from color or motion information may be surpassed in accuracy by a fusion of these two modalities extracted from same video.

iv. Most of the facial video-based fully automatic HR estimation methods, including color-based [3–5, 19] and motion-based [12], assume that the head is static (or close to) during data capture. This means that there is neither internal facial motion nor external movement or rotation of the head during the data acquisition phase. We ascribe internal motion to facial expression and external motion to head pose. However, in real life scenarios there are, of course, both internal and external head motion. Current methods, therefore, provide less accuracy in realistic scenarios.

In this paper we address the aforementioned demands/challenges by proposing a novel Empirical Mode Decomposition (EMD)-based approach to estimate HPL and then HR. Unlike previous methods, the proposed EMD-based decomposition of raw heartbeat traces provides a novel way to look into the heartbeat signal from facial video and generates clearly visible heartbeat peaks that can be used in, among others, further clinical analysis. We estimate the HR from both the number of peaks detected in a time interval and inter-beat distance in a heartbeat signal from HPLs obtained by employing the EMD. We then introduce a multimodal HR estimation from facial video by fusion of color and motion information and demonstrate the effectiveness of such a fusion in estimating HR. We report our results through a publicly available challenging database MAHNOB-HCI [20] in order to demonstrate the success of our system in estimating HR, even when there are voluntary internal and external head motions in the videos.

The rest of the paper is organized as follows. Section 2 describes the proposed system. Section 3 describes HR estimation from HPLs, and an approach to fusing color and motion. Section 4 presents the experimental results. Section 5 contains the conclusions.

2   The Proposed System

This section describes the steps of the proposed EMD-based HPL estimation method from color or motion traces as shown in Fig. 1.

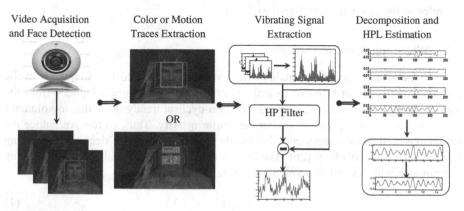

Fig. 1. Steps of the proposed HPL estimation method using skin color or head motion information from facial video.

2.1 Video Acquisition and Face Detection

The first step of the proposed HPL estimation system is face detection from facial video acquired by a simple RGB camera. By following [21, 22], we employ Haar-like feature-based Viola and Jones object detection framework [23] for face detection.

2.2 Facial Color and Motion Traces Extraction

As mentioned earlier, periodic circulation of the blood from the heart through the body causes facial skin to change color, and the head to move or shake in a cyclic motion. The proposed system for HPL estimation can utilize either of the modalities (skin color variations and head motions) as shown in Fig. 1. Recoding RGB values of pixels in facial regions to generate the color trace and tracking some facial feature points to generate the motion trace help to record such skin color variation and head motion from facial video, respectively. In order to obtain the traces of either of the two modalities, we first select a ROI in the detected face. For the color-based approach, the ROI contains 60% of the facial area width (following [2]) detected by the automatic face detection method. We take the average of the RGB values of all pixels in the ROI in each frame instead of only either of the red, green or blue channels to defend the effect of the external lighting condition, as single color channel may prone to increased noise to specific lighting conditions.

For the motion-based approach the ROI (following [14]) contains two areas of forehead and cheek. We divide the ROI into a grid of rectangular regions of pixels and detect some feature points in each region by employing a method called Good Features to Track (GFT) [24, 25]. The GFT works by finding corner points from the minimal Eigen values of the windows of pixels in the ROI. When the head moves due to heart pulse, the feature points also move in the pixel coordinates. We employ a KLT tracker to track the feature points in consecutive video frames and obtain a single trajectory of each feature points in the video by measuring Euclidian distance of the point-locations in consecutive frames. We then fuse all trajectories into a single motion trace.

The next steps follow the same procedure for both color and motion and hereafter we refer to the raw heartbeat signal as $\bar{S}$.

2.3 Vibrating Signal Extraction

The raw heartbeat signal ($\bar{S}$) contains other extraneous high and low frequency cyclic components than heart beat due to ambient color and motion noise induced from the data capturing environment. It also exhibits non-cyclical trendy noise due to voluntary head motion, facial expression, and/or vestibular activity. Thus, to remove/reduce the extraneous frequency components and trends from the signal we decompose it using Hodrick-Prescott (HP) filter [26]. The filter breaks down the signal into the following components with respect to a smoothing penalty parameter, τ:

$$S_{\tau}^{log}(t) = T_{\tau}(t) + C_{\tau}(t) \tag{1}$$

where S_τ^{log} is the logarithm of $\bar{S}$, T_τ is the trend component, and C_τ denotes the cyclical component of the signal with t as the time index (video frame index). We empirically follow two times the decomposition of the trajectory by using two smoothing parameter values $\tau = \infty$ and $\tau = 400$ to obtain all of the cyclic components (C_∞) and high frequency cyclic components (C_{400}), respectively. A detailed description of the HP filter can be obtained from [26]. We completely overlook the trend components (T_τ) because these are not characterized by cyclic pattern of heartbeat. We then obtain the difference between the cyclical components to get the vibrating signal as follows:

$$V(t) = C_\infty(t) - C_{400}(t) \tag{2}$$

2.4   EMD-Based Signal Decomposition for the Proposed HPL Estimation

The vibrating signal (V) cannot clearly depict the heartbeat peaks (which will be shown in the experimental evaluation section). This is because of the contamination of heartbeat information by the other lighting and motion sources, which the HP filter alone cannot restore for visibility. Previous methods in [4, 5, 8, 12, 14, 19] moved to the frequency domain and filtered the signal by different bandpass filters and/or cal-culating the power spectrum of the signal to determine the HR in the frequency domain. However, this cannot provide a heartbeat signal with visible peaks, i.e. no possible HPL estimation, and thus cannot be useful for clinical applications where inter-beat intervals are necessary or signal variation needs to be observed over time. Thus, we employ a derivative of EMD to address the issue. EMD usually decomposes a non-linear and non-stationary time-series into functions that form a complete and nearly orthogonal basis for the original signal [27]. The functions into which a signal is decomposed are all in the time domain and of the same length as the original signal. However, the functions allow for varying frequency in time to be preserved. When a signal is generated as a composite of multiple source signals and each of the source signals may have individual frequency band, calculating IMFs (Intrinsic Mode Func-tions) using EMD can provide illustratable source signals.

In the case of processing the heart signal obtained from facial video, the obtained vibrating signal (V) is a nonlinear and nonstationary time-series that comes as a composite of multiple source signals from lighting change, and/or internal and external head motions along with heartbeat. The basic EMD, as defined by Huang [28], breaks down a signal into IMFs satisfying the following two conditions:

i. In the whole signal, the number of extrema and the number of zero-crossings cannot differ by more than 1.
ii. At any point, both means of the envelopes defined by the local maxima and local minima are zero.

The decomposition can be formulated as follows:

$$V(t) = \sum_{i=1}^{m} M_i + r \tag{3}$$

where M_i presents the mode functions satisfying the aforementioned conditions, m is the number of modes, and r is the residue of the signal after extracting all the IMFs. The procedure of extracting such IMFs (M_i) is called shifting. The shifting process starts by calculating the first mean $(\mu_{i,0})$ from the upper and lower envelopes of the original signal (V in our case) by connecting local maxima. Then a component is calculated as the first component $(I_{i,0})$ for iteration as follows:

$$I_{i,0} = V(t) - \mu_{i,0} \tag{4}$$

The component $I_{i,0}$ is then considered the data signal for an iterative process, which is defined as follows:

$$I_{i,j} = I_{i,j-1} - \mu_{i,j} \tag{5}$$

The iteration stops when a predefined value exceeds the following parameter (δ) calculated in each step:

$$\delta_{i,j} = \sum_{k=1}^{l} \frac{(I_{i,j-1}(k) - I_{i,j}(k))^2}{I_{i,j-1}^2(k)} \tag{6}$$

where l is the number of samples in I (in our case, the number of video frames used).

The basic model of EMD described above, however, exhibits some problems such as the presence of oscillations of very disparate amplitudes in a mode and/or the presence of very similar oscillations in different modes. In order to solve these problems an enhanced model of EMD was proposed by Torres et al. [29], which is called Complete Ensemble Empirical Mode Decomposition with Adaptive Noise (CEEM-DAN). CEEMDAN adds a particular noise at each stage of the decomposition and then computes residue to generate each IMF. The results reported by Torres showed the efficiency of CEEMDAN over EMD. Therefore, we decompose our vibrating signal (V) into IMFs (M_i) by using the CEEMDAN. The total number of IMFs depends on the vibrating signal's nature. As a normal adult's resting HR falls within the frequencies [0.75–2.0] Hz (i.e. [45–120] bpm) [12] and merely 6-th IMF falls within this range, we selected the 6-th IMF as the final uncontaminated (or less contaminated) form of the heartbeat signal of all experimental cases.

We employ a local maxima-based peak detection algorithm on the selected heartbeat signal (the 6-th IMF) to estimate the HPL. The peak detection process was restricted by a minimum peak distance parameter to avoid redundant peaks in nearby positions. The obtained peak locations are the HPLs estimated by the proposed system.

3   HR Calculation Using the Proposed Multi-modal Fusion

The HPLs we obtained in the previous section can be utilized to measure the total number of peaks and peak distances in a heartbeat signal. These can be obtained for either case of the color and motion information from facial video. We calculate the HR in bpm for both approaches in two different ways – from the total number of peaks and average peak distance – as follows:

$$HR_{numPeak} = \left(\frac{N \times F_{rate}}{F_{total}} \right) \times 60 \qquad (7)$$

$$HR_{distPeak} = \left(\frac{F_{rate}}{\frac{1}{(N-1)} \sum_{k=1}^{(N-1)} d_k} \right) \times 60 \qquad (8)$$

where N is the total number of peaks detected, F_{rate} is the video frame rate per second, F_{total} is the total number of video frames used to generate the heartbeat signal, and d_k is the distance between two consecutive peaks.

As we stated in the first section of this article, facial video contains both color and motion information that denote heartbeat. Along with the proposed EMD-based method, the applications of color information for HR estimation were shown in [4, 5, 8, 9, 19], and the applications of motion information were shown in [12, 14]. None of these methods exploited both color and motion information. We assume that, since color and motion information have different notions of heartbeat representation, a fusion of these two modalities in estimating HR can include more deterministic characteristics of heart pulses in the heartbeat signal.

There are three levels that can be considered for the fusion of modalities: raw-data level, feature level, and decision level [30]. Although the extracted raw heartbeat signals in color and motion-based approaches have the same dimensions, they are mismatched due to the nature of the data they present. Thus, instead of raw-data level and feature level fusion, we propose a rule-based decision level (HR estimation results) fusion in this paper for exploiting the HR estimation results from both modalities. For each of the modalities, we obtain two results using the total number of peaks and average peak distance. Thus, we have four different estimates of the HR: $HR_{numPeak}^{color}$, $HR_{distPeak}^{color}$, $HR_{numPeak}^{motion}$, and $HR_{distPeak}^{motion}$. We employed four feasible rules, listed in Table 1, to find the optimal fusion.

Table 1. Fusion rules investigated

Parameter	Definition	Parameter	Definition
$HR_{numPeak}^{Fuse}$	$mean\left(HR_{numPeak}^{color}, HR_{numPeak}^{motion} \right)$	HR_{all}^{Fuse}	$mean\left(\begin{matrix} HR_{numPeak}^{color}, HR_{distPeak}^{color}, \\ HR_{numPeak}^{motion}, HR_{distPeak}^{motion} \end{matrix} \right)$
$HR_{distPeak}^{Fuse}$	$mean\left(HR_{distPeak}^{color}, HR_{distPeak}^{motion} \right)$	$HR_{nearestTwo}^{Fuse}$	$mean_{nearesttwo}\left(\begin{matrix} HR_{numPeak}^{color}, HR_{distPeak}^{color}, \\ HR_{numPeak}^{motion}, HR_{distPeak}^{motion} \end{matrix} \right)$

4   Experiments and Obtained Results

4.1   Experimental Environment

The proposed methods were implemented in Matlab (2013a). Most of the previous methods provided their results on local datasets, which makes the methods difficult to compare with the other methods. In addition, most of the previous methods did not report the results on a challenging database that includes realistic illumination and motion changes. In order to overcome such problems and show the competency of our methods, we used the publicly available MAHNOB-HCI database [20] for the experiment. The database is recorded in realistic Human-Computer Interaction (HCI) scenarios which was treated as a realistic and highly challenging dataset in the literature [8] because it contains facial videos recorded in realistic scenarios, including challenges from illumination variation and internal and external head motions. It contains videos of 491 sessions with 25 subjects that are longer than 30 s, and subjects who consent attribute 'YES'. Both males and females participated; they were between 19 and 40 years of age. Among the sessions, 20 sessions of subject '12' do not have ECG ground truth data and 20 sessions of subject '26' are missing video data. Excluding these sessions, we used the remainder as the dataset for our experiment. As the original videos are of different lengths, we use 30 s (frame 306 to 2135) of each video and the corresponding ECG from EXG3 sensor for the ground truth defined in [20].

We show the experimental results for HPL estimation in a qualitative manner and HR estimation through four statistical parameters used in the previous literature [4, 8]. The first one is mean error, defined as follows:

$$M_E = \frac{1}{N} \sum_{k=1}^{N} \left(HR_k^{video} - HR_k^{groundTruth} \right) \tag{9}$$

where HR_k^{video} is the calculated HR from the k-th video of a database, $HR_k^{groundTruth}$ is the corresponding HR from the ECG ground truth signal, and N is the total number of videos. The second parameter is the standard deviation of M_E, defined as follows:

$$SD_{M_E} = \sqrt{\left(\frac{1}{N} \sum_{k=1}^{N} \left(HR_k^{video} - M_E \right)^2 \right)} \tag{10}$$

The third parameter is the root mean squared error, defined as follows:

$$RMS_E = \sqrt{\left(\frac{1}{N} \sum_{k=1}^{N} \left(HR_k^{video} - HR_k^{groundTruth} \right)^2 \right)} \tag{11}$$

The fourth parameter is the mean of error rate in percentage, defined as follows:

$$M_{ER} = \frac{1}{N} \sum_{k=1}^{N} \left(\frac{HR_k^{video} - HR_k^{groundTruth}}{HR_k^{groundTruth}} \right) \times 100 \tag{12}$$

4.2   Experimental Evaluation

The proposed method tracks color change and head motion due to heartbeat in a video. Figure 2 shows the average trajectory ($\bar{S}$) calculated from the individual trajectories of the feature points and corresponding vibrating signal (V in Eq. (2)) obtained after

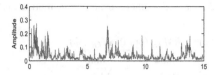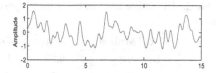

Fig. 2. Vibrating signal extraction by HP filtering: average signal from motion trajectories (left) and filtered vibrating signal (right).

employing the HP filter for a video. We observe that the vibrating signal is less noisy than the previous signal due to the application of the HP filter. We obtain similar results in the color-based approach.

The CEEMDAN, a derivative of EMD, decomposes the vibrating signal into IMFs (M_i) to provide an uncontaminated form of heartbeat signal. Figure 3 shows first eight IMFs obtained from the signal by Eqs. (3)–(6). The IMFs are separated by different frequency components as discussed in Sect. 2.4, and we selected M_6 as the final heartbeat signal to employ the peak detection algorithm. The result of peak detection on M_6 is also shown in Fig. 3. One can observe that the final heartbeat signal has more clearly visible beats than the raw heartbeat signal obtained from motion traces. After employing peak detection we obtained all HPL that can be used in further medical analysis. The qualitative and quantitative comparison of the estimated HPL with the beat locations in ground truth ECG is shown in the performance comparison section.

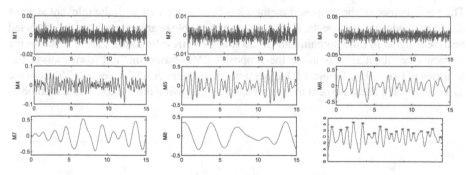

Fig. 3. Obtained IMFs (M_1–M_8) after employing CEEMDAN on the vibrating signal (V) and the detected heartbeat peaks in M_6.

We count the number of peaks and measure average peak distance from HPLs. The associated results are shown in Table 2. We have indicated some of the best cases in **bold**. From the results we observe that counting the number of peaks provides better results than measuring peak distance for both color and motion information. This is because, unlike counting peaks, heartbeat rate variability in the signal can contribute negatively to the average peak distance. The overall error rates (M_{ER}) are less than 10% for HR estimation by counting the number of peaks for both motion and color signals. The fusion results show that, similar to the individual use of motion or color

Table 2. HR estimation results of the proposed EMD-based methods using color, motion and fusion

No.	Method	M_E (bpm)	SD_{M_E} (bpm)	RMS_E (bpm)	M_{ER} (%)
1.	$HR^{motion}_{numPeak}$	**−0.90**	**8.28**	**8.32**	**8.65**
2.	$HR^{motion}_{distPeak}$	−1.33	10.77	10.84	11.51
3.	$HR^{color}_{numPeak}$	**0.21**	**8.55**	**8.54**	**9.26**
4.	$HR^{color}_{distPeak}$	0.95	10.25	10.29	11.59
5.	$HR^{Fuse}_{distPeak}$	−0.19	10.08	10.07	11.00
6.	HR^{Fuse}_{all}	−0.27	9.04	9.03	9.79
7.	$HR^{Fuse}_{nearestTwo}$	−0.29	8.47	8.46	8.92
8.	$HR^{Fuse}_{numPeak}$	**−0.35**	**8.08**	**8.08**	**8.63**

information, the number of peaks fusion generates the best results out of the four fusion rules. Simple arithmetic mean in decision level fusion, as we used, shows a strong correlation with the corresponding color and motion-based results. While comparing the results to the individual motion and color-based estimations, the fusion shows greater accuracy.

4.3   Performance Comparison

The performance of the proposed method has been compared with the relevant state of the art methods in two respects: (i) presentation of visible heartbeat peaks in the extracted heartbeat signal in time domain and (ii) accuracy of HR estimation. Figure 4 shows the extracted heartbeat signals using the proposed EMD-based method from motion trajectories of two videos (subject ID-1, session 14 and subject ID-20, session 26) from the MAHNOB-HCI database next to the extracted signals using the motion-based method of

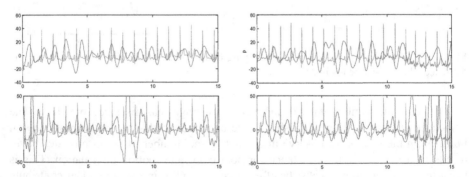

Fig. 4. Illustrating heartbeats obtained by different methods for two facial videos from two different subjects for normal case (left) and challenging case with voluntary motion (right): first row presents the results for our method and the second row presents the results for Guha and Ramin [12, 14]. Ground truth ECG is shown in green. (Color figure online)

[12]. The second video represents the case of voluntary facial motions (right). We have also included ground truth ECG for these videos in overlapping plots. From the figures we observe that the final time domain signal extracted by [12] is not plausible to comprehend for visual analysis. This is also true for the other methods because of similar filters and ICA/PCA/DCT-based decomposition. On the other hand, the final time domain signal generated by the proposed method not only shows heartbeat peaks but also preserves a good correspondence to the ECG ground truth.

We compare the accuracy of the proposed method with state of the art color and the motion-based methods of [4, 5, 12, 19]. The results of the accuracy comparison are summarized in Table 3. From the results, it is clear that the proposed EMD-based methods for both color and motion provide a better estimation of HR than the other state of the art methods. The proposed methods outperformed the other methods in both RMS_E and M_{ER} because EMD can decompose the signal in a better way than the filters and ICA/PCA-based decomposition used in the previous methods. The results of the proposed method demonstrate a high degree of consistency in estimating HR in comparison to the other methods. This, in turn, validates our peak location estimation as well because the peak locations have been used to estimate HR.

Table 3. Performance comparison of the proposed methods with the previous methods for HR estimation

No.	Method	M_E (bpm)	SD_{M_E} (bpm)	RMS_E (bpm)	M_{ER} (%)
1.	Poh (ICA) [4]	−8.95	24.3	25.90	25.00
2.	Kwon (ICA) [5]	−7.96	23.8	25.10	23.60
3.	Guha (PCA) [12]	−14.4	15.2	21.00	20.70
4.	Poh (ICA) [19]	2.04	13.5	13.60	13.20
5.	**Proposed_color (CEEMDAN)**	**0.21**	**8.55**	**8.54**	**9.26**
6.	**Proposed_motion (CEEMDAN)**	**−0.90**	**8.28**	**8.32**	**8.65**
7.	**Proposed_Fusion**	**−0.35**	**8.08**	**8.08**	**8.63**

5 Conclusions

This paper proposed methods for estimating HPL and HR from color and motion information from facial video by a novel use of an HP filter and EMD decomposition. The paper also proposed a fusion approach to exploit both color and motion information together for multi-modal HR estimation. The contributions of these methods are as follows: (i) provided the notion of visually analyzing heartbeat signal in time domain with clearly visible heartbeat peaks for clinical applications, (ii) provided better estimations of HR for separate color and motion traces, (iii) a decision level fusion further improved the result, and (iv) provided a highly accurate HPL and HR estimations method from facial video in the presence of challenging situations due to illumination change and voluntary head motions. The proposed method, however, also imposed some limitations when generating the results. We assume that the camera will be placed in close proximity to the face (about one meter away). Moreover, we did not employ any sophisticated ROI detection and tracking methods, illumination rectification methods, or extraneous motion filtering. Future work should address these points.

References

1. Klonovs, J., et al.: Distributed Computing and Monitoring Technologies for Older Patients, 1st edn. Springer International Publishing, Heidelberg (2015)
2. Haque, M.A., Nasrollahi, K., Moeslund, T.B.: Heartbeat signal from facial video for biometric recognition. In: Paulsen, R.R., Pedersen, K.S. (eds.) SCIA 2015. LNCS, vol. 9127, pp. 165–174. Springer, Cham (2015). doi:10.1007/978-3-319-19665-7_14
3. Takano, C., Ohta, Y.: Heart rate measurement based on a time-lapse image. Med. Eng. Phys. **29**(8), 853–857 (2007)
4. Poh, M.-Z., McDuff, D.J., Picard, R.W.: Non-contact, automated cardiac pulse measurements using video imaging and blind source separation. Opt. Express **18**(10), 10762–10774 (2010)
5. Kwon, S., Kim, H., Park, K.S.: Validation of heart rate extraction using video imaging on a built-in camera system of a smartphone. In: 2012 Annual International Conference of the IEEE Engineering in Medicine and Biology Society (EMBC), pp. 2174–2177 (2012)
6. Wei, L., Tian, Y., Wang, Y., Ebrahimi, T., Huang, T.: Automatic webcam-based human heart rate measurements using Laplacian Eigenmap. In: Lee, K.M., Matsushita, Y., Rehg, J.M., Hu, Z. (eds.) Computer Vision – ACCV 2012, pp. 281–292. Springer, Berlin Heidelberg (2012)
7. Cheng, J., Chen, X., Xu, L., Wang, Z.J.: Illumination variation-resistant video-based heart rate measurement using joint blind source separation and ensemble empirical mode decomposition. IEEE J. Biomed. Health Inform. **99**, 1 (2016)
8. Li, X., Chen, J., Zhao, G., Pietikainen, M.: Remote heart rate measurement from face videos under realistic situations. In: IEEE Conference on Computer Vision and Pattern Recognition (CVPR), pp. 4321–4328 (2014)
9. Monkaresi, H., Calvo, R., Yan, H.: A machine learning approach to improve contactless heart rate monitoring using a webcam. IEEE J. Biomed. Health Inform. **18**(4), 1153–1160 (2014)
10. Shagholi, A., Charmi, M., Rakhshan, H.: The effect of the distance from the webcam in heart rate estimation from face video images. In: 2015 2nd International Conference on Pattern Recognition and Image Analysis (IPRIA), pp. 1–6 (2015)
11. Lam, A., Kuno, Y.: Robust heart rate measurement from video using select random patches. In: IEEE International Conference on Computer Vision (ICCV), pp. 3640–3648 (2015)
12. Balakrishnan, G., Durand, F., Guttag, J.: Detecting pulse from head motions in video. In: IEEE Conference on Computer Vision and Pattern Recognition (CVPR), pp. 3430–3437 (2013)
13. Bouguet, J.: Pyramidal implementation of the Lucas Kanade feature tracker. Intel Corporation Microprocessor Research Labs
14. Irani, R., Nasrollahi, K., Moeslund, T.B.: Improved pulse detection from head motions using DCT. In: 9th International Conference on Computer Vision Theory and Applications (VISAPP), pp. 1–8 (2014)
15. Shan, L., Yu, M.: Video-based heart rate measurement using head motion tracking and ICA. In: 2013 6th International Congress on Image and Signal Processing (CISP), vol. 01, pp. 160–164 (2013)

16. Haque, M.A., Nasrollahi, K., Moeslund, T.B.: Can contact-free measurement of heartbeat signal be used in forensics? In: Presented at the 23rd European Signal Processing Conference (EUSIPCO), pp. 769–773 (2015)

17. Nasrollahi, K., Haque, M.A., Irani, R., Moeslund, T.B.: Contact-free heartbeat signal for human identification and forensics. In: Tistarelli, M., Champod, C. (eds.) Handbook of Biometrics for Forensic Science. ACVPR, pp. 289–302. Springer, Cham (2017). doi:10.1007/978-3-319-50673-9_13

18. Haque, M.A., Irani, R., Nasrollahi, K., Moeslund, T.B.: Heartbeat rate measurement from facial video. IEEE Intell. Syst. **31**(3), 40–48 (2016)

19. Poh, M.-Z., McDuff, D.J., Picard, R.W.: Advancements in noncontact, multiparameter physiological measurements using a webcam. IEEE Trans. Biomed. Eng. **58**(1), 7–11 (2011)

20. Soleymani, M., Lichtenauer, J., Pun, T., Pantic, M.: A multimodal database for affect recognition and implicit tagging. IEEE Trans. Affect. Comput. **3**(1), 42–55 (2012)

21. Haque, M.A., Nasrollahi, K., Moeslund, T.B.: Real-time acquisition of high quality face sequences from an active pan-tilt-zoom camera. In: 10th IEEE International Conference on Advanced Video and Signal Based Surveillance (AVSS), pp. 443–448 (2013)

22. Haque, M.A., Nasrollahi, K., Moeslund, T.B.: Quality-aware estimation of facial landmarks in video sequences. In: IEEE Winter Conference on Applications of Computer Vision (WACV), pp. 1–8 (2015)

23. Viola, P., Jones, M.J.: Robust real-time face detection. Int. J. Comput. Vis. **57**(2), 137–154 (2004)

24. Shi, J., Tomasi, C.: Good features to track. In: IEEE Conference on Computer Vision and Pattern Recognition (CVPR), pp. 593–600 (1994)

25. Haque, M.A., Irani, R., Nasrollahi, K., Moeslund, T.B.: Facial video based detection of physical fatigue for maximal muscle activity. IET Comput. Vis. **10**(4), 323–329 (2016)

26. McElroy, T.: Exact formulas for the Hodrick-Prescott filter. Statistical Research Division, U. S. Census Bureau, September 2006

27. Ren, H., Wang, Y., Huang, M., Chang, Y., Kao, H.: Ensemble empirical mode decomposition parameters optimization for spectral distance measurement in hyperspectral remote sensing data. Remote Sens. **6**, 2069–2083 (2014)

28. Huang, N.E.: An adaptive data analysis method for nonlinear and nonstationary time series: the empirical mode decomposition and Hilbert spectral analysis. In: Qian, T., Vai, M.I., Xu, Y. (eds.) Wavelet Analysis and Applications, pp. 363–376. Birkhäuser, Basel (2006)

29. Torres, M.E., Colominas, M.A., Schlotthauer, G., Flandrin, P.: A complete ensemble empirical mode decomposition with adaptive noise. In: 2011 IEEE International Conference on Acoustics, Speech and Signal Processing (ICASSP), pp. 4144–4147 (2011)

30. Boulgouris, N.V., Plataniotis, K.N., Micheli-Tzanakou, E.: Biometrics: Theory, Methods, and Applications. Wiley, Hoboken (2009)

Segmentation of Multiple Structures in Chest Radiographs Using Multi-task Fully Convolutional Networks

Chunliang Wang[✉]

School of Technology and Health (STH),
KTH Royal Institute of Technology, Stockholm, Sweden
chunliang.wang@sth.kth.se

Abstract. Segmentation of various structures from the chest radiograph is often performed as an initial step in computer-aided diagnosis/detection (CAD) systems. In this study, we implemented a multi-task fully convolutional network (FCN) to simultaneously segment multiple anatomical structures, namely the lung fields, the heart, and the clavicles, in standard posterior-anterior chest radiographs. This is done by adding multiple fully connected output nodes on top of a single FCN and using different objective functions for different structures, rather than training multiple FCNs or using a single FCN with a combined objective function for multiple classes. In our preliminary experiments, we found that the proposed multi-task FCN can not only reduce the training and running time compared to treating the multi-structure segmentation problems separately, but also help the deep neural network to converge faster and deliver better segmentation results on some challenging structures, like the clavicle. The proposed method was tested on a public database of 247 posterior–anterior chest radiograph and achieved comparable or higher accuracy on most of the structures when compared with the state-of-the-art segmentation methods.

Keywords: Multi-task deep neural network · Fully convolutional network · Image segmentation

1 Introduction

Chest radiography is one of the most common medical imaging procedures for screening and diagnosis of pulmonary diseases thanks to its low radiation and cost. Although the interpretation of chest radiography is deemed as a basic skill of a certified radiologist, the inter- and intra-observer performance is highly variable due to the subjective nature of the reviewing process [1]. To assist in the diagnosis of chest radiography, a number of computer-aided diagnosis/detection (CAD) systems have been developed to provide a second opinion on a selective set of possible pathological changes, such as lung nodules [1] or tuberculosis [2]. In such a system, segmentation of various structures from the chest radiograph is often performed as an initial step. The accuracy of the segmentation often has a strong influence on the performance of the following steps such as lung nodule detection, or cardiothoracic ratio quantification, therefore a robust and accurate lung filed and heart segmentation method is essential for these systems. A number of

© Springer International Publishing AG 2017
P. Sharma and F.M. Bianchi (Eds.): SCIA 2017, Part II, LNCS 10270, pp. 282–289, 2017.
DOI: 10.1007/978-3-319-59129-2_24

dedicated segmentation methods have been proposed to target some specific organs, most commonly the lungs in chest radiographs [3–6]. However, the segmentation problem is still a great challenge, even for human observers, due to large variation of anatomical shape and appearance, inadequate boundary contrast and inconsistent overlapping between multiple organs (e.g. the relative positions between bones, muscles and mediastinum) which are hard to be modeled with the common statistical models. In [7, 8], we proposed a hierarchical-shape-model guided multi-organ segmentation method for CT images, that outperformed some dedicated single organ segmentation methods. This gave some support to the hypothesis that solving multiple organ segmentation simultaneously may be better than solving a single organ segmentation problem as the algorithm gets more context information. In this study, we extended the same philosophy to the chest radiography segmentation with a totally different segmentation framework, based on convolutional neural networks (CNN).

CNN, or deep neural networks in general, has gained popularity in recent years due to the outstanding performance on a number of challenging image analysis problems, such as image classification, object detection and semantic segmentation [9], as well as a variety of medical applications [10]. In contrast to conventional machine learning approaches that use handcrafted features designed by a human observer, CNNs automatically adjust the weights of convolutional kernels to create data-driven features that optimize the learning objectives at the end of the neural network. Recently, an increasing number of reports have suggested that adding multiple objectives or combining features trained for different objectives, so-called multi-task CNNs, can deliver better results than the single-task models [11, 12]. In this study, we implemented a multi-task fully convolutional network (FCN) to simultaneously segment multiple anatomical structures, namely the lung fields, the heart, and the clavicles, in standard posterior-anterior chest radiographs. When tested on a public database of chest x-ray images, the proposed method achieved comparable or higher accuracy on most of the structures than the state-of-the-art segmentation methods.

2 Method

2.1 Fully Convolutional Network

In general, a CNN consists of a number of convolutional layers followed by a number of fully connected layers. This setup requires the input images/image patches to share the same size. When used for image segmentation, it requires the input image to be converted to a series of largely overlapping patches around each pixel. This makes the computation very inefficient. FCN can be seen as an extension of the classical CNN, where the fully connected layers are removed or replaced by convolutional layers [9]. This allows FCNs to be applied to images of any size and output label maps proportional to the input image. Combined with "skips" and up-sampling or deconvolution layers [9], the output map can have the same size as the input image. This design eliminates redundant computation on overlapping patches and makes both the training and testing processes more efficient than the patch-based CNN approaches. In this study, we implemented a variation of FCN, called U-Net, which was proposed by

Ronneberger et al. [13]. The overall architecture of the U-Net used in this study is illustrated in Fig. 1. The left arm of the "U" shape consists of four repeating steps of convolution and max pooling. In the convolution steps, we used two consecutive 3×3 convolutional kernels. The exact number of features at each layer is given in the figure. At each max pooling step, the feature maps are reduced to half the size. The right arm of the "U" shape consists of four repeating steps of up-sampling and convolution, which allows the network to output a segmentation mask of the same size as the input image. Right after each up-sampling, the feature maps from the corresponding layers on the left arm are merged with the up-sampled feature maps before the following convolution operations. This allows the network to combine the context information

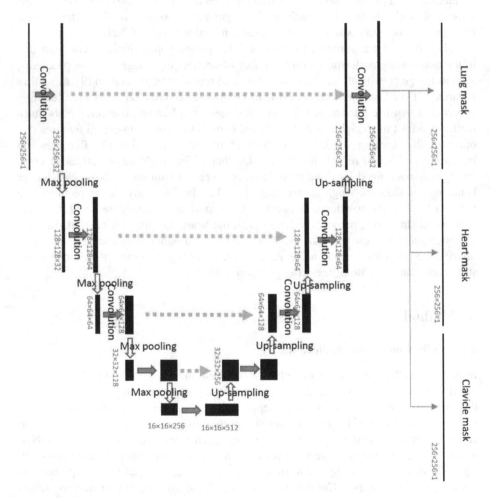

Fig. 1. The U-Net architecture used in this study. Black lines represent the images and feature maps (The height and width of the lines symbolize the size of the maps and the number of features at each step respectively, numbers on the side indicate the exact dimension of the features maps). Blue arrows represent convolutional operations. Dashed green arrows represent the 'skips' (Color figure online)

from the coarse layer and the detailed image features at the finer scale. The final segmentation masks are usually generated with a convolution layer with kernel size of 1×1 that combines the multiple feature maps with a softmax or sigmoid activation function. Compared with the original FCN method reported in [9], U-Net does not required the up-sampling layers to be trained from coarse to fine in multiple stages, but to train all layers in a single round, therefore it is easier to use in practice.

2.2   Multiple Single-Class FCN Vs. Multi-class FCN Vs. Multi-task FCN

In this study we try to segment the lungs, heart and clavicles from a chest x-ray image (cf. Fig. 2). There are a number of options to achieve multiple-class segmentation with FCN. One of the most trivial approaches is to treat the multi-structures separately and train multiple FCNs to segment different structures one after another. Another relatively straightforward method is to simply treat the structure labels as multi-classes and train a single FCN that outputs multiple probability maps with a single objective function. However, as shown in Fig. 2, the segmentation labels of the clavicles overlap largely with the lung fields. The softmax activation function that is commonly used to output multi-class predictions requires the samples to exclusively belong to a single class, and therefore cannot be directly used in this task. Multi-task FCN is another option which can generate multiple segmentation masks through a shared base FCN. The general principle of a multi-task deep neural network is to hybrid multiple output paths with different objective functions on top of a common base network (e.g. Fig. 1). It can also be interpreted as multiple output paths sharing the same feature pool that can be used for multiple tasks. While training a network that optimizes two tasks, such as landmark detection and segmentation, the feature pool may eventually contain two sets of features that complement each other. In the FCN setup, low-resolution layers are also thought of as context information for the finer output layers. In multi-task FCN, the context information is also enriched. In this study, we added 5 output nodes to the U-Net output layer that are expected to generate segmentation masks for 5 different structures. Each node is associated with its own loss function. In this study, we used the negative Dice coefficient as the loss function for each individual region, which eliminates the need of tuning the class weighting factors in situations where there is a strong imbalance between the number of foreground and background voxels [14], like the clavicles. The weighting factors for different objective functions were all set to 1.0, i.e. all structures are equally important.

2.3   Post-processing

As shown in Fig. 2, the output from FCN sometimes developed holes inside the targeted structure or islands outside. To remove these artifacts, we applied a fast level set method [15] that first shrinks from the border of the image with a relatively high curvature force (curvature weighting set to 0.7) and then expands with a low curvature force (curvature weighting set to 0.3). Because the output from U-Net is a probability map that is in a range from 0 to 1, we simply used 0.5 as the threshold to generate the

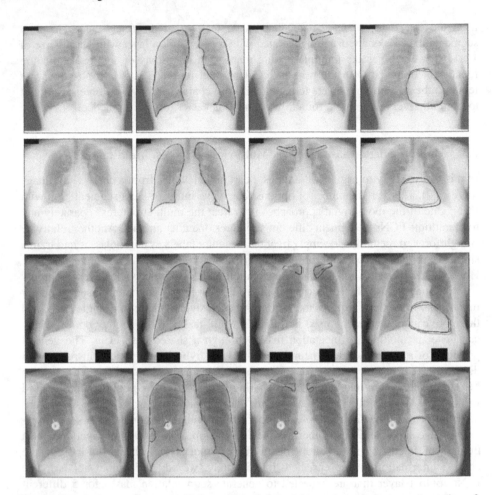

Fig. 2. Four representative cases. Columns from left to right are: input images, segmentation of lungs, segmentation of the clavicle and segmentation of the heart. Green contours represent the ground truth, blue contours represent the segmentation results from the single-task FCN, and red contours represent the segmentation results from the multi-task FCN (Color figure online)

speed-map for the level set method. The shrinking and expanding are similar to mathematical morphology opening and closing operations, but we have found the level set method to be more robust against large holes caused by unexpected objects in the images such as pacemakers or central venous access devices.

3   Experiments and Results

To validate the proposed method, we tested the proposed multi-task FCN approach on the public JSRT dataset of 247 chest radiographs with the segmentation masks from [3]. The original image size was 2048 × 2048, but was down-sampled through linear interpolation to 256×256 as suggested in [3]. Our implementation was based on the Keras framework with Theano backend (http://keras.io). The Adam optimizer with a learning rate of 0.0001 was used. The optimization was stopped at 50 epochs. A single-class FCN and a multi-class FCN were also implemented for comparison. For single-class FCN, five different FCNs were trained for five different structures with the negative Dice coefficient as the loss function. The same optimizer and learning rates were used for the lung segmentation network, but for training the clavicle and heart segmentation networks, a lower learning rate (1×10^{-5}) was needed to make the loss function to decrease and the number epoch was manually determined while monitoring the loss function (around 150 epochs for the heart and 300 epochs for the clavicles). For multi-class FCN, we used a sigmoid function as the activation function, and categorical cross entropy as the objective function. Similar optimizer settings as the multi-task FCN were used. Fivefold cross validation was used to generate the overlap measurements (Jaccard index). In each fold around 200 images were used for training. Using data augmentation technique, we generated 2000 random rotated and scaled image and mask pairs for the actual training. Table 1 compares the segmentation accuracy of three different methods without the post-processing step. In Table 2, the final results from the proposed method after post-processing are shown and compared with state-of-art results found in literature. In addition to the overlap measurement, the average distance (AD) between the manual contour and the segmentation result is also given for comparison.

Table 1. Multiple structure segmentation results using different FCN architectures

Structure name	Jaccard index of single-class FCN	Jaccard index of multi-class FCN	Jaccard index of multi-task FCN	P-value*
Left lung	0.957 ± 0.027	0.951 ± 0.041	0.955 ± 0.026	0.486
Right lung	0.963 ± 0.017	0.957 ± 0.046	0.962 ± 0.015	0.250
Left clavicle	0.845 ± 0.062	0.491 ± 0.171	0.863 ± 0.057	0.001
Right clavicle	0.838 ± 0.061	0.532 ± 0.156	0.862 ± 0.055	<0.001
Heart	0.899 ± 0.044	0.855 ± 0.093	0.898 ± 0.046	0.789

*: P value of student t-test between single-class FCN and multi-task FCN

Table 2. Quantitative comparison of the segmentation accuracy of the proposed method with other state-of-art methods reported on the same dataset

Methods	Lung field		Heart		Clavicle	
	Jaccard	AD (mm)	Jaccard	AD (mm)	Jaccard	AD (mm)
Observer [3]	0.946 ± 0.044	1.64 ± 0.69	0.878 ± 0.054	3.78 ± 1.82	0.896 ± 0.037	0.68 ± 0.26
Ginneken et al. [3]	0.949 ± 0.020	1.62 ± 0.66	0.860 ± 0.056	4.24 ± 1.87	0.736 ± 0.106	1.88 ± 0.93
Shao et al. [4]	0.946 ± 0.018	1.70 ± 0.76	–	–	–	–
Ibragimov et al. [5]	0.953 ± 0.020	1.43 ± 0.85	–	–	–	–
Hogeweg et al. [6]	–	–	–	–	0.860 ± 0.100*	1.09 ± 1.57*
Multi-task FCN	0.959 ± 0.017	1.29 ± 0.80	0.899 ± 0.044	3.12 ± 1.80	0.863 ± 0.045	0.98 ± 0.58

*: The results were reported on a different dataset.

4   Discussion and Conclusion

Chest radiograph segmentation is relative challenging for most conventional methods that are based on handcrafted features, mostly due to its complicated texture pattern. CNNs or FCNs, on the other hand, can cope with the complicate pattern more easily through the data-driven feature composition. As shown in Table 1, for the larger structures, such as the lungs and heart, both single-task and multi-task FCN delivered very promising results that outperform most existing methods summarized in Table 2. The results suggest that the segmentation variability of FCNs is even smaller than the inter-observer variability. However, the segmentation of the clavicles proved to be more challenging due to its size and complex surrounding structures (clavicles overlap with the lung, rib cage, vertebral column and sternum, as well as the soft tissue in the mediastinum). In our experiments, we found that multi-task FCN that segments multiple structures simultaneously, not only reduces the training and running time, but also helps the deep neural network to converge faster and deliver better segmentation results on the clavicles than the single-class FCN trained on single structure masks. One possible explanation is that the context information of the lungs helped the network to determine the boundary of the clavicles. Also the image features that are learned for lung segmentation can be used for the clavicle segmentation, which allows us to use a higher learning rate (10 times higher) when training the multi-task FCN than training a single-class FCN on clavicle alone. These findings are in line with the finding of some other multi-organ segmentation studies [7, 8] and multi-task CNNs studies [11, 12].

It is important to point out that the clavicle segmentation masks used in this study contain only those parts superimposed on the lungs and the rib cage have been indicated as shown in Fig. 2. The reason for this, as explained in [3], is that the peripheral parts of the clavicles are not always visible on a chest radiograph. Due to the overlapping between the clavicle and lung masks, the conventional softmax activation function is inapplicable in this case. In our experiments, we found that a sigmoid activation function combined with the categorical cross entropy objective function gives the best results on all five structures in the multi-class FCN setup. Other objective functions, including negative Dice score, were also tested, but failed to deliver better results. As shown in Table 1, the segmentation accuracy of the multi-class FCN on the clavicles is much worse than the other two methods.

In [10], the author also adapted a multi-task CNN framework for medical image seg-mentation. However, in their work, they trained a network to segment different structure from different image modalities. No comparison of segmentation accuracy between single organ and multi-organ segmentation was made.

In conclusion, we found that FCN-based image segmentation outperformed most conventional methods on lung field, heart and clavicle segmentation in chest radiograph. Multi-task FCN seems to be able to deliver better results on the more challenging structures. Our future works include to test the proposed method on a large dataset and extend it to handle 3D structure segmentation in CT or MRI volumes.

References

1. Kakeda, S., Moriya, J., Sato, H., Aoki, T., Watanabe, H., Nakata, H., Oda, N., Katsuragawa, S., Yamamoto, K., Doi, K.: Improved detection of lung nodules on chest radiographs using a commercial computer-aided diagnosis system. Am. J. Roentgenol. **182**, 505–510 (2004)
2. Melendez, J., Sánchez, C.I., Philipsen, R.H.H.M., Maduskar, P., Dawson, R., Theron, G., Dheda, K., van Ginneken, B.: An automated tuberculosis screening strategy combining X-ray-based computer-aided detection and clinical information. Sci. Rep. **6**, 25265 (2016)
3. van Ginneken, B., Stegmann, M.B., Loog, M.: Segmentation of anatomical structures in chest radiographs using supervised methods: a comparative study on a public database. Med. Image Anal. **10**, 19–40 (2006)
4. Shao, Y., Gao, Y., Guo, Y., Shi, Y., Yang, X., Shen, D.: Hierarchical lung field segmentation with joint shape and appearance sparse learning. IEEE Trans. Med. Imaging **33**, 1761–1780 (2014)
5. Ibragimov, B., Likar, B., Pernuš, F., Vrtovec, T.: Accurate landmark-based segmentation by incorporating landmark misdetections. In: 2016 IEEE 13th International Symposium on Biomedical Imaging (ISBI), pp. 1072–1075. IEEE (2016)
6. Hogeweg, L., Sánchez, C.I., de Jong, P.A., Maduskar, P., van Ginneken, B.: Clavicle segmentation in chest radiographs. Med. Image Anal. **16**, 1490–1502 (2012)
7. Wang, C., Smedby, Ö.: Multi-organ segmentation using shape model guided local phase analysis. In: Navab, N., Hornegger, J., Wells, W.M., Frangi, A.F. (eds.) MICCAI 2015. LNCS, vol. 9351, pp. 149–156. Springer, Cham (2015). doi:10.1007/978-3-319-24574-4_18
8. Jimenez-Del-Toro, O., Muller, H., Krenn, M., Gruenberg, K., Taha, A.A., Winterstein, M., Eggel, I., Foncubierta-Rodriguez, A., Goksel, O., Jakab, A., Kontokotsios, G., Langs, G., Menze, B., Salas Fernandez, T., Schaer, R., Walleyo, A., Weber, M.-A., Dicente Cid, Y., Gass, T., Heinrich, M., Jia, F., Kahl, F., Kechichian, R., Mai, D., Spanier, A., Vincent, G., Wang, C., Wyeth, D., Hanbury, A.: Cloud-based evaluation of anatomical structure segmentation and landmark detection algorithms: VISCERAL anatomy benchmarks. IEEE Trans. Med. Imaging **35**, 2459–2475 (2016)
9. Shelhamer, E., Long, J., Darrell, T.: Fully convolutional networks for semanticsegmentation. IEEE Trans. Pattern Anal. Mach. Intell. **3**, 640–651 (2017)
10. Moeskops, P., Wolterink, J.M., Velden, B.H.M., van der Gilhuijs, K.G.A., Leiner, T., Viergever, M.A., Išgum, I.: Deep learning for multi-task medical image segmentation in multiple modalities. In: Ourselin, S., Joskowicz, L., Sabuncu, M.R., Unal, G., Wells, W. (eds.) MICCAI 2016. LNCS, vol. 9901, pp. 478–486. Springer, Cham (2016). doi:10.1007/978-3-319-46723-8_55
11. Yu, B., Lane, I.: Multi-task deep learning for image understanding. In: 2014 6th International Conference of Soft Computing and Pattern Recognition (SoCPaR), pp. 37–42. IEEE (2014)
12. Li, X., Zhao, L., Wei, L., Yang, M.-H., Wu, F., Zhuang, Y., Ling, H., Wang, J.: DeepSaliency: multi-task deep neural network model for salient object detection. IEEE Trans. Image Process. **25**, 3919–3930 (2016)
13. Ronneberger, O., Fischer, P., Brox, T.: U-Net: convolutional networks for biomedical image segmentation. In: Navab, N., Hornegger, J., Wells, W.M., Frangi, A.F. (eds.) MICCAI 2015. LNCS, vol. 9351, pp. 234–241. Springer, Cham (2015). doi:10.1007/978-3-319-24574-4_28
14. Milletari, F., Navab, N., Ahmadi, S.A.: V-Net: fully convolutional neural networks for volumetric medical image segmentation. In: 2016 Fourth International Conference on 3D Vision (3DV), pp. 565–571 (2016)
15. Wang, C., Frimmel, H., Smedby, O.: Fast level-set based image segmentation using coherent propagation. Med. Phys. **41**, 73501 (2014)

A Novel Method for Automatic Localization of Joint Area on Knee Plain Radiographs

Aleksei Tiulpin[1(✉)], Jerome Thevenot[1], Esa Rahtu[2], and Simo Saarakkala[1,3]

[1] Research Unit of Medical Imaging, Physics and Technology,
University of Oulu, Oulu, Finland
aleksei.tiulpin@oulu.fi
[2] Center for Machine Vision and Signal Analysis,
University of Oulu, Oulu, Finland
[3] Department of Diagnostic Radiology,
Oulu University Hospital, Oulu, Finland

Abstract. Osteoarthritis (OA) is a common musculoskeletal condition typically diagnosed from radiographic assessment after clinical examination. However, a visual evaluation made by a practitioner suffers from subjectivity and is highly dependent on the experience. Computer-aided diagnostics (CAD) could improve the objectivity of knee radiographic examination. The first essential step of knee OA CAD is to automatically localize the joint area. However, according to the literature this task itself remains challenging. The aim of this study was to develop novel and computationally efficient method to tackle the issue. Here, three different datasets of knee radiographs were used (n = 473/93/77) to validate the overall performance of the method. Our pipeline consists of two parts: anatomically-based joint area proposal and their evaluation using Histogram of Oriented Gradients and the pre-trained Support Vector Machine classifier scores. The obtained results for the used datasets show the mean intersection over the union equals to: 0.84, 0.79 and 0.78. Using a high-end computer, the method allows to automatically annotate conventional knee radiographs within 14–16 ms and high resolution ones within 170 ms. Our results demonstrate that the developed method is suitable for large-scale analyses.

Keywords: Knee radiographs · Medical image analysis · Object localization · Proposal generation

1 Introduction

Hip and knee osteoarthritis (OA) are globally ranked as the 11th highest contributor to disability [4]. Typically, OA is diagnosed at a late stage when there is no cure available anymore and when joint replacement surgery becomes the only remaining option. However, if the disease could be diagnosed at an early stage, its progression could be slowed down or even stopped. Such diagnostics is

© Springer International Publishing AG 2017
P. Sharma and F.M. Bianchi (Eds.): SCIA 2017, Part II, LNCS 10270, pp. 290–301, 2017.
DOI: 10.1007/978-3-319-59129-2_25

currently possible; however, it involves the use of expensive imaging modalities, which is clinically unfeasible in the primary health care.

X-ray imaging is the most popular and cheap method for knee OA diagnostics [3,6]. However, the diagnostics at an early stage based on this modality has still some limitations due to several factors. A visual evaluation made by a practitioner suffers from subjectivity and is highly dependent on the experience. Eventually it has been reported that radiologists misdiagnosed OA in 30% of the cases, and that in 20% of the cases a specialist disagrees with his/her own previous decision after a period of time [2,17]. Therefore, to make the diagnostic process more systematic and reliable, computer-aided diagnostics (CAD) can be used to reduce the impact of the subjective factors which alter the diagnostics.

In the literature, multiple attempts have been made to approach knee OA CAD from X-ray images [1,20,22–24]. These studies indicate existence of two parts in the diagnostic pipeline: the region of interest (ROI) localization and, subsequently, a classification – OA diagnostics from localized ROI(s). It has been reported (see, section below), that this problem remains challenging and requires a better solution.

The aim of our study was to propose a novel and efficient knee joint area localization algorithm, which is applicable for large-scale knee X-ray studies. Our study has the following novelties:

- We propose a new approach to generate and score knee joint area proposals.
- We report a cross-dataset validation, showing that the our method performs similarly for three different datasets and drastically outperforms the baseline.
- Finally, we show, that the developed method can be used to accurately annotate from hundreds of thousands to millions of X-ray images per day.

2   Related Work

In the literature, multiple approaches have been used to localize ROIs within radiographs: manual [12,26], semi-automatic [8] and in a fully automatic manner [1,19,22]. To the best of our knowledge, only the studies by Anthony et al. [1] and Shamir et al. [22] focused on knee joint area localization. While the problem of knee joint area localization can be implicitly solved by annotating the anatomical bone landmarks using deformable models [15,21], it would be unfeasible to perform large-scale studies since the use of these algorithms is computationally costly. Currently, despite the presence of multiple approaches, no studies have reported so far their applicability among different datasets. This cross-dataset validation is crucial for the development of clinically applicable solutions.

In the recently published large scale study [1], two methods were analyzed: a template matching adapted from a previously published work [22] and a sliding-window approach. Both methods showed limited localization ability, however, the sliding-window approach demonstrated a better performance. In particular, this approach was designed to find knee joint centers in radiographic images using Sobel [7] gradients and a linear Support Vector Machine (SVM) classifier.

For each sliding window, an SVM score was computed and eventually the patch having the best score was selected. Subsequently, a 300 × 300 pixels region was drawn around the selected patch center. As the localization metric, the authors used intersection over the union (IoU), which is also called the Jaccard index, between the drawn region and the manual annotation:

$$IoU = \frac{A \cap B}{A \cup B},$$ (1)

where A is the manual annotation and B is the detected bounding box. While the sliding window approach was better than the template matching (mean IoU over the dataset was 0.36 vs. 0.1), the performance was still insufficient to perform further OA CAD, as it was indicated by the authors themselves. Consequently, there is a need for more effective methods for ROI localization since nowadays large scale studies are possible due to the availability of multiple image cohorts like Osteoarthritis Initiative (OAI) [9] and Multicenter Osteoarthritis study (MOST) [10]. These cohorts contain follow-up data of thousands of normal subjects and subjects with knee, hip and wrist OA.

Finally, it should be mentioned, that the problem of joint area localization is not limited to only knee radiographs. The attention of the research community is also focused on hand radiographs, where OA occurs as well. It has been recently shown how the anatomic structure of the image can be utilized to annotate hand radiographic images [13]. However, to the best of authors' knowledge, there are no studies in the knee domain where such information is used to segment or annotate the image. In this study, we show how such information can be used to annotate knee radiographs.

3   Method

In this section, we describe our joint area localization approach. The method consists of two main blocks: proposal generation and SVM-based scoring. First, we describe a limb anatomy-based proposal generation algorithm. As shown in previous object detection studies [25,27], object proposal approaches can significantly reduce the amount of candidate regions: from hundreds of thousands to a few thousands. Subsequently, we show the proposal scoring step, based on Histogram of Oriented Gradients (HoG) [5] and SVM classifier. Schematically, our approach is presented in Fig. 1.

3.1   Data Pre-processing

To approach the joint area localization problem on knee radiographs, it would be possible to utilize an exhaustive sliding window approach, however, it is time consuming and inadequate for big data analysis. Thus, we propose an anatomically-based proposal generation approach to significantly reduce the amount of checked joint locations. Our method is trained for only one leg, despite the fact that two legs are usually imaged (see the leftmost block in Fig. 1). We obtain the method

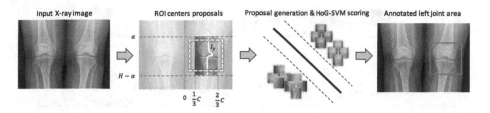

Fig. 1. Schematic representation of the proposed knee joint area localization method. For more details of the method, see the Sects. 3.2 and 3.3.

working for both legs by mirroring the leg which is not used for training the method. Therefore, below we describe the problem for the image containing only one leg and denote it as **I** having size $C \times H$.

3.2 Region Proposal

The core idea and novelty behind our proposal generation are to utilize the anatomic structure of the limb. As the prior information of the knee joint anatomy is known, it can be efficiently utilized. Considering marginal distribution of intensities of a leg, it can be noticed, that around the joint area, there is an intensity increase due to the presence of the patella, and then a sharp intensity drop due to the space between the joints. In our approach, we detect these changes and use their locations as Y-coordinates of region proposals.

At first, we take the middle of an input image, and then sum all the values horizontally to obtain the sequence I_y, which corresponds to a vertical marginal distribution of pixel intensities:

$$I_y[i - \alpha] = \sum_{j=\frac{1}{3}C}^{\frac{2}{3}C-1} \mathbf{I}[i, j], \ \forall i \in [\alpha, H - \alpha). \tag{2}$$

Here, we do not sum all the values – instead, we use a margin α to ignore the outliers which are usually present in the top and the bottom of the image. This also reduces the computational complexity of the method. The next step is to identify the intensity peaks located near the patella. We apply the derivation to the obtained sequence in order to detect the anatomical changes and a moving average as a convolution with a window function $w[\cdot]$ of size s_w to reduce the number of peaks. Eventually, we use a sequence obtained by taking the absolute of each of the I_y values:

$$I_y[i] = |(I'_y * w)[i]|, \forall i \in [0, H - 2\alpha). \tag{3}$$

At the last step, we obtain each k-th index out of the top $\tau\%$ of I_y values. It should be mentioned, that since the margin α was used, it has to be added to each selected index of I_y. A visualization of the procedure described above is given in Fig. 2.

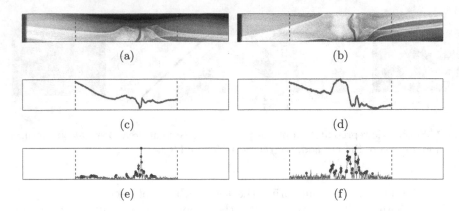

Fig. 2. Visualization of the joint center Y-coordinate proposal procedure for the left and right leg images – left and right columns, correspondingly. (a) and (b) show the sub-areas of the leg images, the red lines indicate the zones which are used for the analysis in Eq. 2. (c) and (d) show the obtained sequences I_y from Eq. 2. (e) and (f) show the result of applying Eq. 3, red dots indicate the values used to obtain Y-coordinates. (Color figure online)

Finally, we take the proposed Y-coordinates and as X-coordinates derive as $x = \frac{1}{2}W + j, \forall j \in [d_1, d_1 + p, d_1 + 2p, \ldots, d_2]$, where p is a displacement step, which can be estimated using a validation set (see, Sect. 4.2). The procedure described above is repeated for each leg.

To generate the proposals at multiple scales, we use a data-driven approach, which requires a training set having manual annotations. Here, we consider a joint area to be within a square of size S_n, where S_n is proportional to an image height H with some factor $\frac{1}{Z_n}$. Using manual annotations of training data we estimate a set $\mathbf{Z}$ having scales Z_n for each training image. Eventually, having an image $\mathbf{I}$ of size $C \times H$, we use a set $\mathbf{S}$ of proposal sizes based on the following estimations: $S_n = \frac{1}{Z_n}H, \forall Z_n \in [\ldots, \overline{\mathbf{Z}} - \sigma(\mathbf{Z}), \overline{\mathbf{Z}}, \overline{\mathbf{Z}} + \sigma(\mathbf{Z}), \ldots]$, where $\overline{\mathbf{Z}}$ is the mean and $\sigma(\mathbf{Z})$ is the standard deviation of the scales set estimated from the training data.

To conclude, the exact amount of generated proposals for each image is calculated as

$$|\mathbf{S}|[R + (R \bmod k \neq 0)] \cdot [d_2 - d_1 + ((d_2 - d_1) \bmod p \neq 0)], \tag{4}$$

where $R = 0.01\tau \cdot (H - 2\alpha)$. Here, we use a product rule and consider that all proposals belong to the image area. To estimate the number of proposals for the whole image with two leg image, their amount has to be multiplied by 2.

3.3 Proposal Scoring

The next step in our pipeline is to train a classification algorithm in order to select the best candidate region among generated proposals. For that, we use

HoG feature descriptor (to compactly describe the knee joint shape) and linear SVM classifier. This combination has been successfully used for human detection and has demonstrated excellent detection performance [5]. Additionally, the extraction of HoG features and SVM-based scoring are relatively fast pipeline blocks, which significantly supported our choice of the proposal scoring approach.

The HoG-SVM pipeline is implemented as follows: for each half of the image we generate the proposals and downscale each of them into a patch of 64 × 64 pixels. Then, we compute HoG features using the following parameters: 16 blocks with 50% overlapping, 4 cells in a block and 9 bins [5]. Finally, we classify vectors of 1,764 values (7 blocks vertically times 7 blocks horizontally times 4 cells times 9 bins). Basic maximum search is used to find the best scored proposal.

4   Experiments

4.1   Data

In this study we used several data sources: 1,574 high resolution knee radiographs from MOST cohort [10], 93 radiographs from [16] (Dataset obtained from Central Finland Central Hospital, Jyväskylä. The dataset is called "Jyväskylä dataset" in this paper), and 77 from OKOA dataset [18]. The images in all datasets contain knees with different severity of OA as well as implants. The images from MOST were used to create training, validation and test sets, while the remaining were only used to assess the generalization ability of the developed algorithm. More detailed information about the data and their usage is presented in Table 1.

Table 1. Description of datasets used in this study. The training data were used to train the algorithm, validation data to find hyperparameters for the algorithm, and the testing set to assess the localization and detection performance of the proposed method.

Dataset	Training set	Validation set	Test set	Average image size
MOST	991	110	473	3588 × 4279
Jyväskylä	-	-	93	2494 × 2048
OKOA	-	-	77	2671 × 2928

We converted the original X-ray data from 16 bit DICOM files to a 8 bit format for standardization: we truncated their histograms between the 5th and the 99th percentiles to remove corrupted pixels. Then we normalized the images by scaling their dynamic range between 0 and 255. After that, we manually annotated all used images using ad-hoc MATLAB-based tool. Main criterion for the annotation was, that the ROI should include the joint itself and the fibula bone.

To create a dataset for training a HoG-SVM proposal scoring, we processed the original training data as follows: for each knee on the training images, we generated the proposals and marked them as positive if the IoU of the manual annotation and the generated proposal was greater or equal than 0.8. Such a strict threshold was selected to make positive proposals be as close to manual annotations as possible. Examples of positive and negative proposals are given in Fig. 3.

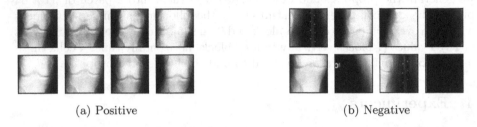

(a) Positive   (b) Negative

Fig. 3. Examples of positive and negative train patches. These examples were generated automatically using the proposal generation algorithm and manual annotations.

To augment the training set for more robust training, we performed the following transformations: the amount of positive was increased by a factor of 5 using rotations in the range [−2,2] degrees with a step 0.8. Here, we trained the classifier for only one leg, since the legs on the image are symmetrical. At the classification step, we used flipped proposals of the left leg image before extracting HoG features.

4.2   Implementation Details

To extract HoG, we used OpenCV [14] implementation and for SVM training we used a dual-form implementation from LIBLINEAR [11] package. To find the appropriate regularization constant C_s for SVM scorer, we used a validation set described earlier. We tried to scale the data before SVM to improve the classification results as well as hard-negative mining. However, neither of these approaches did not provided any improvement. Eventually, we found that $C_s = 0.01$ without data scaling and hard-negative mining gives the best precision and recall on the validation set.

In our pipeline, we fixed the smoothing window width $s_w = 11$ pixels, the displacement range to be $[d_1 = -\frac{1}{4}C, d_2 = \frac{1}{4}C]$ pixels, $k = 10$ and $\tau = 10$. Based on the manual annotations of the training data we estimated the set of scales $\mathbf{Z} = [3.0, 3.2, 3.4, 3.6, 3.8, 4.0, 5.0]$. After that, using the same validation set as for SVM parameters tuning, we adjusted the step p. Our main criterion was to find the best IoU having fast computational time. We found $p = 95$ pixels (IoU=0.843, time per image – 162 ms).

4.3   Localization Performance Evaluation

Before performing the evaluation of our pipeline on the testing datasets, we estimated the quality of the generated proposals for each of them. The evaluations are presented in Fig. 4.

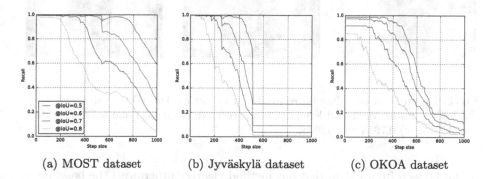

(a) MOST dataset   (b) Jyväskylä dataset   (c) OKOA dataset

Fig. 4. Proposals quality evaluation for each analyzed dataset – recall depending on the step value p and different IoU thresholds. Analysis shows, that it is feasible to reach a recall above 80% in every analyzed dataset having IoU threshold 0.8.

We varied the step of displacement p from 5 to 1,000 and evaluated each generated proposal by calculating IoU with the manual annotations. The best IoU score was used as a measure of quality. Eventually, we used different IoU thresholds to evaluate the best possible recall. It can be seen from Fig. 4, that on all testing datasets for each given IoU threshold our proposal algorithm reaches at least 80% recall. Using the pre-trained SVM classifier described above, we also reached high mean IoU for all analyzed datasets (see, Table 2). Examples of localization are given in Fig. 5.

Table 2. Localization performance evaluation. The computations were parallelized on Intel i7 5820k CPU. Time benchmarks were averaged over 3 runs.

Dataset	# of images	Mean IoU [%]	Average time [s]	Average time/image [ms]
MOST	473	0.8399	79.9922	169
Jyväskylä	93	0.7878	6.4195	14
OKOA	77	0.7761	7.5264	16

Apart from our own implementation, we also adopted the method described in [1] as a baseline (see, Sect. 2). The only difference in our approach was that we used a larger window as a joint center patch – 40 × 40 pixels instead of 20 × 20, since with the latter, the baseline method did not perform well with our images.

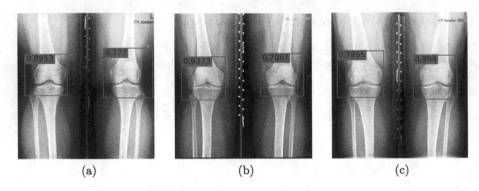

(a)   (b)   (c)

Fig. 5. Examples of bounding boxes produced by our method (OKOA dataset). Here, the IoU values for right and left joint are presented: (a) 0.8953 and 0.77, (b) 0.9373 and 0.7867, (c) 0.7955 and 0.984.

It can be seen from Fig. 6, that our method clearly outperforms the baseline on each analyzed dataset. The mean IoU values for the baseline were 0.1, 0.33 and 0.26 for MOST, Jyväskylä and OKOA datasets, respectively.

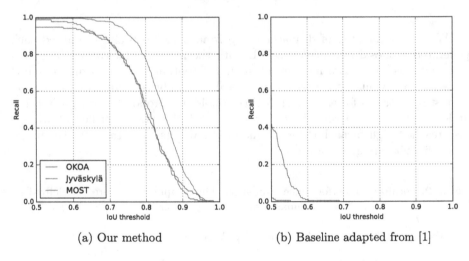

(a) Our method   (b) Baseline adapted from [1]

Fig. 6. Evaluation of recall depending on different IoU thresholds. The curves indicate a significant advantage of our method (a) in comparison to the baseline (b) for each of the testing datasets.

5   Discussion and Conclusions

In this study, we presented a novel automatic method for knee joint localization on plain radiographs. We demonstrated our proposal generation approach, which allows avoiding using exhaustive sliding window approaches. Here, we also showed a new way to use the information about the anatomical knee joint structure for proposal generation. Moreover, we showed that the generated proposals in average are highly reliable, since a recall of above 80% can be reached for IoU thresholds 0.5, 0.7 and 0.8. We showed that our method significantly outperforms the baseline. In the presented results, we showed that the baseline method performs comparatively similar on Jyväskylä dataset as on OAI data in [1] (reported mean IoU was 0.36), however, the detector fails on MOST dataset. This can, most probably, be explained by the presence of artifacts – parts of knee positioning frame, which are detected as joint centers.

We demonstrated, that our method generalizes well: the trained model can be used for other datasets than the one used during the training stage. Moreover, the proposed method neither requires a complex training procedure nor much computational power and memory. The developed method is mainly designed to be used for large scale knee X-ray analysis – especially for a CAD of OA from plain knee radiographs. However, the applications are not limited to this domain only. Our approach can also be easily adapted, for example, to the hand radiographic images.

Nevertheless, some limitations of this study remain to be addressed. First of all, our results are biased due to the manual annotations – only one person annotated the images. Secondly, the used data augmentation might include some of the false positive regions in the positive set for training the scoring block of the pipeline, which can have a negative effect on the detection. Finally, our method can be can be computationally optimized. For example, bias in the HoG-SVM block can explain a slight performance drop on Jyväskylä and OKOA datasets.

The method can be improved by applying the following optimizations. At first, the detection can be done on the downscaled images and then the detected joint area coordinates just need to be upscaled. We believe this optimization will significantly speed up the computations, since there is more than 10 times difference in performance between Jyväskylä dataset and MOST. However, this might require to find new hyperparameters. The second optimization would be to reuse Sobel gradients for the overlapping ROI proposals before computing HoG features, since in our current implementation, we recomputed them for each of the proposals. Furthermore, the following post-processing step can be applied for a possible improvement of the localization performance: localization regions could be centered at the joint center – this can be done by using a classifier from the baseline method [1]. However, the effect of these optimizations on the localization performance needs to be further investigated.

To conclude, despite the limitations, our method scales with a number of cores and can also be even efficiently parallelized on GPU to achieve a high-speed detection, due to the presence of loops. For example, it can be parallelized over X and Y coordinates of joints locations. It can be calculated using the values given in Table 2, that our data-parallel CPU implementation written in Python 2.7 already will allow to annotate more than 6,000,000 images of the size 2500×2000 per day (here, Jyväskylä dataset is taken as a reference), which makes the large-scale knee X-ray studies analysis possible. Eventually, our novel approach may enable the reliable and objective knee OA CAD, which will significantly benefit OA research as well as clinical OA diagnostics. The implementation of our method will be soon released on GitHub.

Acknowledgements. MOST is comprised of four cooperative grants (Felson – AG18820; Torner – AG18832, Lewis – AG18947, and Nevitt – AG19069) funded by the National Institutes of Health, a branch of the Department of Health and Human Services, and conducted by MOST study investigators. This manuscript was prepared using MOST data and does not necessarily reflect the opinions or views of MOST investigators.

The authors would also like to acknowledge the strategic funding of University of Oulu.

References

1. Antony, J., McGuinness, K., O'Connor, N.E., Moran, K.: Quantifying radiographic knee osteoarthritis severity using deep convolutional neural networks. In: Proceedings of 23rd International Conference on Pattern Recognition, ICPR (2016)
2. Berlin, L.: Malpractice issues in radiology. Perceptual errors. AJR Am. J. Roentgenol. **167**(3), 587–590 (1996)
3. Cibere, J.: Do we need radiographs to diagnose osteoarthritis? Best Pract. Res. Clin. Rheumatol. **20**(1), 27–38 (2006)
4. Cross, M., Smith, E., Hoy, D., Nolte, S., Ackerman, I., Fransen, M., Bridgett, L., Williams, S., Guillemin, F., Hill, C.L., et al.: The global burden of hip and knee osteoarthritis: estimates from the global burden of disease 2010 study. Ann. Rheum. Dis. **73**, 1323–1330 (2014). doi:10.1136/annrheumdis-2013-204763
5. Dalal, N., Triggs, B.: Histograms of oriented gradients for human detection. In: 2005 IEEE Computer Society Conference on Computer Vision and Pattern Recognition (CVPR 2005), vol. 1, pp. 886–893. IEEE (2005)
6. Demehri, S., Hafezi-Nejad, N., Carrino, J.A.: Conventional and novel imaging modalities in osteoarthritis: current state of the evidence. Curr. Opin. Rheumatol. **27**(3), 295–303 (2015)
7. Duda, R.O., Hart, P.E., Stork, D.G., et al.: Pattern Classification, vol. 2. Wiley, New York (1973)
8. Duryea, J., Li, J., Peterfy, C., Gordon, C., Genant, H.: Trainable rule-based algorithm for the measurement of joint space width in digital radiographic images of the knee. Med. Phys. **27**(3), 580–591 (2000)
9. Eckstein, F., Hudelmaier, M., Wirth, W., Kiefer, B., Jackson, R., Yu, J., Eaton, C., Schneider, E.: Double echo steady state magnetic resonance imaging of knee articular cartilage at 3 tesla: a pilot study for the osteoarthritis initiative. Ann. Rheum. Dis. **65**(4), 433–441 (2006)
10. Englund, M., Guermazi, A., Roemer, F.W., Aliabadi, P., Yang, M., Lewis, C.E., Torner, J., Nevitt, M.C., Sack, B., Felson, D.T.: Meniscal tear in knees without surgery and the development of radiographic osteoarthritis among middle-aged and elderly persons: The multicenter osteoarthritis study. Arthritis Rheum. **60**(3), 831–839 (2009)
11. Fan, R.E., Chang, K.W., Hsieh, C.J., Wang, X.R., Lin, C.J.: LIBLINEAR: a library for large linear classification. J. Mach. Learn. Res. **9**, 1871–1874 (2008)

12. Hirvasniemi, J., Thevenot, J., Immonen, V., Liikavainio, T., Pulkkinen, P., Jämsä, T., Arokoski, J., Saarakkala, S.: Quantification of differences in bone texture from plain radiographs in knees with and without osteoarthritis. Osteoarthritis Cartilage **22**(10), 1724–1731 (2014)
13. Huo, Y., Vincken, K.L., Viergever, M.A., Lafeber, F.P.: Automatic joint detection in rheumatoid arthritis hand radiographs. In: 2013 IEEE 10th International Symposium on Biomedical Imaging, pp. 125–128. IEEE (2013)
14. Itseez: Open source computer vision library (2016). https://github.com/itseez/opencv
15. Lindner, C., Thiagarajah, S., Wilkinson, J., Wallis, G., Cootes, T., The arcOGEN Consortium, et al.: Development of a fully automatic shape model matching (FASMM) system to derive statistical shape models from radiographs: application to the accurate capture and global representation of proximal femur shape. Osteoarthritis Cartilage **21**(10), 1537–1544 (2013)
16. Multanen, J., Heinonen, A., Häkkinen, A., Kautiainen, H., Kujala, U., Lammentausta, E., Jämsä, T., Kiviranta, I., Nieminen, M.: Bone and cartilage characteristics in postmenopausal women with mild knee radiographic osteoarthritis and those without radiographic osteoarthritis. J. Musculoskelet. Neuronal Interact. **15**(1), 69–77 (2015)
17. Pitman, A.: Perceptual error and the culture of open disclosure in Australian radiology. Australas. Radiol. **50**(3), 206–211 (2006)
18. Podlipská, J., Guermazi, A., Lehenkari, P., Niinimäki, J., Roemer, F.W., Arokoski, J.P., Kaukinen, P., Liukkonen, E., Lammentausta, E., Nieminen, M.T., et al.: Comparison of diagnostic performance of semi-quantitative knee ultrasound and knee radiography with MRI: Oulu knee osteoarthritis study. Sci. Rep. **6** (2016)
19. Podsiadlo, P., Wolski, M., Stachowiak, G.: Automated selection of trabecular bone regions in knee radiographs. Med. Phys. **35**(5), 1870–1883 (2008)
20. Podsiadlo, P., Cicuttini, F., Wolski, M., Stachowiak, G., Wluka, A.: Trabecular bone texture detected by plain radiography is associated with an increased risk of knee replacement in patients with osteoarthritis: a 6 year prospective follow up study. Osteoarthritis Cartilage **22**(1), 71–75 (2014)
21. Seise, M., McKenna, S.J., Ricketts, I.W., Wigderowitz, C.A.: Double contour active shape models. In: BMVC. Citeseer (2005)
22. Shamir, L., Ling, S.M., Scott, W., Hochberg, M., Ferrucci, L., Goldberg, I.G.: Early detection of radiographic knee osteoarthritis using computer-aided analysis. Osteoarthritis Cartilage **17**(10), 1307–1312 (2009)
23. Stachowiak, G.W., Wolski, M., Woloszynski, T., Podsiadlo, P.: Detection and prediction of osteoarthritis in knee and hand joints based on the x-ray image analysis. Biosurface Biotribology **2**, 162–172 (2016)
24. Thomson, J., O'Neill, T., Felson, D., Cootes, T.: Automated shape and texture analysis for detection of osteoarthritis from radiographs of the knee. In: Navab, N., Hornegger, J., Wells, W.M., Frangi, A.F. (eds.) MICCAI 2015. LNCS, vol. 9350, pp. 127–134. Springer, Cham (2015). doi:10.1007/978-3-319-24571-3_16
25. Uijlings, J.R., van de Sande, K.E., Gevers, T., Smeulders, A.W.: Selective search for object recognition. Int. J. Comput. Vis. **104**(2), 154–171 (2013)
26. Woloszynski, T., Podsiadlo, P., Stachowiak, G., Kurzynski, M.: A signature dissimilarity measure for trabecular bone texture in knee radiographs. Med. Phys. **37**(5), 2030–2042 (2010)
27. Zitnick, C.L., Dollár, P.: Edge boxes: locating object proposals from edges. In: Fleet, D., Pajdla, T., Schiele, B., Tuytelaars, T. (eds.) ECCV 2014. LNCS, vol. 8693, pp. 391–405. Springer, Cham (2014). doi:10.1007/978-3-319-10602-1_26

Semi-automatic Method for Intervertebral Kinematics Measurement in the Cervical Spine

Anne Krogh Nøhr[1], Louise Pedersen Pilgaard[1], Bolette Dybkjær Hansen[1],
Rasmus Nedergaard[2], Heidi Haavik[2], Rene Lindstroem[1],
Maciej Plocharski[1(✉)], and Lasse Riis Østergaard[1]

[1] Department of Health Science and Technology, Aalborg University, Aalborg,
Denmark
mpl@hst.aau.dk
[2] Centre for Chiropractic Research, New Zealand College of Chiropractic, Auckland,
New Zealand

Abstract. Cervical spine injuries, such as whiplash or disk herniation, are a worldwide health problem. Digital videofluoroscopic is often used to examine spine movement by means of manual identification and marking of vertebral landmarks, which is a complicated and time-consuming process. The aim of our study was to develop a fast, semi-automatic cervical vertebrae tracking method to accurately calculate the inter vertebral rotation in C1–C7 vertebrae. We manually defined templates for each cervical vertebra, so that these templates would be automatically tracked throughout neck movement. Subjects performed extension and flexion in the sagittal plane, which was recorded with digital videofluoroscopy. We implemented cross-correlation and Kalman filters for template tracking, and validated our method by comparing our results with a manual method, where a trained clinician manually marked the vertebrae. Our method provided higher intraobserver repeatability for C2/C3 to C6/C7 segments. Accordingly, the intraobserver repeatability was also comparable to other methods developed to track the lumbar vertebrae.

Keywords: Intervertebral kinematics · Cervical vertebrae · Digital videofluoroscopy · Template matching · Vertebrae tracking

1 Introduction

The cervical spine is a complex system, which consists of vertebrae, intervertebral discs, as well as muscles and ligaments. Cervical spine injuries are a major health concern, as trauma in the neck represents the majority of all spinal lesions [9]. Real-time motion assessments of the cervical spine provide means for understanding the natural neck motion and reveal movement abnormalities associated with spinal injuries or medical conditions, such as whiplash or disc herniation. However, no widely accepted standards exist for diagnosing cervical spine injuries in soft tissue. Such injuries are closely related to mechanical factors and thereby in-vivo studies of the spine's motion can lead to an improved understanding of

© Springer International Publishing AG 2017
P. Sharma and F.M. Bianchi (Eds.): SCIA 2017, Part II, LNCS 10270, pp. 302–313, 2017.
DOI: 10.1007/978-3-319-59129-2_26

these injuries [5,9]. Digital videofluoroscopy (DVF) analysis techniques, which involve recording multiple X-ray images as video frames, have been demonstrated to successfully investigate the spinal motion [1,2,6,7,15,16,20]. Intervertebral kinematics serve as means to describe and quantify the spinal motion. A commonly used method for calculation of intervertebral kinematics is Distortion-Compensated Roentgen Analysis (DCRA) [8,11,14,16]. DCRA is based on landmarks, which are defined in each of the DVF frames [11]. Definition of landmarks is time-consuming and subjective when performed manually, and small errors in the definition of the landmarks can potentially cause large errors in computation of the intervertebral kinematics [13].

Several semi-automatic methods have been successfully developed to track the lumbar vertebrae [7,10,19,20]. Lumbar vertebrae tracking methods have been based on template matching [7,10,20] or active contours [19]. Template matching technique involves defining a template for each vertebra, so that this template is continuously tracked throughout the video frames, using cross-correlation as a similarity measure [3,7,10,12,20]. These methods consider the vertebrae to be rigid bodies and thereby assume that no out-of-plane coupled motion is present. On the other hand, active contour methods do not assume vertebral rigidity [8]. Wong et al. [19] proposed an active contour method based on feature learning, feature detection, and tracking using a Kalman filter. The semi-automatic tracking methods for the lumbar part of the spine are applied in research [16,17] and in clinical practice [7,20].

The methods published in literature for tracking the cervical vertebrae using DVF are limited. Reinartz et al. [15] proposed a method for tracking C0 to C6 based on template matching and normalized gradient field. Cervical vertebrae tracking is more complicated than tracking the lumbar vertebrae, because the former are smaller and more difficult to identify in the videos than the latter, due to the anatomy of the C1 and C2. C1 and C2 vertebrae have unique shapes, and thus they often appear as one single vertebra in DVF frames. What is more, the C7 vertebra is often partly shadowed by the shoulder. The aim of this study was to develop an efficient, semi-automatic method for tracking the cervical vertebrae in order to accurately calculate the intervertebral rotation. Our method required very little initial input and did not require an expertise in radiography. It was designed to continuously track the C1–C7 cervical vertebrae during flexion and extension movements, recorded in the sagittal plane using digital videofluoroscopy.

2   Method

The Phillips BV Libra mobile diagnostic X-ray image acquisition and viewing system was used to record the neck movements at a rate of 25 frames per second, with a frame size of 576 × 720 pixels, and image resolution of 24 bits per pixel. The system consisted of two main components: (1) the C-arm containing the CCD camera, the image intensifier, the collimator and an X-ray tube; and (2) the mobile viewing station. Videos from two subjects were used for design and implementation of the method and videos from one subject was used for the proof of concept validation. For each subject two videos were recorded: one while the subject performed a flexion movement, and one during an extension movement.

The method we developed consisted of four steps: video preprocessing, manual input and template definition, tracking using template matching and Kalman filtering, and calculation of intervertebral rotation. We validated our method by comparing the results with a digitalized manual method using two fluoroscopic videos, where vertebral corners had been manually defined as landmarks by a trained clinician, which were used for calculation of intervertebral rotation based on DCRA.

2.1 Video Preprocessing

The fluoroscopic video frames were bandpass-filtered in order to enhance the contrast between the vertebral bodies and the surrounding tissue, and to remove the high frequency noise. The applied bandpass filter had a lower cut-off frequency of 5, and an upper cut-off frequency of 71, as proposed by Teyhen et al. [16]. Subsequently, a 10×10 Canny edge detection filter with the lower threshold of 0.02 and the upper threshold of 0.05 was implemented to detect the edges of the vertebral bodies. Figure 1 shows each of the steps in the preprocessing.

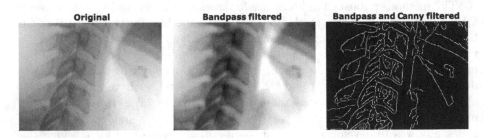

Fig. 1. A segment of an original frame, a bandpass filtered frame and a fully preprocessed frame.

2.2 Manual Input and Template Definition

A template for each of the cervical vertebrae was created by manual definition of landmarks in the first frame in the video, as illustrated in Fig. 2. These landmarks were used to define the templates in the first preprocessed frame and consequently each template was binary. Three separate methods were used for the template definition: one for C1, one for C2 and C7, and one for C3–C6.

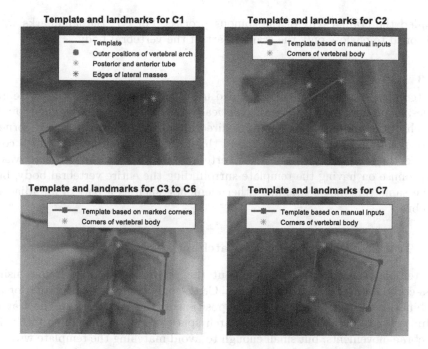

Fig. 2. Illustration of the manually defined landmarks used for template definition in the first frame.

2.2.1   C1

The template for C1 covered the vertebral arch which was visible in all anatomical positions during both flexion and extension movements. Landmarks were defined at the posterior and the anterior tube and at the posterior and anterior edges of the lateral masses to enable calculation of the intervertebral rotation, as illustrated in Fig. 2. The template was defined using the landmark at the posterior edge of the lateral masses and three additional landmarks positioned on the most cranial point, the most caudal point, and the most posterior point of the vertebral arch.

2.2.2   C2 and C7

The cranial part of the C2 vertebral body was not visible in all anatomical positions during movements due to shadowing by C1. C7 was also partially shadowed in the caudal part in some subjects by their shoulder. The corners of the vertebral bodies were defined for the computation of the intervertebral rotation and another four points were defined to enclose the template. The landmarks used for the definition of the template for C2 were positioned so the template enclosed the caudal part of the vertebral body and the cranial part of the vertebral arch. The anterior point was positioned just outside the anterior edge of the vertebral body and the cranial posterior point was positioned so that the template included the cranial part of the spinous process. For C7, the landmarks were positioned to enclose only the cranial part of the vertebral body.

2.2.3   C3–C6

The templates for C3 to C6 were defined using only the corners of the vertebral bodies. Four new points were automatically computed by adding eight pixels to each of the corners in the opposite direction of the diagonally placed corner. Those points were then connected, and this formed a quadrilateral, which constituted the template for the C3–C6 vertebrae. The size of the template was a compromise on having the template surrounding the entire vertebral body, but at the same time not containing other elements, such as a part of an adjacent vertebra.

2.3   Tracking Using Template Matching

The templates were tracked throughout the preprocessed video frames using cross-correlation and Kalman filtering. Cross-correlation was calculated for all combinations of rotation and spatial position of the templates in the frames within a limited search space. The search space was large enough to track the vertebrae movements, but small enough to avoid matching the template with an adjacent vertebra. The rotational search space was defined as the angle of the vertebra's midplane in the previous frame $\pm 4°$, with a step size of $0.01°$. The spatial search space was defined by adding 10 pixels to the position of the template in the previous frame in all directions. Subsequently, the maximum cross-correlation was found and used to determine the rotation and position resulting in the best match for the templates in the frame. Unrealistic high changes in the rotation and position of the templates between two adjacent frames were observed. Kalman filters have earlier been used to estimate the position of objects in images [18]. This inspired us to implement two Kalman filters in the tracking of the templates. One filter estimated the position of the templates by modelling the position and the velocity of the templates in both x and y dimensions. The other Kalman filter estimated the rotation by modelling the angle of the templates and their angular velocity. The acceleration was assumed to be constant for both the position and the rotation of the templates.

2.4   Intervertebral Rotation

The intervertebral rotation was calculated by means of the DCRA [11]. For the C2–C7 vertebrae, the corners of their vertebral bodies were used to calculate the anterior and posterior midpoints which defined midplanes of each vertebrae as illustrated in Fig. 3. The intervertebral rotation was calculated as the angle between the midplanes for two adjacent vertebrae. [11]. The corners of the C1 were detected in a different manner, as the C1 does not have a vertebral body. Instead, a midplane was defined as a line passing through the landmarks positioned at the anterior and the posterior tube (Fig. 2, upper left image).

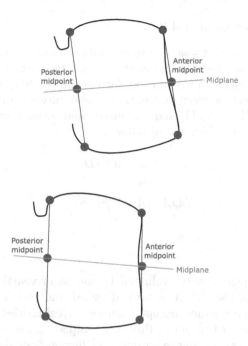

Fig. 3. An illustration of the landmarks on a pair of adjacent vertebrae for the calculation of the intervertebral rotation. The corners of the vertebral bodies are marked with purple points. (Color figure online)

2.5   Validation

We validated our semi-automatic method for vertebrae tracking through a comparison with a manual method using two fluoroscopic videos. The videos were recorded in the sagittal plane, and the subject performed movements of a full flexion and full extension. In the manual marking method, a trained clinician manually defined vertebral landmarks in 11 frames, constituting the 10% epochs of the total range of motion of movement. The midplanes for C3–C6 were defined in the same manner as in our semi-automatic method, whereas the midplanes for C1, C2, and C7 had been defined using only two landmarks. This entailed a modified version of the DCRA in the computation of the intervertebral rotation for these vertebrae in the manual method. The landmarks had been defined three times to enable examination and comparison of the variability both in our semi-automatic and in the manual method, which resulted with three datasets for each method. These three sets of manually defined landmarks in the first frame were used for the template definitions, thus the templates were defined three times. The intervertebral rotation was computed in all frames in the video and compared to the corresponding rotation computed by the manual method.

2.6   Intraobserver Repeatability

The Bland-Altman method was used to evaluate the intraobserver repeatability, and the differences between the intervertebral rotation were compared pairwise for both methods. The mean (d) and standard deviation (SD) were calculated for each set of differences between two datasets, and subsequently the Coefficient of Repeatability (CR) (Eq. 1), and the upper and lower Limits of Agreement (LOA) (Eq. 2) [4] were calculated as follows:

$$CR = 1.96 \cdot SD \tag{1}$$

$$LOA = d \pm 1.96 \cdot SD \tag{2}$$

3   Results

The tracking was qualitatively evaluated by means of visual inspection of the template tracking in the videos performed by our method. The evaluation did not reveal any tracking issues, except for minor irregularities in the end of the extension movement for two out of three C1 templates. The continuously intervertebral rotation both during extension and flexion, calculated by our semiautomatic method, is illustrated in Fig. 4 (red curves). The black asterisks illustrate the intervertebral rotation calculated from the manually marked vertebral landmarks by a trained clinician. The intervertebral rotation calculated by our method during an extension movement was comparable to the rotation calculated by the manual method for the C2/C3 to C6/C7 segments. However, differences were observed for the C1/C2, which might be due the irregularities in the tracking of the template for C1. For the flexion movement, the intervertebral rotation was comparable between the two methods for the C3/C4 to C6/C7 segments, but for C1/C2 and C2/C3 the rotation changed differently for the two methods throughout the movement. No issues were observed for the tracking of the templates for C2 and C3 and consequently, the observed differences might be a result of the midplane for C2 being differently defined in the manual method and our semi-automatic method.

3.1   Intraobserver Analysis

The Bland-Altman plots were made for all pairs of adjacent vertebrae for intervertebral rotation for both our cervical vertebrae tracking method (abbreviated as CVTM in Fig. 5 and Tables 1 and 2), and the digital manual method, abbreviated as DMM. The Bland-Altman plots for intervertebral rotation between C1/C2 and C2/C3 are illustrated in Fig. 5. These plots illustrate the pair of adjacent vertebrae having the highest coefficient of repeatability. The Bland-Altman statistics for all pairs of adjacent vertebrae are presented in Table 1 for our semi-automatic approach and the manual method involving vertebrae marking by a trained clinician. In five out of six cases, the mean of differences for

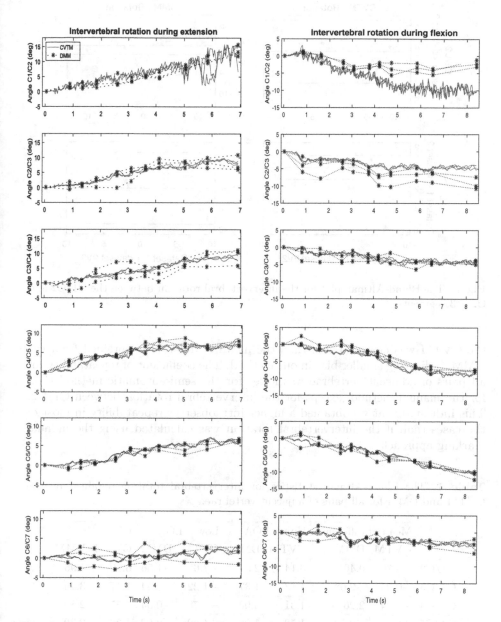

Fig. 4. Intervertebral rotation for all pairs of adjacent cervical vertebrae during an extension (left column) and a flexion movement (right column). The red lines illustrate the intervertebral rotation obtained by our semi-automatic cervical spine tracking method for all frames, and the black asterisks illustrate the intervertebral rotation obtained with the manual method for the 11 manually marked frames. (Color figure online)

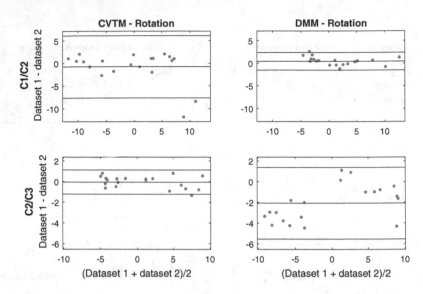

Fig. 5. The Bland-Altman plot for the intervertebral rotation between the C1/C2 and C2/C3 segments.

the CVTM was closer to zero than the mean of differences obtained from the DMM, indicating smaller bias in our method. The coefficient of repeatability for all pairs of adjacent vertebrae was lower for the semi-automatic method than for the manual method, except for the intervertebral rotation between C1/C2. This indicated that we obtained a higher intraobserver repeatability in most of the cases than if the intervertebral rotation was calculated using the manual marking approach.

Table 1. The Bland-Altman statistics for intervertebral rotation obtained by the CVTM and DMM for all pairs of adjacent vertebrae.

	Mean of difference		Upper LOA		Lower LOA		CR	
	CVTM	DMM	CVTM	DMM	CVTM	DMM	CVTM	DMM
C1/C2	−0.76	0.40	6.14	2.37	−7.65	−1.56	6.89	1.96
C2/C3	−0.05	−2.08	1.13	1.37	−1.22	−5.54	1.17	3.46
C3/C4	−0.24	2.26	1.31	5.06	−1.78	−0.54	1.55	2.80
C4/C5	0.28	−0.32	1.58	1.98	−1.02	−2.62	1.30	2.30
C5/C6	0.00	0.09	0.53	3.22	−0.52	−3.04	0.52	3.13
C6/C7	0.02	−0.76	1.08	2.44	−1.05	−3.96	1.07	3.20

4 Discussion

The purpose of this paper was to develop a semi-automatic tracking method for cervical vertebrae using template matching, in order to accurately and efficiently calculate intervertebral rotation of the cervical spine. The method enabled the tracking of C1 to C7 vertebrae in fluoroscopic videos during neck flexion and extension in the sagittal plane. Our semi-automatic method was validated as proof of concept through a comparison of the results with a digitalized manual method of vertebral marking performed by a trained clinician. The two methods showed comparable intervertebral kinematics. Additionally, our approach had a lower coefficient of repeatability than the manual method for all pairs of adjacent vertebrae, except for the intervertebral rotation of C1/C2. We acknowledge the need for a larger data set for validation, but the promising results obtained with our method indicate it to be a valid proof of concept for a fast and accurate calculation of intervertebral kinematics, with a potential of including intervertebral translation.

The user interaction with our semi-automatic is limited only to the initial step involving a definition of landmarks in the first video frame, making it far less time-consuming than the digitalized manual method. Our method had a low coefficient of repeatability, and accordingly was robust to small errors in the definition of landmarks for the templates, allowing successful template definition and intervertebral rotation calculations without an extensive expertise in radiography. The manual method of vertebrae marking requires a skilled clinician to define the landmarks in each frame to be analysed. Moreover, the manual method is susceptible to a variety of external factors, such as the consistency of clinician's precision in marking the vertebrae.

No studies evaluating the repeatability for tracking of the cervical vertebrae were found in the literature. Reinartz et al. [15] tracked C0–C6 by a semi-automatic method and stated that the tracking performed by the algorithm was as accurate as manual template matching, however, no results for the validation of the method was presented. Thus, our method was compared to existing methods for calculation of intervertebral kinematics of the lumbar spine. Frobin et al. [11] calculated the intervertebral rotation between L3/L4 and L5/S1 in X-ray images. They defined the landmarks manually in all frames and obtained a CR of 1.94°. Teyhen et al. [16] calculated intervertebral rotation for L3/L4 to L5/S1 using a semi-automatic method for DVF. The method was developed to optimize the precision of manually defined landmarks in all frames, and it resulted in a CR of 2.31°. Penning et al. [14] proposed a method based on image registration for calculation of intervertebral rotation between L1/L2 to L5/S1 in X-ray images and this method had a CR at 0.86–0.94°. Yeager et al. [20] evaluated a vertebral tracking algorithm, called KineGraph Vertebral Motion Analysis (VMA), which is used to calculate intervertebral rotation between L1/L2 to L5/S1 in DVF and the coefficient of repeatability of this method was 1.53°. An overview of the repeatability coefficients is presented in Table 2, where we compared our method to the aforementioned studies found in literature. However, methods for tracking lumbar and cervical vertebrae cannot be directly compared, since the cervical vertebrae are smaller than the lumbar vertebrae. Consequently, the intervertebral rotation for the cervical vertebrae are more sensitive to variability in the definition of the landmarks and thereby it is more difficult to obtain a low CR.

Table 2. CR and methods for calculation of intervertebral rotation for the CVTM and the methods found in the literature.

Study	Vertebrae	MCIR	CR
CVTM	C1/C2	DCRA	6.89°
CVTM	C2/C3-C6/C7	DCRA	1.19°
Frobin et al. [11]	L3/L4-L5/S1	DCRA	1.94°*
Teyhen et al. [16]	L3/L4-L5/S1	DCRA	2.31°*
Penning et al. [14]	L1/L2-L5/S1	Registration	0.86–0.94°*
Yeager et al. [20]	L1/L2-L5/S1	VMA	1.53°

MCIR: Methods for Calculation of Intervertebral Rotation; * indicates that the CR was calculated from the SD presented in the studies.

The semi-automatic method for cervical vertebrae tracking presented in our study had a lower CR for the C2/C3 to C6/C7 segments than Frobin et al. [11], Teyhen et al. [16], and Yeager et al. [20]. The low CR for our semi-automatic method compared to other methods found in literature might be due to our implementation of the Kalman filters in vertebrae tracking. The Kalman filters estimate a more precise position and rotation of the templates. The high CR for the intervertebral C1/C2 rotation is primarily a result of irregularities in the tracking of the template for C1 during the extension movement. The irregularity in the C1 tracking was likely caused by a change in the shape of C1 during the movement, which resulted in a deviation from the assumption of rigidity. This may be due to a rotation around the longitudinal axis. Consequently, a differently defined template may improve the tracking of C1 and thereby decrease the CR for the intervertebral rotation between C1/C2.

References

1. Anderst, W.J., Donaldson, W.F., Lee, J.Y., Kang, J.D.: Cervical spine intervertebral kinematics with respect to the head are different during flexion and extension motions. J. Biomech. **46**(8), 1471–1475 (2013)
2. Anderst, W.J., Donaldson, W.F., Lee, J.Y., Kang, J.D.: Cervical motion segment contributions to head motion during flexion\extension, lateral bending, and axial rotation. Spine J. **15**(12), 2538–2543 (2015)
3. Bifulco, P., Cesarelli, M., Allen, R., Sansone, M., Bracale, M.: Automatic recognition of vertebral landmarks in fluoroscopic sequences for analysis of intervertebral kinematics. Med. Biol. Eng. Comput. **39**, 65–75 (2001)

4. Bland, J.M., Altman, D.G.: Statistical methods for assessing agreement between two methods of clinical measurement. Lancet **1**, 307–10 (1986)
5. Bogduk, N., Yoganandan, N.: Biomechanics of the cervical spine part 3: minor injuries. Clin. Biomech. **16**, 267–275 (2001)
6. Breen, A.C., Allen, R., Morris, A.: Spine kinematics: a digital videofluoroscopic technique. J. Biomed. Eng. **11**(3), 224–228 (1989)
7. Breen, A.C., Muggleton, J.M., Mellor, F.E.: An objective spinal motion imaging assessment (OSMIA): reliability, accuracy and exposure data. BMC Musculoskelet. Disord. **7**, 1–10 (2006)
8. Breen, A.C., Teyhen, D.S., Mellor, F.E., Breen, A.C., Wong, K.W.N., Deitz, A.: Measurement of intervertebral motion using quantitative fluoroscopy: report of an international forum and proposal for use in the assessment of degenerative disc disease in the lumbar spine. Adv. Orthop. **2012**, 1–10 (2012). doi:10.1155/2012/802350. Article ID 802350
9. Cerciello, T., Bifulco, P., Cesarelli, M., Romano, M., Allen, R.: Automatic vertebra tracking through dynamic fluoroscopic sequence by smooth derivative template matching. In: 9th International Conference on Information Technology and Applications in Biomedicine (2009)
10. Cerciello, T., Romanoa, M., Bifulcoa, P., Cesarelli, M., Allen, R.: Advanced template matching method for estimation of intervertebral kinematics of lumbar spine. Med. Eng. Phys. **33**, 1293–1302 (2011)
11. Frobin, W., Brinckmann, P., Leivseth, G., Briggemann, M., Reikerås, O.: Precision measurement of segmental motion from flexion-extension radiographs of the lumbar spine. Clin. Biomech. **11**(8), 457–465 (1996)
12. Muggleton, J.M., Allen, R.: Automatic location of vertebrae in digitized videofluoroscopic images of the lumbar spine. Med. Eng. Phys. **19**(1), 77–89 (1997)
13. Panjabi, M.: Centers and angles of rotation of body joints: a study of errors and optimization. J. Biomech. **12**, 911–920 (1979)
14. Penning, L., Irwan, R., Oudkerk, M.: Measurement of angular and linear segmental lumbar spine flexion-extension motion by means of image registration. Eur. Spine J. **14**, 163–170 (2005)
15. Reinartz, R., Platel, B., Boselie, T., Mameren, H., Santbrink, H., Haar Romeny, B.: Cervical vertebrae tracking in video-fluoroscopy using the normalized gradient field. In: Yang, G.-Z., Hawkes, D., Rueckert, D., Noble, A., Taylor, C. (eds.) MICCAI 2009. LNCS, vol. 5761, pp. 524–531. Springer, Heidelberg (2009). doi:10.1007/978-3-642-04268-3_65
16. Teyhen, D.S., Flynn, T.W., Bovik, A.C., Abraham, L.D.: A new technique for digital fluoroscopic video assessment of sagittal plane lumbar spine motion. SPINE **30**(14), 406–413 (2005)
17. Teyhen, D.S., Flynn, T.W., Childs, J.D., Kuklo, T.R., Rosner, M.K., Polly, D.W., Abraham, L.D.: Fluoroscopic video to identify aberrant lumbar motion. SPINE **32**(7), 220–229 (2007)
18. Vidal, F.B., Casanova Alcalde, V.H.C.: Window-matching techniques with Kalman filtering for an improved object visual tracking. In: Third Annual IEEE Conference on Automation Science and Engineering, IEEE CASE 2007, vol. 22–25, pp. 829–834. IEEE (2007)
19. Wong, K.W., Luk, K.D., Leong, J.C., Wong, S.F., Wong, K.K.: Continuous dynamic spinal motion analysis. Spine **31**(4), 414–419 (2006)
20. Yeager, M.S., Cook, D.J., Cheng, B.C.: Reliability of computer-assisted lumbar intervertebral measurements using a novel vertebral motion analysis system. Spine J. **14**, 274–281 (2013)

Memory Effects in Subjective Quality Assessment of X-Ray Images

Victor Landre, Marius Pedersen$^{(\boxtimes)}$, and Dag Waaler

Norwegian University of Science and Technology, Gjøvik, Norway
{marius.pedersen,dag.waaler}@ntnu.no
http://www.colourlab.no

Abstract. Experiments with human observers is considered as the most precise way for the assessment of image quality. Although widely used, such experiments have its pitfalls and hazards. In this work we investigate if the quality rating of previously viewed images influence the rating given to the current image, which we refer to as the rating memory effect. A subjective experiment with a group of observers rating x-ray images of different radiation dose was used for the basis of the analysis. The results indicate a memory effect, meaning that the rating of an image can be influenced by the ratings given in previously judged images.

Keywords: X-ray · Memory · Subjective experiments · Psychometrics · Quality assessment

1 Introduction

Assessment of image quality is very important in many applications, such as medical imaging, printing, image enhancement, etc. Assessment of image quality can be done objectively, through image quality metrics [1], or subjectively, by consulting human observers [2]. Subjective assessment of image quality is considered as the most precise way of assessing quality, and is seen as the ground truth [3].

Various methods are used when assessing quality with human observers. In psychophysics there are traditional methods for measuring thresholds such as method of adjustment, method of limits, and method of constant stimuli [4]. These methods are used in detection or discrimination experiments. Psychometric scaling methods [2] have been proposed to obtain the relationship between physical changes and perceived changes. Common experimental methods include paired comparison, rank order and category judgement [2].

V. Landre—This research has been supported by the Research Council of Norway through project no. 247689 IQ-MED: Image Quality enhancement in MEDical diagnosis, monitoring and treatment. The authors would like to acknowledge the Color in Science and Industry master program (Jean Monnet University, Norwegian University of Science and Technology, University of Granada and University of Eastern Finland).

P. Sharma and F.M. Bianchi (Eds.): SCIA 2017, Part II, LNCS 10270, pp. 314–325, 2017.
DOI: 10.1007/978-3-319-59129-2_27

In most, if not all, psychometric experiments the observers are shown a set of images that are rated. In this work we focus on category judgement experiment, in which the observer is shown an image and is instructed to rate it according to a scale. Then a new image is shown, and the observer is asked to rate it according to the same scale. This is repeated until all images in the set have been rated. We investigate if the rating given to one image is influenced by the ratings of the previously rated images. We refer to this as the memory effect.

Most, if not all, statistical analysis of the results from subjective experiments are based on the assumption that the ratings are independent. If there is a memory effect, then this assumption does not hold, making the standard statistical analysis "useless" [5,6]. In quality control testing it is common to include visual assessment [7], and if memory effects are present, this might influence the outcome of such tests.

The paper is organized as follows: first relevant background, then we present the experimental setup, before results, and conclusion.

2 Background

We start by introducing work related to still images, then work related to videos, and at last work on decision making.

2.1 Still Images

Hoßfeld et al. [8] carried out an experiment with observers to study the impact of quality changes in web browsing. They found that the memory effect is a relevant quality of experience factor.

Short-term memory plays an important role in for example pair comparison experiments, where two images are shown at the time. An observer cannot scrutinise both images at the same time, since viewing one image will automatically place the other in the peripheral field making it substantially less detailed [9]. Because of this, one needs to rely on short-term memory when judging the quality the images, resulting in that a limited quantity of information from the two images can be compared at a time [10].

There has also been work related to long-term memory effects and its influence on image quality [11–13]. However, in this paper we focus on the short-term memory effects. Work has also shown that there is a difference between expert and non-expert observers in psychometric experiments [14,15]. Le Moan et al. [16] compared two different setups for paired comparison experiments, showing that the results between the two setups were significantly different.

2.2 Video

For video the working memory, or recency, and its impact on video quality has been studied by several researchers [17,18]. Alridge et al. [17] evaluate this effect on video quality on subjective experiment. Their experimental method consists of showing 30 s of a video with increasing or decreasing the quality. They asked the observers to rate the overall quality judgement on a 5 point scale at the end of the video. They evaluated the data using a Mean Difference Score between the ratings and the reference video. The results of this work showed that subjects tended to forgive the bad section by averaging the quality over all the period of the test, and were strongly influenced by what they see in the last section of 10 s. In their study they found that the prior information 20 or 30 s to the end seem to not influencing the weighting of the final rating.

Seferidis et al. [19] tried to quantify the "forgiveness effect", like the recency memory they evaluated it using videos. The authors introduced a forgiveness factor that adjusted the results obtained from short sessions 10–30 s to real-world viewing condition.

Hands and Avons [18] showed 30 s sequence films with poor-quality at the beginning or at the end; and asked the observers to rate continuously the quality during the observation. They found that the ratings changed more slowly following an improvement in quality than following a sudden impairment. They also investigated the duration of the impairment and its effect on observers, and they found that the duration is not cared by the participants, this effect is call "duration neglect".

A study from Pinson and Wolf [20] using videos showed that perception can be affected by the time spending looking the sample. They showed evidence that the human memory effect for quality estimation is limited to about 15 s. No significant memory effect occurred after 8 to 9 s, there is a low correlation between 1 and 3 s, and that the correlation were high between 7 and 9 s.

Huynh-Thu et al. [21] investigated the difference between discrete and continuous scales. According to their results based on videos they found that most observers tend to align their ratings with the labels, but some observers appear to distribute their ratings more evenly across the scale. This indicates the existence of individual response styles. In conclusion the authors assume that the absolute category rating method, with careful design and proper instruction produce very repeatable subjective results even across different scales.

Ickin et al. [22] studied the challenges in assessing the perceived quality of mobile-phone based video. They found that the extremely good quality videos are remembered better, even if there are intermediate parts with varying qualities.

There is some evidence that viewers have non-symmetrical memory in that they are quick to criticize degradations in video quality but slow to reward improvements [23].

2.3   Decision Making

Quality evaluation is a decision making process. Klapproth [24] reviewed the principal literature related to duration of the expected delay for realizing the options, and the time available to reach a decision. They show the relationship between decision making and time in different aspects. They show that physical (objective) time matters and has been related to decision making often and extensively. Also that psychological (subjective) time affect decision. Moreover, they show that anticipating delays in realizing an option not only reduces the value of that option, but also alters its mental representation. The time available for making choices has an impact on the amount of information that is processed and the quality of the final decision.

3   Experimental Setup

Double stimulus continuous quality scale (pair comparison) method is claimed to be less sensitive to context (i.e., subjective ratings are less influenced by the severity and ordering of the impairments within the test session). Single stimulus continuous quality evaluation (category judgement) method is claimed to yield more representative quality estimates. In this work we will use the single stimulus continuous quality evaluation method [25].

3.1   Viewing Conditions

The experiment was done in a controlled environment and with a medium ambient illumination at around 60 Lux. We used a single monitor, an Eizo ColorEdge CG246 display (24.1" - 61 cm), with a resolution of 1920 × 1080, calibrated with the ColorNavigator 6 software for a color temperature of 6500 K, a gamma of 2.20, a luminous intensity of 100 cd/m^2, and a black level of 0.15 cd/m^2. We did not set any restrictions for participants regarding the proximity of the screen to the user during the experiments. The users were allowed to take flexible and comfortable position; the position they are accustomed to take while doing this observation in daily work life. The viewing distance was around 50–60 cm from the monitor.

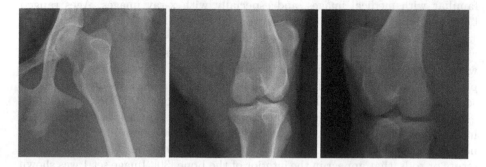

Fig. 1. Experimental X-ray images; image 1 on the left, image 2 in the center, and image 3 on the right.

3.2   Data

The images used for the experiment were three X-ray images of a lamb femur phantom (Fig. 1), image set 1 and image set 2 from Precht et al. [26] and image set 3 from Precht et al. [27]. The images in a set are different according to the variation of the X-ray dose and software optimization (Table 1). This resulted in three versions of image 1, three version of image 2, and four versions of image 3.

Table 1. Details of the images used in the experiment. The table shows the version of the images, x-ray dose in milliampere-seconds (mAs), and software optimization for the three different set of images.

Image set	Version	Dose (mAs)	Software optimization
1	1	16	Canon MLT(M)
1	2	6.3	Canon MLT(S)
1	3	2	Canon MLT(S)
2	1	16	Canon MLT(M)
2	2	6.3	Canon MLT(S)
2	3	2	Canon MLT(S)
3	1	8	Canon MLT(S)
3	2	3.2	Canon MLT(S)
3	3	0.5	Canon MLT(S)
3	4	8	Canon MLT(M)

3.3   Observers and Task

A total of 20 radiography students were used as observers. All of observers were familiar with medical images, and especially with x-ray images. Ages ranged between 19 and 25, and all observers were Norwegian.

The visual grading assessment was implemented using the "QuickEval" software [28]. Identical instructions were given to all users prior to the experiments with a special training session where we "calibrated" the observers with examples of best quality image and worst quality image. The scale used was a 5 point scale as recommended by ITU [29]: −2 (Bad), −1 (Poor), 0 (Fair), +1 (Good), +2 (Excellent). For the experiment one image was shown at a time and the instructions was "Rate the images on a 5 points scale according to the sharpness of the trabeculae", the trabeculae are small elements in the form of beams, struts or rods, that appear in the interior of the bone [30]. Image set 1 was shown first (three versions), then image set 2 (three versions), and at last image set 3 (4 versions). Within each set the images were shown in a random order to the observers. Each version of the image was shown twice to the observers, resulting in a total of 20 images shown to each of the 20 observers, which in total gave 400 ratings for all images.

4 Results and Discussion

4.1 Memory Effect

Figure 2 shows a histogram of the ratings given by the observers, and we can notice that all categories have been used, but that observers have mostly used categories -1, 0 and 1. To study and evaluate the data and the memory effect, the autocorrelation function [31] is an often used mathematical tool for finding repeating patterns. We define the autocorrelation function by the division of the covariance by the variance [32]:

$$r_k = \frac{c_k}{c_0}, \tag{1}$$

where k is the lag, c_0 is the sample variance and

$$c_k = \frac{1}{T-1} \sum_{t=1}^{T-k} (y_t - \bar{y})(t_{t+k} - \bar{y}). \tag{2}$$

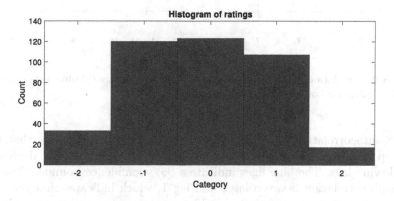

Fig. 2. Histogram of ratings given by the observers.

For the further analysis, we subtracted the average rating given to an image by all observers to the ratings given an observer to this image, and divided it by the standard deviation for all observers for that image:

$$RC_{ij} = \frac{R_{ij} - \frac{1}{N}\sum_{j=1}^{N} R_{ij}}{\sqrt{\frac{1}{N-1}\sum_{j=1}^{N} \left| R_{ij} - \left(\frac{1}{N}\sum_j R_{ij}\right)\right|^2}}, \tag{3}$$

where RC is the standard score rating for image i given by observer j, N the total number of observers, and R the raw rating given by an observer.

The autocorrelation function was calculated for all observers for 20 lags, since the observers each gave 20 ratings. To reduce the impact of the order in which in the observers carried out the experiment 1000 permutations of the observer order were carried out, and the average of the autocorrelation for these 1000 permutations were taken. In a theoretical case, where there is no presence of a memory effect, i.e. that the current rating given by an observer is influenced by the previous rating(s), the autocorrelation function should be close to zero all lags except 0 (which should be 1). The sample distribution can also influence the autocorrelation, for example if most ratings were identical, but as shown in Fig. 2 there is more or less an even distribution between category −1, 0 and 1, and some ratings in categories −2 and 2.

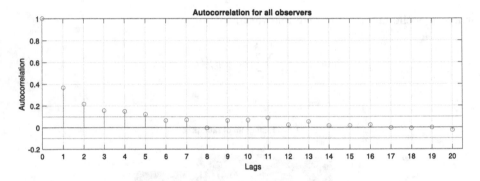

Fig. 3. Autocorrelation for all observers for 20 lags. The horizontal blue lines represents the 95% confidence bounds (Color figure online).

The autocorrelation function is shown in Fig. 3. We can notice that there is a drop from 1 to just below 0.4 at lag 1, then there is a steady decrease in the following lags. The blue lines indicate a 95% confidence bounds. There is a fairly high correlation between lag 0 and lag 1, which indicates that the rating given for the current image is influenced by previous ratings when analyzing the results for all observers.

Figure 4 shows the difference in score (raw score) between the duplicates of images (since each version of the image was shown twice). We notice that for most images the original version and its duplicate is rated with the same score. However, there are cases with a difference between the original version and its duplicate. If there are memory effects there would differences in the ratings given by the observers between the original version and its duplicate.

We also analyze the results also for individual observers, as there might be individual differences. Figure 5 shows the results for every single observer in the experiment. The confidence bounds are of course larger as the number of data points are fewer. However, we can notice that there is a large variation between the observers, but that the autocorrelation is reduced with increasing lag and that it is converging.

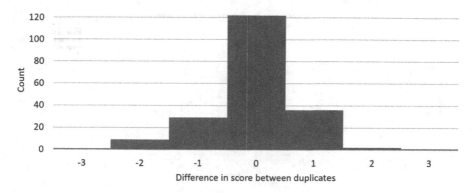

Fig. 4. Difference in score between original version and duplicate for all observers.

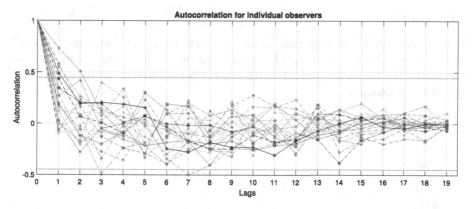

Fig. 5. Autocorrelation for individual observers. Each curve indicate a single observer. The horizontal blue lines represents the 95% confidence bounds (Color figure online).

The content of an image can also influence the ratings given by observers, and therefore we also analyzed the autocorrelation per image set. We also carried out 1000 permutations as described above. Figure 6 shows the results for the three different images, and we can notice that for image 2 there seems to be a higher autocorrelation than for the two others. Given the reduced number of images and number of observers, it is clear that further investigation with regards to the content of images should be done.

One aspect that was not investigated in this work was whether the observers recognized that the same image was shown twice and tried to remember what they scored earlier. If this occurred it might influence the results. This aspect should be investigated in the future.

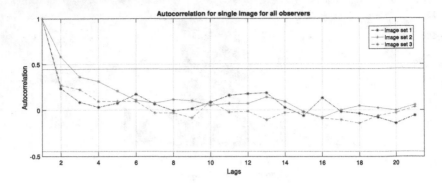

Fig. 6. Autocorrelation per image in the experiment.

4.2   Time

Time is also an interesting parameter in psychometric experiments. Figure 7 shows a boxplot of the time the observers spent on the three image sets. We can notice that the average and median time spent is reduced as the observers rated the different image sets.

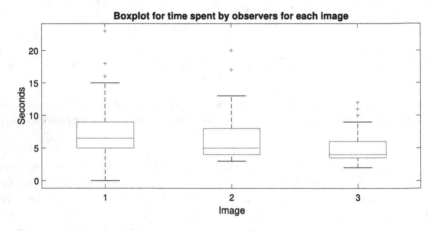

Fig. 7. Boxplot representing the time spent by the observers evaluating the images for the three image sets.

Furthermore, we also analyze the time spent by the observers when giving the different ratings for each category (-2, -1, 0, 1 and 2). Figure 8 shows the average time in seconds for each category in the experiment. We can notice that the observers used less time when they gave a high score compared to giving a low score. The effect is not very dominant though: on average the observers use approximately 20% less time than average to assign a top score, and approx.

20% more time than average to assign the lowest score. This result is similar to [22], who found observers to respond faster when giving a high score compared to low scores.

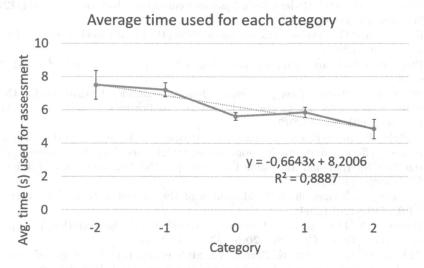

Fig. 8. Average time in seconds for each category represented with a 95% confidence interval.

5   Conclusion and Future Work

In this study we want to evaluate the potential presence of memory effects on in subjective experiments. A set of x-ray images were shown to observers, and the autocorrelation function was used to analyze the influence of previous ratings on the current rating. The results indicate that there is a correlation between previous ratings and the rating currently given by an observers.

Additional experiments with more images and more observers should be carried out to verify the results found. One should also carry out experiments where the order of the image sets are randomized to see if this influences the results.

Acknowledgments. We would like to thank Dr. Helle Precht, who provided the different images for the experiment.

References

1. Pedersen, M., Hardeberg, J.Y.: Full-reference image quality metrics: classification and evaluation. Found. Trends® Comput. Graph. Vis. **7**(1), 1–80 (2012)
2. Engeldrum, P.G.: Psychometric Scaling: A Toolkit for Imaging Systems Development. Imcotek press, Winchester (2000)

3. Hamid, H.R., Sabir, M.F., Bovik, A.C.: A statistical evaluation of recent full reference image quality assessment algorithms. IEEE Trans. Image Process. **15**(11), 3440–3451 (2006)

4. Ehrenstein, W.H., Ehrenstein, A.: Psychophysical methods. In: Windhorst, U., Johansson, H. (eds.) Modern Techniques in Neuroscience Research, pp. 1211–1241. Springer, Heidelberg (1999)

5. Engeldrum, P.G.: Psychometric scaling: avoiding the pitfalls and hazards. In: PICS, pp. 101–107 (2001)

6. Engeldrum, P.G.: Image quality modeling: Where are we? In: PICS, pp. 251–255 (1999)

7. International Atomic Energy Agency: Quality Assurance Programme for Computed Tomography: Diagnostic and Therapy Applications. Number 19 in IAEA Human Health Series (2012)

8. Hoßfeld, T., Biedermann, S., Schatz, R., Platzer, A., Egger, S., Fiedler, M.: The memory effect and its implications on web QoE modeling. In: Proceedings of the 23rd International Teletraffic Congress, pp. 103–110. International Teletraffic Congress (2011)

9. Freeman, J., Simoncelli, E.P.: Metamers of the ventral stream. Nat. Neurosci. **14**(9), 1195–1201 (2011)

10. Cohen, M.A., Dennett, D.C., Kanwisher, N.: What is the bandwidth of perceptual experience? Trends Cogn. Sci. **20**(5), 324–335 (2016)

11. Aideyan, U.O., Berbaum, K., Smith, W.L.: Influence of prior radiologic information on the interpretation of radiographic examinations. Acad. Radiol. **2**(3), 205–208 (1995)

12. Hardesty, L.A., Ganott, M.A., Hakim, C.M., Cohen, C.S., Clearfield, R.J., Gur, D.: "Memory effect" in observer performance studies of mammograms1. Acad. Radiol. **12**(3), 286–290 (2005)

13. Haygood, T.M., Liu, M.A.Q., Galvan, E.M., Bassett, R., Devine, C., Lano, E., Viswanathan, C., Marom, E.M.: Memory for previously viewed radiographs and the effect of prior knowledge of memory task. Acad. Radiol. **20**(12), 1598–1603 (2013)

14. Dugay, F., Farup, I., Hardeberg, J.Y.: Perceptual evaluation of color gamut mapping algorithms. Color Res. Appl. **33**(6), 470–476 (2008)

15. Bonnier, N., Schmitt, F., Brettel, H., Berche, S.: Evaluation of spatial gamut mapping algorithms. In: Color and Imaging Conference, vol. 2006, pp. 56–61. Society for Imaging, Science and Technology (2006)

16. Le Moan, S., Pedersen, M., Farup, I., Blahová, J.: The influence of short-term memory in subjective image quality assessment. In: 2016 IEEE International Conference on Image Processing (ICIP), pp. 91–95, September 2016

17. Aldridge, R., Davidoff, J., Ghanbari, M., Hands, D., Pearson, D.: Recency effect in the subjective assessment of digitally-coded television pictures. In: 5th International Conference on Image Processing and its Applications, pp. 336–339, July 1995

18. Hands, D.S., Avons, S.E.: Recency and duration neglect in subjective assessment of television picture quality. Appl. Cogn. Psychol. **15**(6), 639–657 (2001)

19. Seferidis, V., Ghanbari, M., Pearson, D.E.: Forgiveness effect in subjective assessment of packet video. Electron. Lett. **28**(21), 2013–2014 (1992)

20. Pinson, M.H., Wolf, S.: Comparing subjective video quality testing methodologies. In: Visual Communications and Image Processing 2003, pp. 573–582. International Society for Optics and Photonics (2003)

21. Huynh-Thu, Q., Garcia, M.-N., Speranza, F., Corriveau, P., Raake, A.: Study of rating scales for subjective quality assessment of high-definition video. IEEE Trans. Broadcast. **57**(1), 1–14 (2011)

22. Ickin, S., Janowski, L., Wac, K., Fiedler, M.: Studying the challenges in assessing the perceived quality of mobile-phone based video. In: 2012 Fourth International Workshop on Quality of Multimedia Experience (QoMEX), pp. 164–169. IEEE (2012)

23. Hamberg, R., de Ridder, H.: Time-varying image quality: modeling the relation between instantaneous and overall quality. SMPTE J. **108**(11), 802–811 (1999)

24. Klapproth, F.: Time and decision making in humans. Cogn. Affect. Behav. Neurosci. **8**(4), 509–524 (2008)

25. CIE: Guidelines for the evaluation of gamut mapping algorithms. Commision Internationale De L'eclairage, vol. 153, pp. D8–6 (2003)

26. Precht, H., Gerke, O., Rosendahl, K., Tingberg, A., Waaler, D.: Digital radiography: optimization of image quality and dose using multi-frequency software. Pediatr. Radiol. **42**(9), 1112–1118 (2012)

27. Precht, H., Gerke, O., Rosendahl, K., Tingberg, A., Waaler, D.: Large dose reduction by optimization of multifrequency processing software in digital radiography at follow-up examinations of the pediatric femur. Pediatr. Radiol. **44**(2), 239 (2014)

28. Van Ngo, K., Storvik, J. Jr., Dokkeberg, C.A., Farup, I., Pedersen, M.: Quickeval: a web application for psychometric scaling experiments. In: Larabi, M.-C., Triantaphillidou, S. (eds.) Image Quality and System Performance XI, vol. 9396, p. 93960O (2015)

29. ITUR Rec.: Bt. 500–11 (2002). Methodology for the subjective assessment of the quality of television pictures. International Telecommunication Union, 39 (2009)

30. Martensen, K.M.Q.: Radiographic Image Analysis. Elsevier Health Sciences, Amsterdam (2013)

31. Box, G.E., Jenkins, G.M.: Time Series Analysis: Forecasting and Control, revised edn. Holden-Day, San Francisco (1976)

32. Box, G.E., Jenkins, G.M., Reinsel, G.C., Ljung, G.M.: Time Series Analysis: Forecasting and Control Revised Ed. Wiley, Hoboken (2015)

Classification of Fingerprints Captured Using Optical Coherence Tomography

Ctirad Sousedik[1]($\boxtimes$), Ralph Breithaupt[2], and Patrick Bours[1]

[1] Norwegian University of Science and Technology (NTNU), Gjøvik, Norway
{ctirad.sousedik,patrick.bours}@ntnu.no
[2] Federal Office for Information Security (BSI), Bonn, Germany
ralph.breithaupt@bsi.bund.de

Abstract. We propose a technique for analysis of fingerprints scanned free-air (not pressed against a glass) with Optical Coherence Tomography (OCT). Fingerprints from the surface and subdermal parts of the finger are extracted from a 2 GB volumetric scan in cca. 2 s using our specialized technique and GPU acceleration on GeForce GTX 980. The technique provides fingerprints that perform with promising error rates that demonstrate the potential of the OCT for improved fingerprint identification, as well as its potential for prevention of biometric spoofing (PAD).

1 Introduction

In the recent years, biometrics are on the rise as a convenient alternative authentication mechanism. Unlike passwords, which can be easily forgotten, and access cards or keys, which can be easily lost, biometrics provide a means of authentication that is always readily available.

Among the many existing biometric modes, such as iris, face, retina and others, fingerprint stands among the best known and most widely applied.

However, despite the three-decade-long history, fingerprint sensing solutions still struggle with a number of challenges, which limit their applicability especially for unsupervised scenarios such as border control:

- Worn out fingers - For persons whose fingertip skin has been subjected to a lot of stress, the fingerprint can be abraded or significantly damaged (guitar players, construction workers, chemists etc.)
- Wet or greasy fingers - liquids on the surface of the fingerprint tend to diffuse into the fingerprint valleys when the finger is pushed against the sensor surface, which makes the acquisition difficult
- Dry fingers - If the fingerprint skin is too dry, it does not come into good contact with the fingerprint sensor surface, which results in low quality
- Infant fingerprints - the fingerprint skin of infants is very soft, and if pressed against a sensor surface, the fingerprint pattern will not be observable - this limits the usage of fingerprinting in fight against child trafficking

P. Sharma and F.M. Bianchi (Eds.): SCIA 2017, Part II, LNCS 10270, pp. 326–337, 2017.
DOI: 10.1007/978-3-319-59129-2_28

Similarly to other biometric modes, a significant challenge comes also with the susceptibility of fingerprinting to spoofing attacks [10,14].

A large body of research exists addressing this challenge of Presentation Attack Detection (PAD) for 2D fingerprint sensors. However, no single solution as of yet provides for a good level of security, especially considering resistance against novel materials and production techniques regarding the artefact fingerprints [14]. The existing approaches in the industry typically focus on combining a larger number of additional single-purpose sensors and features extracted from the 2D image to take the PAD decision. Considering the variability of properties of genuine human fingers, this requires machine learning approaches, which inherently depend on the training data, and as such are vulnerable against novel approaches not considered before [14].

1.1 OCT Fingerprinting

Due to the above mentioned shortcomings of the existing fingerprint sensors, the community has been looking for an alternative solution.

A promising path is offered by the Optical Coherence Tomography (OCT). The OCT is a light-beam-based scanning technology that is capable of penetrating the fingerprint skin up to the depth of 2–3 mm and acquiring a 3D volumetric representation of the light-scattering properties of the scanned sample - the surface fingerprint along with the sub-surface structure - Figs. 2 and 3. Contact with a surface is not necessary, and as such many of the challenges such as dry, wet, greasy or infant fingers can be easily overcome. In addition, the OCT is able to spot a second representation of the fingerprint in the subsurface data - the master template responsible for the stability of fingerprint during a person's lifetime. In addition to the inner fingerprint, presence of which could readily be used for the PAD purposes, the OCT is also able to spot sweat glands - fine spiral structures which end as sweat pores on the surface on the fingers. And last but not least, the volumetric measurements from the OCT readily provide a general scattering profile of the underlying material.

However, along with the significant promise, a rather significant challenge comes also associated with the step up from 2D to 3D scanning. The amounts of data generated by the OCT are very significant, and require novel scanning and processing approaches in order to achieve the practical speeds of a few seconds as required by many applications, such as border control.

2   Related Work

Some initial research has been carried out regarding the application of the OCT for the fingerprint sensing scenario.

Cheng and Larin [2] performed fingerprint Liveliness Detection by using autocorrelation analysis applied to the OCT. They scanned a small 2D in-depth part of the fingerprint tip in both the x-direction (2.4 mm) and in the z-direction (2.2 mm). In the depth direction they applied autocorrelation analysis.

They proved that autocorrelation values resulting from living fingerprints differ substantially from the values resulting from fake fingerprints. The same authors also obtained 3D density representations of fake and living fingerprints. They were among the first to use OCT technology to do this. A 3D OCT scan of a fake fingerprint surface on a real finger was published in their paper.

Peterson and Larin [12] classified OCT scans into fake and real samples. For this they used various neural network based approaches. The features that were used were based on first-order image statistics as well as on Gabor filter responses. Self-organizing maps (SOM) were used to reduce the dimensionality of the vector of Gabor filter responses.

Nasiri-Avanaki et al. have also shown in [11] that OCT scanning can be used to separate between genuine fingerprints and fake ones. In [1], Bossen et al. have shown that standard fingerprint comparison methods can be used for classification when applied on both the inner and outer fingerprint extracted from an OCT scan. Menrath [9] came up with a method to use OCT scans for detection of both sweat glands and extraction of the outer and inner fingerprint. His results are based on rather small fingerprint areas of 4×4 mm.

Khutlang et al. [7] have been capturing a partial fingerprint area using a commercially available OCT scanner and experimenting with the detection of the inner fingerprint and its extraction.

Despite the fact that a number of initial studies exists regarding the application of the OCT for fingerprint sensing and genuine/fake finger detection, very few studies take the speed of the processing into account. OCT fingerprint scans represent volumetric data of very significant sizes if one aims for standard resolutions of 500 dpi or 1000 dpi in 2D, which easily exceeds 1 GB per finger ($1024 \times 1024 \times 1024$ at 8 bit). This fact calls for an approach where the speed of the processing techniques is taken seriously into account as an actual research challenge, rather than disregarded as simply a matter of faster processing hardware.

In addition, majority of the studies assume a finger pressed against a flat surface (e.g. glass) during the scanning, which greatly simplifies the processing challenges (since the fingerprint surface is flattened) but introduces a number of disadvantages shared with standard 2D fingerprint sensors, such as difficulty to scan wet, greasy, dry or soft-skinned fingers, which could otherwise be easily overcome by the OCT [3,8].

Last but not least, the free-air scanning approach taken in our work has the potential for touch-less fingerprint sensing applications, where the subjects do not have to come into contact with any potentially unclean surface of the fingerprint sensor, increasing acceptability and convenience.

Regarding the work that satisfies the challenging constraints mentioned above, we are aware only of the promising works by Darlow et al. [3–5].

3   Fingerprint Surface Extraction in 3D

Our approach is to scan the fingerprint free-air, such that it is undistorted by pressing against a flat surface (such as glass). This comes with the challenge of precise 3D segmentation of the volumetric fingerprint OCT scan, so that further analysis is possible. A well segmented OCT fingerprint scan could be further used for extracting the fingerprints both from the surface and from underneath the skin and it could also serve well for further analysis of the skin regarding PAD (Fig. 2).

3.1   Database

We utilized the following OCT fingerprint scan dataset collected by our earlier work [13]:

- $1408 \times 1408 \times 1024$ voxels
- 8 bit per voxel
- 72 participants
- all 10 fingers
- 720 OCT fingerprint scans in total
- 2×2 cm scanning area

Along with the OCT fingerprints, we also collected standard 2D fingerprints from an identical set of participants, in order to enable testing of compatibility of the OCT fingerprints and the standard 2D fingerprints. This dataset contains 500 2D fingerprints, of all 10 fingers from 50 participants.

3.2   Efficient Edge Detection

In order to address the challenge of processing 2 GB OCT fingerprint scans in a matter of a few seconds, as required by numerous applications including border control, we took a combined approach of GPU acceleration and our specifically developed fast approximate filtering technique. Utilizing the following mathematical equality,

$$\int_{-\infty}^{\infty} f(\tau) \int \cdots \int g(t - \tau)dt \cdots dtd\tau = \int \cdots \int \int_{-\infty}^{\infty} f(\tau)g(t - \tau)d\tau dt \cdots dt, \quad (1)$$

it is possible to perform a convolution of a signal and a convolution core by deriving the convolution core, performing the convolution, and integrating the result. The advantage comes with the fact that a wide filtering core, which would take a significant computational effort to perform convolution with, can in specific cases derive into a very sparse representation that is very efficient to perform convolution with. Following this train of thought, we designed the following filter in order to detect edges in the scan-lines of the OCT fingerprint data - Fig. 1:

$$G(n) = \begin{cases} -height_2 + (n + 1) \cdot slope & \text{if } n \in [-\frac{size}{2}, -1] \\ height_2 - n \cdot slope & \text{if } n \in [0, \frac{size}{2} - 1] \\ 0 & \text{otherwise} \end{cases} \quad (2)$$

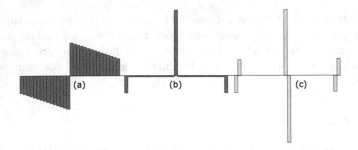

Fig. 1. Fast approximate edge detection filter and its derivatives for $height_1 = 15$, $height_2 = 29$, $size = 30$: (a) filter core $G(x)$; (b) first derivative $G'(x)$; (c) second derivative $G''(x)$ ($slope = \frac{height_2 - height_1}{size/2 - 1}$)

Second derivation of such a filtering core results in the following core:

$$
G''(n) = \begin{cases}
-height_1 & \text{if } n = -\frac{size}{2} \\
height_1 - slope & \text{if } n = -\frac{size}{2} + 1 \\
2 \cdot height_2 + slope & \text{if } n = 0 \\
-(2 \cdot height_2 + slope) & \text{if } n = 1 \\
-height_1 + slope & \text{if } n = \frac{size}{2} \\
height_1 & \text{if } n = \frac{size}{2} + 1 \\
0 & \text{otherwise}
\end{cases} \tag{3}
$$

Notably, such core has always only 6 non-zero coefficients, which come in 3 pairs of 2 coefficients positioned immediately next to each other - Fig. 1. This allows for a very efficient implementation of the convolution on the GPU, since only 3 memory reads per voxel are necessary to perform the filtering along the scan-lines, as the other 3 reads can be implemented using 3 delay variables reusing the results from a previous voxel. Integrating the results twice along the way allows to obtain the convolution of the scan-lines with the core, $G(n)$, in very efficient manner in a single GPU thread per scan-line. In addition, the technique can calculate the results for each of the scan-lines without using any extra intermediary memory buffer, which supports both the speed and the flexibility of the approach.

The actual edge detection is performed by identifying the position of a maximum during the process of convolution of the OCT scan-lines and the above discussed convolution core.

3.3   Outer Fingerprint Surface

However, it is highly sub-optimal to simply perform the above discussed edge detection for each of the (x, y) in-depth scan lines and treat the result as the detected surface. The OCT scan inherently contains large amounts of noise, and such an approach would result in a highly noisy surface, full of holes, where the detection failed due to noise. In addition, the strength of the finger's response

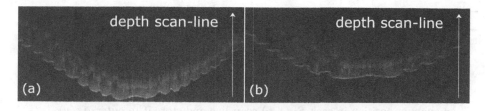

Fig. 2. OCT finger 2D slices - inner and outer fingerprint edges visualization

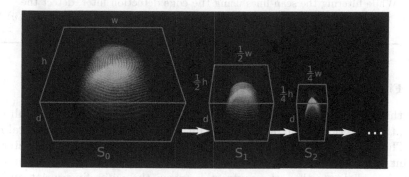

Fig. 3. Downsampling of the OCT scan along the width and height dimensions

diminishes with depth, which poses further challenges - Fig. 2. In order to address the challenges, we took inspiration from the approach by Darlow et al. [5] regarding performing the detection using a 3D equivalent of the image down-sampling pyramid.

Unlike Darlow et al. [5], who suggest building the pyramid by downsampling the OCT scan in all three dimensions, we have taken an approach where the pyramid is built by downsampling the scan always to 1/2 in width and height, leaving the depth dimension unaltered as illustrated by Fig. 3. The detection of the outer fingerprint surface, $outer_0(x, y)$, is then performed using the Algorithm 1.1.

The rationale is that it is far more likely to correctly detect the position of the fingerprint surface in a more downsampled scan, since the down-sampling heavily reduces the noise levels. If the surface, $outer_n(x, y)$, is assumed only in proximity of the surface, $outer_{n+1}(x, y)$, detected in a more down-sampled version of the scan, the likelihood of a correct detection is greatly increased.

In addition, detecting the surface, $outer_n(x, y)$, only in a distance, d_n, from a lower resolution surface $outer_{n+1}(x, y)$, allows for a significant speed-up since the entire scan does not need to be processed on any but the most downsampled level.

Algorithm 1.1. Outer fingerprint detection - 531 ms - GeForce GTX 980

---

1 Copy data from RAM to the GPU;

2 From the original OCT fingerprint scan, $S_0(x, y, z)$, generate N additional
  versions of the OCT scan, $S_1(x, y, z)...S_N(x, y, z)$, each downsampled to 1/2 of
  the previous one along width and height dimensions;

3 While filtering the scan-lines using the edge detection filter, detect the
  maximum position in each $S_N(x, y)$ scan-line and store it to $outer_N(x, y)$;

4 **for** *each n in N − 1 to 0* **do**

5   Up-sample $outer_{n+1}(x, y)$ by a factor of two into $region_n(x, y)$

6   While filtering the scan-lines using the edge detection filter, detect the
      maximum position in each $S_n(x, y)$ scan-line and store it to $outer_n(x, y)$ IF
      at a distance d_n from $region_n(x, y)$;

3.4 Fingerprint Flattening

After the outer fingerprint surface, $outer_0(x, y)$, has been identified at full resolution, the detection of the inner fingerprint surface constitutes the natural next step. The inner fingerprint is more difficult to detect, compared to the outer fingerprint, since the contrast between the inner fingerprint and and surrounding tissue is much lower than the contrast between the outer fingerprint and the empty air - Fig. 2.

In addition, the depth at which the inner fingerprint appears in various people varies widely (range wider than 1–5x in our experience). If one attempts to detect the inner fingerprint simply as a second surface around the outer surface, $outer_n(x, y)$, the outer fingerprint easily interferes with the inner fingerprint on the more downsampled versions of the scan and the detection is highly unreliable, especially for fingers where the distance between the inner and outer layer is rather small.

In order to address this issue, we perform full flattening of the original OCT scan according to the identified outer fingerprint surface. By flattening we mean re-arranging of the data along the z-axis using the following equation:

$$S_0(x, y, z) \leftarrow S_0(x, y, z - outer_0(x, y)) \tag{4}$$

3.5 Inner Fingerprint Surface

The re-arranged data naturally put any remnants of the outer fingerprint close to the bottom of the scan. This prevents interference of the outer fingerprint during the inner fingerprint detection.

Even though the algorithm, used for extracting the inner fingerprint surface, is similar to the algorithm for extracting the outer fingerprint surface, there are two important differences.

The inner fingerprint surface is not searched using a constant filter size, but rather an adaptive filter size. This is necessary to handle the cases of very thin inner layers, where the inner fingerprint is very close to the outer one. It also improves performance for the fingerprint where the distance between the outer and inner fingerprint surface is large, and as such a larger filter size can perform much more reliably.

The inner fingerprint surfaces, detected even in the most down-sampled versions of the flattened OCT scan, can still contain significant number of errors. This appears to be caused by interference of the inner fingerprint with itself, if a significant amount of down-sampling is considered. The problem is mitigated by a guessing procedure, where any detected points that deviate too much from the average depth range are replaced by a position that gained maximum response in a histogram of the detected positions - by the most likely depth encountered. Although this could seem to damage the inner fingerprint surface, in practice, it provides a precise-enough estimate for the higher-resolution detection steps, and the technique can recover even from severe failures encountered at the most down-sampled level.

The concept is described by Algorithm 1.3.

3.6   Surface Conversion to 2D Fingerprint

We extract the fingerprint into 2D representation by utilizing the identified 3D surface, masking out the random noise that appears where the finger was not present - Fig. 6. The concept is described by Algorithm 1.2.

Algorithm 1.2. 2D fingerprint extraction - 59 ms - GeForce GTX 980

---

1 Process the 3D fingerprint surface, inner or outer, $s(x, y)$, by a 2D Gaussian core and store to $s_{blurred}(x, y)$;

2 $s_{subtracted} \leftarrow (s - s_{blurred}) \cdot fingerprint\_height\_ratio$;

3 $s_{fingerprint}(x, y) \leftarrow s(x, y)$ IF $s(x, y) \in [0, 1]$ or 1 IF $s(x, y) > 1$ or 0 IF $s(x, y) < 0$;

4 Threshold the confidence map $confidence(x, y)$ into fingerprint and empty area, median filter to remove holes, and utilize to mask out the noisy areas;

---

4   Results

Direct comparison with the method of Darlow et al. [3] is difficult, since we were not able to obtain the codes/executables from the authors and we do not have any opportunity to share the data either due to their sensitive nature. However, we believe our method clearly outperforms the method of Darlow et al. [3] for the following reasons:

- Our approach using adaptive filter size allows processing of the cases where the inner fingerprint is at very small depth - something that is not possible using a simple 3-point edge detector on a scan down-sampled in all 3 dimensions as in [3]

Algorithm 1.3. Inner fingerprint detection - 328 ms - GeForce GTX 980

---

1 Flatten the fingerprint according to Eq. 4;
2 Re-compute the image pyramid;
3 From the original OCT fingerprint scan, $S_0(x, y, z)$, generate N additional
   versions of the OCT scan, $S_1(x, y, z)...S_N(x, y, z)$, each downsampled to 1/2 of
   the previous one along width and height dimensions (re-calculate the pyramid
   for the flattened scan);
4 $current\_maxima = -INF$;
5 **for** *each tested_filter_size in l to h with step s* **do**
6   While filtering the scan-lines using the edge detection filter, detect the
    maximum position in each $S_N(x, y)$ scan-line and store the maxima
    positions to $inner_N(x, y)$ and the values of the maxima (the response of
    the filter at the maxima positions) to $confidence_N(x, y)$;
7   Calculate a weighted histogram $h(d)$ of $inner_N(x, y)$ using
    $confidence_N(x, y)$ as weights;
8   Gaussian blur the $h(d)$ as a 1D function;
9   Detect maxima m in $h(d)$;
10 **if** $m > current\_maxima$ **then**
11   set $current\_maxima \leftarrow m$;
12   set $optimal\_filter\_size \leftarrow tested\_filter\_size$;
13 set $optimal\_filter\_size$ as the edge detection filtering size for all further
   operations;
14 While filtering the scan-lines using the edge detection filter, detect the
   maximum position in each $S_N(x, y)$ scan-line and store the maxima positions
   to $inner_N(x, y)$ and the value of the maxima (the response of the filter at the
   maxima positions) to $confidence_N(x, y)$;
15 **for** *each n in N − 1 to 0* **do**
16 **if** $n > ec\_depth\_limit$ **then**
17   Calculate a weighted histogram $h(d)$ of $inner_n(x, y)$ using
     $confidence_n(x, y)$ as weights;
18   Calculate a histogram $h(d)$ of $inner\_weighted_n$;
19   Gaussian blur the $h(d)$ as a 1D function;
20   Detect the maxima, m, and its position, p, in $h(d)$;
21   From the position, p, search to the right and to the left in $h(d)$ until a
     number smaller than $m \cdot ratio$ is encountered (intuitively, measure the
     width of the peak in the histogram);
22   Utilize the identified positions to calculate the acceptable level of
     deviation, $inner\_layer\_deviation$;
23   Process $inner_N(x, y)$ such that any value further than
     $inner\_layer\_deviation/2 \cdot d\_ratio$ from m is replaced by m (Replace
     values too far from the average by the average value)
24 Up-sample $inner_{n+1}(x, y)$ by a factor of two into $region_n(x, y)$;
25 While filtering the scan-lines using the edge detection filter, detect the
    maximum position in each $S_n(x, y)$ scan-line and store the maxima
    positions to $inner_n(x, y)$ and the value of the maxima (the response of the
    filter at the maxima positions) to $confidence_n(x, y)$ IF at a distance d_n
    from $region_n(x, y)$;

- Our approach uses a filter with a wide support, which allows obtaining a much higher quality surface than in [3]
- Darlow et al. [3] did not follow the path of extracting the 2D fingerprints directly from the 3D surface, and instead utilized the surface simply as an estimate of the fingerprint positions. The fingerprint is read off from the original scan by averaging the data above and below the estimated surface. Our method does not require this, in fact, we were able to obtain complete fingerprints using the layer information alone.

4.1   Compatibility with 2D Fingerprints

To further prove the robustness of our approach, we performed a $N : N$ comparison between the OCT fingerprints and the 2D fingerprints in the above mentioned dataset: The metrics considered are miss-classified comparisons of identical fingerprints (FNMR) and miss-classified comparisons of non-identical fingerprints (FMR). The failure-to-extract (FTX) metric expresses the percentage of fingerprints the comparison software failed to extract fingerprint features from [6]. The outer fingerprints compared to the 2D fingerprints with an equal-error-rate (EER) of 0.7% and a FTX rate of 11%. The inner fingerprints compared to the 2D fingerprints with EER of 1% and a FTX rate of 3.5%. The 2D

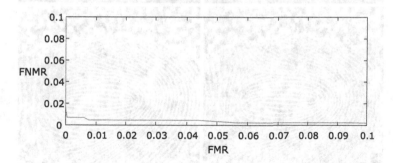

Fig. 4. ROC curve for the outer fingerprints and the 2D fingerprints comparisons

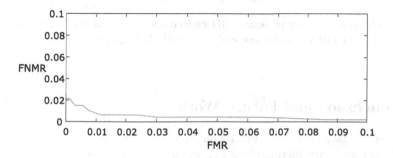

Fig. 5. ROC curve for the inner fingerprints and the 2D fingerprints comparisons

fingerprints from a commercial sensor resulted in FTX of 3.2%. A commercial fingerprint identification software Verifinger 9.0 was used (Figs. 4 and 5).

The average durations of the above proposed algorithms are stated in their heading (Algorithms 1.1, 1.2 and 1.3). The speed was measured as the CPU time since the start of the stage until its end, synchronizing with the GPU to make sure the GPU computation has finished as well. In total, our algorithm processes a 2 GB scan (including additional data management) in cca. 2 s on GeForce GTX 980.

4.2 Fingerprint Surfaces

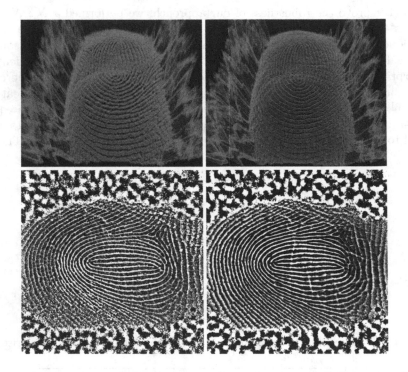

Fig. 6. Left: inner fingerprint surface 3D surface and extracted 2D fingerprint; Right: outer fingerprint surface 3D surface and extracted 2D fingerprint

5 Conclusion and Future Work

Based on the fact that the inner fingerprint seems to perform better than the outer fingerprint, considering the FTX, we believe it is indeed very promising to expect significant improvements from the OCT. At the same time, the fingerprints already do perform close to practical error rates - in spite of the significant potential of our custom-built OCT scanner for improvements. Our technique can process 2 GB in cca. 2 s, which shows promise regarding the speed challenges.

The future work would involve further improvements of the underlying scanner design. In addition, the method for the detection of the outer fingerprint could probably benefit from further improvements.

References

1. Bossen, A., Lehmann, R., Meier, C.: Internal fingerprint identification with optical coherence tomography. IEEE Photonics Technol. Lett. **22**(7), 507–509 (2010)
2. Cheng, Y., Larin, K.V.: Artificial fingerprint recognition by using optical coherence tomography with autocorrelation analysis. Appl. Opt. **45**(36), 9238–9245 (2006)
3. Darlow, L.N., Connan, J., Singh, A.: Performance analysis of a hybrid fingerprint extracted from optical coherence tomography fingertip scans. In: 2016 International Conference on Biometrics (ICB), pp. 1–8, June 2016
4. Darlow, L.N., Connan, J.: Efficient internal and surface fingerprint extraction and blending using optical coherence tomography. Appl. Opt. **54**(31), 9258–9268 (2015)
5. Darlow, L.N., Connan, J., Akhoury, S.S.: Internal fingerprint zone detection in optical coherence tomography fingertip scans. J. Electron. Imaging **24**(2), 023027 (2015)
6. ISO/IEC 2382-37: Information technology - Vocabulary - Part 37: Biometrics (2012)
7. Khutlang, R., Khanyile, N.P., Makinana, S., Nelwamondo, F.V.: High resolution feature extraction from optical coherence tomography acquired internal fingerprint. In: 2016 17th IEEE/ACIS International Conference on Software Engineering, Artificial Intelligence, Networking and Parallel/Distributed Computing (SNPD), pp. 637–641, May 2016
8. Meissner, S., Breithaupt, R., Koch, E.: Defense of fake fingerprint attacks using a swept source laser optical coherence tomography setup. In: Proceedings of SPIE, Frontiers in Ultrafast Optics: Biomedical, Scientific, and Industrial Applications, vol. 8611 (2013)
9. Menrath, M.: Fingerprint with OCT. Master's thesis, Fern-Universität Hagen in Cooperation with Bundesamt für Sicherheit in der Informationstechnik (BSI) (2011)
10. Mura, V., Ghiani, L., Marcialis, G.L., Roli, F., Yambay, D.A., Schuckers, S.A.: LivDet 2015 fingerprint liveness detection competition 2015. In: 2015 IEEE 7th International Conference on Biometrics Theory, Applications and Systems (BTAS), pp. 1–6, September 2015
11. Nasiri-Avanaki, M.R., Meadway, A., Bradu, A., Khoshki, R.M., Hojjatoleslami, A., Podoleanu, A.G.: Anti-spoof reliable biometry of fingerprints using en-face optical coherence tomography. Opt. Photonics J. **1**(3), 91–96 (2011)
12. Peterson, L.E., Larin, K.V.: Image classification of artificial fingerprints using Gabor wavelet filters, self-organising maps and Hermite/Laguerre neural networks. Int. J. Knowl. Eng. Soft Data Paradigms **1**(3), 239–256 (2009)
13. Sousedik, C., Breithaupt, R.: Full-fingerprint volumetric subsurface imaging using fourier-domain optical coherence tomography. In: 2017 International Workshop on Biometrics and Forensics (IWBF) (2017, accepted)
14. Sousedik, C., Busch, C.: Presentation attack detection methods for fingerprint recognition systems: a survey. IET Biometrics **3**(4), 219–233 (2014)

Interpolation from Grid Lines: Linear, Transfinite and Weighted Method

Anne-Sofie Wessel Lindberg$^{(\boxtimes)}$, Thomas Martini Jørgensen,
and Vedrana Andersen Dahl

Department of Applied Mathematics and Computer Science,
Technical University of Denmark, 2800 Kongens Lyngby, Denmark
{awli,tmjq,vand}@dtu.dk

Abstract. When two sets of line scans are acquired orthogonal to each other, intensity values are known along the lines of a grid. To view these values as an image, intensities need to be interpolated at regularly spaced pixel positions. In this paper we evaluate three methods for interpolation from grid lines: linear, transfinite and weighted. Linear method does not preserve the known values along the grid lines. Transfinite method, known from mesh generation, preserves the known values but might cause overshoot. The weighted method, which we propose, is designed to combine the desired properties of transfinite method close to grid lines, and the stability of the linear method. We perform an extensive evaluation of the three interpolation methods across a range of upsampling rates for two data sets. Depending on the upsampling rate, we show significant difference in the performance of the three methods. We find that the transfinite interpolation works well for small upsampling rates and the proposed weighted interpolation method performs very well for all relevant upsampling rates.

Keywords: Interpolation · Image processing · Performance analysis · Line scans · Medical image analysis

1 Introduction

Scanning along a set of parallel lines is a common setting in medical imaging. The modality that motivated our investigation is optical coherence tomography (OCT) [7], well established in ophthalmology for obtaining volumetric images of the retina. Using OCT, the retina is scanned in depth (z) and along a line (x) with a high depth and transversal resolution, resulting in a single xz cross-section of the retina (a so-called B-scan). Collecting a number of images by scanning along parallel lines results in a volumetric xyz data set. Since scanning speed of the commercial systems is limited, and prolonged scanning is unpleasant for the patient, the distance between the recorded B-scans is often large compared to the transverse resolution of the B-scans. Therefore, if an xy cross-section (*en face*) is of interest, the resolution is much coarser in y direction and pixels

© Springer International Publishing AG 2017
P. Sharma and F.M. Bianchi (Eds.): SCIA 2017, Part II, LNCS 10270, pp. 338–349, 2017.
DOI: 10.1007/978-3-319-59129-2_29

are non-square. This makes it difficult to distinguish the anatomical structures, especially evident with blood vessels running parallel to scan lines.

To reveal additional anatomical structures, another OCT scan may be performed, along the lines orthogonal to the first scan. It is our goal to compute a volumetric data which combines the information from the two volumetric scans. Several problems emerge in connection to this. The eye might move in the z direction during scanning, and this needs to be accounted for. Furthermore, the intensity might vary significantly between the scans and images. And most importantly, how to combine two volumetric data covering the same area, one with high resolution cross-sections in xz planes, and the other in yz planes?

In this paper we address the interpolation problem when merging two OCT volumes. As we have a high resolution in z direction (around 5 microns), we practically sample at any height and our problem reduces to a 2D case. Furthermore, given a high resolution along the scan lines we ignore the discrete sampling in this direction. Therefore, our problem is image interpolation from grid lines.

The problem is illustrated in Fig. 1 (a). The information is available along the two sets of parallel lines, and it needs to be sampled at regularly spaced pixel positions.

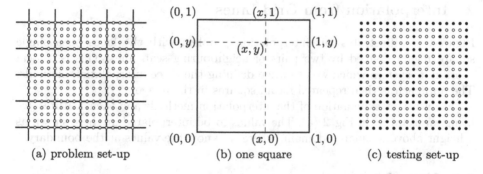

(a) problem set-up   (b) one square   (c) testing set-up

Fig. 1. Interpolation from the grid lines. (a) two sets of scan lines with known intensities are shown in black, this need to be sampled at regularly spaced pixel positions illustrated as white dots. (b) shows one square region defined by four scan lines and its local coordinate system. (c) is a testing set-up, black are the known and white are the unknown pixels, here shown with an upsampling rate of 4.

To the best of our knowledge, the problem of interpolating information from grid lines while preserving boundary values, has not been addressed in the context of image interpolation. In the context of mesh generation for finite element modeling a related problem is often solved using transfinite interpolation [5,6], a method for constructing a smooth function over a planar domain given the values on the boundary. Transfinite interpolation has been used for solving problems where information on boundaries should be preserved. It has been used in more recently studies e.g. [12] for solving time-dependent changes of volumetric material properties in heterogeneous volumes and in [10] for solving elliptic boundary value Poisson problems in arbitrary shaped 2D domains.

In this work we employ the transfinite interpolation for image interpolation from grid lines, and we compare it against an approach based on linear upscaling. Furthermore, we propose a weighted interpolation which preserves the desirable properties of the transfinite and the linear method.

A quality measure for merging two OCT scans should relate to the ease of distinguishing the anatomical structures present in the volume, their sharpness and precision. While sharpness may be quantified, it is difficult to asses the precision of the interpolation. Central to our problem is that we need to determine information where it is lacking. This aspect is similar to image upscaling and single-image super-resolution approaches. Therefore, when it comes to evaluation the performance of interpolation algorithms we turn to the conventional approach [11,16] which tests each method on a set of downsampled images and uses peak signal to noise ratio (PSNR) metric.

During testing, we change upsampling rates and statistically evaluate the results from the three interpolation methods. This allow us to conclude on the methods' performance and to provide guidelines for different upsampling rates.

2   Interpolation from Grid Lines

Figure 1 (b) illustrates our interpolation problem with the focus on a single square region defined by two pairs of neighbouring scan lines. This is a local coordinate system which we use when defining the three interpolation methods. The approach is then repeated for all squares in the image.

For a better explanation of the interpolation methods and their features, we bring an example in Fig. 2 (a). The values to be interpolated are here shown as a height above a squared domain, where we know the values at the boundary.

2.1   Linear Interpolation

A naive approach of combining the two scans involves linearly upsampling each scan independently and averaging the results. Over one square domain we have

$$L_x(x,y) = (1-y)S(x,0) + yS(x,1),$$
$$L_y(x,y) = (1-x)S(0,y) + xS(1,y),$$
$$L = \frac{1}{2}(L_x + L_y),$$

where S are the known values along the boundary of the square domain, L_x and L_y are linearly upsampled boundaries in x and y direction, and L is the interpolant which we in this context denote *linear*. Construction of linear interpolation is demonstrated on Fig. 2 (a)–(d).

Let us point out two properties of linear interpolation. First, every value $L(x,y)$ is a convex combination of four values from S. As a result, L does not

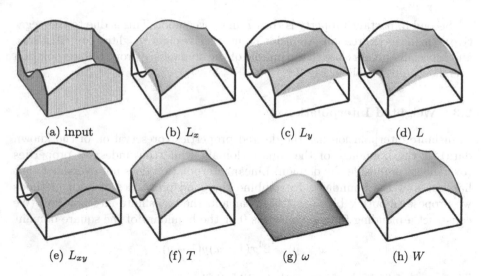

(a) input   (b) L_x (c) L_y (d) L

(e) L_{xy} (f) T (g) ω (h) W

Fig. 2. Interpolation over one squared domain. (a) know values from one direction in red and from the other direction in blue, (b) linear interpolation from one pair (blue) of rectangle sides, (c) linear interpolation from other pair (red) of sides, (d) mean of two linear contributions, (e) bilinear interpolation from corners, (f) transfinite interpolation, (g) weighting scheme (h) weighted interpolation (Color figure online).

produce undesirable overshoot. Secondly, for a point on the boundary, the underlying known data contributes only with a half of its value, the other half coming from the values at two corners. As a result, L does not agree to the known data along the boundaries of the domain. Those two properties combined mean that when used on images, linear interpolation results in smeared-out appearance.

2.2   Transfinite Interpolation

Transfinite interpolation is used for functions given on the boundary of a domain which can be parameterized as a square. For our purposes this reduces to

$$T = L_x + L_y - L_{xy}$$

where

$$L_{xy}(x,y) = (1-x)(1-y)S(0,0) + (1-x)yS(0,1) + x(1-y)S(1,0) + xyS(1,1).$$

Here L_{xy} is the bilinear interpolant from the values at the corners of the domain, and T is the final transfinite interpolant. Those are shown in Fig. 2 (e) and (f).

The most important property of the transfinite interpolation is that it preserves the known values at the boundary of the domain. To perceive how this property is achieved by the construction of T, note in Fig. 2 that at the boundary of the domain, L_{xy} differs from the known values exactly twice as much as L does.

Second important property is that T may overshoot. This is due to the negative term in the expression. All inside points receive eight weighted contributions, and especially points close to the middle of the domain are prone to interpolation overshoot, also visible in Fig. 2 (f).

2.3   Weighted Interpolation

Transfinite interpolation has the desired properties (preservation of the known data) at the boundary of the square domain while the undesired properties (overshoot) are inside the domain. Linear interpolation does not overshoot, but has issues at the boundary. To combine the good properties of both methods we propose smoothly blending the linear and the transfinite interpolation. We construct a blending function which is 0 at the boundary of the square domain

$$\omega(x, y) = 2^4 x (1 - x) y (1 - y).$$

The constant 2^4 is chosen such that $\omega(0.5, 0.5) = 1$.

We define our novel interpolation, which we denote $weighted$, as

$$W = \omega L + (1 - \omega)T.$$

This also evaluates to

$$W = (2 - \omega)L - (1 - \omega)L_{xy}.$$

The blending function and the weighted interpolation are shown on Fig. 2 (g) and (h). The illustrated example confirms desirable properties of the weighted interpolation. Like transfinite, the weighted interpolant matches the exact values at the boundaries of the domain. However, thanks to blending with the linear interpolant, the overshoot from inside of the domain is reduced. Finally, smooth blending function maintains a smooth appearance of the interpolant.

3   Evaluation of Interpolation Methods

Both objective and subjective tests are used [4] for evaluating interpolation methods. Subjective tests measure a perceived image quality, while objective tests use a defined metrics for quantifying image quality or interpolation error. The choice of the tests and the quality measures depends on the intended use.

For our motivating example, interpolation from grid lines is a step towards merging two OCT line scans. We plan to use the merged volume for automatic detection of anatomical structures and quantification of abnormalities in the eye. In this upcoming work we will evaluate the three interpolation methods in terms of the detection and quantification results.

In the work presented here, we bring a more meticulous and general evaluation of the interpolation methods based on measuring interpolation error for a specific upsampling rates. The ground truth is constructed by downsampling an image, which is then upsampled using the three methods, and the results are compared against the original image. For downsampling, we keep image columns and rows at a certain distance, which corresponds to upsampling rate s. See Fig. 1 (c) for our testing set-up.

For a certain upsampling rate, the fraction of the unknown pixel is

$$u = \frac{(s-1)^2}{s^2}.$$

For example, when $s = 2$ we keep every second row and every second column, and fraction of unknown pixels is only 0.25.

To demonstrate the properties of the interpolation methods, we conduct tests for upsampling rates from 2 to 30. However, the high upsampling rates (above 10) are of limited practical value due to high degeneration of image quality.

3.1   Data Sets

We evaluate the three interpolation methods on two data sets. The first contains 200 images from the Berkley Segmentation data set [9], which is widely used for evaluations of image upscaling and super-resolution algorithms [2]. The images depict scenes from nature such as landscapes, people and animals, covering a wide range of image patterns at all scales. We converted all images into grayscale with an intensity range between 0 and 1 prior to processing.

The second data set is ophthalmologic data in form of 72 funduscopies. Fundoscopy is an imaging technique for examination of fundus obtained using a light source and a ophthalmoscope. This was chosen because of image content which is similar to OCT scans, and will allow us to assess the performance of the interpolation methods in a setting which resembles to our application. Figure 3 shows some examples of the funduscopies.

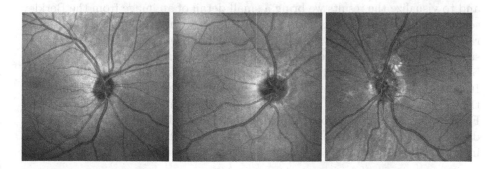

Fig. 3. Three images from the fundoscopy data set. Images depict anatomical structures at the fundus of the eye.

3.2   Performance Measurement

The interpolation quality is assessed by the pixelwise difference between the ground truth image and the interpolated image. The interpolation error can be evaluated as the Root Mean Square Error (RMSE) which is defined as [8]

$$\text{RMSE} = \sqrt{\frac{1}{N} \sum_{i=1}^{N} (\hat{I}(i) - I(i))^2}$$

where summation runs over all pixels i from the original image I and the interpolated image $\hat{I}$.

More often the Peak Signal-to-Noise Ratio (PSNR) is used [16] as an interpolation quality measure. The PSNR is measured in dB and defined as [14]

$$\text{PSNR} = 20 \log_{10} \left(\frac{\text{MAX}_\text{I}}{\text{RMSE}} \right)$$

where MAX_I is the maximum pixel intensity value, in our case 1.

4   Results

First, we present some results from the interpolations for upsampling rate 3 and 6. Second, we present the performance of each interpolation method for upsampling rates varying between 2 and 30. Last, we present the results from a statistical analysis of the methods' performance.

4.1   Interpolated Results

The differences in performance of the three interpolation methods are subtle, and to visualize the results we bring a small detail of an image from the Berkley segmentation data set in Fig. 4 (a), and we also show the grid lines for upsampling rate 3 and 6 in Fig. 4 (b) and (c). The interpolated results for this image, the two sets of grid lines and the three interpolation methods are shown on Fig. 5. We also bring the pixelwise error between the interpolated images and the original image. It can be seen that the error is zero along the grid lines for the transfinite and the weighted interpolation, while this is not the case for the linear interpolation. Furthermore, for all methods, the interpolation quality in form of the PSNR decreases when the upsampling rate is increased. For this example, the weighted interpolation outperforms the transfinite when using upsampling rate 6.

4.2   Performance Analysis

Figure 6 shows the PSNR values for the three interpolation methods for 6 randomly chosen images from the Berkley segmentation data set, interpolated with upsampling rate 6. We see a big variance in performance across the images,

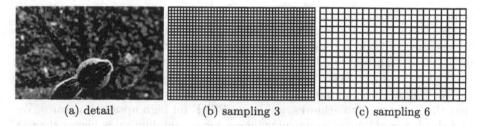

(a) detail   (b) sampling 3   (c) sampling 6

Fig. 4. Testing example. (a) zoom on a detail in one of the images from the Berkley data set. (b) and (c) are grid lines with an upsampling rate of 3 and 6.

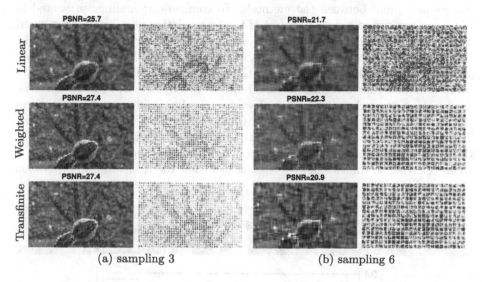

(a) sampling 3   (b) sampling 6

Fig. 5. The interpolated results for the detail and the grid lines shown on Fig. 4. The interpolation methods are presented with linear on top, weighted in the middle and transfinite in the bottom row. Columns (a) and (b) bring the results for an upsampling rate of 3 and 6. The PSNR value for each interpolated image is listed above it. Next to each interpolated image is the pixelwise difference between the interpolated image and the original image, with red/blue color indicating positive/negative difference, and white indicating zero difference. (Color figure online)

compared to relatively small variance between the three interpolation methods. However, weighted interpolation obtains the best performance for 5 out of 6 images, while transfinite method is best for the last image. Linear method is not the best for any of the images, but is still superior to transfinite in 4 out of 6 images.

To evaluate an overall performance of the interpolation methods, we computed the mean PSNR for the whole Berkley segmentation data set, for each interpolation method and for a range of upsampling rates. We conducted a similar experiment for the Fundoscopy data set. Figure 7 shows a plot of the obtained values with upsampling rates varying between 2 and 19. We notice the same performance pattern for both data sets. The transfinite method has the largest mean PSNR for smallest upsampling rates, while linear method has the largest mean PSNR for highest upsampling rates. In the interval around the point where transfinite and linear method cross, the weighted method achieves the highest mean PSNR.

We conducted experiments for upsampling rate up to 30, and we confirm that the linear method achieves best mean PSNR for high upsampling rates. We find this being of limited practical value, as for upsampling rates higher than 18 we interpolate over 89% of the pixels in the image.

As already shown in Fig. 6 the variance of performances is big between the images and small between the methods. To confirm our findings presented in Fig. 7, we performed a statistical test of the interpolation performance measured by the PSNR value. We set up a regression model for correlation between the PSNR value and the categorical variables for image and for method, for each sampling rate. F-values indicated that the method is the main descriptor. Moreover, we found that a significant difference between the three methods exists. Therefore, we tested the methods pairwise to check for difference between them at each upsampling rate and moreover, to find out which method performs best. The results are listed in Table 1(a) for the Berkley data set and in Table 1(b) for the Fundoscopy data set. The results show similar trend, and we use notation a/b when referring to the two data sets. It is seen that the transfinite interpolation

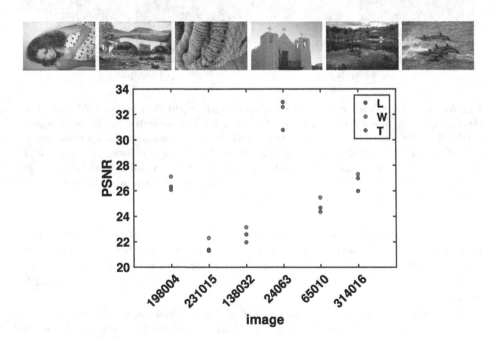

Fig. 6. A set of 6 randomly chosen images from the Berkley segmentation data set, and the resulting PSNR for three interpolation methods, linear (L), weighted (W) and transfinite (T). Upsampling rate is 6.

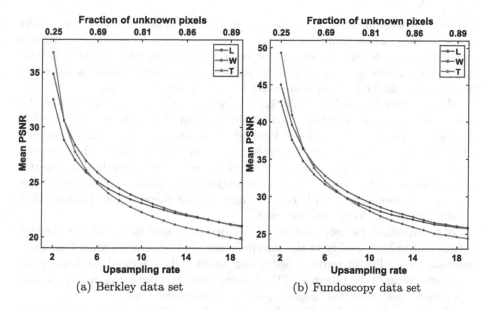

Fig. 7. Comparison of the linear (L), weighted (W) and transfinite (T) interpolation. The mean PSNR value for (a) Berkley and (b) Fundoscopy data set at different upsampling rate. The second x-axis indicates the fraction of unknown pixels.

Table 1. Results from statistical analysis of interpolation performance for three methods at different upsampling rates, for Berkley data set (a) and Fundoscopy data set (b). The methods are linear (L), weighted (W) and transfinite (T), and upsampling rates are shown in intervals between 2 and 30. Number 1 indicates the method that performs best for a given upsampling rate, while 3 indicates the method that performs worst. The star indicates that no significant difference was found between the two methods for the given upsampling rate.

(a) Berkley data set								(b) Fundoscopy data set						
	2	3	4-5	6	7-14	15-20	21-30		2-3	4	5-6	7-8	9-19	20-22	23-30
L	3	3	3	2*	2	1*	1	L	3	3	3	2*	2	1*	1
W	2	1*	1	1	1	1*	2	W	2	1*	1	1	1	1*	2
T	1	1*	2	2*	3	3	3	T	1	1*	2	2*	3	3	3

performs best for upsampling rates below 3/4 and the weighted interpolation performs best for sampling rates above 4/5 and below 15/20. The linear method performs best for upsampling rates above 20/22.

5 Discussion

Our experiments and the statistical evaluation of the three interpolations are in alignment with the previously demonstrated properties of the methods. Prior to experiments, we knew that the transfinite methods performs best close to the grid lines containing known information, while the linear method performs best in areas further away from the grid lines. Therefore we expected the transfinite method to achieve superior results for small upsampling rates where grid lines cover a large fraction of the image. Our results confirm this hypothesis. Likewise, we show that linear method is superior at high upsampling rates.

We designed the weighted method to combine the good properties of the linear and transfinite method. Our results confirm that weighted method has superior performance for a large interval of upsampling rates, and especially where the transfinite and linear method perform equally badly. This happens at upsampling rate of 6/8 and the interval where weighted method is superior extends from 4/5 to 14/19. For our application of merging OCT images we are aiming at upsampling ratio between 5 and 10 and these results indicate that weighted interpolation should be used.

It is important to note that image quality measured as PSNR not directly correspond to high quality of the interpolation. Further investigation that measures image quality in terms of sharpness should be performed for finding the most suitable method. Likewise, if the images are to be used for visual inspection, a perceived quality of the images should be measured.

The three interpolation methods presented here only use the known intensity information along the grid lines. We would expect the performance to improve significantly if a prior knowledge about the appearance of images is incorporated in the method. A significant work in this line has been conducted for single-image super resolution [3] or image inpainting [1], for example using image patches [15] and sparse representation [13]. We believe that those methods might be adapted to solve the problem of interpolation from the grid lines.

6   Conclusion

In conclusion, the contribution of our present work is twofold. First, we introduce the problem of interpolation from grid lines and suggest three possible solutions: a linear, a transfinite and a weighed interpolation. Secondly, we provide systematically test the three methods and conclude that transfinite method is superior for very small upsampling rates, while weighted method should be considered for a broad range of upsampling rates.

References

1. Bertalmio, M., Sapiro, G., Caselles, V., Ballester, C.: Image inpainting. In: Proceedings of the 27th Annual Conference on Computer Graphics and Interactive Techniques, pp. 417–424 (2000)
2. Dong, W., Zhang, L., Shi, G., Wu, X.: Image deblurring and super-resolution by adaptive sparse domain selection and adaptive regularization. IEEE Trans. Image Process. **20**(7), 1838–1857 (2011)
3. Elad, M., Feuer, A.: Restoration of a single superresolution image from several blurred, noisy, and undersampled measured images. IEEE Trans. Image Process. **6**(12), 1646–1658 (1997)

4. Giachetti, A., Asuni, N.: Real-time artifact-free image upscaling. IEEE Trans. Image Process. **20**(10), 2760–2768 (2011)
5. Gordon, W.J., Hall, C.A.: Construction of curvilinear co-ordinate systems and applications to mesh generation. Int. J. Numer. Methods Eng. **7**(4), 461–477 (1973)
6. Gordon, W.J., Hall, C.A.: Transfinite element methods: blending-function interpolation over arbitrary curved element domains. Numer. Math. **21**(2), 109–129 (1973)
7. Huang, D., Swanson, E.A., Lin, C.P., Schuman, J.S., Stinson, W.G., Chang, W., Hee, M.R., Flotte, T., Gregory, K., Puliafito, C.A., Fujimoto, J.G.: Optical coherence tomography. Science **254**(5035), 1178 (1991)
8. Loia, V., Sessa, S.: Fuzzy relation equations for coding/decoding processes of images and videos. Inf. Sci. **171**(1–3), 145–172 (2005)
9. Martin, D., Fowlkes, C., Tal, D., Malik, J.: A database of human segmented natural images and its application to evaluating segmentation algorithms and measuring ecological statistics. In: Proceedings of the 8th International Conference in Computer Vision, vol. 2, pp. 416–423, July 2001
10. Provatidis, C.G.: Solution of two-dimensional Poisson problems in quadrilateral domains using transfinite coons interpolation. Commun. Numer. Methods Eng. **20**(7), 521–533 (2004)
11. Romano, Y., Protter, M., Elad, M.: Single image interpolation via adaptive nonlocal sparsity-based modeling. IEEE Trans. Image Process. **23**(7), 3085–3098 (2014)
12. Sanchez, M., Fryazinov, O., Adzhiev, V., Comninos, P., Pasko, A.: Space-time transfinite interpolation of volumetric material properties. Trans. Vis. Comput. Graph. **21**(2), 278–288 (2015)
13. Shen, B., Hu, W., Zhang, Y., Zhang, Y.J.: Image inpainting via sparse representation. In: Proceedings of the IEEE International Conference on Acoustics, Speech, and Signal Processing, pp. 697–700, April 2009
14. Sung, M.M., Kim, H.J., Kim, E.K., Kwak, J.Y., Yoo, J.K., Yoo, H.S.: Clinical evaluation of JPEG2000 compression for digital mammography. Nucl. Sci. **49**(3), 827–832 (2002)
15. Xu, Z., Sun, J.: Image inpainting by patch propagation using patch sparsity. Trans. Image Process. **19**(5), 1153–1165 (2010)
16. Yang, C.-Y., Ma, C., Yang, M.-H.: Single-image super-resolution: a benchmark. In: Fleet, D., Pajdla, T., Schiele, B., Tuytelaars, T. (eds.) ECCV 2014. LNCS, vol. 8692, pp. 372–386. Springer, Cham (2014). doi:10.1007/978-3-319-10593-2_25

Automated Pain Assessment in Neonates

Ghada Zamzmi[1], Chih-Yun Pai[1], Dmitry Goldgof[1], Rangachar Kasturi[1],
Yu Sun[1], and Terri Ashmeade[2(✉)]

[1] Department of Computer Science and Engineering,
University of South Florida, Tampa, FL 33620, USA
{ghadh,chihyu,goldgof,r1k,yusun}@mail.usf.edu
[2] Department of Pediatrics, USF Health, University of South Florida,
Tampa, FL 33620, USA
tashmead@mail.usf.edu

Abstract. The current practice of assessing infants' pain is subjective
and intermittent. The misinterpretation or lack of attention to infants'
pain experience may lead to misdiagnosis and over- or under-treatment.
Studies have found that poor management and treatment of infants' pain
can cause permanent alterations to the brain structure and function. To
address these shortcomings, the current practice can be augmented with
an automated system to monitors various pain indicators continuously
and provide a quantitative assessment. In this paper, we present methods to analyze infants' crying sounds, and other pain indicators for the
purpose of developing an automated multimodal pain assessment system. The average accuracy of estimating infants' level of cry was around
88%. Combining crying sounds to facial expression, body motion, and
vital signs for classifying infants' emotional states as no pain or severe
pain yielded an accuracy of 96.6%. The reported results demonstrate the
feasibility of developing an automated system that integrates multiple
pain modalities for pain assessment in infants.

Keywords: Crying analysis · Emotion recognition · Pain monitoring

1 Introduction

On average, infants receiving care in the Neonatal Intensive Care Unit (NICU)
experience fourteen painful procedures per day [7]. Inadequate management
of infants' pain due to the poor assessment during this vulnerable developmental period can cause serious long-term impacts. Studies have found that
the experience of pain in preterm and post-term infants might be associated
with permanent neuroanatomical changes, behavioral, developmental and learning disabilities [9]. Additionally, inadequate treatment of infants' pain may
increase avoidance behaviors and social hypervigilance [10]. Therefore, assessing pain accurately using valid, standardized, and reliable pain assessment tools
is critical.

Current assessment of infants' pain involves observing objective measures
(e.g., heart rate and blood pressure) along with subjective indicators (e.g., facial

© Springer International Publishing AG 2017
P. Sharma and F.M. Bianchi (Eds.): SCIA 2017, Part II, LNCS 10270, pp. 350–361, 2017.
DOI: 10.1007/978-3-319-59129-2_30

expression and crying) by bedside caregivers. This practice has two main short-comings. First, it relies on caregivers' subjective interpretation of multiple indicators and fails to meet rigorous psychometric standards. The inter-observer variation in the use of pediatric pain scales may result in an inconsistent assessment and treatment of pain. Second, the current pain assessment is intermittent, which can potentially lead the caregivers to miss pain or delay their ability to promptly detect and treat pain. To mitigate these shortcomings and provide standardized, yet continuous assessment of infants' pain, we proposed an automated pain assessment system in [19]. This paper extends our previous work [19] that has preliminary results of assessing pain using facial expression, body motion, and vital sign data to include crying sounds and state of arousal[1] as behavioral measures for pain.

Crying is one of the most widely used pain indicators in infants. Several studies [8, 14, 18] reported that crying sound is one of the most specific indicators of pain and emphasized the importance of including it as a behavioral measure when assessing infants' pain. As such, most pediatric pain scales include crying sound as a main indicator of pain. Additionally, clinical studies [5, 8] have found that premature infants (i.e., gestational age 32–34 weeks) have limited ability to sustain facial actions associated with pain, since their facial muscles are not well-developed, and reported that premature infants communicate their pain mainly through crying sounds. Hence, it is important to include crying sounds as a main indicator of pain when developing an automated pain assessment system. State of arousal is another behavioral measure that is often included as a dimension of valid pain scales. Analyzing crying sounds and state of arousal along with other pain indicators allows us to develop an automated multimodal assessment system comparable to the current pediatric scales.

This paper makes two main contributions.

First, it is the first paper to present a completely automated version of the current multimodal pediatric scales that is continuous and standardized. The proposed system can be easily integrated into clinical environments since it uses non-invasive devices (e.g., RGB cameras) to monitor infants. The continuous assessment of pain is important because infants might experience pain when they are left unattended. The multimodal nature allows for a reliable assessment of pain during circumstances when not all data points are available owing to developmental stage, clinical condition, or level of activity.

Second, this paper presents the first automatic analysis of infants' sounds for the purpose of estimating the level of pain. Particularly, we used signal processing and machine learning methods to analyze infants' sounds and classify them into no cry, whimper cry, and vigorous cry. We also combined infants' crying sounds to other pain indicators to classify the infants' emotional state as no pain or severe pain.

In the next section, we briefly discuss current automated methods to assess infants' pain, followed by a description of the study design and data collection. In Sect. 4, we present our automated system for assessing infants' pain. Section 5

---

[1] The individual degree of alertness to a stimulus.

includes the experimental results for infants' pain assessment. Finally, we conclude and discuss possible future directions in Sect. 6.

2 Overview of Current Work

Although much effort has been made to assess pain using computer vision and machine learning methods, the vast majority of the current work in this area focus on adults' pain assessment. Those works that investigate machine assessment of infants' pain are discussed next.

Crying sound is a behavioral indicator that has been used commonly to classify infants' emotional states. One of the first studies to analyze infants' crying sounds was introduced in [14]. To extract features for classification, Mel Frequency Cepstral Coefficients (MFCC) method was applied in segmented crying signals to extract sixteen coefficients as features. The accuracy of classifying crying sounds for sixteen infants as pain cry, fear cry, or anger cry was 90.4%. Similarly, Vempada et al. [18] investigated the use of thirteen MFCC coefficients along with other time-domain features for recognizing infants' crying sounds. A total of 120 hospitalized premature infants were recorded undergoing different emotional states, namely pain, hunger, or wet-diaper. For classification, a score-level and feature-level fusion were performed to classify infants' cries as one of the three classes. The weighted accuracies for feature-level and score-level fusion were observed to be around 74% and 81%, respectively.

Another behavioral indicator that is commonly used to assess infants' pain experience is facial expression. Automated methods to analyze infants' facial expression and classify them as no pain or pain expression can be found in [1,3]. In case of physiological pain indicators, studies found that there is an association between pain and physiological measures such as vital signs [2] and changes in cerebral oxygenation [15,16].

The failure to record a specific pain indicator is common in clinical environment due to several reasons such as developmental stage (e.g., an infant's facial muscles are not well-developed), physical exertion, (e.g., exhaustion), specific disorders (e.g., paralysis), clinical conditions (e.g., occlusion by oxygen mask), and individual differences. Therefore, combining multiple indicators can provide a more reliable pain assessment.

Pal et al. [12] presented a bimodal method to classify infants' emotional states as pain, hunger, anger, sadness, and fear based on analysis of facial expression and crying sounds. Similarly, Zamzmi et al. [19] introduced an automated approach to assess infants' pain based on analysis of facial expression, body motions, and vital signs measures. The results of these works showed that combining multiple modalities might provide reliable assessment of infants' emotional states. We refer the reader to our survey paper [20] for a comprehensive review and discussion of existing machine-based pain assessment methods.

3 Study Design

3.1 Subjects

The data for a total of eighteen infants were recorded during acute episodic painful procedure. Infants' average gestational age was 36 weeks; seven infants were Caucasian, three Hispanic, three African American, two Asian, and three others. Any infant born in the range of 28 and 41 weeks gestation was eligible for enrollment after obtaining informed consent from the parents. Infants with cranial facial abnormalities and neuromuscular disorders were excluded.

3.2 Apparatus

Video and audio recordings were carried out using GoPro Hero3+ camera to capture the infant's facial expression, body motion, and record crying sound. Since digital output of vital signs (i.e., vital signs that are shown in Fig. 1D) was not readily available from the monitor, we placed another GoPro camera in front of a Philips MP-70 cardio-respiratory monitor to record it and Optical Character Recognition (OCR) was performed to convert the recorded vital signs into data. All study recordings were carried out in the normal clinical environment that is only modified by the addition of the cameras.

3.3 Data Collection and Ground Truth Labeling

Data were recorded for infants who were undergoing routine acute painful procedure such as heel lancing and immunization over seven epochs. Specifically, the infants were recorded for five minutes prior the painful procedure (i.e., epoch 1) to get the baseline state. Then, they were recorded at the start of the painful procedure (i.e., epoch 2) and every minute for five minutes after the completion of painful procedure (i.e., epochs 3 to 7). To get the ground truth labels, trained nurses scored the infants and estimated their level of pain at the beginning of each epoch using a pediatric pain scale known as the Neonatal Infant Pain Scale (NIPS).

NIPS is a multimodal pain scale that takes into account both subjective and objective measures (i.e., behavioral and physiological indicators) when assessing infants' pain. The scale involves observing facial expression, crying sound, body motion (i.e., arms and legs), and state of arousal data along with vital signs readings to estimate the level of pain. Table 1 presents the score ranges for each indicator/measure of NIPS scale. After scoring each of these measures individually, caregivers add these scores together to generate the total pain score. The computed total score is then used to classify the infant's pain state as: no pain (0–2 score), moderate pain (3–4 score), or severe pain (>4 score). We refer the interested reader to [4] for further description about this pediatric scale and its measures. The performance of our algorithms is measured by comparing their output to the ground truth.

Table 1. The range for NIPS scores

NIPS measure	Score of 0	Score of 1	Score of 2
Facial expression	Neutral or relaxed	Grimace	—
Body motion	Relaxed	Flexed arms/legs, kicking	—
State of arousal	Sleeping/calm	Uncomfortable	—
Vital signs	Normal	Different than baseline	—
Crying sounds	No cry	Whimper	Vigorous

4   Methodology

We divided the pain analysis into two groups: Those that analyze behavioral measures and those that analyze physiological measures. An overview of the methodology for the entire assessment system is depicted in Fig. 1.

4.1   Methods that Analyze Behavioral Measures

We analyzed four behavioral measures and used them to assess infants' pain. These measures are crying sound, facial expression, body motion, and state of arousal. The proposed methods to analyze each of these measures and extract pain-relevant features for classification are presented below.

Facial Expression. We used the method presented in our previous work [19] to analyze infants' facial expression. As shown in Fig. 1A, the method consists of three main steps, which are preprocessing, feature extraction, and classification.

In the preprocessing stage, we detected the infants face in each frame using Viola-Jones object detector that is trained to specifically detect faces of infants in different positions and with varying degrees of facial obstruction. The detector was able to detect faces with frontal and near-frontal views but failed to detect faces with significant changes in position or facial obstruction; these faces were excluded from further analysis. We also excluded the frames where the infant's face was out of sight from further analysis. After detecting the face, we applied facial landmark points algorithm [21] to extract 68 facial points. These points were then used to align the face, crop it, and divide it into four regions.

To extract pain-relevant features, we used the strain-based expression segmentation algorithm presented in [19]. For classification, the extracted features (i.e., the strain magnitude computed for each region from I to IV) are used to train different machine-learning classifiers such as Support Vector Machine (SVM) and Random Forest trees. We performed LOSOXV to evaluate the trained model and estimate the generalization performance. For each training fold, 10-fold cross-validation was performed for parameter estimation.

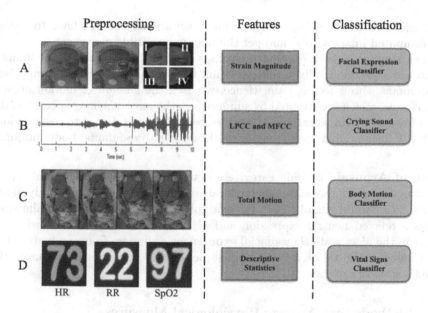

Fig. 1. Overview of the Methodology for the Assessment System. A. The face in each frame is detected, cropped, and divided into four regions; optical flow is applied in each region and used to estimate the strain. B. The input crying signal is analyzed using LPCC and MFCC. C. The motion image is computed for the cropped body area. D. Statistics are computed for the input HR, RR, and SpO2 data.

Crying Sound. We employed Yang's speech recognition method [11] to extract pain-relevant features from infants' sounds. As shown in Fig. 1B, the method consists of three main steps: preprocessing, feature extraction, and classification.

The preprocessing step involves dividing the audio signals into several consecutive overlapped windows. After segmenting the signals into small windows, Linear Prediction Cepstral Coefficients (LPCC) and MFCC coefficients were extracted from each window as features. The window to extract LPCC was 32 ms Hamming window with 16 ms overlapping; 30 ms Hamming window with 10 ms shift was used to extract MFCC. After extracting LPCC and MFCC coefficients, we compressed them by employing vector quantization method. The compressed vector was then used to train a LS-SVM for classification. To evaluate the classifier and estimate its generalization performance, we performed leave-one-subject-out cross-validation (LOSOXV). For each training fold, another level of cross-validation was performed for parameter estimation.

Body Motion. To extract pain-relevant features from infants' body, we computed the motion image between consecutive video frames after cropping the exact body area, as shown in Fig. 1C. The motion image is a binary image that has values of one to represent pixels that move and zero to represent pixels that do not move. The frames where the infant's body was out of sight or occluded were excluded from further analysis. Then, we applied median filter to reduce the computed image's noise and get the maximum visible movement.

The amount of body motion presents a good indication about the infant's emotional state. Generally, a relaxed infant has less body motion in comparison to an infant who is feeling pain. Hence, we used the amount of motion in each video frame, which is computed by summing up the motion's image pixels, as the main feature for classification. The interested reader is referred to our previous work [19] for more details about our method to analyze infants' body motion.

State of Arousal. Bedside caregivers determine the state of arousal for an infant by observing the infant's behavioral responses (e.g., facial and body movements) to a painful stimulus [17]. For example, an infant in a sleep or calm state shows a relaxed/neutral expression and a little or no body activity while an infant in the alert state shows facial expressions and more frequent body activity. Consequently, we used facial expression and body motion to estimate the infant's state of arousal.

4.2   Methods that Analyze Physiological Measures

To assess infants' pain using objective measures, vital signs (i.e., heart rate [HR], respiratory rate [RR] and oxygen saturation levels [SpO2]) numeric readings were extracted from short period epochs. Then, we applied median filter to remove outliers. For classification, we calculated different descriptive statistics (e.g., mean and standard deviation) from the extracted vital signs readings and used them to train different machine-learning classifiers (e.g., Random Forest trees). For classifier's evaluation and estimation, we performed the same model evaluation discussed under facial expression.

5   Pain Assessment Results

This section presents the results of estimating the scores for NIPS pain measures (see Table 1). Also, it presents the results of classifying the emotional states of an infant as no pain or severe pain using each pain modality and combinations of multiple pain modalities. Along with the score estimation and pain classification, a statistical comparison for pain classification performance is provided. Before we proceed any further, we would like to note that NIPS has three levels of pain, namely no pain, moderate pain, and severe pain (see Sect. 3.3). However, due to the small number of instances for moderate pain in our limited dataset, we decided to exclude instances of this pain level and classify the infant's emotional state as no pain or severe pain.

5.1   Score Estimation

Score estimation is the process of generating NIPS scores presented in Table 1. To generate the score for each pain modality, we employed the methods presented in Sect. 4 to extract pain-relevant features. Next, we used the extracted features of these modalities to classify the scores as either 0 or 1, except for crying which is scored as 0, 1 or 2. Our previous work [19] reports the results of estimating the scores for facial expression, body motion, and vital signs readings.

The score estimation for crying sound involves classifying infants' sounds into vigorous cry (score of 2), whimper cry (score of 1), or no cry (score of 0). To generate these scores, we used the extracted LPCC and MFCC coefficients with k-nearest neighbors (k-NN) classifier and LOSOXV. The accuracy of estimating the infants' crying sounds as vigorous cry, whimper cry, or no cry in comparison with the ground truth was around 84%. Table 2 shows the confusion matrix.

We would like to mention that the achieved accuracy was obtained using crying sounds recorded in a realistic clinical environment that has ambient noises such as human speech, machine sounds, and other infants' crying. Excluding the episodes that have significant noise from analysis (i.e., a clean dataset) yielded an accuracy of 90% as discussed in [11].

Table 2. Confusion matrix for sounds score estimation

N = 85	No cry	Whimper	Vigorous
No cry	62	0	4
Whimper	3	1	4
Vigorous	3	0	8

5.2 Pain Classification

Pain classification is the process of classifying the infant's emotional state as severe pain or no pain. To classify infants' pain, we conducted two sets of experiments. The first experiment involves recognizing the emotional state of infants using a single pain indicator. In the second experiment, we combined different pain indicators for pain classification. The results of both are reported and discussed below.

In unimodal pain classification, the features from each pain modality were used individually to classify infants' pain. Particularly, features of crying sound, facial expression, body motion, and vital signs were used separately to classify the emotional state of an infant as no pain or severe pain.

The *One Measure* column of Table 3 summarizes the unimodal pain classification performance. As is evident from the table, facial expression has the highest classification accuracy. This result is consistent with our previous work [19] and supports previous findings [5,6] that facial expression is the most specific and common indicator of pain. As such, most pediatric pain scales utilize facial expression as the main indicator for assessment. Crying sound and body motion have similar classification performance that are notably higher than the performance of vital signs. A possible explanation for the low accuracy of classifying pain using vital signs can be attributed to the fact that vital signs readings are less specific for pain since they are more sensitive to other conditions such as loud noise or underlying disease.

Table 3. Summary of unimodal and multimodal pain classification

	One measure				Two measures						Three and above				
C	✓				✓		✓		✓		✓		✓	✓	✓
F		✓			✓	✓		✓			✓	✓	✓		✓
B			✓			✓	✓			✓	✓	✓		✓	✓
V				✓				✓	✓	✓		✓	✓	✓	✓
ACC	87.7	93.2	87.5	73.6	93.5	90.2	93.9	92.4	90.6	87.0	96.1	89.9	94.5	90.3	96.7

Note: **C** stands for crying sounds, **F** for facial expression, **B** for body motion, **V** for vital signs, and **ACC** for the accuracy averaged across all subjects of LOSOXV method.

To combine pain modalities together for the multimodal assessment, we performed two schemes: thresholding and majority voting. For the thresholding scheme, we added the scores of crying sound, facial expression, body motion, state of arousal, and vital signs together and then preformed a thresholding on the total pain score to classify the infant's pain state (see Sect. 3.3); this scheme follows the exact process of assessing pain by bedside caregivers but it is automated. Classifying infants' pain using the thresholding scheme achieves an average accuracy of around 94%.

The majority-voting scheme is a decision-level fusion method to predict the final outcome of different modalities by combining the outcomes (i.e., class labels) of these modalities together and choosing the major label in the combination as the final outcome. For example, the final assessment of pain would be severe pain for a combination of two measures with severe pain labels and one measure with no pain label. In case these labels make a tie, we break this tie by choosing the class label with the highest confidence score as the final assessment.

The **Two Measure** and **Three and Above** columns in Table 3 present the results of assessing pain using the majority voting scheme for eleven combinations of pain indicators. As can be seen, the performance of pain classification using combination of behavioral pain indicators (**CFB**) was around 96%. Adding vital signs to crying sounds, facial expression, and body motion (**CFBV**) slightly increases the pain classification performance. These results indicate that combining multiple indicators might provide a better and more reliable pain assessment.

5.3   Statistical Comparison

Referring to Table 3, we note that different combinations of pain indicators have similar performance. For example, the difference of performance between (**CFB**) and (**CFBV**) is around 0.6%. Also, the difference between (**F**) performance and (**CF**) performance is around 0.3%. The purpose of this section is to determine if there is a significant difference in the performance of two combinations using a statistical significance test known as Mann Whitney U test. U test is a non-parametric test to measure the central tendencies of two groups. We decided to use U test for comparison instead of T-test because our data (i.e., subjects' accuracies) are not normally distributed.

Using U-test, we compared the pain classification performance of vital signs, which is the lowest in Table 3, with the performance of all other combinations. U test indicated that the performance of classifying infants' pain was significantly higher for a combination of behavioral and physiological measures (**CFBV**) than for physiological measures alone **V**, (U = 33.5, p < 0.05 two-tailed). Also, we found that the performance of classifying infants' pain using behavioral indicators (**CFB**) is significantly higher than using physiological measures (**V**), (U = 35.5, p < 0.05 two-tailed). These results support previous finding [13] that physiological changes such as an increase in heart rate are less specific for pain and thus are not sufficient for pain assessment.

We also compared the pain classification of facial expression (**F**), which has the highest accuracy under *One Measure* column in Table 3, with the pain classification of all other combinations under *Two Measures* and *Three and Above* columns. U test indicated that there is no significant difference between the pain classification using facial expression individually and the pain classification using combinations of different pain indicators. Although the pain classification accuracy using facial expression is not statistically different than the accuracies of multiple indicators, we still believe it is important to consider multiple pain indicators when assessing infants' pain. The multimodal nature can provide more reliable assessment system that is able to function in case of missing data due to developmental stage, occlusions, noise, or level of activity (e.g., exhaustion).

The final statistical comparison we performed was to compare the pain predication using the thresholding scheme with the pain predication of (**CFBV**) using the voting scheme. No significant difference was found between the pain classification performance of these two schemes.

6   Conclusions and Future Research

This paper expands upon our previous work that presented an approach for assessing infants' pain using facial expression, body movement, and vital signs by including crying sounds and state of arousal. The results of assessing infants' pain presented in this paper are encouraging and, if further confirmed on a larger dataset, would influence and ultimately improve the current practice of assessing infants' pain.

In terms of future research possibilities, we would like to evaluate our method on a larger dataset and expand our system to include chronic pain. Another research direction is to perform a feature-level fusion of all modalities' features to generate a single feature vector for classification. A final research direction would be to include clinical data such as the gestational age to the assessment process.

References

1. Brahnam, S., Chuang, C.-F., Shih, F.Y., Slack, M.R.: Machine recognition and representation of neonatal facial displays of acute pain. Artif. Intell. Med. **36**(3), 211–222 (2006)
2. Faye, P.M., De Jonckheere, J., Logier, R., Kuissi, E., Jeanne, M., Rakza, T., Storme, L.: Newborn infant pain assessment using heart rate variability analysis. Clin. J. Pain **26**(9), 777–782 (2010)
3. Fotiadou, E., Zinger, S., a Ten, W.T., Oetomo, S.B. et al.: Video-based facial discomfort analysis for infants. In: IS&T/SPIE Electronic Imaging, pp. 90290F–90290F. International Society for Optics and Photonics (2014)
4. Gallo, A.-M.: The fifth vital sign: implementation of the neonatal infant pain scale. J. Obstet. Gynecol. Neonatal Nurs. **32**(2), 199–206 (2003)
5. Grunau, R.V.E., Craig, K.D.: Pain expression in neonates: facial action and cry. Pain **28**(3), 395–410 (1987)
6. Grunau, R.V.E., Johnston, C.C., Craig, K.D.: Neonatal facial and cry responses to invasive and non-invasive procedures. Pain **42**(3), 295–305 (1990)
7. Hummel, P., van Dijk, M.: Pain assessment: current status and challenges. Semin. Fetal Neonatal Med. **11**, 237–245 (2006). Elsevier
8. Johnston, C.C., Stevens, B., Craig, K.D., Grunau, R.V.E.: Developmental changes in pain expression in premature, full-term, two-and four-month-old infants. Pain **52**(2), 201–208 (1993)
9. American Academy of Pediatrics, Fetus, Newborn Committee, et al.: Prevention and management of pain in the neonate: an update. Pediatrics **118**(5), 2231–2241 (2006)
10. Page, G.G.: Are there long-term consequences of pain in newborn or very young infants? J. Perinat. Educ. **13**(3), 10–17 (2004)
11. Pai, C.-Y.: Automatic pain assessment from infants' crying sounds (2016)
12. Pal, P., Iyer, A.N., Yantorno, R.E.: Emotion detection from infant facial expressions and cries. In: 2006 IEEE International Conference on Acoustics Speech and Signal Processing Proceedings, vol. 2, pp. II-II. IEEE (2006)
13. Pereira, A.L.D.S.T., Guinsburg, R., Almeida, M.F.B.D., Monteiro, A.C., Santos, A.M.N.D., Kopelman, B.I.: Validity of behavioral and physiologic parameters for acute pain assessment of term newborn infants. São Paulo Med. J. **117**(2), 72–80 (1999)
14. Petroni, M., Malowany, A.S., Johnston, C.C., Stevens, B.J.: Identification of pain from infant cry vocalizations using artificial neural networks (ANNs). In: SPIE's 1995 Symposium on OE/Aerospace Sensing and Dual Use Photonics, pp. 729–738. International Society for Optics and Photonics (1995)
15. Slater, R., Cantarella, A., Gallella, S., Worley, A., Boyd, S., Meek, J., Fitzgerald, M.: Cortical pain responses in human infants. J. Neurosci. **26**(14), 3662–3666 (2006)
16. Slater, R., Fabrizi, L., Worley, A., Meek, J., Boyd, S., Fitzgerald, M.: Premature infants display increased noxious-evoked neuronal activity in the brain compared to healthy age-matched term-born infants. Neuroimage **52**(2), 583–589 (2010)
17. Thoman, E.B.: Sleeping and waking states in infants: a functional perspective. Neurosci. Biobehav. Rev. **14**(1), 93–107 (1990)
18. Vempada, R.R., Kumar, B.S., Rao, K.S.: Characterization of infant cries using spectral and prosodic features. In: 2012 National Conference on Communications (NCC), pp. 1–5. IEEE (2012)

19. Zamzmi, G., et al.: An approach for automated multimodal analysis of infants' pain. In: 2016 23rd International Conference on Pattern Recognition (ICPR). IEEE (2016)
20. Zamzmi, G., Pai, C.-Y., Goldgof, D., Kasturi, R., Sun, Y., Ashmeade, T.: Machine-based multimodal pain assessment tool for infants: A review. arXiv preprint. arXiv:1607.00331 (2016)
21. Zhang, Z., Luo, P., Loy, C.C., Tang, X.: Facial landmark detection by deep multi-task learning. In: Fleet, D., Pajdla, T., Schiele, B., Tuytelaars, T. (eds.) ECCV 2014. LNCS, vol. 8694, pp. 94–108. Springer, Cham (2014). doi:10.1007/978-3-319-10599-4_7

Enhancement of Cilia Sub-structures by Multiple Instance Registration and Super-Resolution Reconstruction

Amit Suveer[1(✉)], Nataša Sladoje[1,2], Joakim Lindblad[1,2], Anca Dragomir[3], and Ida-Maria Sintorn[1,4]

[1] Centre for Image Analysis, Uppsala University, Uppsala, Sweden
amit.suveer@it.uu.se
[2] Mathematical Institute, Serbian Academy of Sciences and Arts, Belgrade, Serbia
[3] Department of Surgical Pathology, Uppsala University Hospital, Uppsala, Sweden
[4] Vironova AB, Stockholm, Sweden

Abstract. Ultrastructural analysis of cilia cross-sectional images using transmission electron microscopy (TEM) assists the pathologists to diagnose Primary Ciliary Dyskinesia, a genetic disease. The current diagnostic procedure is manual and difficult because of poor signal-to-noise ratio in TEM images. In this paper, we propose an automated multi-step registration approach to register many cilia cross-sectional instances. The novelty of the work is in the utilization of customized weight masks at each registration step to achieve good alignment of the specific cilium regions. Registration is followed by super-resolution reconstruction to enhance the substructural information. Landmarks matching based evaluation of registration results in pixel alignment error of 2.35 ± 1.82 pixels, and the subjective analysis of super-resolution reconstructed cilium shows a clear improvement in the visibility of the substructures such as dynein arms, radial spokes, and central pair.

Keywords: Non-rigid registration · Transmission Electron Microscopy · Super-resolution · Cilia ultrastructures · Dynein arms · Radial spokes

1 Introduction

Transmission Electron Microscopy (TEM) can reveal information about very fine structures (~ 1 nm) in biological tissue sections. Pathologists analyze the morphology of such ultrastructures to diagnose certain clinical conditions. One such example is Primary Ciliary Dyskinesia (PCD). PCD is an autosomal recessive genetic disease in which specialized cell structures, *cilia*, do not function normally. Cilia are hair-like organelles protruding from cells, and dysfunctional motile cilia results in serious problems, e.g. long term respiratory infection and infertility in males and females. Therefore, pathologists analyze cilia substructures like dynein arms (DA), radial spokes, nexin links, and central pair, see Fig. 1. Though the outer and inner dynein arms (ODA and IDA) are of particular interest for PCD diagnosis [18], defects in other substructures could also be the result of the PCD condition and must not be ignored.

© Springer International Publishing AG 2017
P. Sharma and F.M. Bianchi (Eds.): SCIA 2017, Part II, LNCS 10270, pp. 362–374, 2017.
DOI: 10.1007/978-3-319-59129-2_31

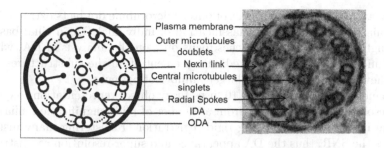

Fig. 1. Example image of perpendicularly cut cilium and related terminology.

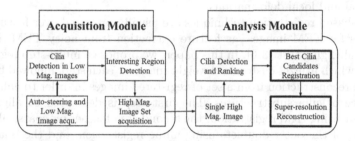

Fig. 2. Automated TEM imaging and analysis workflow. (The steps in focus in this paper are highlighted.)

In current practice, a large number ($\geq$50) of perfectly perpendicularly cut cilia need to be detected in the sample and visually analyzed at high magnification. The procedure is time consuming and therefore costly, it takes around two hours for a proficient pathologist to perform analysis of a single case. The poor signal-to-noise ratio (SNR) in the TEM images also makes image interpretation difficult. The automation of processes at different levels of the diagnostic procedure is hence highly desirable. The work-flow of a proposed automated approach is illustrated in Fig. 2.

In this multiscale approach, we search for cilia-like objects at low-resolution, on detecting a sufficient number of objects we acquire relatively high-resolution images of specific regions and perform analysis to create super-resolution cilium. The collage of high-resolution images along with the enhanced reconstructed cilium is presented to a pathologist for final diagnosis. We have successfully addressed the main challenge, cilia detection at low magnification [16]. We have also presented preliminary results for super-resolution reconstruction of cilium at mid magnification [10], using an automated approach.

Some work related to enhanced cilium reconstruction for PCD diagnosis has previously been reported. It involves manual or user guided selection, segmentation, and alignment of good cilia instances from digitized high-resolution images [1,4,5]. As mentioned, these approaches are manual and don't take local deformations into account, unlike the one proposed in this work. In addition, alternative automated approaches for PCD diagnosis involves enhancement of

ODA by averaging a large number of outer microtubule doublets extracted from many cilia instances [12], and classification of individual DA in cilia, based on their lengths [13]. These approaches focus only on the analysis of DA, whereas our technique performs automated enhancement of all the substructures which could allow PCD diagnosis in a larger number of patients.

The method we proposed in [10] uses fuzzy object representation to create a super-resolution reconstruction from a number of mid-magnification cilia cross-sectional instances aligned using rigid registration. Registering many instances improves the SNR, thus the DA appearance and super-resolution reconstruction further boost the DA representation. It is observed that the quality of the super-resolution image would benefit from an improved registration, which could allow both global and local deformations.

Image-based registration techniques are usually preferred over feature-based techniques for TEM images, as feature extraction from noisy TEM images is quite a challenging task [3,6]. In this paper, we propose a multi-step registration approach which utilizes customized weight masks at each step, followed by super-resolution reconstruction from a set of registered images in order to enhance the appearance of cilia substructures. The first step is dedicated to aligning the central pair using rigid registration, the second step focuses on coarse alignment of the ring of the outer doublets using affine registration, and the final step is dedicated to the fine alignments of substructures using non-rigid registration. Utilizing optimized weight mask for each registration step is the novelty of this work. Mask derivation is discussed in details in Sect. 3.3.

2   Methodology

2.1   Image Registration

We use standard multi-step registration strategy adapted to our application. In each step, we performed pair-wise registration where one image acts as a static image (I_s) and the second image as a moving image (I_m), and the goal is to find the displacement field which defines the mapping between the co-ordinates of the two images. Thus the registration of image-pair can be defined as energy minimization problem to find the displacement field:

$$D = \arg\min_{\phi} E_{reg}(\phi; I_s, I_m), \tag{1}$$

where ϕ is current displacement field and D is the final displacement field. $E_{reg}(\phi)$ is the energy function, which is defined as:

$$E_{reg}(\phi, I_s, I_m) = -wNCC(I_s, T(I_m, \phi)) + \lambda_{reg}R(\phi), \tag{2}$$

where $T(I_m, \phi)$ is the result of transformation of I_m using the displacement field ϕ, wNCC is weighted normalized cross-correlation used as similarity measure, and R is regularization term with regularization parameter λ_{reg}. For non-rigid registration ($\lambda_{reg} > 0$) and for the rigid and affine registrations ($\lambda_{reg} = 0$).

The wNCC used in the registration and at later stage for choosing a set of registered images for super-resolution reconstruction, is defined as:

$$wNCC(I_s, I_v) = \frac{\sum W_n \cdot [I_s - \overline{I}_{sw}] \cdot [I_v - \overline{I}_{vw}]}{\left(\sum W_n \cdot [I_s - \overline{I}_{sw}]^2 \sum W_n \cdot [I_v - \overline{I}_{vw}]^2\right)^{0.5}}, \qquad (3)$$

where W_n is a normalized weight mask such that $\sum W_n = 1$. $\overline{I}_{sw}$ and $\overline{I}_{vw}$ are the weighted means of the static and variable image calculated using W_n as:

$$\overline{I}_{sw} = \sum W_n \cdot I_s, \quad \overline{I}_{vw} = \sum W_n \cdot I_v \quad and \quad W_n(x, y) = \frac{W(x, y)}{\sum W}, \qquad (4)$$

where W is the user defined weight mask, and where $(A \cdot B)$ means point-wise multiplication, and $(\sum A)$ means sum over all elements of A.

2.2   Multi-step Registration

Rigid Registration: The focus of the first step is on aligning the central pair. The rotational symmetry of the ring of the outer doublets makes this task challenging, as there is a high possibility of getting stuck in local minima. Therefore, we follow multi-position initialization, using 9 rotations of the moving image as starting positions in the range $[0–320°]$ with a step size of $40°$, and selecting the one with the best-achieved wNCC score. The weighting mask (W_{rig}) only covers the central pair with high weights at the central region, as later described in Sect. 3.3. The resulting transformation matrix is used as the initial transformation in the next registration step.

Affine Registration: The second step focuses on aligning the ring of the outer doublets using affine transformation without affecting the alignment of the central pair attained by rigid registration. This is achieved by using another experimentally derived weight mask (W_{aff}), as later described in Sect. 3.3. The mask covers all the rings but assigns relatively high weights to the central region in comparison to the periphery. The resulting transformation matrix is used as the initial transformation in the next registration step.

Non-rigid Registration: The final step focuses on the fine adjustments in local regions. This step uses a free form deformation (FFD) model based on B-splines [8,9] which is often used in medical image registration [15]. The FFD model transforms an image based on the transformation of a grid placed over an image where the grid nodes act as control points. A set of B-splines is used to guide the transformation where each B-spline transformation is influenced by a set of control points. Each individual control point is tuned iteratively to lead the local transformation until convergence or a specified exit condition is reached. The initial displacement field is generated from the affine transformation, and the final displacement field is a combination of multiple transformed B-splines updated iteratively under the influence of their respective control points [13]. Let the image domain is $\Omega = \{(m,n) \mid 0 \leq m < M, 0 \leq n < N\}$ and Φ denotes a $[P_m \times P_n]$ mesh of control points with constant uniform spacing of δ in x and y-Cartesian direction. Let $\phi_{i,j}$ is the value of the control point located at (i,j), with $-1 \leq i < P_m$, $-1 \leq j < P_n$ and $(P_m = M + 2, P_n = N + 2)$. Then the approximate transformation using cubic B-splines function that represents the local deformation can be defined as [7,9]:

$$T_{local}(m,n) = \sum_{r=0}^{3} \sum_{s=0}^{3} B_r(u) B_s(v) \phi_{(i+r,j+s)} . \tag{5}$$

Here, $i = \lfloor m/\delta \rfloor - 1$, $j = \lfloor n/\delta \rfloor - 1$, $u = (m/\delta) - \lfloor m/\delta \rfloor$ and $v = (n/\delta) - \lfloor n/\delta \rfloor$. The functions B_r and B_s are the cubic B-spline polynomials as defined in [7,9], where $0 \leq z < 1$:

$$B_0(z) = (1-z)^3/6, \qquad\qquad B_1(z) = (3z^3 - 6z^2 + 4)/6,$$
$$B_2(z) = (-3z^3 + 3z^2 + 3z + 1)/6, \qquad B_3(z) = z^3/6.$$

The control points determine the degrees of freedom (DoF) and the amount of non-rigid deformation, depending on the resolution of the control points grid. A low-resolution grid performs coarse non-rigid alignment and on increasing the grid resolution the alignment gets finer. However, many DoF come with a high computational cost. To achieve a good balance, we employed a pyramidal multi-resolution approach [9], where both the image and the control grid resolutions are increased from coarse-to-fine at each level. Let the local transformation at any pyramid level be denoted as T_{local}^{pl}, then the final non-rigid registration displacement field is defined as, $D = \sum_{pl=1}^{L} T_{local}^{pl}$.

Regularization: In order to constrain the deformation to avoid unrealistic transformations, a 2D bending energy of a thin-plate of metal [17] is used as penalty term to regularize the deformation [14] is defined as:

$$R(\phi) = \frac{1}{|\Omega|} \iint_{\Omega} \left[\left(\frac{\partial^2 \phi}{\partial x^2} \right)^2 + 2 \left(\frac{\partial^2 \phi}{\partial x \partial y} \right)^2 + \left(\frac{\partial^2 \phi}{\partial y^2} \right)^2 \right] dx dy , \tag{6}$$

The amount of penalty is controlled by regularization parameter λ_{reg} as stated in Eq. (2). The larger the regularization parameter $\lambda_{reg} > 0$, the smoother the deformation field will be.

2.3   Super-Resolution Reconstruction

The set of registered cilia images is used to reconstruct a super-resolution (SR) image. The approach is inspired by the work presented in [10], but with a slightly adjusted regularization, based on experiences from [2]. To reduce disturbance from possible misalignments, we exclude the 25% lowest scoring registered images based on their wNCC scores. We formulate the SR reconstruction as a regularized energy minimization problem, where the reconstructed image h is estimated as:

$$h = \arg\min_{u} E_{sr}(u). \tag{7}$$

We utilize an energy function that includes the robust ℓ_1 norm to ensure noise insensitive adherence to the observed images, in combination with a Huberized TV-regularization which provides noise reduction while preserving edges. The energy function is of the form

$$E_{sr}(u) = \frac{1}{c} \sum_{i=1}^{c} \|S(u) - T(I_i; D_i)\|_1 + \lambda_{sr} \Phi_H(|\nabla(u)|), \tag{8}$$

where $S(\cdot)$ is a factor 2 subsampling operation, $T(I_i; D_i)$ is the i-th registered observed image and D_i is the displacement field estimated for I. $\Phi_H(t)$ is the Huber potential function

$$\Phi_H(t) = \begin{cases} \frac{t^2}{2\omega}, & t \leq \omega \\ t - \frac{\omega}{2}, & t > \omega, \end{cases}$$

and ∇ is the discrete image gradient. The two regularization parameters λ_{sr} and ω are empirically tuned to $\lambda_{sr} = 0.05$, $\omega = 0.05$. Equation (7) is minimized using spectral projected gradient optimization, see [10] for details.

3 Experiments

The focus of the experiments is to derive the different optimized weight masks suitable for each registration step and the evaluation of the accuracy of the proposed registration technique. This approach is referred as the multiple-mask strategy from here on.

3.1 Image Data and Ground Truth

All the experiments are performed on a dataset of 20 representative cilia instance image patches chosen from a total of 30, which were detected using the automated detection technique presented in [11]. Each image patch is of size 128×128 pixels, extracted from a 2048×2048 pixels image, acquired at mid-magnification using the MiniTEM[1] system. To evaluate registration algorithm performance, 20 landmarks were placed for each cilium instance by author IMS at the approximate centre of the microtubuli of the central pair (2) and the outer doublets (18). An example of a cilium image patch and corresponding marked landmarks is shown in Fig. 3.

---

[1] Vironova AB, Stockholm, Sweden.

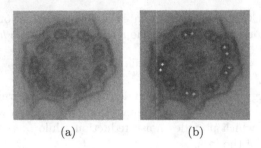

(a)   (b)

Fig. 3. (a) Cilium instance, (b) landmarks manually placed on (a).

3.2   Average Pixel Alignment Error (PAE)

The registration accuracy is measured as the pixel alignment error defined as the average Euclidean distance between the landmarks from the reference image to its closest landmark in the registered image. Let L_{ref} and L_{reg} be the set of 20 landmarks in the reference and the registered images, then the PAE for that image-pair is:

$$PAE = \frac{1}{N_L} \sum_{p \in L_{ref}} \min_{q \in L_{reg}} d(p,q), \tag{9}$$

where $d(\cdot)$ is the Euclidean distance and N_L is the number of landmarks considered while computing the PAE. For the central pair $N_L = 2$, for outer rings $N_L = 18$, and for all rings $N_L = 20$.

3.3   Weight Masks

Each weight mask is tuned for two parameters, the size, and the weight distribution. We evaluate weight masks with sizes ranging from those covering only the central pair up to those completely covering the outer ring. For weight distribution, uniform distribution and variations of the Hann window are tested. H1 denotes the Hann window defined by a radial profile $h(r) = 0.5(1 + cos(\frac{\pi r}{R_{sz}}))$, where r is the radial distance from the center of the mask, and R_{sz} is the total radius of the mask, and H2, H3, and H4 denote the windows defined by $(h(r))^2$, $(h(r))^3$, and $(h(r))^4$, respectively.

Figure 4 shows the coverage of circular mask of radius 1 r and the corresponding weights over cilium regions for different distributions. For the non-rigid registration, only the uniform distribution is considered while performing size optimization, as all regions in the cilium are equally important. The approximate radius (r) of cilium, which is the distance from the cilium centre to the plasma membrane is considered as 55 pixels. This is chosen based on the observations from 30 initially detected cilia instances.

Table 1 shows the PAE_{cl} for the central landmarks, measuring the performance of weight masks tested for the rigid registration (W_{rig}) step, in the size range [0.4 r–0.7 r]. The performance for (H2, 0.6r) clearly indicates that to achieve

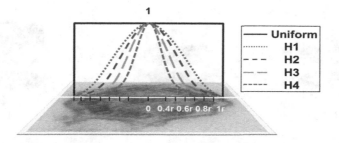

Fig. 4. Illustration of weight mask size extent and distributions (mask size $= 1\,$r)

Table 1. PAE_{cl} for W_{rig}

	0.4 r	0.5 r	0.6 r	0.7 r
Uniform	3.2(1.5)	5.1(2.4)	4.2(2.3)	4.5(2.3)
H1	3.2(2.0)	2.7(1.4)	3.1(1.5)	3.6(1.8)
H2	5.3(3.1)	3.1(1.9)	**2.6(1.5)**	2.8(1.4)
H3	5.6(2.9)	5.2(3.2)	2.7(1.6)	2.6(1.5)
H4	5.4(2.8)	5.1(3.1)	3.6(2.4)	2.7(1.6)

Table 2. PAE_{ol} for W_{aff}

	0.8 r	0.9 r	1.0 r
Uniform	3.65(1.86)	4.08(2.79)	4.14(2.63)
H1	5.30(3.55)	4.32(3.05)	**3.27(1.75)**
H2	6.54(3.68)	5.86(3.74)	5.15(3.64)
H3	7.20(3.61)	6.70(3.65)	6.15(3.69)
H4	7.18(3.68)	7.20(3.67)	6.69(3.65)

Table 3. PAE for W_{nrr}

	0.90 r	1.00 r	1.1 r
PAE_{cl}	2.83(1.72)	**2.79(1.84)**	2.89(1.63)
PAE_{ol}	2.34(1.74)	**2.31(1.81)**	2.44(1.80)
PAE_{al}	2.39(1.74)	**2.35(1.82)**	2.48(1.78)

Table 4. Weight masks details

	Rigid	Affine	Non-rigid
Radius	0.6 r	1.0 r	1.0 r
Mask size	66 × 66	110 × 110	110 × 110
Distribution	H2	H1	Uniform

a good alignment of the central pair, high weight must be given to the central region. Table 2 shows the PAE_{ol} for the outer landmarks, measuring the performance of weight masks tested for the affine registration (W_{aff}) step, in the size range [0.8 r–1 r]. The performance for (H1, 1 r) indicates that to achieve good alignment of the outer rings without disturbing the central ring alignment, a good balance of weight is important for the central and the outer regions with high weight at the central region to compensate for the 9 outer doublets. Table 3 shows the PAE_{cl}, PAE_{ol} and PAE_{al} for the central, outer and all landmarks respectively, measuring the performance of weight masks tested for the non-rigid registration (W_{nrr}) step, in the size range [0.9 r–1.1 r]. The performance for (Uniform, 1 r) indicates that to achieve good local-alignment, especially for the outer rings, we should not consider the plasma membrane as it could mislead the outer region registration. Table 4 summarizes our recommendation for suitable masks at each registration step, and Fig. 5 illustrates the extent and weight distribution of the recommended masks.

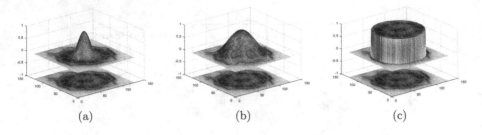

Fig. 5. Recommended weight masks (a) W_{rig}, (b) W_{aff}, (c) W_{nrr}.

3.4 Algorithm

The algorithm takes an image-pair and multiple weight masks (W_{rig}, W_{aff} and W_{nrr}) as input. In the rigid registration step, input images are pre-processed with a Gaussian filter and downsampled to half of the original size. In the affine registration step also images are pre-processed using Gaussian smoothing filter, but registration is performed on the original image sizes. In the non-rigid registration step a three level resolution pyramid is used. At the highest level, images are smoothened with a Gaussian filter and resized to half of the original image size, and coarse non-rigid registration is performed using a low resolution grid. The generated displacement field is further refined at the middle level, where images are smoothened with Gaussian filter and processed at their original sizes. The grid resolution used is twice that of the previous level. At the lowest pyramid level registration is performed on the original images and a grid resolution is set close to the original image size. The resulting displacement field defines the mapping between the image-pair and when applied on the moving image results in the registered image I_r.

The parameter details are shown in Table 5. G_σ and G_{sz} represents the sigma value and the Gaussian filter size while, I_{sz} and Gr_{sz} are the image size and the grid size. P_{sp} is the control point spacing and the regularization weight is ($\lambda_{reg} = 8e - 4$). The values for G_σ and λ_{reg} were chosen based on experiments on synthetic data for G_σ and on real data for λ_{reg}. The algorithm takes $\approx$60 s per image-pair in MATLAB on a 2.3 GHz Intel Core i7 CPU.

Table 5. Registration algorithm parameter details

	Rigid	Affine	Non-Rigid		
			PL = 1	PL = 2	PL = 3
G_σ	4.0	2.7	1.0	1.0	0
G_{sz}	11×11	11×11	7×7	7×7	-
I_{sz}	64×64	128×128	64×64	128×128	128×128
Gr_{sz}	-	-	36×36	68×68	132×132
P_{sp}	-	-	[2,2]	[2,2]	[1,1]

4   Results

In this section results for cilia registration using the proposed multiple-masks strategy and super-resolution reconstruction are presented. We also presented the cilia registration performance of two obvious approaches, first, registering without using any mask, and second, registering using the uniform weight mask with radius $1\,r$, which only covers the region within the plasma membrane of the cilium. We referred to the former as no-mask and the later as a constant-masks strategy. In Table 6, the mean of average pixel alignment error $(\overline{PAE})$ calculated using all the image-pairs is reported for before registration, and after registration using different weight mask(s) strategies. The registration performance is computed separately for the central, outer and all landmarks. Results clearly indicate that the central pair is best aligned using the multiple-masks, whereas the alignment of the ring of the outer doublets is best for the constant-masks, closely followed by the multiple-masks. The overall performance of the multiple-masks is slightly better than the constant-masks. An example of the results for each registration step using multiple-masks is shown in Fig. 6. Figure 7 shows the SR images reconstructed using different weight masks strategies. Here, 15 cilia instances with the highest wNCC scores in the registration were used to create the respective SR image.

Table 6. Mean PAE summary

Landmarks	Before	No-Mask	Constant	Multiple
Central	4.32(1.87)	3.29(2.03)	3.78(2.14)	**2.79(1.84)**
Outer	5.75(3.73)	4.98(3.73)	**2.20(2.03)**	2.30(1.80)
All	5.61(3.61)	4.81(3.63)	2.36(2.09)	**2.35(1.82)**

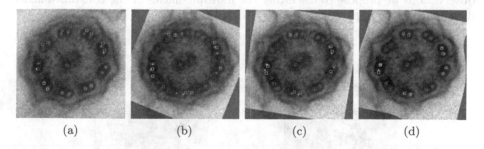

| (a) | (b) | (c) | (d) |

Fig. 6. Step-wise performance of registration using the multiple-masks approach. (a) Relative position of landmarks in the static (+) and the moving (□) image before registration, (b) after rigid registration, (c) after affine registration and (d) after non-rigid registration. Landmarks are used for evaluation only.

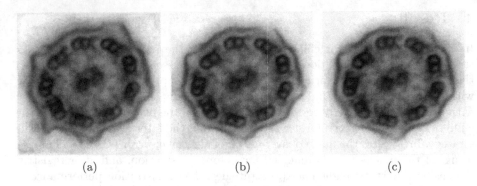

(a)   (b)   (c)

Fig. 7. SR cilia with (a) no-mask, (b) constant-masks, (c) multiple-masks.

5   Discussion and Conclusion

In this paper, we present a technique to enhance the cilia substructures by registering multiple instances of cilia cross-sections followed by SR reconstruction.
Cilium instance registration is achieved using a multi-step registration strategy
where we use different weight masks, with the aim to align different regions at
each registration step. Using the proposed multiple-masks results in a $\overline{PAE}$ of
2.35 ± 1.82 for all the landmarks, which is better than using no-masks or constant-
masks, see Table 6 (All). The constant-masks, however, performed slightly better
for aligning the outer rings but failed for the central pair, see Table 6 (Outer and
Central). The subjective results of reconstructed SR cilium in Fig. 7c also support
our quantitative results where the central pair, radial spokes and dynein arms
have relatively better visibility than the other two results. As the constant-masks
performed better in aligning the outer doublets, the corresponding SR cilium in
Fig. 7b has good visibility of DA, but poor for radial spokes and the central
pair. The same is validated by our expert pathologist, author AD. With these
observations we propose to use either the multiple-masks or the constant-masks
depending of the requirements of the problem at hand.

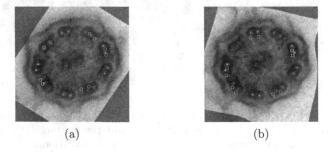

(a)   (b)

Fig. 8. Exceptions in alignment, (a) stuck in local minima, (b) case of poor result from
multi-position initialization.

Figure 8a shows a case when the algorithm achieves good alignment of the central pair after the rigid registration step but fails to keep the alignment after the affine registration step. Figure 8b shows the best multi-position initialization which was achieved for a position that was unfavorable for the ring of the outer doublets, making the total registration poor. To avoid disturbance from failure cases, the SR reconstruction is only performed from well-registered instances. Future work involves evaluation of the method on a larger dataset.

Acknowledgments. This work is supported by the Swedish Innovation Agency through the MedTech4Health program, and the Ministry of Science of the Republic of Serbia through projects ON174008 and III44006 (authors NS and JL).

References

1. Afzelius, B.A., Dallai, R., Lanzavecchia, S., Bellon, P.L.: Flagellar structure in normal human spermatozoa and in spermatozoa that lack dynein arms. Tissue Cell **27**(3), 241–247 (1995)
2. Bajić, B., Lindblad, J., Sladoje, N.: Restoration of images degraded by signal-dependent noise based on energy minimization: an empirical study. J. Electron. Imaging **25**(4), 043020 (2016)
3. Berkels, B., et al.: Optimized imaging using non-rigid registration. Ultramicroscopy **138**, 46–56 (2014)
4. Carson, J.L., Hu, S.S., Collier, A.M.: Computer- assisted analysis of radial symmetry in human airway epithelial cilia: assessment of congenital ciliary defects in primary ciliary dyskinesia. Ultrastruct. Pathol. **24**(3), 169–174 (2000)
5. Escudier, E., et al.: Computer-assisted analysis helps detect inner dynein arm abnormalities. Am. J. Respir. Crit. Care Med. **166**, 1257–1262 (2002)
6. Jones, L., et al.: Smart align-a new tool for robust non-rigid registration of scanning microscope data. Adv. Struct. Chem. Imaging **1**(1), 1 (2015)
7. Jorge-Peñas, A., et al.: Free form deformation-based image registration improves accuracy of traction force microscopy. PLoS One **10**(12), e0144184 (2015)
8. Lee, S., Wolberg, G., Chwa, K.Y., Shin, S.Y.: Image metamorphosis with scattered feature constraints. IEEE Trans. Vis. Comput. Graph. **2**(4), 337–354 (1996)
9. Lee, S., Wolberg, G., Shin, S.Y.: Scattered data interpolation with multilevel b-splines. IEEE Trans. Vis. Comput. Graph. **3**(3), 228–244 (1997)
10. Lindblad, J., et al.: High-resolution reconstruction by feature distance minimization from multiple views of an object. In: 5th International Conference on Image Processing Theory, Tools and Applications (IPTA), pp. 29–34. IEEE, Orleans (2015)
11. Lindblad, J., Sladoje, N.: Linear time distances between fuzzy sets with applications to pattern matching and classification. IEEE Trans. Image Process. **23**(1), 126–136 (2014)
12. O'Toole, E.T., Giddings, T.H., Porter, M.E., Ostrowski, L.E.: Computer-assisted image analysis of human cilia and chlamydomonas flagella reveals both similarities and differences in axoneme structure. Cytoskeleteon (Hoboken) **69**(8), 577–590 (2012)
13. Palm, C., et al.: Interactive computer-assisted approach for evaluation of ultra-structural cilia abnormalities. In: SPIE Medical Imaging. p. 97853N. International Society for Optics and Photonics (2016)

14. Rueckert, D., et al.: Nonrigid registration using free-form deformations: application to breast MR images. IEEE Trans. Med. Imaging **18**(8), 712–721 (1999)
15. Sotiras, A., Davatzikos, C., Paragios, N.: Deformable medical image registration: a survey. IEEE Trans. Med. Imaging **32**(7), 1153–1190 (2013)
16. Suveer, A., et al.: Automated detection of cilia in low magnification transmission electron microscopy images using template matching. In: IEEE International Symposium on Biomedical Imaging (ISBI), pp. 386–390. IEEE (2016)
17. Wahba, G.: Spline Models for Observational Data. SIAM, Philadelphia (1990)
18. Zariwala, M., Knowles, M., Omran, H.: Genetic defects in ciliary structure and function. Annu. Rev. Physiol. **69**, 423–450 (2007)

Faces, Gestures and Multispectral Analysis

Residual vs. Inception vs. Classical Networks for Low-Resolution Face Recognition

Christian Herrmann[1,2(✉)], Dieter Willersinn[2], and Jürgen Beyerer[1,2]

[1] Vision and Fusion Lab, Karlsruhe Institute of Technology KIT,
Karlsruhe, Germany
[2] Fraunhofer IOSB, Karlsruhe, Germany
{christian.herrmann,dieter.willersinn,
juergen.beyerer}@iosb.fraunhofer.de

Abstract. When analyzing surveillance footage, low-resolution face recognition is still a challenging task. While high-resolution face recognition experienced impressive improvements by Convolutional Neural Network (CNN) approaches, the benefit to low-resolution face recognition remains unclear as only few work has been done in this area. This paper adapts three popular high-resolution CNN designs to the low-resolution (LR) domain to find the most suitable architecture. Namely, the classical AlexNet/VGG architecture, Google's inception architecture and Microsoft's residual architecture are considered. While the inception and residual concept have been proven to be useful for very deep networks, it is shown in our case that shallower networks than for high-resolution recognition are sufficient. This leads to an advantage of the classical network architecture. Final evaluation on a downscaled version of the public YouTube Faces Database indicates a comparable performance to the high-resolution domain. Results with faces extracted from the SoBiS surveillance dataset indicate a superior performance of the trained networks in the LR domain.

1 Introduction

Forensic analysis of video data can help to solve crimes and identify the offenders. The struggle to evaluate increasing amounts of video footage can be addressed by automated solutions. A key component for automated video analysis is a robust face recognition system for person identification. Recent Convolutional Neural Network (CNN) approaches [19,21,28] which lead to significant performance improvements on this task, mainly came from internet companies such as Google or Facebook. Consequently, the addressed target domain are high quality face shots such as celebrity photos, selfies or personal profile photos. To address the low-resolution (LR) surveillance domain, at least an adaptation of such networks is required.

Several different types of network architectures are successfully applied for CNNs. Here, we select three different state-of-the-art architecture types (inception [21], residual [5] and classical [19]) and evaluate their suitability for low-resolution (LR) face recognition. An analysis of the respective architectures is performed to identify the necessary adjustments and potential bottlenecks of

© Springer International Publishing AG 2017
P. Sharma and F.M. Bianchi (Eds.): SCIA 2017, Part II, LNCS 10270, pp. 377–388, 2017.
DOI: 10.1007/978-3-319-59129-2_32

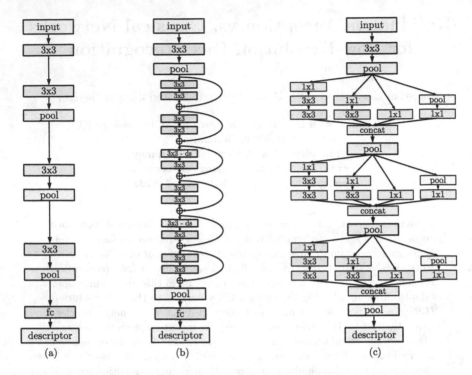

Fig. 1. Adapted LR networks for different architecture types: classical (a), residual (b) and inception (c). Green background denotes downsampling layers. (Color figure online)

the networks for the LR task. According to the analysis, each architecture is implemented with regard to the LR domain. After a systematical optimization of the meta-parameters, the final evaluation is performed on a LR version of the YouTube Faces Database (YTF) [30] as well as surveillance domain faces extracted from the SoBiS person identification database [22].

2   Related Work

Face recognition is a common topic in computer vision and recent state-of-the-art solutions for high-resolution (HR) faces are usually all based on CNNs [3,14, 15,19,21,28,29]. While groups from companies usually train on their own very large scale datasets containing up to several hundred million face images [21,28], other groups are limited to smaller public datasets [4,13,17–19,31,32] usually containing less or slightly above one million face images. Data augmentation strategies [15] can compensate for this but might also introduce unwanted effects.

In comparison to HR approaches, recent LR face recognition strategies are usually still based on conventional strategies consisting of a combination of local features and learned representations such as metric learning [16], dictionary learning [23] or manifold learning [11]. Currently, only few attempts to apply CNN approaches exist in this domain [7,8]. In summary, this gives the impression that LR face recognition is still moving towards the deep learning age.

Especially, unlike the HR face recognition task where it is possible to fine tune pre-trained networks [19], a LR face recognition network has to be trained from scratch due to the lack of general LR networks. One key choice which has to be made is the architecture type of the network. The most recent CNN architectures are usually first developed and proposed for the ImageNet challenge [2]. The three currently most distinctive and widespread architecture types are the inception architecure [26,27], the residual architecture [5,6] and the rather classically designed VGG architecture [25], which is quite similar to the well known AlexNet [12]. In the following sections, we will analyze and adjust these three architecture types to the LR scenario and finally compare their performance.

3   Network Architecture

When designing a CNN for LR face recognition, several issues have to be considered. For the ease of presentation, we focus on a face resolution of 32×32 pixels.

The main issue when adapting the different architectures to LR scenarios are the downsampling layers usually implemented as pooling layers. They group spatial information by propagating the average or maximum of a $k \times k$ region. For the common choice of $k = 2$ this involves a downsampling by a factor of two. If some spatial information should be kept in the feature maps, a maximum of 4 pooling layers is acceptable, for 32×32 pixel input size. The 5th pooling layer would condense the remaining 2×2 feature map into a 1×1 feature map and destroy the respective spatial information.

This conflicts one general rule of thumb for creating CNNs stating that deeper networks are preferred over wider networks. When depth is limited by the maximum number of possible downsampling steps, it will be necessary to compromise by using wider networks than for HR applications.

3.1   Classical Architecture

This traditional philosophy became popular with the famous AlexNet [12] and was extended with minor adaptions in the VGG Face network [19]. Some guidelines how to transfer this architecture to the LR task are given in [8].

Conceptually, the beginning of the network consists of alternating convolutional and pooling layers as shown by Fig. 2a. After these layers, a set of fully connected layers is appended to classify the resulting feature maps of the first part.

As motivated by the authors of the VGG network [25], in this architecture, consecutive convolutional layers can be understood as a replacement of a single convolutional layer with a larger filter size. This means for example that two 3×3 convolutional layers are a replacement for one 5×5 convolutional layer.

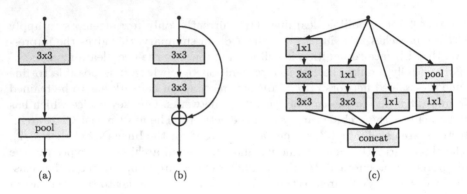

Fig. 2. Key components for each architecture type. Alternating convolutional and pooling layers for classical type (a), residual block (b) and inception module (c).

Following this motivation, the number of layers in such a network has to remain limited for LR applications because filter sizes have to represent the small content size in the LR image. Large filter sizes in the shape of many consecutive convolutional layers are unnecessary or even counterproductive.

Design choices for this architecture type include the number of convolutional, pooling and fully connected layers as well as the layer meta parameters itself. These include the number of filters per convolutional layer C, kernel size $P \times P$ of pooling layers or the number of neurons N in fully connected layers. To ease optimization, the basic structure of the network starts with a 3×3 convolutional layer and continues with a varying number of groups each consisting of one 3×3 convolutional layer and one pooling layer.

3.2   Residual Architecture

The main difference to the classical architecture is a kind of bypass of two convolutional layers (Fig. 2b). The benefit according to He et al. [5] is the better trainability for very deep networks. Instead of learning a function mapping of the input x to the output $F(x)$ where F represents the learned function, this strategy has to learn only a small offset or residuum $F(x)$ of the identity, leading to the output $F(x) + x$. Using this trick, this architecture type currently allows the deepest networks with about 1000 layers in certain applications [6]. Because the potentially lower number of layers in the application of this paper limits training time, the full residual blocks instead of the reduced bottleneck ones are employed. A specialty of the residual architecture are the missing pooling layers for downsampling. This is instead performed by specific convolutional layers which have a stride of 2 resulting also in a downsampling by factor 2. Once again, the amount of these special layers limits network depth for LR scenarios. Calling a fixed set of consecutive residual blocks a group, one of these downsampling layers is inserted at the beginning of each group.

Key design choices for this architecture include the number of residual groups, their sizes, as well as the number of trailing fully connected layers, besides the layer meta parameters itself.

3.3 Inception Architecture

The key concept of the inception architecture is a kind of meta layer called inception module [26, 27] shown by Fig. 2c. It includes several parallel data processing paths which are motivated by multi-scale processing. Efficient usage of computing resources is implemented by 1×1 convolutional layers which reduce data dimension.

Besides the layer meta parameters, the key design choice is the number of inception modules and trailing fully connected layers. Filter numbers within an inception module are fixed to a ratio of 1:4:2:1 for the double 3×3, single 3×3, 1×1 and pooling path respectively. The filter number ratio between 1×1 and consecutive 3×3 layers is 1:2. Following the same argumentation about filter sizes as for the classical architecture, we add a pooling layer after each inception module to avoid overly large filter sizes and call the combination of inception module and pooling layer a group.

3.4 General Configuration and Training Setup

As shown by the examples in Fig. 1, the basic structure of each network begins with a 3×3 convolutional layer followed by a pooling layer in case of residual and inception architectures. Afterwards, a modifiable number of layer groups of the respective architecture kind is added. The number of convolutional filters is doubled after each group in the network. In all cases, downsampling is performed between layer groups. The networks end with a varying number of fully connected layers in front of the output layer which includes the descriptor.

All networks are trained using a Siamese setup which has two key advantages compared with a conventional softmax classification strategy. First, the descriptor dimension, which equals the number of neurons in the output layer, is independent of the number of person identities in the dataset and can be chosen arbitrarily. In practice, this allows smaller descriptors. Second, it allows to combine several datasets without any effort because no consistent identity labels between datasets are required. As loss function, the max-margin hinge loss proposed in [8] is employed.

With this setup, the network can be understood as a projection from the image space into a discriminative feature space with respect to face recognition. The dimension of this feature space, which equals the target descriptor size, is set to 128 motivated by a number of further approaches [8,21,24]. Batch normalization [9] is incorporated in all cases.

4   Experiments

Exploiting the advantage of the Siamese setup, the networks are trained on a combination of several large-scale face datasets including Celebrity-1000 [13],

FaceScrub [18], MegaFace [17], MSRA [32], TV Collection (TVC) [8], and VGG Face Dataset [19]. This results in about 9M training images. The recent MS-Celeb-1M [4] as well as the CASIA WebFace [31] databases are omitted, because we found that they do not improve the results. All networks are trained from scratch using the Caffe framework [10] and Nvidia Titan X GPUs. Evaluation is performed on a downscaled version of the YouTube Faces Database (YTF) [30] (1,595 persons, 3,425 sequences) as well as faces collected from the SoBiS surveillance dataset [22]. Faces from the SoBiS dataset are extracted by a Viola-Jones based face tracker and registered by eye locations [20]. This results in 1,559 face sequences from 45 persons.

4.1   Architecture Optimization

Some of the training datasets include identities also present in the YTF database used for the final evaluation. These are removed from the training set and serve as validation set in this section. Table 1 lists all datasets including the relevant split sizes. Due to memory limitations, architecture optimization in this section is performed without the MegaFace and VGG Face dataset. After determination of a potential parameter optimum for each architecture type, a systematic optimization is performed in each case. The batch size is set to 100 for the majority of the networks and reduced to 20 for the largest networks. Each batch contains samples from all datasets. Table 2 shows selected results for some key parameters. The filter number C is given for the first 3×3 convolutional layer.

Having a detailed look at the impact of the different meta parameters, one can note a few things in Table 2. Regarding the number of layer groups, either 3 or 4 perform best (case 1–3). Larger differences were observed for the number of fully connected layers at the end of the network (case 11–13). At least one is required in the classical, and exactly one in the residual architecture. The results for the inception architecture are inconclusive with no or two fully connected layers performing comparably and, in the sense of reducing parameters,

Table 1. Training and validation datasets. Image datasets have the same number of images and sequences (each image can be understood as single sequence), video datasets have multiple images per sequence. Differences to official dataset sizes might occur in case of images being no longer downloadable from internet links.

Dataset name	Train			Validation		
	#images	#sequences	#persons	#images	#sequences	#persons
Celebrities-1000 [13]	2, 117, 837	145, 751	930	210, 154	13, 981	70
FaceScrub [18]	51, 162	51, 162	451	10, 299	10, 299	79
MegaFace [17]	4, 741, 425	4, 741, 425	672, 957	–	–	–
MSRA [32]	163, 018	163, 018	1, 372	39, 704	39, 704	208
TV collection [8]	1, 151, 545	15, 427	604	–	–	–
VGG face [19]	834, 375	834, 375	2, 558	–	–	–
Combination	9, 059, 362	5, 951, 158	678, 872	260, 157	63, 984	357

Table 2. Optimization of the three architectures on the validation set at 32×32 pixels face size. Baseline values are indicated. Results are mean accuracy and standard deviation for a 10-fold cross-validation. Blanked settings were omitted due to limited GPU memory.

Parameter	Case	Value	Classical		Residual		Inception	
			Accuracy	Std	Accuracy	Std	Accuracy	Std
#groups	1	2	0.771	0.013	0.664	0.006	0.754	0.011
	2	3	*0.776*	0.010	*0.758*	0.009	*0.772*	0.013
	3	4	0.789	0.010	0.770	0.009	0.657	0.008
#filters C	4	64	0.687	0.014	*0.758*	0.009	0.728	0.008
	5	128	0.771	0.013	0.762	0.007	0.755	0.015
	6	192	0.784	0.018	0.760	0.012	*0.772*	0.013
	7	256	*0.776*	0.010	0.762	0.009	0.778	0.011
	8	384	0.784	0.009	-	-	0.737	0.112
	9	512	0.786	0.013	-	-	-	-
	10	768	0.795	0.012	-	-	-	-
#fully connected layers	11	0	0.767	0.010	0.719	0.004	*0.772*	0.013
	12	1	*0.776*	0.010	*0.758*	0.009	0.732	0.013
	13	2	0.779	0.012	0.641	0.009	0.777	0.008
#neurons per fully connected layer N	14	1024	0.782	0.012	0.761	0.010	0.732	0.013
	15	2048	*0.776*	0.010	*0.758*	0.009	0.693	0.015
	16	4096	0.781	0.011	0.765	0.006	0.636	0.022
	17	8192	0.759	0.014	0.700	0.010	0.581	0.018

it is preferred to choose none. The different impact of the fully connected layers can be explained by the varying capabilities of the respective architecture blocks in front. The results for the number of filters in the convolutional layers behave as expected (case 4–10). In each case, increasing the number of filters improves the results up to a range where saturation is observed. The upper limit is caused by GPU memory in all cases. The number of neurons in any fully connected layer is rather irrelevant in the tested range, except for inception where performance decreases with increasing number of neurons (case 14–17). All in all, the classical architecture performs best with the inception architecture coming second. This indicates that architectures allowing deeper networks are inferior for LR applications compared to the classical concept. It can be suspected that the LR images contain too few information to feed a very deep network. Note that the differences in validation accuracy near the optimum are mostly below the measured standard deviation. So from a statistical viewpoint, the measurement noise on the accuracy justifies the choice of a parameter setting which seems

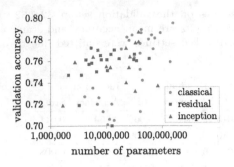

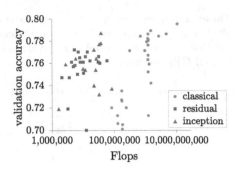

Fig. 3. Validation accuracy vs. number of network parameters.

Fig. 4. Validation accuracy vs. network Flops

Table 3. Final network configuration and properties for each architecture type.

	Classical	Residual	Inception
Configuration (cases)	2,9,12,16	2,5,12,16	2,7,11
Validation accuracy	0.786	0.766	0.787
Parameters	74.1 M	32.0 M	29.2 M
Flops	1,279 G	512 G	986 G
Prediction time (ms)	2.1	4.2	6.4
Memory footprint (MB)	338	230	239

to be slightly off the maximum. This becomes relevant if runtime and memory consumption has to be considered along with the performance.

One of the main differences between the classical architecture and the modern residual and inception one is the number of parameters necessary to achieve a certain performance. The modern architectures require less parameters to achieve a comparable performance as shown by Fig. 3 which includes all trained networks. Similarly, Fig. 4 shows a comparison between performance and necessary computational power for all the trained networks. But the stated number of Flops is only a theoretical value. In practice, execution times increase significantly with network depth due to relatively higher memory usage by the large amount of intermediate layers.

Table 3 shows the properties for the best network of each architecture including measured execution times on a Titan X. Note that despite having the most Flops, the classical network's forward pass is the fastest due to less required memory operations. These three networks will serve for final training and evaluation.

4.2   In-the-Wild Results

Final training is performed using the best setup for each architecture type as denoted by Table 3 and visualized by Fig. 1 with all training data listed in Table 1

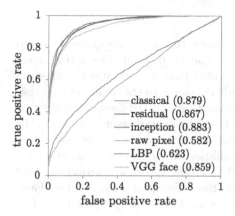
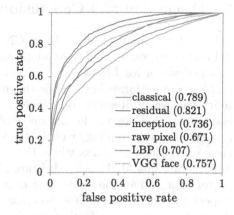

Fig. 5. Example scenes and faces from the SoBiS Face dataset.

(left plot legend)
classical (0.879)
residual (0.867)
inception (0.883)
raw pixel (0.582)
LBP (0.623)
VGG face (0.859)

true positive rate / false positive rate

(right plot legend)
classical (0.789)
residual (0.821)
inception (0.736)
raw pixel (0.671)
LBP (0.707)
VGG face (0.757)

true positive rate / false positive rate

Fig. 6. ROC curve and accuracy (in brackets) comparison for low-resolution YTF dataset.

Fig. 7. ROC curve and accuracy (in brackets) comparison for SoBiS Face dataset.

until convergence. For comparison, raw pixel, local binary patterns (LBP) [1], and the HR VGG Face descriptors [19] are employed. The appropriate vector distance is combined with each descriptor: Hellinger distance for LBP because of its histogram character, cosine distance for raw pixel and VGG Face because of the softmax training setup and Euclidean distance for LrfNet and the trained descriptors because the loss function was chosen to minimize this distance. Aggregation of multiple frames in a consecutive face sequence is performed by averaging the respective face descriptors.

A 10-fold cross-validation is performed on both the YTF and SoBiS Face dataset (Fig. 5). The results are reported for all face descriptors as ROC curves and the respective classification accuracy in Figs. 6 and 7. While all CNN-based descriptors significantly outperform the LBP descriptor in the case of higher quality data represented by the YTF dataset, the high-quality trained VGG Face descriptor performs nearly as bad as LBP in the case of the SoBiS surveillance data. The proposed face descriptors with exception of the inception one outperform the VGG Face descriptor by a significant margin in this case. We are unsure about the reason for the significant performance drop of the inception architecture compared to the high-quality YTF data. Potential explanations include a too strong focus on fine grained details by the multi-scale inception blocks or domain overfitting by this powerful architecture on the training data which is more similar to the YTF than the SoBis data. Comparing the results on both datasets, it can be observed that the order of the architectures is inverse in the YTF and SoBiS experiments. This supports the overfitting hypothesis where the residual architecture generalizes best, classical coming second and the inception being the worst in this respect. All in all, the achieved accuracy of 0.883 on YTF is near the HR accuracy 0.916 of VGG Face on the same dataset.

5   Discussion and Conclusion

All in all, the analysis of the three CNN architecture concepts classical, residual and inception allowed their adaptation to the LR domain. The optimization and evaluation for LR face recognition revealed that in this case the classical architecture performs best. While this result is inverse to the architecture capabilities shown in literature on image categorization and HR face recognition, it can be explained by the low image resolution. The spatial resolution limits the amount of downsampling layers in the network which makes it unreasonable to create very deep networks where the capabilities of the residual and inception architecture unfold. Training dedicated LR face descriptors was proven to yield better results than the application of a HR face descriptor such as VGG Face. Especially when surveillance data has to be analyzed, the proposed LR face descriptors outperform further descriptors significantly.

References

1. Ahonen, T., Hadid, A., Pietikainen, M.: Face description with local binary patterns: application to face recognition. Pattern Anal. Mach. Intell. **28**(12), 2037–2041 (2006)
2. Deng, J., Dong, W., Socher, R., Li, L.J., Li, K., Fei-Fei, L.: Imagenet: a large-scale hierarchical image database. In: Computer Vision and Pattern Recognition, pp. 248–255. IEEE (2009)
3. Ding, C., Tao, D.: Trunk-Branch Ensemble Convolutional Neural Networks for Video-based Face Recognition. arXiv preprint arXiv:1607.05427 (2016)
4. Guo, Y., Zhang, L., Hu, Y., He, X., Gao, J.: MS-Celeb-1M: challenge of recognizing one million celebrities in the real world. In: Imaging and Multimedia Analytics in a Web and Mobile World (2016)
5. He, K., Zhang, X., Ren, S., Sun, J.: Deep Residual Learning for Image Recognition. arXiv preprint arXiv:1512.03385 (2015)
6. He, K., Zhang, X., Ren, S., Sun, J.: Identity mappings in deep residual networks. arXiv preprint arXiv:1603.05027 (2016)
7. Herrmann, C., Willersinn, D., Beyerer, J.: Low-quality video face recognition with deep networks and polygonal chain distance. In: Digital Image Computing: Techniques and Applications, pp. 1–7. IEEE (2016)
8. Herrmann, C., Willersinn, D., Beyerer, J.: Low-resolution convolutional neural networks for video face recognition. In: Advanced Video and Signal Based Surveillance. IEEE (2016)

9. Ioffe, S., Szegedy, C.: Batch normalization: accelerating deep network training by reducing internal covariate shift. In: International Conference on Machine Learning (2015)

10. Jia, Y., Shelhamer, E., Donahue, J., Karayev, S., Long, J., Girshick, R., Guadarrama, S., Darrell, T.: Caffe: Convolutional Architecture for Fast Feature Embedding. arXiv preprint arXiv:1408.5093 (2014)

11. Jiang, J., Hu, R., Wang, Z., Cai, Z.: CDMMA: coupled discriminant multi-manifold analysis for matching low-resolution face images. Sig. Process. **124**, 162–172 (2016)

12. Krizhevsky, A., Sutskever, I., Hinton, G.E.: Imagenet classification with deep convolutional neural networks. In: Neural Information Processing Systems, pp. 1097–1105 (2012)

13. Liu, L., Zhang, L., Liu, H., Yan, S.: Toward large-population face identification in unconstrained videos. IEEE Trans. Circuits Syst. Video Technol. **24**(11), 1874–1884 (2014)

14. Liu, X., Kan, M., Wu, W., Shan, S., Chen, X.: VIPLFaceNet: An Open Source Deep Face Recognition SDK. arXiv preprint arXiv:1609.03892 (2016)

15. Masi, I., Tran, A.T., Leksut, J.T., Hassner, T., Medioni, G.: Do we really need to collect millions of faces for effective face recognition? arXiv preprint arXiv:1603.07057 (2016)

16. Mudunuri, S.P., Biswas, S.: Low resolution face recognition across variations in pose and illumination. IEEE Trans. Pattern Anal. Mach. Intell. **38**(5), 1034–1040 (2016)

17. Nech, A., Kemelmacher-Shlizerman, I.: Megaface 2: 672,057 Identities for Face Recognition (2016)

18. Ng, H.W., Winkler, S.: A data-driven approach to cleaning large face datasets. In: International Conference on Image Processing, pp. 343–347. IEEE (2014)

19. Parkhi, O.M., Vedaldi, A., Zisserman, A.: Deep face recognition. In: British Machine Vision Conference, vol. 1(3), p. 6 (2015)

20. Qu, C., Gao, H., Monari, E., Beyerer, J., Thiran, J.P.: Towards robust cascaded regression for face alignment in the wild. In: Computer Vision and Pattern Recognition Workshops (2015)

21. Schroff, F., Kalenichenko, D., Philbin, J.: FaceNet: a unified embedding for face recognition and clustering. In: Computer Vision and Pattern Recognition, pp. 815–823 (2015)

22. Schumann, A., Monari, E.: A soft-biometrics dataset for person tracking and re-identification. In: IEEE Conference on Advanced Video and Signal Based Surveillance (AVSS), August 2014

23. Shekhar, S., Patel, V.M., Chellappa, R.: Synthesis-based robust low resolution face recognition. IEEE Trans. Image Process. (under review) (2014)

24. Simonyan, K., Parkhi, O.M., Vedaldi, A., Zisserman, A.: Fisher vector faces in the wild. In: British Machine Vision Conference, vol. 1, p. 7 (2013)

25. Simonyan, K., Zisserman, A.: Very deep convolutional networks for large-scale image recognition. In: International Conference on Learning Representations (2015)

26. Szegedy, C., Liu, W., Jia, Y., Sermanet, P., Reed, S., Anguelov, D., Erhan, D., Vanhoucke, V., Rabinovich, A.: Going deeper with convolutions. In: Computer Vision and Pattern Recognition, pp. 1–9 (2015)

27. Szegedy, C., Vanhoucke, V., Ioffe, S., Shlens, J., Wojna, Z.: Rethinking the inception architecture for computer vision. arXiv preprint arXiv:1512.00567 (2015)

28. Taigman, Y., Yang, M., Ranzato, M., Wolf, L.: Deepface: closing the gap to human-level performance in face verification. In: Computer Vision and Pattern Recognition, pp. 1701–1708 (2014)
29. Wen, Y., Zhang, K., Li, Z., Qiao, Y.: A discriminative feature learning approach for deep face recognition. In: Leibe, B., Matas, J., Sebe, N., Welling, M. (eds.) ECCV 2016. LNCS, vol. 9911, pp. 499–515. Springer, Cham (2016). doi:10.1007/978-3-319-46478-7_31
30. Wolf, L., Hassner, T., Maoz, I.: Face recognition in unconstrained videos with matched background similarity. In: Computer Vision and Pattern Recognition (2011)
31. Yi, D., Lei, Z., Liao, S., Li, S.Z.: Learning face representation from scratch. arXiv preprint arXiv:1411.7923 (2014)
32. Zhang, X., Zhang, L., Wang, X.J., Shum, H.Y.: Finding celebrities in billions of web images. IEEE Trans. Multimedia **14**(4), 995–1007 (2012)

Visual Language Identification
from Facial Landmarks

Radim Špetlík[✉], Jan Čech, Vojtěch Franc, and Jiří Matas

Department of Cybernetics, Center for Machine Perception (CMP), Czech Technical University in Prague, Karlovo nám. 13, 121 35 Prague, Czech Republic
{spetlrad,cechj,xfrancv,matas}@cmp.felk.cvut.cz
http://cmp.felk.cvut.cz

Abstract. The automatic Visual Language IDentification (VLID), i.e. a problem of using visual information to identify the language being spoken, using no audio information, is studied. The proposed method employs facial landmarks automatically detected in a video. A convex optimisation problem to find jointly both the discriminative representation (a soft-histogram over a set of lip shapes) and the classifier is formulated. A 10-fold cross-validation is performed on dataset consisting of 644 videos collected from *youtube.com* resulting in accuracy 73% in a pairwise discrimination between English and French (50% for a chance). A study, in which 10 videos were used, suggests that the proposed method performs better than average human in discriminating between the languages.

1 Introduction

By Visual Language IDentification (VLID), a recognition of language a subject speaks in a video (with no audio track) is meant.

Successful VLID is useful in scenarios in which conventional audio processing is ineffective (very noisy environments), or impossible (no audio signal is available) [1]. It has been shown that visual speech cues may improve human's ability to understand a speech under noisy conditions [2]. There are two tasks related to the VLID, lip-reading and speaker authentication using visual lip information. Lip-reading means understanding the content of the uttered speech. There is a significant body of research, see [3] for a review. Speaker authentication is a biometric problem, where the identity of a person is recognized based on the way of speaking [4]. Unlike these tasks, we are only concerned about the language being spoken.

Automatic Language Identification (LID), i.e. a recognition of language from an audio channel, is a mature technology achieving a high identification accuracy from a few seconds of a speech [5]. In contrast, VLID is still an unexplored area of research that is considered both an interesting research topic by means of practical usability [1].

© Springer International Publishing AG 2017
P. Sharma and F.M. Bianchi (Eds.): SCIA 2017, Part II, LNCS 10270, pp. 389–400, 2017.
DOI: 10.1007/978-3-319-59129-2_33

Related Work. Newman and Cox [5]
present a preliminary study on an automatic VLID. The authors describe a
system that adopts a model successfully used in audio-based speech recognition systems. Bigram language models are built for each language using a
highly speaker-dependent representation
extracted from the videos. The database
used to perform the experiments consists of 21 multilingual subjects reading a UN Declaration of Human Rights
in all languages in which they are proficient. The dataset is thus also highly
content-dependent. Despite the fact that

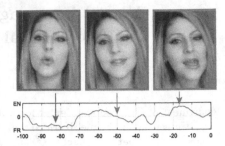

Fig. 1. An example of a VLID. The output score of our method over 100 frames.
From left to right expressions producing:
strong French score, neutral score and
strong English score.

the database consists of 21 speakers, they present only results for two bilingual
speakers (English - Arabic, English - German) and for one trilingual speaker
(English, French and German).

Newman and Cox continue their research in [6]. They fully articulate some
of the details of the method presented in [5] and they try to extend it so it
is capable of a speaker-independent VLID. The authors manually transcribe
the audio track at word level and automatically expand this transcription to
units of sound (phonemes) and then, via a custom designed mapping table, to
units of visual communication (visemes). A tied-state mixture triphone HMMs
are used to model the probability of a given viseme sequence being a sequence
of a particular language. They classify with an RBF kernel SVM built on the
outcomes of the individual HMMs. The results for five English/French bilingual
speakers are presented.

In the latest work [1], Newman and Cox repeat both the described speaker-
dependent and speaker-independent experiments on a larger dataset. Again,
the videos are highly content-dependent and recorded in a controlled environment. The authors perform a new speaker-independent experiment with features based only on the shape of the lips. After 30 s of speech, ≈22% mean
error-rate is achieved in speaker-independent experiments using only lip-shape
features and ≈10% mean error-rate using a combination of the lip-shape and lip-
appearance features. In the final discussion, the authors point out that the results
of their speaker-independent experiments, in which the combined features are
used, seems to be dependent on the difference between the colour of the speakers' skin. This comes from the fact that the experiments were performed with
nineteen Arabic and English speakers in total. Also, a speaking-rate is suspected
to affect both the speaker-dependent and speaker-independent experiments.

In [7] an automatic LID in music videos is explored. A "bag-of-words" approach based on audio-visual features and linear SVM classifiers is presented.
Using a combination of audio and video features, 48% accuracy (compared to
4% for chance) is obtained in experiments with 25000 music videos and 25

languages. Using only video features, 14.3% accuracy is achieved. Visual features consist of 8×8 hue-saturation histogram, several statistics derived from the face detector and textons computed for each video frame.

All the presented VLID experiments of Newman and Cox are done with the videos recorded in a studio-like environment with all speakers reading the same text. In [5], highly speaker-dependent features are used. In [6], the experiments are performed using only five bilingual speakers. Despite low mean error rates presented in [1], there is a high standard deviation present in all the results.

To the best of our knowledge, there are no other works on VLID.

In this paper, an approach to the VLID using facial landmarks is described (see Fig. 1 for an example output of the method). A hypothesis is examined that a language being spoken may be deduced by observing the lip-shapes of the speaker. A convex optimisation problem is formulated in which simultaneously a linear classifier is learnt and the most discriminative representation of the lip-shapes is found.

Our contribution is the introduction of a speaker and content-independent VLID method. Unlike the existing approaches, our method was validated on realistic videos that were not recorded specifically for the purpose of the research. Additionally, we collected a dataset *CMP-VLID* consisting of videos downloaded from *youtube.com* of native English and French speakers, mostly bloggers. The dataset is publicly available at the following website: http://cmp.felk.cvut.cz/~spetlrad/cmp-vlid/.

The rest of the paper is structured as follows: Sect. 2 describes the proposed method, implementation details are given in Sects. 3 and 4 presents the datasets, Sect. 5 covers experiments settings and their results. Sect. 6 discusses the results and concludes the paper.

2 The Proposed Method

The basic idea of the proposed method is to perform a VLID by observing the shape of the speaker's lips. The recognition of a language is done by measuring the frequency of the most discriminative lip-shapes in a given video. We model a single lip-shape as a subset of standard facial landmarks automatically detected in every frame. The most discriminative lip-shapes are found jointly together with the classifier.

Let the training set be

$$T = \{(V_i, y_i)\}_{i=1}^{N} \tag{1}$$

where V_i is a video with a single speaker, y_i is the label of one of two possible languages of the speaker in the video, $y_i \in \{-1, 1\}$, and N is the number of the videos.

Facial landmarks in all frames in all videos are detected. Every video has a set of associated facial landmarks, i.e.

$$V_i = \{\mathbf{s}_i^t\}_{t=1}^{L_i} \tag{2}$$

where $\mathbf{s}_i^t$ are the facial landmarks detected in the t-th frame of the i-th video, $\mathbf{s}_i^t \in \mathbb{R}^D$, D is twice the number of the landmark points, because we get $\frac{D}{2}$ coordinates for x and $\frac{D}{2}$ coordinates for y axis, and L_i is the number of frames in video i.

The i-th video is represented by a soft-assignment variant of bag-of-words. A histogram with K bins

$$\mathbf{x}_i = \begin{bmatrix} x_i^1 \cdots x_i^K \ 1 \end{bmatrix}^T \tag{3}$$

is used where for $k \in \{1, \ldots, K\}$ the bin values are given as

$$x_i^k = \frac{1}{L_i} \sum_{t=1}^{L_i} \exp\left(-\sigma \|\mathbf{s}_i^t - \mu_k\|_2^2\right), \tag{4}$$

L_i is the number of frames with detected facial landmarks in a video i, σ is a hyperparameter. Centroids μ_k are constructed by clustering the set $\{s\} = \bigcup_{i=1}^{N} \{\mathbf{s}_i^t\}_{t=1}^{L_i}$, which contains all landmarks from all videos, by the k-means algorithm using the Euclidean norm. The result

$$\{\mu_k\}_{k=1}^{K} \tag{5}$$

is a set of K centroids, $\mu_k \in \mathbb{R}^D$.

Finally, to discriminate between the languages, a linear binary classification over the representation

$$\hat{y}_i = \text{sign}(\mathbf{w}^T \mathbf{x}_i) \tag{6}$$

is performed, where the $\mathbf{w}$ is an unknown weight vector $\mathbf{w} = \begin{bmatrix} w_1, \cdots, w_K, b \end{bmatrix}^T$.

To train the classifier, the optimisation problem

$$\min_{\mathbf{w}} \left\{ \lambda_1 \|\mathbf{w}\|_1 + \lambda_2 \|\mathbf{w}\|_2^2 + \frac{1}{N} \sum_{i=1}^{N} \max\{0, 1 - y_i \cdot \mathbf{w}^T \mathbf{x}_i\} \right\} \tag{7}$$

is formulated. The last term is the hinge loss. The other two terms are the regularizers. Hyperparameters λ_1 and λ_2 control the strength of the regularization.

Problem (7) is convex and can be converted into a standard quadratic programming problem for which fast and efficient solvers are available.

Let us now shortly discuss the ℓ_1 norm in Eq. (7). By omitting the norm, a standard linear hinge loss SVM is obtained. In that case, a solution $\mathbf{w}^*$ assigning an optimal weight to each soft histogram bin x_i^k is produced. By using the ℓ_1 norm, a sparsity of the solution $\mathbf{w}^*$ is enforced, thus a subset of the soft histogram bins is used. In other words, a simultaneous search for the optimal bin weights and a selection of the most informative bins, i.e. the learning of the optimal representation, is undertaken. In Sect. 5, we show that ℓ_1 norm improves the results of the experiments and thus a subset of centroids (having non-zero weight) is selected.

In theory, landmarks from all videos may be used as the centroids μ_k in the representation (4). The most discriminative centroids would still be found by solving (7). However, the problem would be too large and computationally intractable. Therefore the data are pre-clustered with the k-means algorithm.

Also note that the method has the following properties. First, although a formulation of a binary decision problem is presented, an extension to multiple class scenarios is straightforward. Second, the fixed-length representation (3) enables classification of an arbitrarily long sequence of images. We expect the representation to be more stable for longer image sequences. Third, a simple linear classifier is used, thus applications in real-time environments are at hand. Fourth, the representation (3) is inherently invariant to any permutation of the input sequence frames.

3   Implementation Details

In this section the details concerning the implementation of the method are specified. The details are described in the order following the structure of Sect. 2.

Before the landmarks detection a face bounding box is detected. A commercial implementation of *WaldBoost* [8] based detector[1] was used to perform this task. The detector provides a head yaw estimate. All detected faces having $|yaw| > 15°$ were discarded. Also all faces for which the width of the face bounding box was <150 pixels were discarded. These two steps were taken to ensure that the landmarks detector has good enough input to provide precise results.

When the face bounding boxes were detected, a facial landmarks detection was performed with the *IntraFace* detector [9]. Each detected set of landmarks was translated to the origin and it was normalized by the interocular distance.

After the facial landmarks were detected, the non-speaking parts of the video were removed. That was done by computing a location variance for each landmark using a sliding temporal window of length 75 frames (3 s). That was done separately for each video. Then the *mean temporal window variance* was computed for each frame. Frames where the mean temporal window variance was less than 10^{-4} were discarded.

The quadratic problem defined in Eq. (7) was solved by the *IBM ILOG CPLEX Optimization Studio* quadratic problem solver using the default settings.

4   Data

In this section, the experimental datasets are described. First, it is explained, how the *CMP-VLID* dataset was collected. Then the *CMP-VLID Test* dataset, used for the experiment with human participants and for the testing, is presented.

Table 1. The *CMP-VLID* database.

	English	French
Videos count	832	485
Mean frames count	4714	6508
Median frames count	4291	4461
Mean width × height	1189 × 673	1174 × 669

---
[1] Eyedea Recognition Ltd. http://www.eyedea.cz/.

Fig. 2. *CMP-VLID Test* dataset. The first and the second row contains randomly selected frames from videos with five English (EN1, ..., EN5) and five French (FR1, ..., FR5) speakers respectively. The third row contains the expressions yielding strong French score and the fourth row strong English score for every subject in the set.

CMP-VLID Dataset. Both the English and French parts of the *CMP-VLID* dataset were collected from *youtube.com*. The majority of the dataset was collected semi-automatically. There were two main techniques used in the collection of the data. First, a script was created that automatically downloaded all results of a single full-text search at the site. The search terms were similar to the terms "English blogger" or "English YouTuber", and their adequates in the French language, i.e. "Une Blogueuse Française", "Un Blogueur Français", "Une YouTubeuse Française" and "Un YouTubeur Français". The downloaded videos were manually checked for to be sure that the persons in the videos speak the expected language. Second, a Wikipedia site containing a list of English and French film and theatre actors was crawled, and the youtube full-text search engine was used to find the interviews with the individuals. One interview per each actor was then manually selected from each result of the search. Table 1 contains statistics describing the collected database. Two non-overlapping datasets, the *CMP-VLID Test* dataset and the *Cross-validation* dataset, were created from the *CMP-VLID*.

Cross-Validation Dataset. Consists of 322 videos for each language. Each video was trimmed to the first 500 frames. The *Cross-validation* dataset was used for training and for introspection experiments.

CMP-VLID Test Dataset. Consists of 5 videos for each language. It is used for an experiment with human participants. The selection was performed in the following manner. A video was randomly picked and then manually checked. The single speaker videos with neutral surroundings were selected. The videos were demultiplexed, the audio track was removed, and they were cut to the length of 60 s. The proposed method, trained on the *CMP-VLID* dataset, was evaluated on the *CMP-VLID Test* dataset to provide a comparison between human and the algorithm.

5   Experiments

This section contains the description of the performed experiments. First, an experiment with human participants is presented, then a series of experiments demonstrating the accuracy and properties of the proposed method follows.

5.1   Experiment with Human Participants

This experiment was intended to uncover the extent to which a human can distinguish between the English and French speakers if only the visual information is available. Also, techniques used by people to make the guess were investigated. An on-line form was created using the *Test* dataset described in Sect. 4. The participant's task was to guess which language a person in a video (with no audio track) speaks. One hundred people participated. Every participant was also asked if he or she is familiar with the French language. The experiment was not carried out in controlled conditions, but since the participation was voluntary, we believe that the participants were interested in their own performance rather than in "getting it all right".

Figure 3(a) shows the mean guess accuracy for each of the ten videos. There were two videos with the mean guess accuracy lower than fifty percent and the accuracy in the remaining cases fluctuates around the same average. The mean guess accuracy is 72.6% and it is depicted by the green dashed plot. The first

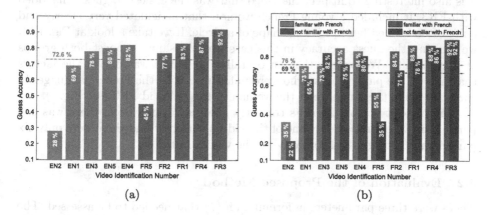

(a)   (b)

Fig. 3. VLID accuracy of 100 human participants on *CMP-VLID Test* dataset videos with 5 English (EN · columns) and 5 French (FR · columns) speakers. Green dashed line in (a) is the average accuracy. Light and dark green dashed lines in (b) is the average accuracy of the people familiar and unfamiliar respectively. (Color figure online)

video that was hard to guess contained a young fast speaking English man. The second hard to guess video included a French middle-aged man with a slow speaking pace. These results indicate that there were some expectations amongst people who participated about how fast a speaker of a given language should speak. If we take a glimpse at Fig. 3(b), we see that the results pictured there, i.e. the results where the familiarity with the French language is taken into account, support our hypothesis. People familiar with French were more accurate in guessing the language of the speaker in both cases than the people not familiar with French.

Figure 3(b) shows the average guess accuracy for each of the ten videos with the results displayed for both groups of participants, i.e. the participants familiar and participants not familiar with the French language. In the case of the video with id EN2 and FR5, there is a ten percent and twenty percent points difference between the two groups of participants. Also, there are two other videos with the difference ten percent points or higher. The video FR1 contained clearly articulating French teenage man and a video FR2 included young French man with fast speaking pace. The results show that people who participated in our experiment and who are familiar with French are able to uncover the language of the French speakers better than the people not familiar with French.

We asked some of the participants to give us a more detailed feedback on the way they tried to guess the language of the speaker. The participants mentioned the suspected qualities, i.e. the pace of the speech and the shapes the speakers lips took most frequently. But they also mentioned something other – their guesses were also based on something which we call "an overall appearance". We heard things like: "This must be a French girl, look at the way she dresses. . . ", or "This is totally a guy from Algeria. . . ". In other words, also the things like the colour of the skin, the shape of the head or the way one dresses support the decision whether the subject speaks one language or another. In the light of this finding, it is also interesting to inspect the video that was the easiest to guess, the video with FR3. It included a young French woman that articulated very clearly and quite often formed her lips into a shape of a circle. If we take a look at Fig. 3(b), we see that the guess accuracy in this case was the same amongst both groups of participants. We believe that an overall appearance was one of the factors that helped the participants, who were unfamiliar with the French language, to correctly guess the language of the woman in the video id FR3 (see Fig. 1).

The majority of participants correctly guessed seven videos, there was one participant who guessed only one of the videos correctly and there were only six participants that guessed correctly all the videos in the set.

5.2   Evaluation of the Proposed Method

There were three parameters in formulation (7) that needed to be assessed. The weight of the ℓ_1 norm λ_1, the weight of the ℓ_2 norm λ_2 and the width of the soft-histogram bin σ. An exhaustive grid search with $\lambda_1, \lambda_2 \in \{10^{-10}, 10^{-9}, \ldots, 10^6\}$ and $\sigma \in \{10^{-3}, 10^{-2}, \ldots, 10^2\}$ was performed. A 10-fold cross-validation was used on the *Cross-validation* dataset defined in Sect. 4. The parameters $\lambda_1 = 10^{-3}$,

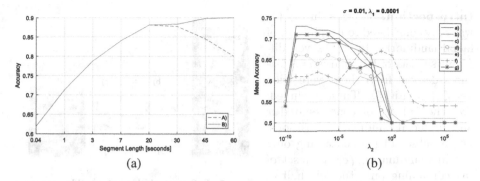

(a)   (b)

Fig. 4. (a) Accuracy of the method on *CMP-VLID Test* dataset as a function of a video segment length. *(A)* the accuracy when the representation (4) is build using the whole input sequence, *(B)* the accuracy when an ensemble of classifiers is used. (b) Average accuracy as a function of different choices of s_i^t and λ_2 for fixed σ and λ_1. The same colours as in Fig. 5 were used to distinguish between different choices of s_i^t. Cross-validation results. (Color figure online)

$\lambda_2 = 10^{-6}$ and $\sigma = 1$ were found resulting in 73% accuracy with 6 % points standard deviation on the validation data. The parameter K in (5) of the k-means algorithm was set to 500. This choice was empirically proven to give the best results when the training time was taken into account. Solving problem (7) takes approximately 3 s.

Accuracy Evaluation. The proposed method was evaluated on the *CMP-VLID Test* dataset described in Sect. 4. The sequences of landmarks detected in each video were divided sequentially and exhaustively to give test durations of 60, 45, 30, 20, 7, 3, 1 and $\frac{1}{25}$ s (assuming frame rate 25 frames per second). Figure 4(a) presents the results of the evaluation. The blue plot depicts the accuracy of the representation (4) build from all available frames. The best accuracy is achieved at 20 s, it drops for longer durations. Let us now discuss this result. The described data partitioning leads to a situation, in which the number of test sequences for shorter test durations greatly exceeds the number of sequences for longer test durations. This means, that the results for longer durations may not be statistically significant. Thus, we performed an additional experiment in which a sliding temporal window of length 500 frames was applied on each sequence of length $L > 500$ leading to another $L - 500 + 1$ sequences. The label of the whole sequence was obtained by a vote of $L - 500 + 1$ classifiers, each having the same weight. The results represented in Fig. 4(a) suggest that the descending trend witnessed for durations longer than 20 s in case of a single classifier may be indeed caused by a small size of the testing dataset. The classifier ensemble represented by the orange plot yields the best accuracy 90% at the longest duration.

Results of both presented experiments lead to a conclusion that the proposed method is better than an average human in discriminating between the French and English language.

Introspection. So far, $\mathbf{s}_i^t$ in Eq. (4) was only considered to be a full set of facial landmarks. In fact, seven different subsets of landmarks and features derived from them were tried as $\mathbf{s}_i^t$ (see Fig. 5). These included: *(a)* the full set of concatenated x and y facial landmarks positions, *(b)* a subset of (a) containing only the lips landmarks, *(c)* a subset of (a) containing only the left half of the lips landmarks, *(d)* a subset of (a) containing only the lips contour landmarks, i.e. the lips landmarks with the inner lips landmarks excluded. Saitoh and Konishi [10] compute the radius feature r, which is the distance r_0, r_1, $\ldots$, r_L from the centre of gravity of the lip to the contour, where L is the number of facial lips landmarks. This method was used to compute: *(e)* a set of the features r computed for left eye, right eye, left eyebrow, right eyebrow

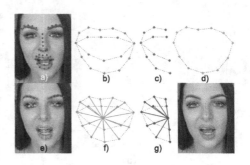

Fig. 5. Different choices of $\mathbf{s}_i^t$. The top row (from left to right): *(a)* full set of facial landmarks, *(b)* subset of (a) – inner and outer lips, *(c)* subset of (a) – left half of the lips, *(d)* subset of (a) – lip contours. The bottom row (from left to right): *(e)* features proposed in [10] computed for left and right eye, left and right eyebrow and lips landmarks, *(f)* subset of (e) – features computed from inner and outer lips, *(g)* subset of (e) – features computed from the left half of the lips, a full set of facial landmarks highlighted by green lines. (Color figure online)

and lips landmarks separately, *(f)* a subset of (e) containing only the lips features r, *(g)* a subset of (e) containing only the left half of the lips features r. The overview of how the different choices of $\mathbf{s}_i^t$ affected the average accuracy is given in Fig. 4(b). Particularly, the graph shows accuracies obtained in a 10-fold cross-validation when the parameters σ and λ_1 were fixed and the λ_2 parameter spanned. A grid search discussed in Sect. 5.2 was performed that included the search over different choices of $\mathbf{s}_i^t$. The best results were obtained for the concatenated x and y positions of the lips landmarks, i.e. for type *(b)*.

In Figs. 6(a), (b) and 7(a), the dependency of the mean accuracy on the parameters λ_1, λ_2 and σ is shown. Besides the average accuracy, Fig. 6(a) shows the number of the non-zero bins for different λ_1 settings. We see, that the highest average accuracy is obtained when $\lambda_1 = 10^{-3}$. For this particular choice, the ℓ_1 norm reduces the number of the non-zero bins to 99, effectively selecting the most informative centroids μ_k defined in (5). The representation is indeed learnt simultaneously with the classifier. Setting the λ_1 to some smaller values makes the ℓ_1 norm ineffective, i.e. we leave the regularization purely on the ℓ_2 norm, setting it to some high values leads to a too sparse weight vector producing a low classification accuracy on the validation set.

Figure 7(a) presents dependency of average accuracy on σ parameter. As σ represents the width of the soft-histogram bin defined in (4), setting its value too high leads to the bins too wide, spreading the representation of a single

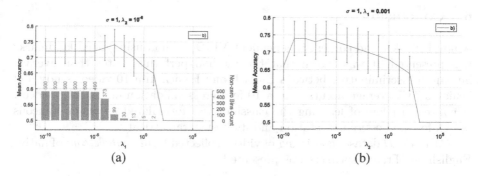

Fig. 6. (a) Dependency of average accuracy (orange plot) and the non-zero bins count (blue bars) on λ_1 for fixed σ and λ_2. The vertical bars represent the standard deviation. Cross-validation results. (b) Dependency of average accuracy on λ_2 for fixed σ and λ_1. The vertical bars represent the standard deviation. Cross-validation results. (Color figure online)

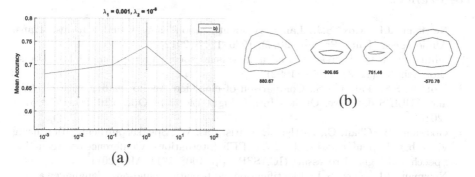

Fig. 7. (a) Dependency of average accuracy on σ for fixed λ_1 and λ_2. Vertical bars represent the standard deviation. Cross-validation results. (b) English (orange) and French (blue) centroids. The two shapes on the left are the centroids with the highest corresponding weight, the next two have the second highest corresponding weight. (Color figure online)

lip-shape across many bins. Setting it too low leads to the bins too thin, therefore only a small number of bins is activated for a given lip-shape. Both situations negatively affect the generalization capabilities of the representation, so a proper σ must be selected. In our case the highest average accuracy occurs for $\sigma = 1$.

Two centroids for each language with the highest corresponding weights are visualised in Fig. 7(b). These centroids have the highest impact on the classification. A tendency of English and French speakers to deform their lips in a particular way, visible also in Fig. 2, is captured. Notice, that both the top two English and French centroids represent opened and closed mouth, only in a reversed order. The most discriminative English lip-shape corresponds to a fully opened mouth slightly rotated to the side, while the French one reminds of the lips formed into a shape of a circle. This suggests that the typical English and French lip-shapes were found.

6   Conclusion

A novel speaker and content-independent VLID method using facial landmarks was presented. It was shown that the method performs better than average human in discriminating between English and French in a 10 videos study and achieves 73% average accuracy in a 10-fold cross-validation on realistic videos. A convex problem of learning the classifier was formulated as a simultaneous search for the best representation and its best parametrization.

CMP-VLID dataset consisting of videos collected from *youtube.com* of native English and French speakers was presented.

Acknowledgement. The research was supported by The Czech Science Foundation Project GACR P103/12/G084, by the project ERC-CZ LL1303 and by CTU student grant SGS17/185/OHK3/3T/13.

References

1. Newman, J.L., Cox, S.J.: Language identification using visual features. Trans. Audio Speech Lang. Process. **20**(7), 1936–1947 (2012)
2. Summerfield, Q.: Lipreading and audio-visual speech perception. Philos. Trans.: Biol. Sci. **335**(1273), 71–78 (1992)
3. Morade, S.S., Patnaik, S.: Comparison of classifiers for lip reading with CUAVE and TULIPS database. Optik - Int. J. Light Electron. Opt. **126**(24), 5753–5761 (2015)
4. Goswami, B., Chan, C., Kittler, J., Christmas, W.: Speaker authentication using video-based lip information. In: 2011 IEEE International Conference on Acoustics, Speech and Signal Processing (ICASSP), pp. 1908–1911, May 2011
5. Newman, J.L., Cox, S.J.: Identification, automatic visual-only language: a preliminary study. In: Proceedings of the 2009 IEEE International Conference on Acoustics, Speech and Signal Processing, ICASSP 2009, pp. 4345–4348. IEEE Computer Society, Washington, DC, USA (2009)
6. Newman, J.L., Cox, S.J.: Speaker independent visual-only language identification. In: 2010 IEEE International Conference on Acoustics, Speech and Signal Processing, pp. 5026–5029, March 2010
7. Chandrasekhar, V., Emre Sargin, M., Ross, D.A.: Automatic language identification in music videos with low level audio and visual features. In: 2011 IEEE International Conference on Acoustics, Speech and Signal Processing (ICASSP), pp. 5724–5727, May 2011
8. Sochman, J., Matas, J.: WaldBoost - learning for time constrained sequential detection. In: 2005 IEEE Computer Society Conference on Computer Vision and Pattern Recognition (CVPR 2005), vol. 2, pp. 150–156, June 2005
9. Xiong, X., De la Torre, F.: Supervised descent method and its applications to face alignment. In: 2013 IEEE Conference on Computer Vision and Pattern Recognition, pp. 532–539, June 2013
10. Saitoh, T., Konishi, R.: Lip reading based on sampled active contour model. In: Kamel, M., Campilho, A. (eds.) ICIAR 2005. LNCS, vol. 3656, pp. 507–515. Springer, Heidelberg (2005). doi:10.1007/11559573_63

HDR Imaging Pipeline for Spectral Filter Array Cameras

Jean-Baptiste Thomas[1,2](✉), Pierre-Jean Lapray[3], and Pierre Gouton[1]

[1] Le2i, FRE CNRS 2005, Université de Bourgogne, Franche-Comté, Dijon, France
[2] The Norwegian Colour and Visual Computing Laboratory,
NTNU - Norwegian University of Science and Technology, Gjøvik, Norway
jean.b.thomas@ntnu.no
[3] MIPS Laboratory, Université de Haute Alsace, Mulhouse, France

Abstract. Multispectral single shot imaging systems can benefit computer vision applications in needs of a compact and affordable imaging system. Spectral filter arrays technology meets the requirement, but can lead to artifacts due to inhomogeneous intensity levels between spectral channels due to filter manufacturing constraints, illumination and object properties. One solution to solve this problem is to use high dynamic range imaging techniques on these sensors. We define a spectral imaging pipeline that incorporates high dynamic range, demosaicing and color image visualization. Qualitative evaluation is based on real images captured with a prototype of spectral filter array sensor in the visible and near infrared.

Keywords: Multispectral imaging · Spectral filter arrays · High dynamic range · Imaging pipeline

1 Introduction

Spectral filter arrays (SFA) technology [22] provides a compact and affordable mean to acquire multispectral images (MSI). Such images have been proven to be useful in countless applications, but their extended use to general computer vision application was limited due to complexity of imaging set-up, calibration and specific imaging pipelines and processing. In addition, spectral videos are not easily handled either. SFA, however, is developed around a very similar imaging pipeline than color filter arrays (CFA), e.g. RGB, which is rather well understood and already implemented in many solutions. Indeed, SFA, similarly to CFA, is a spatio-spectral sampling of the scene captured in a single shot of a solid-state, single, image sensor. In this sense, SFA may provide a conceptual solution that improves vision systems.

Up to recently, only simulations of SFA camera were available, which made its experimental evaluation and validation difficult. Recent works on optical filters [8,29,45] in parallel to the development of SFA camera prototypes in the visible electromagnetic range [15], in the near infrared (NIR) [9] and in combined visible and NIR [20,43] permitted the commercialization of solutions, e.g. Imec [12], Silios [41], Pixelteq [31]. In addition, several color cameras include

© Springer International Publishing AG 2017
P. Sharma and F.M. Bianchi (Eds.): SCIA 2017, Part II, LNCS 10270, pp. 401–412, 2017.
DOI: 10.1007/978-3-319-59129-2_34

custom filter arrays that are in-between CFA and SFA (e.g. [13,28]). We could then consider that the use of SFA technology may reach a large scale of use soon after the development of standard imaging pipelines and drivers.

We address and demonstrate the imaging pipeline in this communication. One remaining limitation of SFA is to preserve the energy balance between channels [21,30] while capturing a scene. Indeed, due to the large number of filters and their spectral characteristics, *i.e.* narrow band sensitivities and inadequacy with the scene and illumination, or large inhomogeneity between filter shapes, it is frequent to observe one or several channels under- or over-exposed for a given integration time, which is common to all filters. This may be solved in theory by optimizing the filters before creating the sensor [21]. But filter realization is not yet flexible enough. Another way to solve this issue would be to develop sensors with by-pixel integration control. This is in development within some 3D silicon sensor concepts [5,16], but this technology is at its very beginning, despite of recent developments.

On the other hand, in gray-level and color imaging, the problem of under and over-exposure of parts of the scene is addressed by means of high dynamic range (HDR) imaging [6,24]. HDR imaging permits to potentially recover the radiance of the scene independently of the range of intensities present in the scene. As the dynamic range of a given sensor is limited, the quantization of the radiance values is a source of problems. The signal detection of very low intensity is limited by the dark noise. On the other hand, high intensities of the input signal can not be completely recovered and are sometimes voluntarily ignored (saturated pixels). To overcome these problems, a low exposure time image could be used to discretize the highest intensities, whereas a longer exposure time allows quantifying well relatively low light signals.

In an ideal configuration, an HDR image is simply obtained by bringing Low Dynamic Range (LDR) images in the same domain by dividing each image by its particular exposure time (normalization), and then by summing the corresponding pixel values. However, due to the effect of electronic circuits, most of the cameras have a non-linear processing regarding to the digitization of intensities into integer values. This non-linear transformation is materialized by the camera response function, denoted by $g(i)$, where i indexes the pixel value. It is assumed that this curve is monotonic and smooth. Some algorithms have been developed to recover this characteristic [6,27,35]. The most common method is the non-parametric technique from Debevec *et al.* [6]. For a given exposure time and intensity value, the relative radiance value is estimated by using $g(i)$ and a weighting function $\omega(i)$. Debevec *et al.* use a "hat" function as weighting function (see Fig. 6(b)), based on the assumption that mid-range pixels (values close to 128 for an 8-bit sensor) are the most reliable and the best exposed pixel for a given scene and integration time. In addition, recent advances have been done on the capture and processing of HDR video with low latency, using hardware-based platforms [18,19,25]. For HDR video, merging images captured at different times could lead to ghost artefacts when there are moving objects. This has been largely studied in recent years [1,3]. So, we argue that such methodology could be embedded in the SFA imaging pipeline without breaking the advantages of SFA technology for computer vision.

HDR multispectral acquisition is already treated by e.g. Brauers *et al.* [4] and Simon [39]. However, they consider the problem of HDR using individual bands acquired sequentially, so each band is treated independently. In the case of SFA, we may consider specific joined processes. Our contribution is to define the imaging pipeline to SFA and to provide results on real experimental data.

In this communication, we first generalize the imaging CFA pipeline to SFA. This new SFA imaging pipeline incorporates HDR concept based on multiple exposure images, in Sect. 2. Then, the experimental implementation is shown in Sect. 3, which is based on real images, acquired by a SFA system spanning the visible and NIR range [43]. Results are discussed in Sect. 4, before we conclude in Sect. 5.

2 Imaging Pipelines

In this section, we describe different imaging pipelines from CF to HDR-SFA.

2.1 CFA Imaging Pipeline

Several CFA imaging pipelines exist. We can classify them in two large groups: one concerns the hardware and real-time processing community [33,44,46], the other concerns the imaging community [14,32,38]. A very general distinction is that the former one often demosaics the raw image in very early steps, conversely the latter demosaics after or jointly with other processings such as white balance. In this work we base our design after the generic imaging pipeline defined by Ramanath *et al.* [32], which is shown in Fig. 1.

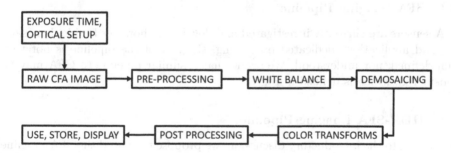

Fig. 1. CFA imaging pipeline similarly defined as in [32]. The pipeline contains pre-processing on raw data, which include for instance a dark noise correction and other denoising. Raw data would be corrected for illumination before to be demosaiced. Images are then projected into an adequate color space representation and followed by some post-processing, e.g. image enhancement, before to get out of the pipeline.

2.2 HDR-CFA Imaging Pipeline

HDR imaging has been developed mostly within monochromatic sensors for acquisition. However, there is a huge amount of work that developed the tone mapping of HDR color images for visualization, (e.g. [34,40]), the HDR capture is mostly an intensity process performed per channel [39]. We propose to encapsulate a general HDR-CFA imaging pipeline such as shown in Fig. 2. This pipeline is based on sequence of images of the same scene having different integration times. We argue that a HDR pipeline may have two distinguished output: one leads to HDR radiance images, which can be stored and used for automatic applications. The other leads to a display-friendly visualization of color image. Note that the two outputs may overlap in specific applications.

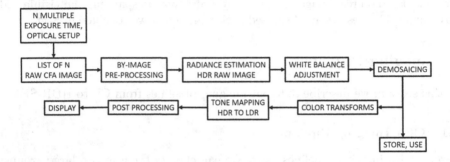

Fig. 2. HDR-CFA imaging pipeline. In this case, the denoising is typically performed per image similarly to the LDR-CFA case. Then, radiance estimation is performed based on the multiple images, providing radiance raw images. White balancing and demosaicing is performed on this data. Then, the HDR image may be used as is, or continue into a visualization pipeline, where a color transform, tone-mapping and image-enhancement may be applied before visualization.

2.3   SFA Imaging Pipeline

SFA sensors are currently investigated and developed, however beside demosaicing and applications dedicated processing, the rest of the pipeline is not very well defined, nor understood. We argue that a similar pipeline to CFA may be considered, which is defined in Fig. 3.

2.4   HDR-SFA Imaging Pipeline

According to the introductory discussion, we propose to extend the SFA pipeline to an HDR version in order to benefit from HDR, in particular towards a better balance between channel sensitivities. We propose to consider the raw image and to treat it as a gray-level image for relative radiance estimation. Thus, we perform all radiance reconstruction prior to any separation between bands. The pipeline is defined in Fig. 4.

3   Implementation of the HDR-SFA Imaging Pipeline

This section explicitly defines which processing is embedded in each of the pipeline boxes. We take well-established and understood methods from the state

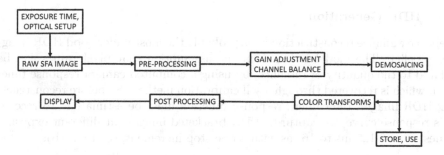

Fig. 3. SFA imaging pipeline. At the instar of CFA, this pipeline defines some illumination discarding process and demosaicing. The spectral image would be typically used for application after demosaicing. However, these data may not be observable as they are, so the pipeline is prolonged for visualization. The color transform is ought to be slightly different than CFAs, for several more channel and NIR information may be present in the spectral image.

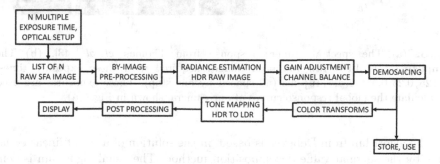

Fig. 4. HDR-SFA imaging pipeline. The radiance estimation is performed on the raw image taken as a whole and not per channel. This leads to a raw HDR image, which may be corrected for illumination and demosaiced. Then, the HDR multispectral image may be stored or used. The visualization process projects the data into a HDR color representation of the data, which is tone-mapped and processed for visualization.

of the art in order to provide benchmarking proposal and analysis. These methods are combined into the pipeline. Our proposal is not exclusive in the sense that any method may be used and different orders may also be considered.

The prototype SFA camera from Thomas *et al.* [43] is used in this study. Sensitivities are shown in Fig. 5(a). Spatial arrangement and other details may be found in their article. The raw images are pre-processed and denoised according to what is performed in this article, which is basically a dark noise removal. Then, following the pipeline, HDR data are computed. Subsection 3.1 covers the HDR data recovering. HDR images are demosaiced, according to Miao *et al.* [26] algorithm, forming the full resolution HDR multispectral image. The part of the pipeline that concerns visualization is developed in Subsect. 3.2.

3.1   HDR Generation

Debevec radiance reconstruction [6] is probably the most understood HDR imaging pipeline. The model is based on the assumption that pixel values can be related to the quantity of radiance, by using a computed camera response function, which is recovered through a self calibration method. So before reconstructing HDR images, the camera response function must be estimated. To recover this response curve, we capture 8 LDR bracketed images[1] at different exposure times, from 0.125 ms to 16 ms with a one-stop increment, see Fig. 5(b).

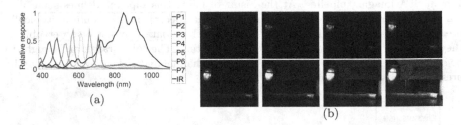

Fig. 5. (a) The spectral camera response from Thomas *et al.* [43]. (b) The complete set of LDR raw mosaiced images acquired with these exposure times: $\{0.125, 0.25, 0.5, 1, 2, 4, 8, 16\}$ ms (all spaced by one stop). These exposures are used to calculate the global response curves of our camera, shown in Fig. 6(a).

The algorithm from Debevec is based on the solution of a set of linear equations by the singular value decomposition method. The usual algorithm is generally applied on RGB cameras, and it recovers 3 different response curves, one by channel. In our case, as we have 8 spectral channels, we recover 8 curves (see Fig. 6(a)). We notice that the dispersion is relatively low among channels, so in the following, we use the median value of these curves for all channels, allowing us to work directly on the raw data at once to generate HDR values.

As described in the pipeline in Fig. 4, we recover relative radiance values directly from preprocessed data (called "raw data"). A number of 3 exposure times are selected. We chose only 3 exposures because it is a number commonly used in the literature [3, 25], as it gives relatively high dynamic range and not too much ghost effects. The radiance values are recovered using the response curve, and by combining the pixel value with the corresponding exposure time (c.f. Debevec equation [6]). A weighted sum of radiance values among all exposure times is done using the hat weighting function (see Fig. 6(b)), to give more contribution to mid range pixel intensities during the HDR reconstruction. The single raw-HDR image is then demosaiced to recover the whole spatial resolution for each HDR band. We have obtained a HDR multispectral image.

---

[1] We could work with three images as later in this work, but it is commonly accepted that using more images than necessary can lead to a better response curve estimation in terms of robustness to noise.

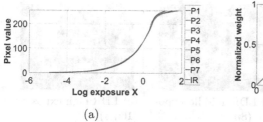 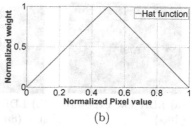

(a) (b)

Fig. 6. (a) Response functions recovered from the image set shown in Fig. 5(b) for each of the bands $P_{1-7,IR}$. (b) The well-exposedness hat function used in our experiment.

3.2 Visualization Procedure

HDR spectral data are projected into an HDR coded CIEXYZ color space according to a color transform based on the 24 Gretag Macbeth color checker patches reflectance and the scene acquisition illumination measured in situ. This colorimetric image may be either transformed into sRGB directly or tone mapped by a more efficient algorithm. In the following, we use two tone mapping as examples, one is a global logarithmic mapping from Duan *et al.* [7], the other is from Krawczyk *et al.* [17], and combines global and local tone mapping processing[2].

3.3 Other Pipeline Components

One obvious gain adjustment would be defined by the sensitivity curves efficiency ratio. Some works are also initiated toward spectral constancy [42]. These works are under development, and for benchmarking purpose, we decided to keep the ratio untouched in this study.

Post-processing, except the tone-mapping, are avoided in this article, but may consider spectral demultiplexing [36,37], image enhancement such as ghost removal [10] or other items.

Spectral LDR data are stored into multiband Tiff files. Spectral HDR data are stored using 32-bit Tiff files (8 channels). To provide a direct visualization with a great number of software implementing tone-mapping solutions, RGB HDR data are also computed, encoded within ".hdr" RGBE [23] format.

We assume color data to be displayed onto a 8 bit display. The pipeline may be adapted to the new generation of HDR displays that rose up on the market.

4 Results

We provide examples of resulting images for two scenes in Figs. 7 and 8. Due to space constraint, the description is embedded into the captions of the figures.

---

[2] We used the code which is implemented in the Matlab HDR Toolbox [2].

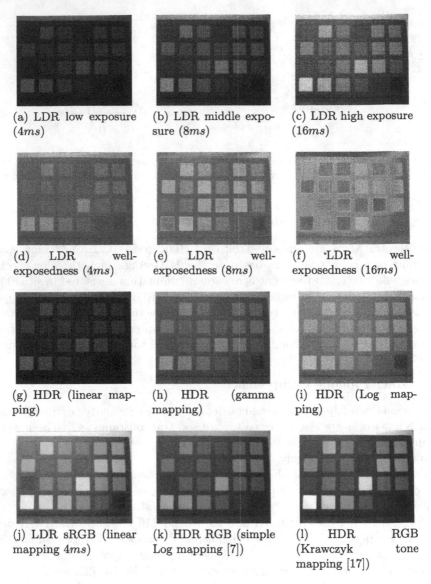

(a) LDR low exposure (4*ms*)

(b) LDR middle exposure (8*ms*)

(c) LDR high exposure (16*ms*)

(d) LDR well-exposedness (4*ms*)

(e) LDR well-exposedness (8*ms*)

(f) ˙LDR well-exposedness (16*ms*)

(g) HDR (linear mapping)

(h) HDR (gamma mapping)

(i) HDR (Log mapping)

(j) LDR sRGB (linear mapping 4*ms*)

(k) HDR RGB (simple Log mapping [7])

(l) HDR RGB (Krawczyk tone mapping [17])

Fig. 7. Results for the MacBeth color chart (typically a low dynamic range scene). (a,b,c) Raw images, (d,e,f) false color well-exposedness representation, (g,h,i) HDR generation results from the 3 exposure set (a,b,c), (j,k,l) sRGB representation of both LDR and HDR data for comparison. The important issue, that we try to solve in this work, is that if we look the raw images at a neutral MacBeth patch, we can clearly distinguish the inherent energy balance problems between pixel values through the 8 channels. This phenomenon is highlighted in Fig. 7(d), (e) and (f), where a pixel position could hold a good intensity for a given exposure time (red color), and a bad exposition in another (blue color). It leads to visual noise when visualizing a single LDR reconstructed image (Fig. 7(j)). Our HDR-SFA imaging pipeline can correct this problem by a certain amount, which is visually appreciated. (Color figure online)

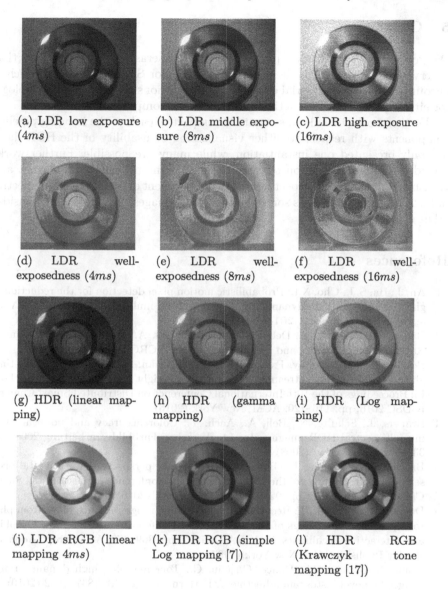

(a) LDR low exposure (4*ms*)

(b) LDR middle exposure (8*ms*)

(c) LDR high exposure (16*ms*)

(d) LDR well-exposedness (4*ms*)

(e) LDR well-exposedness (8*ms*)

(f) LDR well-exposedness (16*ms*)

(g) HDR (linear mapping)

(h) HDR (gamma mapping)

(i) HDR (Log mapping)

(j) LDR sRGB (linear mapping 4*ms*)

(k) HDR RGB (simple Log mapping [7])

(l) HDR RGB (Krawczyk tone mapping [17])

Fig. 8. Results for the CD scene (relatively high dynamic range scene compared to the previous MacBeth scene). (a,b,c) Raw images, (d,e,f) false color well-exposedness representation, (g,h,i) HDR generation results from the 3 exposure set (a,b,c), (j,k,l) sRGB respresentation of both LDR and HDR data for comparison. As for Fig. 7, we show that channels are inequitably affected by signal noise among exposures. Showing different good or bad pixel intensities, there is the necessity to take several exposure times in order to equalize the distribution of noise between channels. Contrary to the MacBeth scene, which is a typical low dynamic range scene, we can see in addition that some areas with high specular reflection have less saturated pixels in the resulting HDR image. We still observe saturated pixels at specular directions, which saturates the lowest integration time.

5   Conclusion

We generalized the imaging pipeline to SFA cameras. We demonstrated that a very similar to CFA architecture can be used for SFA successfully, which is encouraging for the industrial development of solutions based on this technology, for either spectral reconstruction and traditional computer vision tasks.

Further works include to evaluate the impact of each of the imaging pipeline components with respect to either visualization or usability of the HDR data. We only presented one instantiation, while many are possible. Further works include also standardization of camera and pipeline as well as file format and transmission line. Other aspect lies in the development of quality of HDR spectral data. Although there exists some work for HDR images [11], little work consider HDR spectral data.

References

1. An, J., Ha, S.J., Cho, N.I.: Probabilistic motion pixel detection for the reduction of ghost artifacts in high dynamic range images from multiple exposures. EURASIP J. Image Video Process. **2014**(1), 42 (2014)
2. Banterle, F., Artusi, A., Debattista, K., Chalmers, A.: Advanced High Dynamic Range Imaging: Theory and Practice. AK Peters (CRC Press), Natick (2011)
3. Bouderbane, M., Lapray, P.J., Dubois, J., Heyrman, B., Ginhac, D.: Real-time ghost free HDR video stream generation using weight adaptation based method. In: Proceedings of the 10th International Conference on Distributed Smart Camera, ICDSC 2016, pp. 116–120. ACM, New York (2016)
4. Brauers, J., Schulte, N., Bell, A., Aach, T.: Color accuracy and noise analysis in multispectral HDR imaging. In: 14. Workshop Farbbildverarbeitung 2008, pp. 33–42. Shaker Verlag (2008)
5. Brochard, N., Nebhen, J., Ginhac, D.: 3D-IC: new perspectives for a digital pixel sensor. In: Proceedings of the 10th International Conference on Distributed Smart Camera, ICDSC 2016, pp. 92–97. ACM, New York (2016)
6. Debevec, P.E., Malik, J.: Recovering high dynamic range radiance maps from photographs. In: Proceedings of the 24th Annual Conference on Computer Graphics and Interactive Techniques, SIGGRAPH 1997, pp. 369–378. ACM Press/Addison-Wesley Publishing Co., New York (1997)
7. Duan, J., Bressan, M., Dance, C., Qiu, G.: Tone-mapping high dynamic range images by novel histogram adjustment. Pattern Recogn. **43**, 1847–1862 (2010)
8. Eichenholz, J.M., Dougherty, J.: Ultracompact fully integrated megapixel multispectral imager. In: SPIE OPTO: Integrated Optoelectronic Devices, p. 721814. International Society for Optics and Photonics (2009)
9. Geelen, B., Blanch, C., Gonzalez, P., Tack, N., Lambrechts, A.: A tiny VIS-NIR snapshot multispectral camera. In: SPIE OPTO, p. 937414. International Society for Optics and Photonics (2015)
10. Granados, M., Kim, K.I., Tompkin, J., Theobalt, C.: Automatic noise modeling for ghost-free HDR reconstruction. ACM Trans. Graph. (TOG) **32**(6), 201 (2013)
11. Hanhart, P., Bernardo, M.V., Pereira, M., Pinheiro, A.M.G., Ebrahimi, T.: Benchmarking of objective quality metrics for HDR image quality assessment. EURASIP J. Image Video Process. **2015**(1), 39 (2015)

12. IMEC: hyperspectral-imaging. http://www2.imec.be
13. Jia, J., Barnard, K.J., Hirakawa, K.: Fourier spectral filter array for optimal multispectral imaging. IEEE Trans. Image Process. **25**(4), 1530–1543 (2016)
14. Kao, W.C., Wang, S.H., Chen, L.Y., Lin, S.Y.: Design considerations of color image processing pipeline for digital cameras. IEEE Trans. Consum. Electron. **52**(4), 1144–1152 (2006)
15. Kiku, D., Monno, Y., Tanaka, M., Okutomi, M.: Simultaneous capturing of RGB and additional band images using hybrid color filter array. In: Proceedings of the SPIE, vol. 9023, pp. 90230V–90230V-9 (2014)
16. Knickerbocker, J.U., Andry, P., Dang, B., Horton, R., Patel, C.S., Polastre, R., Sakuma, K., Sprogis, E., Tsang, C., Webb, B., et al.: 3D silicon integration. In: 2008 58th Electronic Components and Technology Conference, pp. 538–543. IEEE (2008)
17. Krawczyk, G., Myszkowski, K., Seidel, H.P.: Lightness perception in tone reproduction for high dynamic range images. Comput. Graph. Forum **24**, 635–645 (2005). Wiley Online Library
18. Lapray, P.J., Heyrman, B., Ginhac, D.: HDR-ARtiSt: an adaptive real-time smart camera for high dynamic range imaging. J. Real-Time Image Process., 1–16 (2014)
19. Lapray, P.J., Heyrman, B., Ginhac, D.: Hardware-based smart camera for recovering high dynamic range video from multiple exposures. Opt. Eng. **53**(10), 102–110 (2014)
20. Lapray, P.J., Thomas, J.B., Gouton, P.: A multispectral acquisition system using MSFAs. In: Color and Imaging Conference, vol. 2014(2014), pp. 97–102 (2014-11-03T00: 00: 00)
21. Lapray, P.J., Thomas, J.B., Gouton, P., Ruichek, Y.: Energy balance in spectral filter array camera design. J. Eur. Opt. Soc. (2017)
22. Lapray, P.J., Wang, X., Thomas, J.B., Gouton, P.: Multispectral filter arrays: recent advances and practical implementation. Sensors **14**(11), 21626 (2014)
23. Larson, G.W., Shakespeare, R.: Rendering with Radiance: The Art and Science of Lighting Visualization. Booksurge LLC, New York (2004)
24. Mann, S., Picard, R.: Being undigital with digital cameras: extending dynamic range by combining differently exposed pictures. In: Proceedings of IS&T 46th Annual Conference, pp. 422–428 (1995)
25. Mann, S., Lo, R.C.H., Ovtcharov, K., Gu, S., Dai, D., Ngan, C., Ai, T.: Realtime HDR (high dynamic range) video for eyetap wearable computers, FPGA-based seeing aids, and glasseyes (eyetaps). In: 2012 25th IEEE Canadian Conference on Electrical & Computer Engineering (CCECE), pp. 1–6. IEEE (2012)
26. Miao, L., Qi, H., Ramanath, R., Snyder, W.E.: Binary tree-based generic demosaicking algorithm for multispectral filter arrays. IEEE Trans. Image Process. **15**(11), 3550–3558 (2006)
27. Mitsunaga, T., Nayar, S.K.: Radiometric self calibration. In: 1999 IEEE Computer Society Conference on CVPR, vol. 1, p. 380 (1999)
28. Monno, Y., Kikuchi, S., Tanaka, M., Okutomi, M.: A practical one-shot multispectral imaging system using a single image sensor. IEEE Trans. Image Process. **24**(10), 3048–3059 (2015)
29. Park, H., Dan, Y., Seo, K., Yu, Y.J., Duane, P.K., Wober, M., Crozier, K.B.: Vertical silicon nanowire photodetectors: Spectral sensitivity via nanowire radius. In: CLEO: Science and Innovations, p. CTh3L-5. OSA (2013)
30. Péguillet, H., Thomas, J.B., Gouton, P., Ruichek, Y.: Energy balance in single exposure multispectral sensors. In: CVCS 2013, pp. 1–6, September 2013

31. PIXELTEQ: Micro-patterned optical filters. https://pixelteq.com/
32. Ramanath, R., Snyder, W.E., Yoo, Y., Drew, M.S.: Color image processing pipeline. IEEE Signal Process. Mag. **22**(1), 34–43 (2005)
33. Rani, K.S., Hans, W.J.: FPGA implementation of bilinear interpolation algorithm for CFA demosaicing. In: 2013 International Conference on Communications and Signal Processing (ICCSP), pp. 857–863. IEEE (2013)
34. Reinhard, E., Stark, M., Shirley, P., Ferwerda, J.: Photographic tone reproduction for digital images. ACM Trans. Graph. **21**(3), 267–276 (2002)
35. Robertson, M.A., Borman, S., Stevenson, R.L.: Dynamic range improvement through multiple exposures. In: Proceedings of the 1999 International Conference on Image Processing, ICIP 1999, vol. 3, pp. 159–163. IEEE (1999)
36. Sadeghipoor, Z., Lu, Y.M., Mendez, E., Ssstrunk, S.: Multiscale guided deblurring: chromatic aberration correction in color and near-infrared imaging. In: 2015 23rd European Signal Processing Conference (EUSIPCO), pp. 2336–2340, August 2015
37. Sadeghipoor, Z., Thomas, J.B., Susstrunk, S.: Demultiplexing visible and near-infrared information in single-sensor multispectral imaging. In: Color and Imaging Conference 2016, pp. 76–81 (2016–11–03T00: 00: 00). http://ist.publisher.ingentaconnect.com/contentone/ist/cic/2016/00002016/00000001/art00012
38. Sharma, G., Trussell, H.J.: Digital color imaging. IEEE Trans. Image Process. **6**(7), 901–932 (1997)
39. Simon, P.M.: Single shot high dynamic range and multispectral imaging based on properties of color filter arrays. Ph.D. thesis, University of Dayton (2011)
40. Tamburrino, D., Alleysson, D., Meylan, L., Süsstrunk, S.: Digital camera workflow for high dynamic range images using a model of retinal processing. In: Electronic Imaging 2008, p. 68170J. International Society for Optics and Photonics (2008)
41. Technologies, S.: Micro-optics supplier. http://www.silios.com/
42. Thomas, J.B.: Illuminant estimation from uncalibrated multispectral images. In: CVCS 2015, pp. 1–6. IEEE (2015)
43. Thomas, J.B., Lapray, P.J., Gouton, P., Clerc, C.: Spectral characterization of a prototype SFA camera for joint visible and NIR acquisition. Sensors **16**(7), 993 (2016)
44. Tsin, Y., Ramesh, V., Kanade, T.: Statistical calibration of CCD imaging process. In: Proceedings of the Eighth IEEE International Conference on Computer Vision, ICCV 2001, vol. 1, pp. 480–487. IEEE (2001)
45. Yi, D., Kong, L., Wang, J., Zhao, F.: Fabrication of multispectral imaging technology driven MEMS-based micro-arrayed multichannel optical filter mosaic. In: SPIE MOEMS-MEMS, p. 792711. International Society for Optics and Photonics (2011)
46. Zhou, J.: Getting the most out of your image-processing pipeline. White Paper, Texas Instruments (2007)

Thistle Detection

Søren I. Olsen[✉], Jon Nielsen, and Jesper Rasmussen

University of Copenhagen, Copenhagen, Denmark
ingvor@di.ku.dk

Abstract. Detecting green thistles in yellow mature cereals seems an easy task. In practice however, the two colors may be close and changing over time and place. The total thistle area may be negligible, illumination effects at low sun angles etc. may make a stable detection less easy. This paper describes a method *Weeddetect* for detecting thistles in drone images of fields with mature wheat. Preliminary experiments indicate a classification accuracy of about 95%.

1 Introduction

Pesticide regulations and a relatively new EU directive [5] on integrated pest management create strong incentives to limit herbicide applications. In Denmark, several pesticide action plans have been launched since the late 1980 s with the aim to reduce herbicide use [12]. One way to reduce the herbicide use is to apply site-specific weed management, which is an option when weeds are located in patches, rather than spread uniformly over the field. Site-specific weed management can effectively reduce herbicide use, since herbicides are only applied to parts of the field [2]. This requires reliable remote sensing and sprayers with individual controllable boom sections or a series of controllable nozzles that enable spatially variable applications of herbicides. Preliminary analyses [14] indicate that the amount of herbicide use for pre-harvest thistle (Cirsium arvense) control with glyphosate can be reduced by at least 60% and that a reduction of 80% is within reach. The problem is to generate reliable and cost-effective maps of the weed patches. One approach is to use user-friendly drones equipped with RGB cameras as the basis for image analysis and mapping.

The use of drones as acquisition platform has the advantage of being cheap, hence allowing the farmers to invest in the technology. Also, images of sufficiently high resolution may be obtained from an altitude allowing a complete coverage of a normal sized Danish field in one flight. Importantly, thistles shorter than the cereals may not be visible when viewed obliquely from 2–3 m height, but is clearly visible from above.

To differentiate between different types of weed in drone images, pure color based analysis probably will be insufficient. Here low altitude (high resolution) images allowing leaf shape analysis will be needed. However, since the dominating weed type is thistles, and since these may be detected from larger altitude allowing mapping of a larger area, we will here use the terms *weed* and *thistles* interchangeably and will not try to classify the detections into the different types of weed.

P. Sharma and F.M. Bianchi (Eds.): SCIA 2017, Part II, LNCS 10270, pp. 413–425, 2017.
DOI: 10.1007/978-3-319-59129-2_35

The objective of this paper is to demonstrate that single-image weed- (and in particular thistle-) detection in drone images of cereal fields is possible. The paper includes a short review of previous work, a description of the proposed method, dubbed *Weeddetect*, results of preliminary experiments and suggestions for further work.

2   Previous Work

The amount of previous work on color image processing is enormous. A few relevant textbooks include [6, 7, 10]. Less work exist on drone image analysis of field images. The most relevant contributions include [4, 8, 9, 11, 14]. The latter is directly relevant for the present work. Here each image is split into 64×64 pixel patches, that are converted to a scalar value using the *Excess green-* projection [13] $ExG = 2G - R - B$. Then, the average of the 5% largest ExG-values within the patch is computed and thresholded by a fixed parameter value learned from a set of annotated images. It is shown [14] that this method may achieve a classification accuracy above 95% on many images, but also that a number of problems exist. Most importantly, optimal performance could not be reached without adjusting the threshold parameter to each image capture campaign. This is because of different crop and light conditions makes automatic usage problematic. For some images, far too many or far too few patches were classified correctly as weed. To a large degree this seems to be caused by lack of preprocessing, but also the ExG-projection and the usage of a fixed threshold plays a role.

In [3] texture analysis was attempted in addition to color analysis. Texture analysis may seem useful because cereals and the thistle leaves are different in shape. However, texture based detection is inherently difficult no matter the approach, because even for small sized image patches, the distribution of weed features and cereal features mix. To obtain stable statistics, larger image patches is required but here the fraction of visible weed may be very small. Experiments [3] showed that detection was possible but unstable and less accurate compared to the color-based procedure reviewed above. Also, texture analysis was far more computationally demanding.

In the present work both pre- and post processing is addressed, but the major contribution of this work is a procedure for robustly estimating an optimal threshold for possibly unimodal distributions composed of a cereal and a weed component. Also, a new projection vector, significantly outperforming previous choices are presented.

3   Weed Detection

Previous work [3, 4] shows that simple approaches may work remarkably well. However, too often such approaches fail due to lack of robustness. One practical problem is that change of camera or use of auto-white balancing at image capture changes the color balance. This makes the images more grayish and reduces the difference between yellow (cereals) and green (thistles). In general, the colors change with the lighting conditions. For color based weed detection the more saturated colors, the easier the detection will be. Another problem arises when the angle between the sun and the horizon is below about 60°. In this case, because the view is approximately ortho, the image part opposite to the sun azimuthal angle will show the illuminated side of the flora, while the image part towards the sun will be darker. This effect clearly depend on the shape of the crop leaves, the viewing direction etc. If not corrected for, too many thistles in the affected image area will not be detected.

Depending on the resolution of the spraying equipment, the detections need to be resampled. For an altitude of 50 m and using a standard commercial camera with fixed standard optics, the pixels side length may correspond to a few centimeters or less at ground. This resolution is far too high to be used for sprayers. Typically, the controllable spraying resolution is about 3 m. The resampling should be made accordingly taking altitude and effective focal length into account. Finally, the farmer/user should be able to specify the trade off between minimizing the risk of false positives (causing too much spraying) and false negatives (causing too little spraying).

Previous attempts [3,4] within thistle detection have used a fixed threshold to separate green from yellow. Despite all preprocessing such as color balancing, such non-adaptive approach may not be sufficiently robust and an adaptive procedure for threshold selection is needed.

3.1   Preprocessing

To remedy an undesirable color balance we follow the approach of Cheng et al. [1]. However, instead of adjusting to a white-balance, i.e. the maximum vector $(1,1,1)$ we use the vector $(1,0.9,0.7)$ reflecting that the images show yellow and green colors and vey little blue. Colors in images obtained using auto white balance are restored reasonably well (in the sense that the resulting color distribution becomes similar to what is seen at other image captures). To a large extent, this procedure replaces a more tedious camera color calibration prior to image capture.

To eliminate the effect of uneven illumination obtained when the angle between the horizon and the sun is not large, we first transform the representation to HSV. Here, the effect is concentrated both in the saturation component and the value component, whereas the hue component is hardly affected. The shape and position of the effect depends on the crop type, the sun height and azimutal position, the camera tilt, and would probably be difficult to estimate. We choose to correct both saturation S and value V using a linear fit to each image, i.e.:

$$S(x,y) \longrightarrow \frac{\bar{S}}{a_s + b_s x + c_s y} S(x,y) \quad \text{and} \quad V(x,y) \longrightarrow \frac{\bar{V}}{a_v + b_v x + c_v y} V(x,y) \quad (1)$$

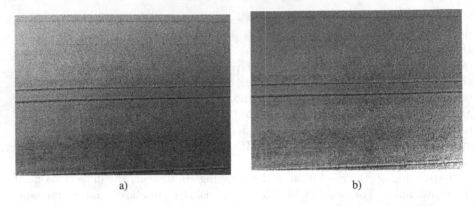

a)   b)

Fig. 1. (a) An original image with uneven illumination. (b) Compensated illumination.

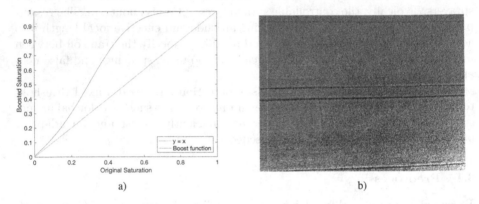

a)   b)

Fig. 2. (a) Saturation boosting function. (b) Corrected and saturation-boosted image (Color figure online)

where $\bar{S}$ and $\bar{V}$ are global averages of saturation and value, and where the denominators show planar fits to the local averages of saturation and value. This simple approach is robust to situations where spots (e.g. of flowers) make the local average deviate significantly from the fit. More complex models (than a planar fit) have shown too sensitive to such cases. An example of a correction is shown in Fig. 1.

To further ease the following separation of the green and the yellow, the saturation values are nonlinearly increased. This is illustrated in Fig. 2. Low values are left almost unchanged whereas medium and in particular large values are increased towards the maximal value.

3.2   Detection

Based on a large number of annotated areas, the preprocessed images were examined and verified to split the color space linearly into two slightly overlapping subspaces. This is illustrated in Fig. 3(a) for a random subset of training points. The best fitting plane separating the two populations had a unit surface normal of $(-0.609, 0.773, -0.178)$. The best separating plane vary from image to image but all had approximately the same normal. Compared to [14] this vector is significantly different from the normalized ExG-vector, putting much less weight on the blue color component. By using the normal to the best separating plane as projection vector, the following threshold estimation is eased.

Separation of the projected (smaller) yellow cereal values from the larger green thistle values is difficult because the amount of the latter is usually only a tiny fraction. Thus, for almost all images the distribution of projected values is unimodal. Often a visual inspection will not reveal the thistle component. Bimodality exceptionally shows up when large image areas are covered with weeds, grass or trees.

If the cereal pixel projection values are modelled as i.i.d. random values, then the central limit theorem suggests that the yellow cereal distribution component should be Normal. Visually, this is confirmed. However, both the mean value and the variance vary from image to image. Also for images with large amount of green plants, the corresponding distribution component interferes with the right part of the normal cereal distribution. This suggest an approach where first the parameters of the Normal cereal distribution is estimated, then the weed distribution is defined as the right residual, and the threshold is selected as the one minimizing the sum of false positives and false negatives. This is illustrated in Fig. 3(b). The procedure is fast because the binning of the histograms may be chosen to limit the number of possible threshold values allowing an exhaustive search.

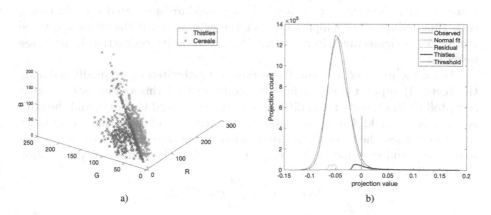

a)   b)

Fig. 3. (a) Projection in red of cereal- and in green of thistle- RGB-values. The two classes are almost linearly separable. (b) A typical distribution of projected values. The black curve is the measured distribution. The red curve is the estimated cereal Normal distribution. The curve in green and blue show the estimated left and right residual weed distribution. The vertical cyan line marks the optimal threshold. (Color figure online)

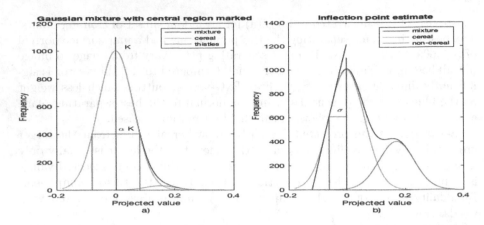

Fig. 4. (a) Estimation based on the central area marked in magenta. Used when the non-cereal distribution shown in red only affects the right tail of the distribution. (b) Estimation based on the position of the left inflection point and the tangent here. Used when the middle and right part of the distribution seriously deviates from a normal distribution.

The challenge is to estimate the mean and variance of the (left) cereal distribution not knowing how severely the right part is corrupted by other components. First, the contribution of the latter distribution component is assumed to be minor. In this case, the mean is estimated from the central data around the mode and with value larger than α (say 0.4) times the mode value. This support area is illustrated in Fig. 4(a). If the mean deviates much from the mode, α is increased and the mean reestimated. This procedure is iterated up to 3 times, each time reducing the support area for the estimation. If the estimated value still deviates significantly from the mode an alternative estimation is used (see below).

As for the mean estimation, the variance is estimated by gradually reducing the (central) support area around the mode up to 3 times. The estimation is accepted if the absolute mean difference of the estimated Gaussian and the data is below a threshold. One problem is that the accuracy of the estimated variance is reduced when the tails of the distribution is discarded—The variance is systematically underestimated. It is easy to show that the variance compensation factor will be:

$$\gamma(\alpha) = \sqrt{2\pi} \int_{-\beta}^{\beta} t^2 e^{-t^2/2} dt$$

where $\beta = \sqrt{-2\ln(\alpha)}$. Thus, γ is independent on the standard deviation of the true distribution. Other expressions for the variance of asymmetric or one-sided truncated normals may be found in [15]. The expression for γ has no closed form solution, but is (for a limited number of values of α) estimated from simulations.

If the estimation of either the mean value or the standard deviation fails, an alternative estimation procedure is attempted. This is applied when a large fraction of the (right part of the) normal cereal distribution is mixed with the distributions of non-cereals. As illustrated in Fig. 4(b), here, only the left part of the empirical distribution is used. After smoothing, the left inflection point I is localized. From the value $v(I)$ in this point and an estimate of the derivative $d(I)$, the variance, and then the mean value may easily be computed by $\hat{\sigma} = v(I)/d(I)$ and $\hat{\mu} = I + v(I)/d(I)$. Since the latter procedure is based on far fewer data, it is invoked as a last resort. Preliminary experiments indicate a robust, but not very accurate estimation. In the experiments reported later the alternative estimation was not in use because all images (with ground truth) only showed cereals and thistles.

3.3 Postprocessing

The result of the pixel-based classification is not appropriate for thistle mapping. A large number of tiny (few pixels large) green areas often may be incorrectly registered as thistles. Mostly, the tiny areas are caused by ground vegetation (most often within the central image area). To eliminate the small detections a morphological opening is applied. This significantly reduces the number of segments even though the structure element is chosen small.

Finally, pixels based detection is not really of use for glyphosate spraying because the image resolution is far higher than the resolution of the sprayer. First, the image patch size in pixels corresponding to the desired resolution in meters is computed. Then, to avoid aliasing and missing detections, a windowed sum of pixel detections within the final resolution map is computed and a final acceptance threshold T applied. This threshold is not determined based on computer science or math, but on agricultural objectives. It specifies the farmers' trade-off between risking not to spray thistles versus unnecessary spraying. Figure 5 shows an example of an original image, the pixel-based classification and the final thistle map.

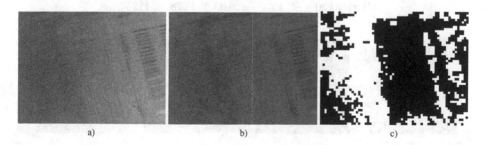

a)   b)   c)

Fig. 5. (a) An original image. (b) Initial pixel based detection. (c) Final $1\,\mathrm{m}^2$-resolution detection with thistles areas marked as white.

4   Experimental Results

To measure the performance of the system, dubbed *Weeddetect*, 97 field images of
mature wheat and 101 field images of mature barley was taken from altitudes of
20, 30 and 50 m. Based on a previous program called *ThistleTool* [14] a number
of image patches were extracted. All patches showed a field area of 3 × 3 m
(corresponding to side lengths of 450, 300 and 180 pixels). Approximately half
the patches showed crop, the remaining thistles. For each patch the position of
the central 1 × 1 m sub-patch, used for performance measurement, was recorded.
The patches were presented to an agricultural expert and classified as showing
either weed or only crop. This forms the basic ground truth. See Table 1 for
details.

In Fig. 6 two patches classified as crop and two patches classified as weed
are shown. Two of the patches are easy to classify (expert or not), while the
remaining two less clearly belong to either of the classes. Even for an expert
in agriculture there is a significant amount of patches where the ground truth
classification is debatable. In addition to the data specified above, the extracted
patches for the 19 images of Wheat observed using an altitude of 50 m was reclas-
sified 3 times by the same expert from agriculture. For the first reclassification
the expert was told to be as liberal in his classification into weed as for the basic
classification. For the second and third reclassification he was told to classify

Table 1. Amount of data used for testing including the manual classification of the
automatically extracted image patches

Type	Height	Images	Patches	Weed	Crop
Wheat	50	19	950	475	475
Wheat	30	39	1574	599	975
Wheat	20	39	1638	663	975
Barley	50	18	900	450	450
Barley	30	38	1900	950	950
Barley	20	45	2103	978	1125

Fig. 6. The leftmost two images patches are classified as crop where the rightmost
two are classified as weed. The classification of the middle two patches is debatable.
The central area used for performance evaluation is indicated by the small magenta
markers. (Color figure online)

Table 2. Patch classification distribution for initial classification and 3 reclassifications of the image patches extracted from the images of Wheat from 50 m altitude.

Classification	Weed	Crop
Basic Liberal	481	419
Liberal 2	494	406
Slightly Conservative	317	583
Conservative	320	580

Table 3. Performance on two types of crop and three altitudes. The columns marked *Within sample* mark the raw results. The columns marked *Estimated total* show the performance after normalization with the true amount of weed seen in the images.

Crop	Height	Estimated total			Within sample		
		TPR	SPC	ACC	TPR	SPC	ACC
Wheat	20	.954	.956	.955	.984	.890	.933
Wheat	30	.992	.929	.943	.997	.948	.912
Wheat	50	.938	.965	.960	.990	.866	.928
Barley	20	.871	.890	.878	.923	.787	.888
Barley	30	.920	.855	.888	.942	.777	.874
Barley	50	.887	.945	.914	.920	.919	.920

slightly conservative and conservative. In Table 2 below the distribution of the initial and the three reclassifications are shown. These data serves as supplementary ground truth used to evaluate the sensitivity of *Weeddetect* as a function of ground truth bias.

Next, *Weeddetect* was applied to all images for a number of different values of the final threshold value T and the final binary classification (with 1 m resolution) was compared to the ground truth. For each altitude, crop type and value of T, the true positive rate (TPR) or sensitivity, and the true negative rate or specificity (SPC), as well as the accuracy (ACC) was computed. In Table 3, these measurements are marked as *Within sample*. Because the fraction of classified weed patches is far larger than observed in general, an estimate of the performance on the total image area is made. These measures marked *Estimated total* more correctly shows the capability of the approach.

Although the results for the different altitudes were slightly different (30 m giving the best results for Wheat and 50 m for Barley), the ranking was constant as a functions of the threshold T. Thus to summarize data, the TPR and SPC were averaged across altitude. Figure 7 shows the ROC-curves for Weed and Barley as functions of the threshold parameter T. The figure shows that *Weeddetect* performs well on Wheat where high values of both TPR and SPC can be achieved simultaneously. For Barley, the performance is significantly worse. This is in agreement with agricultural experience.

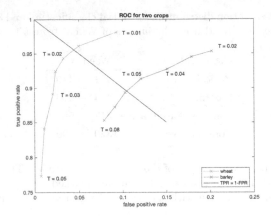

Fig. 7. ROC-curves for "Estimated total" measures for *Weeddetect*. The left curve is for Wheat and the right for Barley. It is clearly more difficult to obtain ideal results for Barley. The curves are obtained by averaging across altitude.

Also shown in Fig. 7 is the line TPR = 1 − FPR corresponding to simultaneously minimal values of false positives and false negatives. This trade-off has been chosen by experts in agriculture to represent the ideal case. The two types of crop clearly require different threshold values to obtain this trade-off. In the following we therefore fix the threshold parameter values to 0.060 for Barley and to 0.015 for Wheat. Important to notice is that this choice does not necessarily correspond to the threshold values for which ACC is maximized. Table 3 show the detailed performance measures for the chosen threshold values. For the chosen values of the threshold T for the two types of crop (but otherwise fixed and identical parameter settings), the accuracy does not change significantly across altitudes, but differs significantly between Wheat and Barley.

Compared with the previous approach *ThistleTool* [14], Table 4 shows the over-all performance when averaging over altitude. The table shows that for Wheat, the TPR is slightly better and SPC is slightly worse for *Weeddetect*. For Barley, the TPR is slightly better for *Weeddetect* while at the same time being significantly better in specificity. Important here is that the numbers for

Table 4. Comparison of *Weeddetect* and *ThistleTool* using Within sample measures and averaging across altitude. Notice that for *ThistleTool* threshold tuning to each field was applied.

Method	Crop	TPR	SPC	ACC
Weeddetect	Wheat	.961	.950	.953
	Barley	.887	.945	.914
ThistleTool	Wheat	.940	.982	.963
	Barley	.872	.724	.823

Table 5. Performance on 900 image patches of Wheat fields viewed from 50 m altitude. The columns marked *Within sample* mark the raw results. The columns marked *Estimated total* show the performance after normalization with the true amount of weed seen in the images. The rows correspond to different ground truth definitions.

Ground truth	Estimated total			Within sample		
	TPR	SPC	ACC	TPR	SPC	ACC
Basic liberal	.938	.965	.960	.990	.866	.928
Liberal 2	.920	.972	.963	.986	.879	.935
Slightly conservative	1.000	.906	.915	1.000	.629	.753
Conservative	1.000	.912	.922	1.000	.641	.762

Weeddetect are obtained for a fixed set of parameters while for *ThistleTool* tuning to each specific field was allowed. Thus, the preprocessing made in *Weeddetect* seems to have eliminated the problems previously observed with *ThistleTool*.

Finally, still keeping all parameters fixed, including the threshold values for the two types of crop, the performance on the three supplementary data sets were measured. Table 5 below show, for comparison, in the first line the results on the initial data set. In the next three lines the result obtained on the supplementary data sets (see Table 2) are shown. Table 5 shows, that bias in the expert classification of patches has a significant impact on the performance measures. If the classification is more conservative (fewer patches are marked to show thistles) then it's easy for *Weeddetect* to reach a perfect TPR. However, for such classification more patches marked as crop will be detected as thistles and the accuracy will be reduced. When the expert classifies patches more eagerly into thistles the performance is better and more balanced although the TPR is lower. The difference between the results using the two liberal classifications indicate the ground truth accuracy here may be about 1%.

Weeddetect is currently implemented in Matlab and it takes about 8 s to process one $3\,K \times 4\,K$ image on a high-end laptop. If ported to say C++, analysis of all images covering a field may be reduced to a few minutes facilitating application in practice.

5 Conclusions

We have presented a fully automatic system *Weeddetect* for detecting weeds, in particular thistles, in drone images of mature wheat. The system automatically compensates for uneven illumination and unbalanced colors. The preliminary experiments indicate that the system performs well on all altitudes from 20 to 50 m but that the performance depends on the type of crop. For Barley the user specified threshold on the minimal necessary amount of detections within one square meter should be increased significantly with respect to the parameter value for Wheat. Although the method is developed for detecting thistles in wheat it may be possible to apply it on other cereals such as Barley. In such cases it is likely that the threshold has to be tuned to obtain the desired tradeoff between false positives and false negatives.

The experiments revealed that a slight bias in the expert classification of ground truth, had a significant impact on the obtained results. This both point at the difficulty of the problem, but also raises fundamental questions on how to test systems when noise and bias in ground truth may be present.

In future work we plan to combine the detections obtained at single images into one field map. Such map will be easy to geo-reference and convert to a spraying-program. Also consistent detection in multiple images may be used to increase the quality of the final map.

Acknowledgements. This work was supported by the Danish Environmental Protection Agency project MST-667-00141.

References

1. Cheng, D.L., Prasad, D., Brown, M.S.: Illuminant estimation for color constancy: why spatial domain methods work and the role of the color distribution. J. Opt. Soc. Am. - A (JOSA) **31**(5), 1049–1058 (2014)
2. Christensen, S., Rasmussen, J., Pedersen, S.M., Dorado, J., Fernandez-Quintanilla, C.: Prospects for site specific weed management. In: Second International Conference on Robotics and Associated High-Technologies and Equipment for Agriculture and Forestry. New trends in Mobile Robotics, Perception and Actuation for Agriculture and Forestry, Madrid, pp. 541–549 (2014)
3. Egilsson G.J.: Detecting weed on images of cereal fields acquired by drones. Master thesis project, Department of Computer Science, University of Copenhagen (2014)
4. Egilsson, G.J., Pedersen, K.S., Olsen, S.I., Nielsen, J., Ntakos, G., Rasmussen, J.: Pre-harvest assessment of perennial weeds in cereals basedon images from unmanned aerial systems (UAS). In: The 17th European Weed Research Society Symposium, Montpellier, France (2015)
5. The Directive on Sustainable Use of pesticides EU 128/EC Directive of the European parliament and of the council establishing a framework for Community action to achieve the sustainable use of pesticides. OJL 309/71, 24 November 2009
6. Fernandez-Maloigne, C. (ed.): Advanced Color Image Processing and Analysis. Springer, New York (2012)
7. Forsyth, D., Ponce, J.: Computer Vision: A modern Approach, 2 edn. Pearson (2012)
8. Franco, C.A., Pedersen, S.M., Papharalampos, H., Ørum, J.E.: An image-based decision support methodology for weed management. Precis. Agric. **15**, 595–601 (2015)
9. Garcia-Ruiz, F.: UAV based imaging for crop, weed and disease monitoring. Ph.D. dissertation, Department of Plant and Environmental Sciences, Faculty of Science, University of Copenhagen (2014)
10. Gonzales, R.C., Woods, R.E.: Digital Image Processing, 3rd edn. Prentice Hall, Upper Saddle River (2008)
11. Hansen, K.D., Garcia-Ruiz, F., Kazmi, W., Bisgaard, M., la Cour-Harbo, A., Rasmussen, J., Andersen, H.J.: An autonomous robotic system for mapping weeds in fields. In: Proceedings of the 8th IFAC Symposium on Intelligent Autonomous Vehicles, Australia, pp. 217–224 (2013)

12. Jørgensen, L.N., Kudsk. P.: Twenty years' experience with reduced agrochemical inputs. In: Proceedings of the Home Grown Cereal Authority, Research and Development Conference. Arable Crop Protection in the Balance Profit and the Environment, pp. 16.1–16.10 (2006)

13. Rasmussen, J., Nielsen, J., Garcia-Ruiz, F., Christensen, S., Streibig, J.C.: Potential uses of small unmanned aircraft systems (UAS) in weed research. Weed Res. **53**, 242–248 (2013)

14. Rasmussen, J., Nielsen, J., Olsen, S.I., Petersen, K.S., Jensen, J.E., Streibig, J.C.: Droner til monitorering af flerårigt ukrudt i korn; (In Danish). Miljøstyrelsen, Miljøministeriet. http://mst.dk/service/publikationer/. Accessed Jan 2017

15. Barr, D.R., Sherill, E.T.: Mean and variance of truncated normal distributions. Am. Stat. **53**(4), 357–361 (1999)

An Image-Based Method for Objectively Assessing Injection Moulded Plastic Quality

Morten Hannemose[1]([✉]), Jannik Boll Nielsen[1], László Zsíros[2],
and Henrik Aanæs[1]

[1] DTU Compute, Technical University of Denmark, Kongens Lyngby, Denmark
mohan@dtu.dk
[2] Department of Polymer Engineering, Budapest University of Technology
and Economics, Budapest, Hungary

Abstract. In high volume productions based on casting processes, like high-pressure die casting (HPDC) or injection moulding, there is a wide range of variables that affect the end quality of produced parts. These variables include production parameters (temperature, pressure, mixture), and external factors (humidity, temperature, etc.). With this many variables it is a challenge to maintain a stable output quality, wherefore massive amounts of resources are spent on quality assurance (QA) of produced parts. Currently, this QA is done manually through visual inspection. We demonstrate how a multispectral imaging system can be used to automatically rate the quality of a produced part using an autocorrelation and a Fourier-based method. These methods are compared with human rankings and achieve good correlations on a variety of samples.

Keywords: Quality inspection · Plastics · Injection moulding · Maximum autocorrelation factor · Multispectral · Fourier transform

1 Introduction

Injection moulding is a versatile manufacturing process, used for producing a wide range of everyday objects. The process is based on injecting material into a mould under high pressure and temperature, whereafter the material will cool and harden, completing the injection moulding cycle. The approach works for a wide range of materials and is very fast, making it one of the most commonly used means of manufacturing today. Figure 1 shows a few examples of everyday objects produced through injection moulding: A computer mouse, a plate, and a pen.

When doing injection moulding it is important to ensure consistent quality of produced products. This both implies the mechanical properties, such as dimensions, roughness, and strength, and the visual appeal, or surface "appearance". For the former, several standards and recommendations exist for ensuring consistent quality objectively, however, for the latter, no guidelines exist and most quality control is today done by manual visual inspection. This need for manual visual quality control, which is in addition often repeated multiple times and averaged for consistency, is a very expensive and slow operation to carry out

© Springer International Publishing AG 2017
P. Sharma and F.M. Bianchi (Eds.): SCIA 2017, Part II, LNCS 10270, pp. 426–437, 2017.
DOI: 10.1007/978-3-319-59129-2_36

Fig. 1. An example of products manufactured using injection moulding. Left: computer mouse. Center: plate. Right: pen.

that scales poorly. It is, however, as the consumers value visual quality highly, a task that can very rarely be neglected. Often today we see a compromise, where humans periodically assess quality of samples and adjust production parameters accordingly to maintain a visual quality that is within bounds. This means that whole batches may be discarded due to late identification of errors. It also often means that the quality is evaluated subjectively, with the potential of having varying qualities over time and different production locations.

It is apparent that being able to *automatically* and *objectively* quantify the visual surface quality of produced parts has the potential of saving manufacturers many resources in production. In this work, we deal with surface quality of injection moulded plastic (acrylonitrile butadiene styrene, ABS), where we demonstrate an objective surface quality measure based on multispectral images captured in a controlled lighting environment. Here, visual quality is synonymous with high color homogeneity, i.e. the produced plastic part has the correct color everywhere, whereas in cases of low quality there may be smaller patches with discoloring. Our main contributions are: (a) Proposing an approach for strongly emphasizing surface discolourings using multispectral imaging, and (b) presenting a Fourier-based, rotation invariant, quality measure that can potentially be translated to current human-based scores.

2   Related Work

It is well known that injection moulding parameters have a significant effect on the visual appearance and quality of moulded parts. Specifically, especially color and gloss, two of the main contributors to the visual appearance, are highly affected by production parameters [6]. Especially mould temperature and packing pressure are identified by Piscotti et al. to have a high impact on color and gloss. Inhomogeneity in color across the material surface is often caused by insufficient dispersion of fillers or colorants, or by injection moulding parameters too [7,8].

The assessment of color quality and consistency itself is a well-addressed topic and is employed in many very different fields of research [12–14]. The CIE calibrated color spaces are a convenient way of obtaining device independent color measurements [9]. Especially, in relation to quality assurance of color, the CIELAB color space has been employed as this space corresponds well with the color perception of humans [4]. Especially, the color distance metric ΔE_{00} calculated in this space is very convenient, as it mimics how humans perceive color differences, and threshold values for "hardly perceptible" differences have been identified [2].

Moving beyond standard three stimulus color measurements, multispectral imaging approaches, e.g. designed for replacing colorimeters that only rely on point-samples, have been proposed in the food industry [12]. Unfortunately these have until now not been ported to the field of quality control in manufacturing.

Recently, convolution based methods have, with some success, been proposed for estimating color-inhomogeneity of samples scanned in a flatbed scanner [15]. Advancing from this, projector-based approaches trying to identify subsurface miscolorings using structured light have also been suggested [3].

3 Data

For the experiments carried out in this paper, a collection of injection moulded unfilled acrylonitrile butadiene styrene (ABS) samples was produced. To add coloring, a 4% by weight of masterbatch (MB) was dry mixed into the matrix. The samples were injection moulded on an Arburg Allrounder Advance 370 S 700-290 machine, using a screw diameter of 30 mm. To obtain varying color homogeneity, i.e. appearance quality, a range of relevant production parameters were varied. We chose to reuse the parameters identified by Zsíros et al. [15]. In Fig. 2, the mould used for producing samples is presented to the left. To the right in the figure, en example of a produced sample, using dark-blue coloring, is shown. Notice that small but distinct color inhomogeneities exist on the sample surface.

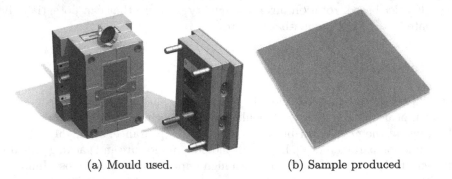

(a) Mould used.   (b) Sample produced

Fig. 2. (a): Mould used for producing the plastic samples that are analyzed in this paper. (b): A plastic sample from the dark blue material. Notice that this sample has many visible inhomogeneities on the surface. (Color figure online)

In total 100 samples were created. The collection covers ten different and commonly used colors, each with ten samples of varying quality. Unfortunately, absolute quality scores obtained by human quality inspectors were not available in this experiment. Instead, an assessment panel was instructed to produce a relative ranking, by ordering samples from the smallest visual error (1) to the largest visual error (10). This was done both for the ten samples within every color, and across all colors using one sample of each color.

For analyzing the produced samples, we utilize a multi-spectral imaging device called VideometerLab 4 to obtain high resolution (2192 × 2192 pixels) images captured at 19 different wavelengths [1]. It captures band-sequential images by diffusely illuminating the sample using LEDs operating in 19 different wavelengths ranging from 365 nm to 970 nm.

As the typical human eye is only sensitive to wavelengths from 390 nm to 700 nm [10], we only use the thirteen bands that fall within this spectrum. Table 1 lists all available wavelengths from the VideometerLab 4, and the ones emphasized in bold writing are the ones used for analysis. This is done because we are in this study not interested in imperfections that are not visible to the human eye. An example of the multi-spectral image of the material sample presented in Fig. 2 can be seen in Fig. 3, where each wavelength is shown as a grayscale image. Note here that the last five bands (near infrared), which fall outside the

Table 1. The wavelengths at which we have acquired images of our samples using the VideometerLab. Due to the limited sensitivity of the human eye, we only use bands two to fourteen. The bands used are highlighted with bold.

Band	1	2	3	4	5	6	7	8	9	10	11	12	13	14	15	16	17	18	19
λ [nm]	365	405	430	450	470	490	515	540	570	590	630	645	660	690	780	850	880	940	970

Fig. 3. Raw spectral images acquired of the sample presented in Figure 2. The wavelengths are in an increasing order from left to right and top to bottom. These correspond to the wavelengths mentioned in Table 1. In the five longest wavelengths (near infrared), there can be seen a faint bright rectangle near the middle of the sample. This is because the sample is slightly transparent at these wavelengths and there is a sticker on the opposite side of the sample.

human visual spectrum, renders the sample slightly transparent as it is possible to faintly see a sticker that is located on the other side of the sample. This shows up as a bright rectangular shape slightly left of the center.

The dark circles seen in the center of each image in Fig. 3 are a result of the samples being very specular and are in fact a reflection of the camera hole itself in the VideometerLab 4 device. This is an unwanted artifact and as a final processing step we therefore produce a mask that masks out this center circle as well as the edge of the samples. This mask ensures that no false color inhomogeneities are introduced during data-processing.

4   Method

The 13 color bands within the human visible spectrum, that were acquired with the multispectral camera, are initially filtered with a 3×3 median kernel to reduce sensor noise and small dust particles that were present on the sample at the time of image acquisition. This is done as dust particles introduce unwanted high frequent noise in the signal, which would complicate later analyses. Afterwards, the mask created during data acquisition, identifying valid pixels in the images, is applied to the images to extract only valid information. After masking, the 1 percentile brightest and darkest pixels are clipped. This clipping is an additional way to reduce the effect of tiny dust particles, as they make up less than 1% of the area in the image.

Every valid pixel in the thirteen channel multispectral image may be treated as an $n = 13$ dimensional observation, $\mathbf{x} \in \mathbb{R}^n$. As 3,246,212 pixels in every channel were identified as being valid using the predefined mask, a total of $m = 3246212$ n-dimensional points have been observed. As the distribution of these points is currently not centered, the mean from each channel is subtracted from the respective channel to ensure proper centering of the distribution:

$$\tilde{\mathbf{x}}_i = \mathbf{x}_i - \mu_i = \mathbf{x}_i - \frac{\sum_{k=1}^{m} x_{ik}}{m} \qquad \forall i \in [1; n] \tag{1}$$

and the centered observations may be gathered in an $m \times n$ sized observation matrix, $\mathbf{X} \in \mathbb{R}^{m \times n}$:

$$\mathbf{X} = [\tilde{\mathbf{x}}_1, \ldots, \tilde{\mathbf{x}}_m]^T \tag{2}$$

Based on this observation matrix, the maximum autocorrelation factor (MAF) transformation can be performed [11]. This is very similar to doing principal component analysis (PCA), but yields results that are in some ways more interpretable in the context of images, as it uses the spatial information.

The autocorrelation-based variance-covariance matrix Σ_Δ is defined as:

$$\Sigma_\Delta = 2\Sigma - \Gamma(\Delta) - \Gamma(-\Delta), \tag{3}$$

where Σ is the covariance matrix of $\mathbf{X}$ and $\Gamma(\Delta)$ is the autocorrelation of the image using a specified spatial shift, Δ, usually 1 pixel both horizontally and vertically.

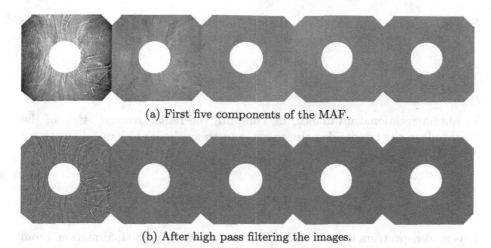

(a) First five components of the MAF.

(b) After high pass filtering the images.

Fig. 4. The first five MAF components of the masked and transformed image before and after high pass filtering the images. The high pass filtering is performed by subtracting a highly blurred version of each image from the image.

The MAF components correspond to the eigenvectors, $\mathbf{W} = [\mathbf{w}_1 \ldots, \mathbf{w}_n]$, of $\mathbf{\Sigma}_\Delta$ with respect to $\mathbf{\Sigma}$.

The resulting components of the MAF transformation can be seen in Fig. 4. Note that there is a low frequency component visible in the first channel. This is because the sample has not been perfectly evenly lit. As we are not interested in this bias, we subtract a highly blurred version of each channel as a means of applying a high pass filter. We use a Gaussian kernel with $\sigma = 30$ to blur the image. The image after this operation can be seen in Fig. 4b. The value of σ has been chosen empirically to reproduce the impurities as perceived by looking at the sample.

Clearly, as can be seen in Fig. 4, both with and without highpass filtering, the first MAF component clearly emphasizes the inhomogeneities that are only faintly visible in the reference image shown in Fig. 2.

4.1   Feature Extraction

At this step we discard everything but the most significant component, i.e. the first, identified by the MAF, as it captures the inhomogeneities in the material sample very well.

Using this single grayscale image we propose two approaches of extracting information about the surface quality: two Fourier-based methods and one auto-correlation based method.

Fourier-based method: Here, we compute the 2D Fourier transform of the image which yields the complex-valued 2D Fourier spectrum,

$$F(u, v) = \sum_{m=0}^{M} \sum_{n=0}^{N} f[m, n] e^{-j2\pi\left(\frac{um}{M} + \frac{vn}{N}\right)}. \tag{4}$$

To obtain rotational invariance, we compute the radial average, $A(r)$, of the amplitude in the Fourier domain, to get something similar to a one-dimensional power spectrum of the image:

$$A(r) = \frac{1}{2\pi} \int_{0}^{2\pi} |F(r \sin \phi, r \cos \phi)| \, d\phi. \tag{5}$$

This power spectrum is shown in Fig. 5. We discard the phase information. From the power spectrum, $A(r)$, we can compute the average radial amplitude.

$$A_{avg} = \int_{0}^{\min(M,N)} A(r) dr \tag{6}$$

The upper limit in the integral is chosen so the A_{avg} is finite because the Fourier transform is cyclic.

Second Fourier-based method: Using the above defined $A(r)$ we compute only the average radial amplitude from $r = 20$ to $r = 80$. This range has been chosen as the ordering of the lines in Fig. 5 in this range visually correlates well with the human based quality ordering. This can be interpreted as a mid-pass filter. The range is depicted by the dashed red vertical lines in Fig. 5.

$$A_{mid} = \int_{20}^{80} A(r) dr \tag{7}$$

Autocorrelation-based method: Here, we compute the correlation length of the 1st MAF component using its weighted autocorrelation. The weighted autocorrelation is given by:

$$\mathbf{\Gamma}_{weighted}(\Delta) = \frac{\mathbf{\Gamma}(\Delta)}{N - |\Delta|}, \tag{8}$$

where N is the length of the discrete signal. The correlation length, l, is then defined as the distance at which the autocorrelation drops below the value $1/e$ for the first time [5]:

$$\mathbf{\Gamma}_{weighted}(l) = \frac{1}{e} \quad \text{(first occurrence)} \tag{9}$$

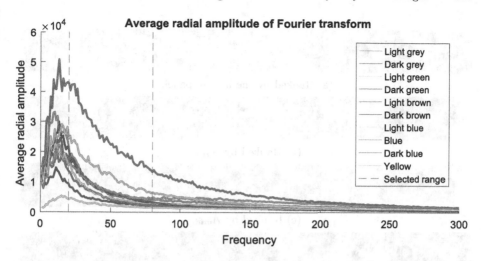

Fig. 5. Radially averaged Fourier transform of the first MAF component. This is shown for one sample of each color. (Color figure online)

5   Results

In the following we summarize the results obtained by the proposed quality features compared to the results obtained by the human quality assessment panel. For compactness, we only summarize the rankings of a subset of the samples, as these were deemed to span variation of our method's performance well. This subset includes the samples "dark blue", "yellow", "dark grey", "light grey", "dark brown", and "one of each color".

We employ each of the three quality measures presented in the previous section and rank the samples according to those. Figures 6, 7, 8 and 9 visualize these rankings, where the first MAF component is overlaid with the human-assigned rank (10 being the worst). The samples are ordered from left to right according to the respective method's ranking. Note that the extremes of the rankings generally conform well with the ones ranked by a human assessment panel. Around the middle of the rankings, we observe a little more variation in the assignments, which would be expected.

For a quantitative evaluation of our method's performance, we employ Spearman's and Kendall rank correlation coefficients to compare our rankings to the human rankings. These correlation coefficients are summarized in Tables 2 and 3. As may be seen from the table all quality features perform decently with an average Spearman's rank correlation coefficient above 0.5 and Kendall rank correlation coefficient slightly above 0.4. What is noteworthy is that the features perform well with different samples, indicating that they may be able to compliment one another. We have not investigated this further but will look into this in future work.

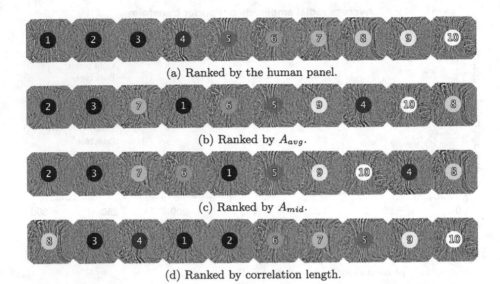

(a) Ranked by the human panel.

(b) Ranked by A_{avg}.

(c) Ranked by A_{mid}.

(d) Ranked by correlation length.

Fig. 6. Comparison of how the dark blue samples were ranked by the human panel and our proposed quality features. The numbers shown in the center of each sample are the ranks assigned by the human assessment panel. The ordering from left to right corresponds to the method's ranking.

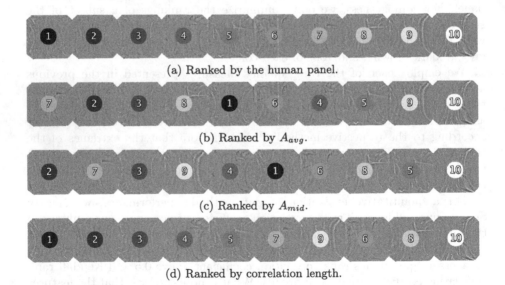

(a) Ranked by the human panel.

(b) Ranked by A_{avg}.

(c) Ranked by A_{mid}.

(d) Ranked by correlation length.

Fig. 7. Comparison of how the light grey samples were ranked by the human panel and our proposed quality features. The numbers shown in the center of each sample are the ranks assigned by the human assessment panel. The ordering from left to right corresponds to the method's ranking.

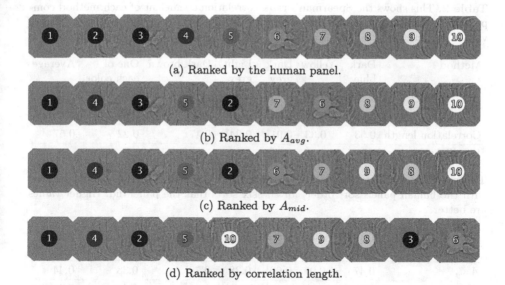

(a) Ranked by the human panel.

(b) Ranked by A_{avg}.

(c) Ranked by A_{mid}.

(d) Ranked by correlation length.

Fig. 8. Comparison of how the dark brown samples were ranked by the human panel and our proposed quality features. The numbers shown in the center of each sample are the ranks assigned by the human assessment panel. The ordering from left to right corresponds to the method's ranking.

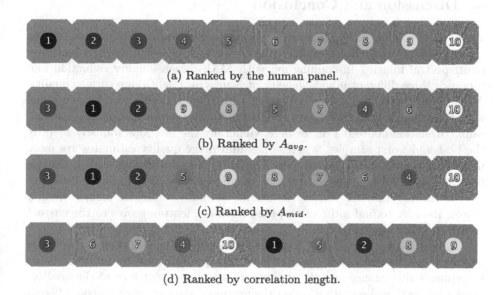

(a) Ranked by the human panel.

(b) Ranked by A_{avg}.

(c) Ranked by A_{mid}.

(d) Ranked by correlation length.

Fig. 9. Comparison of how one sample of each color were ranked by the human panel and our proposed quality features. The numbers shown in the center of each sample are the ranks assigned by the human assessment panel. The ordering from left to right corresponds to the method's ranking.

Table 2. This shows the Spearman's rank correlation coefficient of each method compared with the human panels sort order, for the six sorts that the panel did. Higher values are better.

Method	Dark blue	Yellow	Dark grey	Light grey	Dark brown	One of each color	Average
A_{avg}	0.67†	0.15	0.75†	0.48	0.90†	0.60	0.59
A_{mid}	0.54	−0.24	0.85†	0.43	0.90†	0.66†	0.53
Correlation length	0.53	0.33	0.91†	0.94†	0.47	0.22	0.57

†Significant at $p < 0.05$.

Table 3. This shows the Kendall rank correlation coefficient of each method compared with the human panels sort order, for the six sorts that the panel did. Higher values are better.

Method	Dark blue	Yellow	Dark grey	Light grey	Dark brown	One of each color	All
A_{avg}	0.47	0.07	0.60†	0.38	0.78†	0.38	0.44
A_{mid}	0.38	−0.16	0.69†	0.33	0.78†	0.42	0.41
Correlation length	0.42	0.29	0.78†	0.87†	0.33	0.20	0.48

†Significant at $p < 0.05$.

6   Discussion and Conclusion

We have in this paper demonstrated a method for automatically assessing the visual quality of injection moulded plastic surfaces. The approach is based on multispectral imaging in conjunction with MAF dimensionality reduction and proposes three different inhomogeneity measures for quantifying surface quality. All three methods are capable of correctly ranking according to a human panel with Spearman's rank correlation coefficient above 0.5 on average and Kendall rank correlation above 0.4 on average. Generally the methods robustly identify the best and worst samples, whereas medium-range quality estimates are more uncertain. If only one of our variants should be picked we propose A_{avg}, as this has the most, and highest, statistically significant rank correlation coefficients.

For future work, we will try to obtain absolute values for sample quality scores, in order to find a direct mapping from our feature scores to the current standard in human visual quality control.

Currently, our method only looks at inhomogeneities and does not consider whether the base color of the sample is within specifications. Future work could incorporate this nuance as well, as this too is an important part of the quality.

An interesting observation is that the proposed features work well on different material samples. This could indicate that even better overall ranking performance may be obtained by concatenating the features before predicting material quality. In addition using additional MAF components and/or extracting more than one range from the Fourier spectrum is subject for future research.

References

1. Carstensen, J.M., Folm-Hansen, J.: An apparatus and a method of recording an image of an object (1999). www.google.com/patents/WO1999042900A1?cl=en, WOPatentApp.PCT/DK1999/000,058
2. Hardeberg, J.Y.: Acquisition and Reproduction of Color Images: Colorimetric and Multispectral Approaches. Universal-Publishers, Boca Raton (2001)
3. Lai, W., Zeng, X., He, J., Deng, Y.: Aesthetic defect characterization of a polymeric polarizer via structured light illumination. Polym. Test. **53**, 51–57 (2016)
4. Leon, K., Mery, D., Pedreschi, F., Leon, J.: Color measurement in $L^*a^*b^*$ units from RGB digital images. Food Res. Int. **39**(10), 1084–1091 (2006)
5. Nielsen, A.A.: Geostatistics and analysis of spatial data. Informatics and Mathematical Modelling, Technical University of Denmark, DTU, pp. 7–12 (2009)
6. Pisciotti, F., Boldizar, A., Rigdahl, M., Ariño, I.: Effects of injection-molding conditions on the gloss and color of pigmented polypropylene. Polym. Eng. Sci. **45**(12), 1557–1567 (2005)
7. Santos, R., Pimenta, A., Botelho, G., Machado, A.: Influence of the testing conditions on the efficiency and durability of stabilizers against abs photo-oxidation. Polym. Test. **32**(1), 78–85 (2013)
8. Sathyanarayana, S., Wegrzyn, M., Olowojoba, G., Benedito, A., Giménez, E., Hübner, C., Henning, F.: Multiwalled carbon nanotubes incorporated into a miscible blend of poly (phenylenether)/polystyrene-processing and characterization. Express Polym. Lett. **7**(7), 621 (2013)
9. Smith, T., Guild, J.: The CIE colorimetric standards and their use. Trans. Opt. Soc. **33**(3), 73 (1931)
10. Starr, C., Evers, C., Starr, L.: Biology: Concepts and Applications without Physiology. Cengage Learning, USA (2010)
11. Switzer, P., Green, A.A.: Min/max autocorrelation factors for multivariate spatial imagery. In: Computer Science and Statistics, pp. 13–16 (1984)
12. Trinderup, C.H., Dahl, A., Jensen, K., Carstensen, J.M., Conradsen, K.: Comparison of a multispectral vision system and a colorimeter for the assessment of meat color. Meat Sci. **102**, 1–7 (2015)
13. Wu, D., Sun, D.W.: Colour measurements by computer vision for food quality control - a review. Trends Food Sci. Technol. **29**(1), 5–20 (2013)
14. Yam, K.L., Papadakis, S.E.: A simple digital imaging method for measuring and analyzing color of food surfaces. J. Food Eng. **61**(1), 137–142 (2004)
15. Zsíros, L., Suplicz, A., Romhány, G., Tábi, T., Kovács, J.: Development of a novel color inhomogeneity test method for injection molded parts. Polym. Test. **37**, 112–116 (2014)

Creating Ultra Dense Point Correspondence Over the Entire Human Head

Rasmus R. Paulsen[1]([✉]), Kasper Korsholm Marstal[1,3], Søren Laugesen[2], and Stine Harder[1]

[1] DTU Compute, Technical University of Denmark, Richard Petersens Plads, Building 324, 2800 Kongens Lyngby, Denmark
{rapa,kkmars,sthar}@dtu.dk
[2] Interacoustics Research Unit, DGS Diagnostics A/S, Ørsteds Plads, Building 352, 2800 Kongens Lyngby, Denmark
slau@iru.interacoustics.com
[3] Departments of Radiology and Medical Informatics Erasmus Medical Center, Biomedical Imaging Group Rotterdam (BIGR), Rotterdam, The Netherlands
http://www.compute.dtu.dk

Abstract. While the acquisition and analysis of 3D faces has been an active area of research for decades, it is still a complex and demanding task to accurately model the entire head and ears. Having accurate models would for example enable virtual design of hearing devices. In this paper, we describe a complete framework for surface registration of complete human heads where the result is point correspondence with a very high number of points. The method is based on a volumetric and multi-scale non-rigid registration of signed distance fields. The method is evaluated on a set of 30 human heads and the results are convincing. The output can for example be used to compute statistical shape models. The accuracy of predicted anatomical landmarks is on the level of experienced human operators.

Keywords: Surface registration · Signed distance field · Human head modelling

1 Introduction

Automated analysis and identification from 2D photos is a very well-established area and for the last decade 3D face modelling, analysis, and recognition has become an established technology [6]. Acquisition has typically been done using laser scanning devices that require the subject to be in a fixed position for some time and be careful of eye exposure. Recently, a wide range of acquisition devices has become available. From multi-camera setups using multi-view stereo reconstruction algorithms [2] that can acquire a full face in a single capture to low-cost and accessible systems using the Microsoft Kinect [12]. Most of these systems provide a triangulated surface as output with either a texture map or colour values on each vertex.

P. Sharma and F.M. Bianchi (Eds.): SCIA 2017, Part II, LNCS 10270, pp. 438–447, 2017.
DOI: 10.1007/978-3-319-59129-2_37

Due to the complexity and optical properties of human hair, most systems can not do a realistic acquisition of hair and it is therefore often covered by a cap. 3D face acquisition is for example being used for entertainment where photorealistic faces are captured and used for computer games, in facial surgery for surgery planning [4], and for facial recognition [1]. In most applications, the focus is on the acquisition and analysis of the face.

While the acquisition of the human face is an established method, it is much more complicated to get an accurate surface scan of the entire head including the complex anatomy of the outer ear. 3D scannings of the outer ear have been used in biometrical applications [20]. Recently, accurate 3D scannings of the ear and head have been used for product design. In particular, the acoustical optimisation of hearing devices is an attractive application of head scans. The combination of 3D surface scans and advanced finite element simulations enables the computation of the so-called head related transfer function (HRTF) that enables user-specific hearing device optimisation [8]. We have previously demonstrated a method to acquire accurate full head scans that are well suited for acoustical applications [9]. The data used in this paper is similar to this type of data.

Many approaches to the analysis of human heads are based on a surface registration step where, for example, a template mesh is warped to fit a new unseen face. Surface registration has been a major research area for years [18, 19] and a variety of approaches exists. In the seminal paper on 3D morphable models [3] a modified optical flow method is used to register 3D face scans. In [13] partial scans acquired using the Microsoft Kinect are registered using a novel deformation model that potentially enables multi-level approaches [17], where a sub-sampled coarse model is initially aligned and gradually refined in further steps. However, the difference in complexity between the human face and the entire human head including ears is quite large. In our experiments we did not find an existing framework that could successfully register the entire head. To successfully do the registration, the methods need to be multi-level, so coarse features like the overall head is registered first. While the fine details of the outer ear in are registered at the final fine resolutions. In this paper, we present a method based on non-rigid volumetric registration of signed distance fields to solve the task, where the multi-level properties is given by the volumetric registration method.

2   Data and Preprocessing

The data used in this work consists of 30 3D surface scans of the entire heads including the outer ear. The outer ear is here defined as the concha, pinna, and the entrance to the ear canal. The surface scans are acquired using a Canfield Vectra M3 scanner, which is a dedicated human head scanner typically used for facial restorative surgery. Due to the very complex anatomy of the human ear it is not possible to acquire a full surface scan of the head and ear in a single acquisition. Therefore, each head was scanned from up to ten different angles by placing the person in a rotating chair. For each scan, relevant areas are manually marked to avoid using areas influenced by motion or facial expressions. A set of sparse landmarks are also placed on each sub-scan using the template described in [9]. Using the landmarks, the marked areas are brought into rough alignment. Following the approach described in [15], a combined Markov Random Field surface reconstruction and implicit iterative closest point algorithm is used to create a triangulated surface of the entire head and ear. Finally, the colour values sampled from the scannings are transferred to the vertices of the reconstructed surface. The resulting surfaces consists of approximately 450.000 vertices and 950.000 triangles. An example of one of the 30 entire heads can be seen in Fig. 1. The green areas of the scans indicate where raw colours of the original scans have not been available. Due to the optical properties of the eye it is not possible to acquire the true outer surface of the cornea with an optical scanner. Therefore the eye region will typically be either flat or curve inwards in the used data set.

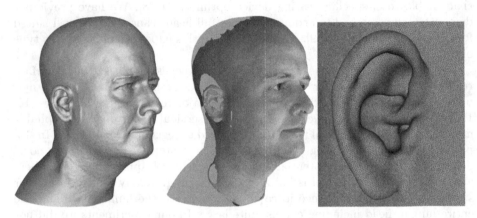

Fig. 1. A reconstructed full head and ear scan. To the left the raw mesh, in the middle with vertex colouring and to the right a close up of the right ear. (Color figure online)

3 Methods

The goal is to register two surfaces. This means creating a dense point correspondence between a source surface ($\mathcal{S}$) and a target surface ($\mathcal{T}$) so each vertex is placed on the exact same anatomical spot on both $\mathcal{S}$ and $\mathcal{T}$. In this work, $\mathcal{S}$ is deformed to fit $\mathcal{T}$. Initially, $\mathcal{S}$ is rigidly aligned to $\mathcal{T}$ using a sparse set of anatomical landmarks manually placed on both surfaces as seen in Fig. 2. The aligned source is $\mathcal{S}_a$. In this work, we use the implicit shape description embedded in signed distance fields to drive the registration. A signed distance field is computed for both $\mathcal{T}$ and $\mathcal{S}_a$ using the method described in [16]. Here the distance field is represented as a voxel volume covering the entire surface, where the value in a voxel is the Euclidean distance to the surface. The surface is implicitly defined as the zero-level iso-surface of the distance field. In order to close holes in the surface and to accommodate for missing data, a weighted

Laplacian regularisation is performed on the distance field. An example of a regularised signed distance field can be seen in Fig. 3, which shows that the shape information is well represented by the iso-curves and that the overall shape is preserved but in a smoother form in the distance field furthest from the surface.

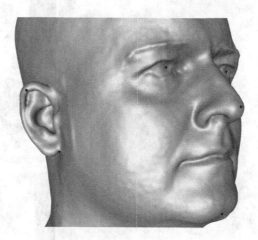

Fig. 2. The set of eight manually placed landmarks. The two landmarks placed the left ear are not shown.

The actual surface is described as the zero-level iso-surface in the distance field, but other iso-surfaces also contain implicit information about the shape. By sub-sampling the field, a coarser description of the surface shape is obtained. Finally, the gradient field of a signed distance field is also well described since the gradient will mostly point towards the zero-level iso-surface. These properties make it attractive to use a well-established and state-of-the-art volumetric registration algorithm to do a non-rigid registration of the signed distance fields. The volumetric image registration is formulated as an optimisation problem,

$$\hat{\mathbf{T}}_\mu = \arg\min_{\mathbf{T}_\mu} = \mathcal{C}(\mathbf{T}_\mu; I_F, I_M), \tag{1}$$

where I_F is the fixed volume and I_M is the moving volume. Here I_F is the signed distance field created from $\mathcal{T}$ and I_M is the signed distance field created from $\mathcal{S}_a$. $\mathbf{T}_\mu$ is a non-rigid volumetric transformation that transforms I_M and it is parameterised by the parameter-vector μ. The goal is to find the values of μ that minimise the cost function $\mathcal{C}$. The `elastix` library [11] is used to perform the volumetric registration. The transformation used is a multi-level cubic B-spline using four resolution levels. The multi-level approach ensures that coarse anatomical structures are aligned first and then finer structures are gradually being registered. In our case, it means that the overall shape of the head is aligned first and the finer details of the ears are registered in the final resolution. Since the two volumes is of the same nature and the scale of the voxel values are very similar, the mean squared voxel value difference (MSD) is chosen as the

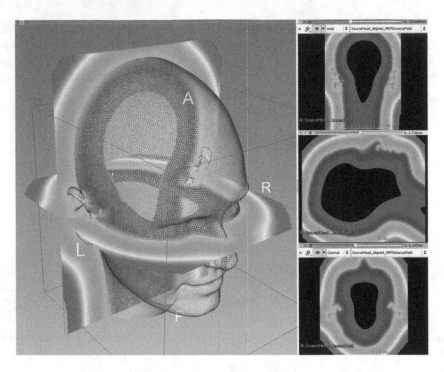

Fig. 3. A signed distance field computed for a full head. To the right the field is projected on three cuts trough the field. The distance field is thresholded to only show distances close to the surface of the head.

similarity metric. The surface of interest is by definition close to the zero-level of the distance field and therefore a binary sampling mask is applied to the moving volume. The mask is generated by only including the voxels that have a value in the range $[-20, 20]$ (measured in mm) in the distance field. Due to the large shape variation around the ears we found it necessary to aid the registration in a few case by adding a set of eight manually placed landmarks as seen in Fig. 2. This is included in the registration by adding

$$\mathcal{S}_{\text{CP}} = \frac{1}{P} \sum_i \| \mathbf{y}_i - \mathbf{T}_\mu(\mathbf{x}_i) \| \tag{2}$$

as a metric that penalises distances between corresponding landmarks. Here $\mathbf{y}_i$ and $\mathbf{x}_i$ are P corresponding points on $\mathcal{S}$ and $\mathcal{T}$. The final similarity metric then becomes

$$\mathcal{S} = \omega_1 \text{MSD}(\mu; I_F, I_M) + \omega_2 \mathcal{S}_{\text{CP}}, \tag{3}$$

where $\omega_1 = 1.0$ and $\omega_2 = 0.001$ are weights that are experimentally set. The degree of smoothness is implicitly regularised by the knot spacing in the B-spline. This means that the final cost function used in the registration is equal to the similarity metric: $C = S$ [11]. The optimisation was done using adaptive stochastic gradient descent [10] with 2048 random samples per iteration for a maximum of 500 iterations.

The result of using the non-linear registration on the signed distance fields is that a transformation, $\hat{\mathbf{T}}_\mu$, that brings the distance field representing S_a in alignment with the distance field representing T. By applying $\hat{\mathbf{T}}_\mu$ to the vertices of S_a, the vertices are propagated to T and thus creates point correspondence. The transformed mesh is S_{NR}.

Since the registration is based on a large part of the distance field it is not guaranteed that the zero-level iso-surfaces match exactly. Therefore some propagated vertices do not fall exactly on T. We apply the point propagation method originally described in [14] to fix the vertices to T. Here the vertices are first projected to the closest position on T. There is now a correspondence vector for each vertex in S_{NR}. The correspondence vector goes from the position computed in the registration step, stored in S_{NR}, to the projected position on T. Together these correspondence vectors represents a correspondence vector field (CVF).

The CVF is now cast into a Markov Random Field regularisation (MRF) framework, where each vector is penalised by deviations from its neighbours. In each iteration the CVF is MRF regularised and each vector is reprojected onto T. The final result of applying this method is that all vertices from S have been propagated directly onto T. In this work, we use a mesh with 600.000 vertices and 1.200.000 triangles as S and thereby creating an ultra dense point correspondence over the entire head and outer ear. This mesh has been selected from the set of full head scannings and remeshed into this mesh resolution.

In order to validate the registration, the template mesh was registered to all the other full heads in the data set, thereby creating a full correspondence over the data set. Following the steps of building a point distribution model [5], a Procrustes alignment of the registered meshes is performed. To avoid the bias introduced by using a specific shape as template, the Procrustes average shape was used as the template in a second registration step.

4   Results

The registration framework is applied to 30 entire head scans. We use a template mesh as the source mesh and apply it to all the other heads (targets). An example can be seen in Fig. 4. The template mesh seen in Fig. 4 is the average shape from the Procrustes alignment [5] of the initial registration of the data set, where an arbitrary head was used as template. As can be seen, the average shape is smooth but contains all important facial features including detailed outer ears. The result of the registration is that the template mesh is placed exactly on top of the target meshes and it is not possible to visualise the potential discrepancies by overlaying the registration results on the target. Instead, the template mesh was manually annotated with 93 landmarks as defined in [9] (green points in Fig. 4). The same set of landmarks was also manually annotated on all the heads

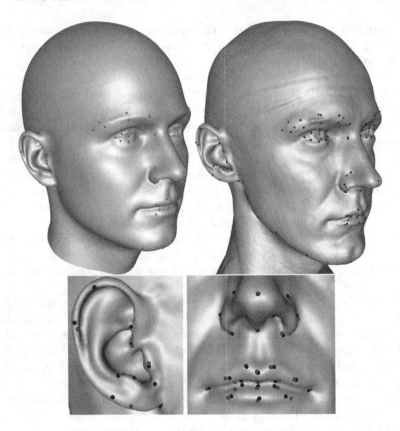

Fig. 4. Upper left:.eps The green landmarks are manually placed on the template. Blue landmarks are the results of the registration and the red landmarks are manually placed on the target. (Color figure online)

in the data set (red points in Fig. 4). In order to validate the accuracy of the registration, the landmarks from the template mesh are propagated to the target mesh (blue points in Fig. 4). An estimate of the accuracy can be computed as the distance between the annotated (red) and predicted (blue) landmarks. However, it is well known that manual annotation is error-prone and that each annotated landmark has a spatial uncertainty [7]. We have therefore chosen to only validate the accuracy with landmarks that can be accurately placed manually - well knowing that this is not a truly neutral estimate of the registration accuracy. The results are shown in Table 1.

It can be seen that the average error is in the range of 1.2–2.6 mm which is comparable to the errors from manual annotators [7]. The landmarks that are difficult to place by a human operator had errors in the range of 4 to 8 mm.

After the template mesh has been applied to the target it is possible to sample the colours from the original scans. In Fig. 5, a registered mesh with a colour assigned to each vertex can be seen. The average vertex colour over the entire set

Table 1. Error (in mm) between annotated and predicted landmarks (LM).

LM	Place	Mean	Median	Standard deviation
17	Right lateral canthus	1.9	1.4	1.4
21	Right medial canthus	1.9	1.8	1.0
25	Left lateral canthus	2.6	2.4	1.4
29	Right medial canthus	2.1	2.0	1.1
40	Columellar connection to upper lip	1.2	1.0	0.8
47	Right oral commissure	1.7	1.8	0.9
53	Left oral commissure	2.0	1.7	1.1
75	Left ear bottom of concha	1.6	1.3	0.9
76	Left ear tragus	1.6	1.6	0.8
77	Left ear crus of helix	2.0	2.0	0.9
85	Right ear bottom of concha	1.3	1.2	0.9
86	Right ear tragus	1.6	1.6	0.9
87	Right ear crus of helix	1.8	1.5	1.0

can then be computed due to the vertex correspondence. Applying the average vertex colours to the Procrustes average can be seen in Fig. 5. It can clearly be seen that the eye colours have been smoothed out and that the outline of the

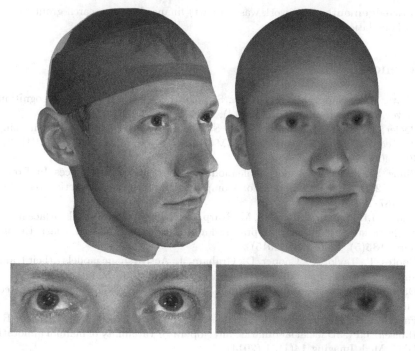

Fig. 5. Left: A registered mesh with vertex colours sampled from the original scans. **Right:** The Procrustes average with average vertex colours.

eye is blurry. This can be caused by the problems in acquiring the shape and texture of the eye correctly and the fact that the eye also changes shape and texture when exposed to light.

The running time of one full head registration is in the order of 20 minutes on a standard Windows-7 laptop computer with 8 GB RAM. Some parts of the algorithm are implemented as parallel processes using the 8 processing cores of the laptop otherwise the algorithm was not optimised for speed.

5   Conclusion

We have presented a method for registration of full heads including ears that successfully computes an ultra dense correspondence from a template mesh to an arbitrary mesh. It is also demonstrated that the method accurately maps anatomical meaningful landmarks from a template mesh to an arbitrary target mesh. The results in terms of accuracy of single landmark placement is comparable to what trained human operators can do. The method is based on matching the pure shape information implicitly described by a signed distance field. It is possible that including the surface colours in the registration could further increase the accuracy, in particular in regions with little shape information. The method can for example be used to build statistical shape models that can further be used in modelling, analysis, and design-related applications.

Acknowledgements. This work was (in part) financed by a research grant from the Oticon Foundation.

References

1. Abate, A.F., Nappi, M., Riccio, D., Sabatino, G.: 2D and 3D face recognition: a survey. Pattern Recogn. Lett. **28**(14), 1885–1906 (2007)
2. Beeler, T., Bickel, B., Beardsley, P., Sumner, B., Gross, M.: High-quality single-shot capture of facial geometry. ACM Trans. Graph. (Proc. SIGGRAPH) **29**(3), 40:1–40:9 (2010)
3. Blanz, V., Vetter, T.: A morphable model for the synthesis of 3D faces. In: Proceedings of 26th Annual Conference on Computer Graphics and Interactive Techniques, pp. 187–194 (1999)
4. Chang, J.B., Small, K.H., Choi, M., Karp, N.S.: Three-dimensional surface imaging in plastic surgery: foundation, practical applications, and beyond. Plast. Reconstr. Surg. **135**(5), 1295–1304 (2015)
5. Cootes, T., Taylor, C., Cooper, D., Graham, J.: Active shape models - their training and application. Comput. Vis. Image Underst. **61**(1), 38–59 (1995)
6. Daoudi, M., Srivastava, A., Veltkamp, R.: 3D Face Modeling, Analysis and Recognition. Wiley, UK (2013)
7. Fagertun, J., Harder, S., Rosengren, A., Moeller, C., Werge, T., Paulsen, R.R., Hansen, T.F.: 3D facial landmarks: inter-operator variability of manual annotation. BMC Med. Imaging **14**(1), 1 (2014)

8. Harder, S., Paulsen, R.R., Larsen, M., Laugesen, S., Mihocic, M., Majdak, P.: A framework for geometry acquisition, 3-D printing, simulation, and measurement of head-related transfer functions with a focus on hearing-assistive devices. Comput. Aided Des. **75**, 39–46 (2016)
9. Harder, S., Paulsen, R.R., Larsen, M., et al.: A three dimensional children head database for acoustical research and development. In: Proceedings of Meetings on Acoustics, vol. 19, p. 050013. Acoustical Society of America (2013)
10. Klein, S., Pluim, J.P., Staring, M., Viergever, M.A.: Adaptive stochastic gradient descent optimisation for image registration. Int. J. Comput. Vis. **81**(3), 227–239 (2009)
11. Klein, S., Staring, M., Murphy, K., Viergever, M.A., Pluim, J.P.: Elastix: a toolbox for intensity-based medical image registration. IEEE Trans. Med. Imaging **29**(1), 196–205 (2010)
12. Li, B.Y., Mian, A.S., Liu, W., Krishna, A.: Using kinect for face recognition under varying poses, expressions, illumination and disguise. In: 2013 IEEE Workshop on Applications of Computer Vision (WACV), pp. 186–192. IEEE (2013)
13. Li, H., Sumner, R.W., Pauly, M.: Global correspondence optimization for non-rigid registration of depth scans. Eurographics Symp. Geom. Process. **27**(5), 1421–1430 (2008)
14. Paulsen, R.R., Hilger, K.: Shape modelling using markov random field restoration of point correspondences. In: Proceedings of Information Processing in Medical Imaging, pp. 1–12 (2003)
15. Paulsen, R.R., Larsen, R.: Anatomically plausible surface alignment and reconstruction. In: Theory and Practice of Computer Graphics, pp. 249–254 (2010)
16. Paulsen, R., Bærentzen, J., Larsen, R.: Markov random field surface reconstruction. IEEE Trans. Vis. Comput. Graph. **16**(4), 636–646 (2010)
17. Sumner, R.W., Schmid, J., Pauly, M.: Embedded deformation for shape manipulation. In: ACM Transactions on Graphics (TOG), vol. 26, p. 80. ACM (2007)
18. Tam, G.K., Cheng, Z.Q., Lai, Y.K., Langbein, F.C., Liu, Y., Marshall, D., Martin, R.R., Sun, X.F., Rosin, P.L.: Registration of 3D point clouds and meshes: a survey from rigid to nonrigid. IEEE Trans. Vis. Comput. Graph. **19**(7), 1199–1217 (2013)
19. Van Kaick, O., Zhang, H., Hamarneh, G., Cohen-Or, D.: A survey on shape correspondence. In: Computer Graphics Forum, vol. 30, pp. 1681–1707. Wiley Online Library (2011)
20. Yan, P., Bowyer, K.W.: Biometric recognition using 3D ear shape. IEEE Trans. Pattern Anal. Mach. Intell. **29**(8), 1297–1308 (2007)

Collaborative Representation of Statistically Independent Filters' Response: An Application to Face Recognition Under Illicit Drug Abuse Alterations

Raghavendra Ramachandra[(✉)], Kiran Raja, Sushma Venkatesh,
and Christoph Busch

Norwegian Biometrics Laboratory,
Norwegian University of Science and Technology (NTNU), Gjøvik, Norway
{raghavendra.ramachandra,kiran.raja,sushma.venkatesh,
christoph.busch}@ntnu.no

Abstract. Face biometrics is widely deployed in many security and surveillance applications that demand a secure and reliable authentication service. The performance of face recognition systems is primarily based on the analysis of texture and geometric variation of the face. Continuous and extensive consumption of illicit drugs will significantly result in deformation of both texture and geometric characteristics of a face and thus, impose additional challenges on accurately identifying the subjects who abuse drugs. This work proposes a novel scheme to improve robustness of face recognition system to address the variations caused by the prolonged use of illicit drugs. The proposed scheme is based on the collaborative representation of statistically independent filters whose responses are computed on the face images captured before and after substance (or drug) abuse. Extensive experiments are carried out on the publicly available Illicit Drug Abuse Database (DAD) comprised of face images from 100 subjects. The obtained results indicate better performance of the proposed scheme when compared with six different state-of-the-art approaches including a commercial face recognition system.

Keywords: Biometrics · Face recognition · Drug abuse · Collaborative representation · Statistical features · Texture features

1 Introduction

Face recognition systems are widely used in various access control applications including law enforcement. The ease of operation and the non-intrusive process of capturing face have further increased their applicability for different authentication applications. However, the face biometric system's performance is known to degrade due to the variation in expression, ageing, illumination, etc. The availability of large-scale face databases and high performing deep learning techniques [5] has enabled a new generation of face recognition systems, which are

© Springer International Publishing AG 2017
P. Sharma and F.M. Bianchi (Eds.): SCIA 2017, Part II, LNCS 10270, pp. 448–458, 2017.
DOI: 10.1007/978-3-319-59129-2_38

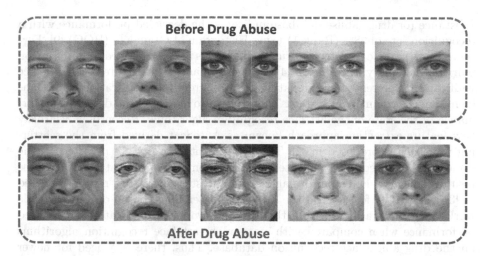

Fig. 1. Face images of the subjects before and after drug abuse [4]. Images compiled from www.facesofmeth.us

robust to these variations. Nevertheless, remaining challenges for face recognition such as a beautification surgery [7], make-up [6], gender discriminablility problem for transgender subjects [8], facial disguise [6] remain due to prominent changes in both texture and geometric facial characteristics which significantly impact the performance of face recognition systems. In spite of the significant research conducted to improve the face recognition systems under covariates as mentioned above, the availability of small scale databases (such as, limited or one sample before and after surgery) makes it more challenging to achieve a robust performance.

Recently, a new challenge to face recognition systems was identified due to the prolonged consumption of illicit drugs that can prominently change the physical structure of the face [3,4]. Figure 1 shows the example of faces before and after illicit substance abuse that caused noticeable changes in appearance of the face characteristic. Furthermore, the availability of only single sample before and single sample after substance abuse, and the fact that such samples are taken in an un-constrained environment makes the problem even more challenging. To achieve good recognition performance on such data, with strong intra-class variation, a robust face recognition needs to be designed which remains a challenge still. The significant change in the facial structure is caused by suppressed appetite that leads to undernourishment which further leads the body to consume muscle tissue and facial fat [4]. This process will result in a gaunt and hollowed-out appearance of the face leading to exhaustive changes in facial geometry.

The early work on illicit drug abuse in the context of face recognition was first introduced by [3,4]. In [3], the performance of the state-of-the-art face recognition system including two different commercial-off-the shelf systems is presented. The recognition performance of the state-of-the-art schemes evaluated on the substance (or drug) abuse database indicates the degraded performance with a recognition rate of 38% (Rank 1) using Histogram of Gradients (HOG) [9]. Further, a new scheme to detect a drug abuse face is also introduced in [3] based on pairwise dictionary learning. In [4], the performance of the state-of-the-art face verification is presented by evaluating eight different face verification algorithms including one commercial system. This study also indicated the degraded performance of the state-of-the-art face recognition algorithms. Recently, a relatively new feature extraction approach known as AutoScat derived from the scattering wavelet was proposed in [2]. The experimental results are reported on the same database used in [3] indicated a recognition rate of 32% at Rank 1. These results indicate a lower performance than the state-of-the-art results reported in [3] using HOG features on the same database. Thus, based on the available work, the results achieved from the available works has indicated the reduced performance when compared with state-of-the-art face recognition algorithms on the drug abuse face recognition database. Thus, there is a need for newer algorithms to significantly address the texture and geometric variations of the face before and after drug abuse to improve the biometric performance.

In this work, we present a novel scheme based on the collaborative representation of the statistically independent filters' response computed on the face images before and after drug abuse. Given the face image, the proposed scheme will first detect and segment the face region which is then normalized to a size of 120×120 pixels. In the next step, we subdivide the normalized face images into six non-overlapping patches such that three patches are obtained horizontally and remaining three patches are obtained vertically as illustrated in Fig. 3. The key motivation for subdivision in non-overlapping region is to obtain at-least few regions that are not physiologically changed to greater extent. We then process each of these patches to extract statistically independent texture features using both multi-scale and multi-bit Binarized Statistical Image Features (BSIF) [10]. The Multi-scale and Multi-bit BSIF has 56 different filters (or kernels) with a varying scale ranging from 5×5, 7×7, 9×9, 11×11, 13×13, 15×15 and 17×17 with varying bits ranging from 5, 6, 7, 8, 9, 10, 11 and 12 according to different filter size. The features obtained for each of these filters on individual patches are processed independently and classified using a probabilistic Collaborative Representation Classifier (Pro-CRC) [11]. Based on the comparison score generated using Pro-CRC, we obtain a ranked list of identities by sorting a comparison score in the descending order. Since we have six patches and 56 different filter kernels from BSIF that are used to extract the features independently, we have $56 \times 6 = 336$ rank lists for each face image. We then perform the rank level fusion using a majority voting rule to obtain the final rank list to identify the person. Thus, it is our assertion that the use of non-overlapping face image patch together with the features extracted independently using 56 different kernels from BSIF filter bank is expected to handle the variations due to both texture and geometric structure of the face to achieve the improved face recognition after drug abuse. The main contributions of this work are as listed below:

- Presents a novel scheme based on the collaborative representation of statistically independent filters whose responses are computed using 56 different kernels (or filters) from BSIF filter banks on the drug abuse face images.
- Presents extensive experiments that are carried out on the publicly available Illicit Drug Abuse Database (DAD) [4] comprised of face images of 100 subjects.
- Reports extensive comparative study by comparing the performance of the proposed scheme with six different state-of-the-art techniques that includes the commercial face recognition software from Neurotechnology, transfer learning approach using Deep Convolutional Neural network (CNN) along with the recently proposed AutoScat features [2].

The rest of the paper is organized as follows: the proposed scheme is discussed in the Sect. 2, the details of the experimental results are presented in the Sects. 3 and 4 draws the conclusion.

2   Proposed Scheme for Drug Abuse Face Recognition

Figure 2 shows the block diagram of the proposed illicit drug abuse face recognition framework that can be structured in two main working components such as: (1) Face detection and normalisation unit (2) Proposed scheme for feature extraction, classification and rank-level fusion to achieve improved face recognition.

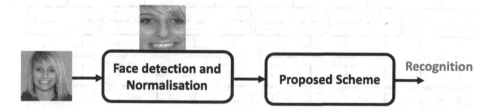

Fig. 2. Block diagram of the proposed face recognition scheme for subjects who abuse illicit drugs.

2.1   Face Detection and Normalisation

Given an image I, the first step is to detect and normalize the face images. The face detection is carried out by employing the Viola-Jones algorithm [1] by considering it's robustness and performance in a real-time scenario. Due to the unconstrained capture of the face images, the use of face detection technique has resulted in a few false detections that are rectified using the technique described in [13]. In the next step, the face image is normalized to compensate rotation using the affine transform as mentioned in [12]. The final normalized face images I_N is of 120×120 pixels.

2.2   Proposed Scheme

Figure 3 illustrates the block diagram of the proposed scheme for robust face recognition of subjects who abuse drugs. The primary challenge with faces of subjects who abuse drugs is to address the variations in textural features that are due to the presence of random moles and acnes, non-uniform deformation of the face structure due to the loss of facial muscles. Since these variations are random in different parts of the face and also across the subjects, based on the metabolic activity as a result of drug consumption, we are motivated to approach this problem using a patch-based paradigm. Given the normalised face image I_N, we obtain six non-overlapping face image blocks $I_{Bi} = \{I_{B1}, I_{B2}, \ldots, I_{B6}\}$ both vertically and horizontally. Figure 3 illustrates the six blocks I_{Bi} obtained on the normalised face image I_N. We then consider each block I_{Bi} at a time to extract the features using 56 different filters (or kernels) from the BSIF filter bank [10].

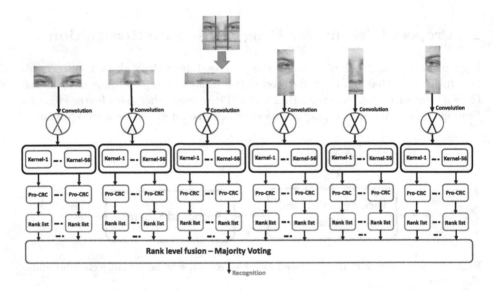

Fig. 3. Block diagram of the proposed scheme

In this work, we employed the open-source BSIF filter bank [10] to extract the features corresponding to each face image block I_{Bi}. The BSIF filters are learned in an unsupervised manner using an Independent Component Analysis (ICA) on the set of image patches extracted from thirteen different natural images. The natural images are first divided to have 50000 images patches which are then mean subtracted followed by dimensionality reduction using Principal Component Analysis (PCA), then used to learn the filters by employing ICA. The learned ICA basis will form the filters that are statistically independent. Thus, depending on the size of the natural image patches and the selection of a top number of basis from ICA, one can learn the bank of filters with different size and length (or bit). For instance, the BSIF filter of size 7×7 with 10 bits (or length) indicates that top 10 basis of ICA is selected that is trained using the natural images of size 7×7. In this work, we have considered the filter with seven different scales such as: 5×5, 7×7, 9×9, 11×11, 13×13, 15×15 and 17×17 and eight different bits (or length) such as: 5, 6, 7, 8, 9, 10, 11, 12 to constitute a filter bank with $7 \times 8 = 56$ different statistically independent filters (or kernels). Thus, given the ith face image block I_{Bi} and the BSIF filter $F_b^{s \times s}$ the response is computed as follows [10]:

$$r_i = \sum I_{Bi} * F_b^{s \times s} \tag{1}$$

Where, I_{Bi} denotes the ith face image block, $*$ denotes the convolution operation and $F_b^{s \times s}$ denotes the filter with the size $s \times s \ \forall \ s = \{5, 7, 9, 11, 13, 15, 17\}$ and b denotes the length (or bits) $\forall \ b = \{5, 6, 7, 8, 9, 10, 11, 12\}$ and r_i indicates the response for the ith face image block I_{Bi}.

In the next step, the obtained response r_i is binarized to obtain the binary string as follows [10]:

$$b_i = \begin{cases} 1, & \text{if } r_i > 0 \\ 0, & \text{otherwise} \end{cases} \tag{2}$$

Finally, the BSIF encoded features are obtained as the histogram of pixel's binary codes that can effectively characterize the texture components in the ith face image block I_{Bi}.

The use of BSIF filter bank comprised of independent filters with different size and different length of filters will provide rich feature representation for the given face image block I_{Bi}. Hence, it is our assertion that the use of the features extracted using independent BSIF filters and combining them at the rank level will further reduce the variations and improve the performance of face recognition after drug abuse.

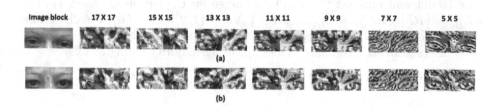

Fig. 4. Illustration of the BSIF features extracted on face image block using different scales with fixed length (or bit) of 8 (a) before drug abuse (b) after drug abuse

Figure 4 illustrates the feature extraction approach when different scale size is used with the fixed length of 8 bit corresponding to the ith face image block before and after drug abuse. It can be observed that, the use of larger scale describes the coarse texture information when compared to that of the small scale size. Figure 5 illustrates the features extracted on both before and after drug abuse face image block I_{Bi} when scale size is fixed to 7×7 and length (or bits) is varied from 5 to 12. Thus, here it can also be observed that, the use of different bits will provide different feature representation. Thus, the application of multi-scale and multi-length filters from the BSIF filter bank can provide robust feature representation to achieve an improved performance for the drug abuse face recognition task.

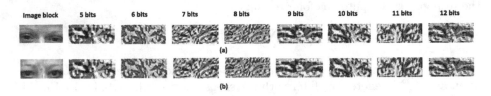

Fig. 5. Illustration of the BSIF features face image block extracted using different length at a fixed scale of 7×7 (a) before drug abuse (b) after drug abuse

In the next step, we employ the probabilistic Collaborative Representation Classifier (Pro-CRC) [11] independently on 56 different feature representations obtained using the BSIF filter bank. We considered to employ the Pro-CRC because of it's robust performance even when only a small training set is available. Such a constraint suits well to our application with one sample per subject available for training. The primary idea of the Pro-CRC is to jointly maximize the likelihood that a test sample belongs to each of the multiple subjects and finally classify the test sample to the subject with maximum likelihood. The features extracted from each of the 56 BSIF filters are classified independently using Pro-CRC to obtain the corresponding comparison scores. Finally, the comparison scores are sorted in the descending order to obtain the ranked list of the possible user identities. Since, we have 56 filters that are used independently, we have 56 different rank list that can be obtained for the face image block I_{Bi} as: $RL_{Bi}^{x} = \left\{ RL_{Bi}^{1}, RL_{Bi}^{2}, RL_{Bi}^{3}, \ldots, RL_{Bi}^{56} \right\}, \forall x = \{1, 2, 3, 4, \ldots 56\}$. Finally, given the test face image I_N, the recognition is performed by combining the rank list of 56 filters from six face image blocks I_{Bi} using majority voting as follows:

$$ Fu_{RL} = MJ \left\{ RL_{B1}^{1:56}, RL_{B2}^{1:56}, RL_{B3}^{1:56}, RL_{B4}^{1:56}, RL_{B5}^{1:56}, RL_{B6}^{1:56} \right\} \tag{3} $$

Where, MJ indicates the majority voting, Fu_{RL} indicates the final fused rank list and $\left\{ RL_{B1}^{1:56}, \ldots RL_{B6}^{1:56} \right\}$ represents the rank list obtained for each face image block corresponding to 56 different BSIF filters.

3 Experiments and Results

In this section, we present the performance of the proposed scheme on the publicly available Drug Abuse face Database (DAD) [4]. This database is similar to that of the Illicit Drug Abuse Face Database (IDAF) [3] and both of these databases are collected through the internet, especially from the 'Face of Meth' webpage [15]. Since IDAF is not publicly available, we have used the Drug Abuse Database for the experiments. The DAD is comprised of frontal face images (mostly) captured from 101 subjects before and after illicit drug abuse. But, in this work, we have selected 100 subjects owing to the reasonable quality of the face images in the DAD databases. Thus, in this work, we have used 100 *subjects* × 2 *samples* = 200 *images*. The results are presented in terms of the Recognition Rate % at Rank-1 to Rank-5. Thus, the higher the value of the Recognition Rate, the better is the biometric performance.

Table 1. Recognition performance of the proposed scheme on the DAD database

Algorithms	Recognition Rate (%)	
	Rank 1	Rank 5
LBP - features	10	34
HOG - features	26	45
LG - features	27	40
AutoScat - Features [2]	21	45
COTS-Neurotech	26	43
Deep-CNN-AlexNet	22	44
Proposed method	**52**	**55**

The performance of the proposed method is compared with the six different face recognition systems that include the features extracted using: Local Binary Patterns (LBP), Histogram of Gradients (HOG), Log-Gabor (LG), AutoScat features [2] and commercial face recognition system from Neurotechnology. In addition to these face recognition systems, we have also evaluated the performance of the Deep Convolutional Neural Network (CNN) based on transfer learning paradigm. To this extent, we have used the AlexNet [14] in which the last fully connected layer is retrained using the face images from DAD database. Since the DAD database has only one sample for each subject, we have carried out the data augmentation using random crop to retrain the deep CNN. Finally, the features of the last layer are classified using linear Support Vector Machine (SVM) to obtain the performance-quantified using Recognition Rate (%). To the best of our knowledge, the deep CNN is utilized for the first time with the application to recognize the faces affected with illicit drug abuse.

Table 1 presents the quantitative results of the proposed scheme along with the comparative methods in terms of Recognition Rate (%) at Rank-1 and also Rank-5. The results are also presented in terms of the Cumulative Match Characteristics (CMC) plots as shown in the Fig. 7. The following are key findings:

- The proposed method yields the best performance with a Rank-1 recognition rate of 52%. Thus, compared to the state-of-the-art face recognition

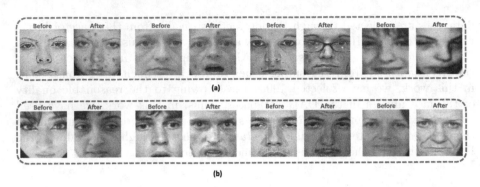

Fig. 6. Sample results demonstrating the performance of the proposed scheme with (a) correct Rank 1 recognition (b) in-correct Rank 1 recognition

techniques including the AutoScat [2], the commercial face recognition system and deep-CNN, the proposed scheme has indicated the best performance.
- The proposed scheme has indicated the improvement of Recognition Rate (%) at Rank-1 with 26% and Rank-5 with 10% when compared to the second best performing technique based on the HOG-features.
- The CMC curves indicate the performance of the proposed scheme together with state-of-the-art algorithms for varying ranks from 1 to 7. It can be observed here that, the proposed system has demonstrated high performance consistently across all ranks when compared with six different state-of-the-art schemes.

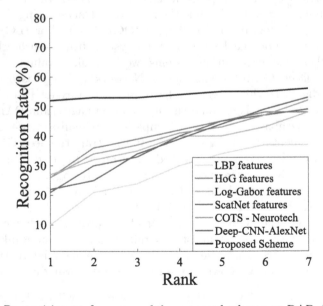

Fig. 7. Recognition performance of the proposed scheme on DAD database

- Figure 6 illustrates the example image pairs that are correctly (see Fig. 6(a)) and incorrectly (see Fig. 6(b)) recognised at Rank-1 using the proposed scheme. Based on the obtained results, the proposed system is capable of correctly identifying the subjects with major physiological variations, occlusion (presence of spectacles), little expressions along with varying image quality. However, most of the failed identification cases are due to the larger variations due to fast paced ageing appearance in addition to the variations of the prolonged use of illicit drugs.

4   Conclusion

In this work, we address a face recognition problem especially related to illicit drug abuse variations that will deform both local and global characteristics of the face images. We proposed a novel framework based on the block image processing and the collaborative representation of the statistically independent features extracted using 56 different filters (or kernels) from BSIF filters bank. The proposed method addresses the variations due to the drug abuse by dividing the face image into six non-overlapping image block which is then processed using BSIF filter bank to have 56 different feature representation. Then, each of these 56 different feature representation is classified independently using probabilistic Collaborative Representation Classifier (Pro-CRC) to obtain comparison scores. These scores are then sorted in the descending order to the obtain a rank list corresponding to all user identities. Finally, the rank list corresponding to 56 different feature representation are combined using the majority voting to obtain the recognition accuracy. Extensive experiments are carried out on the publicly available Illicit Drug Abuse Database (DAD) comprised of faces from 100 substance users. The performance of the proposed scheme is compared with six different state-of-the-art systems including a commercial face recognition system and transfer learning using deep CNN. The experimental results have demonstrated the improved performance from the proposed scheme for recognizing the subjects who abuse the drugs.

Acknowledgements. This work is carried out under the funding of the Research Council of Norway (Grant No. IKTPLUSS 248030/O70).

References

1. Viola, P., Jones, M.J.: Robust real-time face detection. Int. J. Comput. Vis. **57**(2), 137–154 (2004)
2. Pandey, P., Singh, R., Vatsa, M.: Face recognition using scattering wavelet under Illicit drug abuse variations. In: International Conference on Biometrics (ICB), Halmstad, pp. 1–6 (2016)
3. Yadav, D., Kohli, N., Pandey, P., Singh, R., Vatsa, M., Noore, A.: Effect of Illicit drug abuse on face recognition. In: IEEE Winter Conference on Applications of Computer Vision (WACV), Lake Placid, NY, pp. 1–7 (2016)

4. Raghavendra, R., Raja, K., Busch, C.: Impact of drug abuse on face recognition systems: a preliminary study. In: Proceedings of the 9th International Conference on Security of Information and Networks (SIN-2016), New York, NY, USA, pp. 24–27 (2016)
5. Zhi-Peng, F., Yan-Ning, Z., Hai-Yan, H.: Survey of deep learning in face recognition. In: 2014 International Conference on Orange Technologies, Xian, pp. 5–8 (2014)
6. Dantcheva, A., Chen, C., Ross, A.: Can facial cosmetics affect the matching accuracy of face recognition systems? In: IEEE Fifth International Conference on Biometrics: Theory, Applications and Systems (BTAS), pp. 1–8 (2012)
7. De Marsico, M., et al.: Robust face recognition after plastic surgery using local region analysis. In: International Conference Image Analysis and Recognition, pp. 1-8. Springer Heidelberg (2011)
8. Kumar, V., Raghavendra, R., Namboodiri, A., Busch, C.: Robust transgender face recognition: approach based on appearance and therapy factors. In: 2016 IEEE International Conference on Identity, Security and Behavior Analysis (ISBA), Sendai, pp. 1–7 (2016)
9. Dalal, N., Triggs, B.: Histograms of oriented gradients for human detection. In: IEEE Computer Society Conference on Computer Vision and Pattern Recognition (CVPR) (2005)
10. Kannala, J., Rahtu, E.: BSIF: binarized statistical image features. In: 21st International Conference on Pattern Recognition (ICPR), pp. 1–5 (2012)
11. Cai, S., Zhang, L., Zuo, W., Feng, X.: A probabilistic collaborative representation based approach for pattern classification. In: 2016 IEEE Conference on Computer Vision and Pattern Recognition (CVPR), Las Vegas, NV, USA, pp. 2950–2959 (2016)
12. Struc, V.: The PhD face recognition toolbox: toolbox description and user manual. Faculty of Electrical Engineering, Ljubljana (2012)
13. Raghavendra, R., Yang, B., Busch, C.: Accurate face detection in video based on likelihood assessment. Norsk Informasjons Sikkerhets Konferanse (NISK 2012) (2012)
14. Krizhevsky, A., Sutskever, I., Hinton, G.E.: Imagenet classification with deep convolutional neural networks. In: Advances in Neural Information Processing Systems, pp. 1097–1105 (2012)
15. Faces of Meth. http://www.facesofmeth.us. Accessed 30 Jan 2017

Multispectral Constancy Based on Spectral Adaptation Transform

Haris Ahmad Khan[1,2](✉), Jean Baptiste Thomas[1,2],
and Jon Yngve Hardeberg[1]

[1] The Norwegian Colour and Visual Computing Laboratory,
NTNU - Norwegian University of Science and Technology, Gjøvik, Norway
{haris.a.khan,jean.b.thomas,jon.hardeberg}@ntnu.no
[2] Le2i, FRE CNRS 2005, Univ. Bourgogne Franche-Comté, Dijon, France

Abstract. The spectral reflectance of an object surface provides valuable information about its characteristics. Reflectance reconstruction
from multispectral images is based on certain assumptions. One of these
assumptions is that the same illumination is used for system calibration and image acquisition. We propose the novel concept of multispectral constancy, achieved through a spectral adaptation transform, which
transforms the sensor data acquired under an unknown illumination to a
generic illuminant-independent space. The proposed concept and methods are inspired from the field of computational color constancy. Spectral
reflectance is then estimated by using a generic linear calibration. Results
of reflectance reconstruction using the proposed concept show that it is
efficient, but highly sensitive to the accuracy of illuminant estimation.

Keywords: Reflectance reconstruction · Multispectral constancy ·
Illuminant estimation · Spectral adaptation transform

1 Introduction

Multispectral imaging refers to the acquisition of image data at specific intervals
in the electromagnetic spectrum by the use of spectral filters. These spectral
filters are sensitive to specific wavelengths and allow acquisition of data in form
of channels. Color images contain three channels while a multispectral image
consists of more than three channels. With advancement in the sensor technology, use of multispectral imaging for indoor scene acquisition under controlled
conditions has also increased. The advantage of multispectral imaging is the
ability to acquire more spectral information of a scene, which can be used for
spectral reflectance reconstruction [1] of the object's surfaces. A constraint associated with such use of multispectral imaging is the need of system calibration
for the imaging environment when performing the spectral reconstruction [2]. A
system being calibrated for one type of conditions needs to be re-calibrated if
it has to be used in another imaging environment. One of the major limitations
is the need for having the same illuminant for spectral reconstruction system's

© Springer International Publishing AG 2017
P. Sharma and F.M. Bianchi (Eds.): SCIA 2017, Part II, LNCS 10270, pp. 459–470, 2017.
DOI: 10.1007/978-3-319-59129-2_39

calibration and the image acquisition. This limitation is a major challenge and hurdle for the use of multispectral imaging in outdoor environments.

In this paper, we introduce the concept of multispectral constancy, as an extension of the computational color constancy [3] but in higher spectral dimension. By multispectral constancy, we refer to the representation of imaged surfaces without the effect of the scene illuminant. In other words, the surfaces appear as if they are taken under a canonical illuminant. To achieve multispectral constancy, we introduce a spectral adaptation transform (SAT). The closest concept to SAT in the literature is spectral adaptation being introduced by Fairchild [4] and is applied on the full spectral reflectances, while we generalize it to the sensor measurements. By attaining multispectral constancy, multispectral imaging can be used for any illuminant without the requirement of re-calibration when imaging environment is changed.

This paper is organised as follows. Section 2 formalises the problem of spectral acquisition, including the calibration and spectral reconstruction as a linear problem. We propose to add a term to the usual calibration, which discards the illumination change. Section 3 defines our simulation and experimental protocol. Results are analyzed in Sect. 4 before we conclude.

2   Spectral Adaptation Transform

In the context of a simplified noiseless imaging model, a pixel captured in the imaging system is the combination of spectral reflectance of the surface $r(\lambda)$, spectral power distribution of illuminant $e(\lambda)$ and spectral sensitivity of the nth spectral channel $m_n(\lambda)$. This formation for the visible wavelength spectrum ω is defined as

$$f_n = \int_\omega r(\lambda)e(\lambda)m_n(\lambda)d\lambda \tag{1}$$

In practice, we can formulate a discrete version of Eq. 1 as $\mathbf{F} = \mathbf{REM}$. Considering the spectral sampling of 10 nm within the wavelength spectrum of 400 nm to 720 nm and N number of spectral filters, $\mathbf{F}$ is $S \times N$ matrix (S is the number of spectral samples being acquired), $\mathbf{R}$ is $S \times 33$ matrix of surface reflectance, $\mathbf{E}$ is the diagonal matrix (33×33) of the scene illuminant and $\mathbf{M}$ is $33 \times N$ matrix, consisting of spectral sensitivities of the filters.

Here we consider two cases of image acquisition. One is with a canonical illumination $\mathbf{E}_c$ and the other case is when an unknown illuminant $\mathbf{E}_{ill}$ is used. We present both cases in parallel in Eq. 2.

$$\mathbf{F}_{ill} = \mathbf{RE}_{ill}\mathbf{M} \quad ; \quad \mathbf{F}_c = \mathbf{RE}_c\mathbf{M} \tag{2}$$

To perform the spectral reflectance estimation $\hat{\mathbf{R}}$ from the imaged data in both of the above mentioned cases, we can apply a generalized inverse, denoted by $^+$, as in Eq. 3.

$$\hat{\mathbf{R}} = \mathbf{F}_{ill}\mathbf{M}^+\mathbf{E}_{ill}^+ \quad ; \quad \hat{\mathbf{R}} = \mathbf{F}_c\mathbf{M}^+\mathbf{E}_c^+ \tag{3}$$

For the reconstruction of spectral reflectance, knowledge of illumination is required. There are many methods to estimate the sensor sensitivity $\mathbf{M}$ for a given camera [5,6]. However, it is not an easy task to measure the scene illuminant every time along with the image acquisition.

The spectral reconstruction is performed by training a calibration matrix $\mathbf{W} = \mathbf{M}^+\mathbf{E}^+$ [7]. This calibration is specific for a given illumination $\mathbf{E}_c$. By using the calibration matrix $\mathbf{W}$, the equations for spectral reflectance reconstruction become $\hat{\mathbf{R}} = \mathbf{F}_{ill}\mathbf{W}_{ill}$ and $\hat{\mathbf{R}} = \mathbf{F}_c\mathbf{W}_c$ for both cases, respectively. We propose to transform the acquired image $\mathbf{F}_{ill}$ into a canonical representation $\mathbf{F}_c$. In this way, $\mathbf{W}_c$ can be used for the spectral reconstruction from a multispectral image, being taken under any illumination. We call this concept as the multispectral constancy. It is achieved through SAT, which is in the form of a diagonal matrix $\mathbf{A}_{SAT}$ and is applied as,

$$\mathbf{F}_c = \mathbf{A}_{SAT}\mathbf{F}_{ill}. \tag{4}$$

In Eq. 4, SAT being applied to the acquired spectral data enables the estimation of reflectance spectra from an image being taken under any illuminant and can be used as,

$$\hat{\mathbf{R}} = \mathbf{W}_c\mathbf{A}_{SAT}\mathbf{F}_{ill} \tag{5}$$

With the use of SAT as in Eq. 5, the requirement of having same illuminant for spectral reconstruction system's calibration and the image acquisition, is no longer required since the acquired spectral image is transformed before the estimation of spectra. Obtaining $\mathbf{A}_{SAT}$ and $\mathbf{W}$ are explained in Sects. 3.3 and 3.4, respectively.

3 Methodology and Experimental Setup

3.1 Simulation Pipeline

Figure 1 shows the pipeline of experimental framework for the spectral reconstruction, based on multispectral constancy. This pipeline consists of sensor simulation, acquisition of spectral data, SAT, spectral reconstruction and the evaluation of results. These blocks are explained in the following sections.

3.2 Sensor

To implement and validate the proposed idea of multispectral constancy, we use reflectance data in the wavelength range of 400 nm to 720 nm, from the Gretag-Macbeth ColorChecker [8]. We apply equi-Gaussian filters [9] for simulation of the spectral filters and use 8 filters in the experiments. By increasing the number of filters, more noise is introduced in the image which affects the spectral reconstruction. This effect was observed by Wang *et al.* [7]. Radiance data is created by using the illuminants E and D65 and then the simulated sensors are used to acquire the multispectral data.

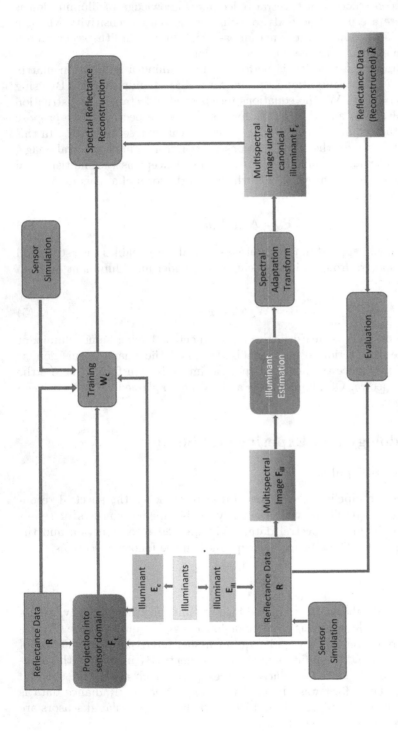

Fig. 1. Pipeline for evaluation of the proposed framework of multispectral constancy. Spectral measurements from Macbeth ColorChecker are used as the reflectance data. Illuminants $\mathbf{E}_c$ and $\mathbf{E}_{ill}$ are used for training and testing, respectively. To achieve multispectral constancy, SAT is applied to the acquired spectral data after estimation of illuminant in the sensor domain. Result from spectral reconstruction of transformed data is evaluated with the measured spectra of the Macbeth ColorChecker.

3.3 Multispectral Constancy Through SAT

In order to achieve multispectral constancy, SAT is applied to the spectral image after the estimation of illuminant in the sensor domain as in Eq. 5. For estimation of the illuminant in spectral images of natural scenes, we propose using the Max-Spectral Algorithm and the Spectral Gray-Edge Algorithm. These algorithms are extensions of Max-RGB Algorithm [10] and the Gray-Edge Algorithm [11,12]. The extension of these algorithms from color to spectral is discussed in detail in [9,13]. Once the illuminant is estimated, then we propose to apply the SAT in form of a diagonal correction to the acquired data, so that it appears as if being taken under a canonical illuminant. Such a diagonal transform was initially proposed by Von Kries [14]. We extend this transform into the spectral domain. For N number of channels, SAT is defined as in Eq. 6.

$$\begin{pmatrix} F_1^c \\ F_2^c \\ \vdots \\ F_N^c \end{pmatrix} = \begin{pmatrix} k_1 & 0 & \dots & 0 \\ 0 & k_2 & \dots & 0 \\ \vdots & \vdots & \ddots & \vdots \\ 0 & 0 & \dots & k_N \end{pmatrix} \cdot \begin{pmatrix} F_1^u \\ F_2^u \\ \vdots \\ F_N^u \end{pmatrix} \tag{6}$$

Here, F_n^u is the pixel of nth channel, taken under an unknown illuminant while F_n^c is the transformed images so that it appears to be taken under a canonical illuminant. k_n is the correction parameter for the channel n, which is obtained from the illuminant estimation (IE).

3.4 Spectral Reflectance Reconstruction

As explained in Sect. 2, a training matrix is required for the spectral reconstruction from camera data. This matrix $\mathbf{W}$ is called the calibration matrix. It is obtained by using measured reflectance spectra $\mathbf{R}_t$ and the acquisition of radiance data $\mathbf{F}_t$, using the same camera. We use reflectance data from the GretagMacbeth ColorChecker [8] $\mathbf{R}_t$ and illuminant E for calibration of $\mathbf{W}$.

There are many methods being proposed for spectral reconstruction. We use linear methods to keep the proposed system simple and robust [2]. Those linear methods include the linear least square regression, principal component analysis [15] and Wiener estimation [16]. We evaluated the performance of these three methods and got similar results. We decided to use the Wiener estimation method because it is robust to noise and fulfills the criteria of being linear. It is defined as

$$\mathbf{W} = \mathbf{R}_t \mathbf{R}_t^T (\mathbf{ME})^T ((\mathbf{ME}) \mathbf{R}_t \mathbf{R}_t^T (\mathbf{ME})^T + \mathbf{G})^{-1} \tag{7}$$

Here, $\mathbf{R}_t \mathbf{R}_t^T$ and $\mathbf{G}$ are the autocorrelation matrices of training spectra and additive noise, respectively. G is in the form of a diagonal matrix consisting of the variance of noise σ^2.

One of the major shortcomings of the linear methods for spectral reconstruction is the assumption that the same illuminant is used for system's calibration and acquisition of data. We are interested in the development of a spectral reconstruction system which does not require the same illuminant for system's calibration and data acquisition. This can be achieved through the illuminant estimation in spectral images and then applying SAT to the acquired data as in Eq. 5. The matrix $\mathbf{W}$ is obtained by using the training reflectance spectra and radiance with the canonical illuminant $\mathbf{E}_c$ (we use illuminant E). This matrix is used for the spectral reconstruction from the spectral data being taken under any lighting conditions.

We evaluate the validity of our proposed idea by performing an experiment on the measured reflectance data of Macbeth ColorChecker for spectral reflectance reconstruction. Those results are provided in Sect. 4.

3.5   Evaluation

To measure the performance of spectral reconstruction, we compare the reconstruction $\hat{\mathbf{r}}$ for each color patch of the Macbeth ColorChecker with the measured reflectance $\mathbf{r}$, through root mean square error (RMSE) and goodness of fit coefficient (GFC) [17] as in Eqs. 8 and 9 respectively.

$$\text{RMSE} = \sqrt{\frac{1}{33}\sum_{j=1}^{N}(r_j - \hat{r}_j)^2} \tag{8}$$

$$\text{GFC} = \frac{\mathbf{r}^T\hat{\mathbf{r}}}{\sqrt{(\mathbf{r}^T\mathbf{r})(\hat{\mathbf{r}}^T\hat{\mathbf{r}})}} \tag{9}$$

Besides that, we also provide spectral reconstruction results for the reflectance patches of the Macbeth ColorChecker in the form of graphs so that the overall performance of spectral reconstruction can be visually analyzed.

4   Results and Discussion

In this section we analyze the validity of our proposed idea of multispectral constancy through SAT. We also investigate the influence of illuminant estimation on the results of spectral reconstruction. We use three different noisy estimates of illuminant for testing the proposed framework to assess the influence of erroneous illuminant estimation.

For training of matrix $\mathbf{W}$, 24 patches of Macbeth ColorChecker are used within the visible wavelength range (400–720 nm). For spectral reconstruction, we use different scenarios, which are given in Table 1.

For testing the influence of error in the illuminant estimation, we use three different noisy estimates. Angular error (ΔA) is calculated in term of radians between the vectors of original illuminant $\mathbf{e}$ and the estimated illuminant $\hat{\mathbf{e}}$ as

$$\Delta A = \arccos \frac{\mathbf{e}^T\hat{\mathbf{e}}}{\sqrt{(\mathbf{e}^T\mathbf{e})(\hat{\mathbf{e}}^T\hat{\mathbf{e}})}} \tag{10}$$

Table 1. Description of the experiments being performed. Adequate calibration of system is performed by using the same illuminant. In no correction case, we use different illuminants and do not apply any correction. In ideal correction, SAT is applied by assuming efficient estimation of the illuminant while in naïve correction, SAT is applied using the illuminant E, while D65 is used for image acquisition. In next three experiments, we apply SAT using the three different noisy estimate of illuminants.

Experiment	Illuminant for training	Illuminant for acquisition	SAT applied	Illuminant for SAT
Adequate calibration	D65	D65	No	–
No correction	E	D65	No	–
Ideal correction	E	D65	Yes	D65
Naïve correction	E	D65	Yes	E
Noisy estimate 1	E	D65	Yes	Estimated illuminant with $\Delta A = 0.0210\,\mathrm{rad}$
Noisy estimate 2	E	D65	Yes	Estimated illuminant with $\Delta A = 0.1647\,\mathrm{rad}$
Noisy estimate 3	E	D65	Yes	Estimated illuminant with $\Delta A = 0.3658\,\mathrm{rad}$

For checking the effect of error in illuminant estimation, use three different randomly generated estimates of illuminant with ΔA of 0.0210, 0.1647 and 0.3658 rad. We apply these erroneous illuminants in the sensor domain to the acquired spectral image of Macbeth ColorChecker according to Eq. 6. First we evaluate spectral reconstruction and then analyze the performance of SAT. Figures 2, 3 and 4 show results obtained from the 24 reflectance patches of the Macbeth ColorChecker in the visible spectrum. These results are discussed in the following sections.

4.1 Reflectance Estimation

With the use of linear method for spectral reconstruction (Wiener estimation [16]), we evaluate the performance of the algorithm using the adequate calibration of system. We measure the performance with both illuminants E and D65. They provide similar results. We show results of adequate calibration with illuminant D65 in Figs. 2, 3 and 4. This is the best reconstruction that can be obtained with this given number of sensors and sensor configuration. We investigate the performance of Wiener estimation when different illuminants are used for training and testing and no SAT is applied. We also perform SAT with illuminant E in sensor domain and the camera data being acquired with illuminant D65. We call this naïve correction. With adequate calibration, the best spectral

Table 2. Mean RMSE and mean GFC from spectral reconstruction of 24 patches of the Macbeth ColorChecker.

Experiment	RMSE	GFC
Adequate calibration	0.0011	0.9993
No correction	0.0118	0.9898
Ideal correction	0.0081	0.9973
Naïve correction	0.0091	0.9939
Noisy estimate 1	0.0106	0.9955
Noisy estimate 2	0.0128	0.9853
Noisy estimate 3	0.0240	0.9437

reconstruction results we could obtain with Wiener estimator, provided an average RMSE of 0.001 and average GFC of 0.9993 over the 24 patches of Macbeth ColorChecker. With no correction being applied and using different illuminants for training and acquisition, average RMSE and GFC were 0.0118 and 0.9898 respectively. To validate the idea of multispectral constancy through SAT, the error in spectral reconstruction must be low as compared to the error in case of applying no correction. We evaluate the performance of SAT in Sect. 4.2.

4.2 SAT Performance

Figures 2, 3 and 4 show the spectral reconstruction results from the multispectral data acquired using eight spectral filters. Although exact spectral reconstruction is not possible with a reduced number of bands, Wiener estimation method is still able to make a close match when adequate calibration is applied. For testing the performance of SAT, we calibrate the spectral reconstruction system for illuminant E and acquire the radiance data under the illuminant D65. It is obvious from the spectral reconstruction results that the overall accuracy of our proposed framework is dependent on the accuracy of illuminant estimation, as would be expected. With efficient illuminant estimation, SAT performs almost the same as in the case of adequate calibration. However, it is interesting to note that although there is overlapping between the spectral reflectance reconstruction curves in the case of adequate calibration and ideal correction as seen in Figs. 2, 3 and 4, the difference in RMSE and GFC is not as close as expected (see Table 2). This leads to opening the discussion about efficiency of SAT and the question of whether SAT should be optimised as well in order to get better results. Another problem to be investigated is the required efficiency of both SAT and the illuminant estimation for applications like object detection and classification on the basis of their spectral properties. However, the closeness in result proves that if efficient illuminant estimation is performed and SAT is applied, we can attain close results as compared with adequate calibration. The advantage of using multispectral constancy is that there is no requirement of measuring the scene illuminant explicitly. The only factor which remains important in our proposed idea is the efficient estimation of illuminant in the spectral image.

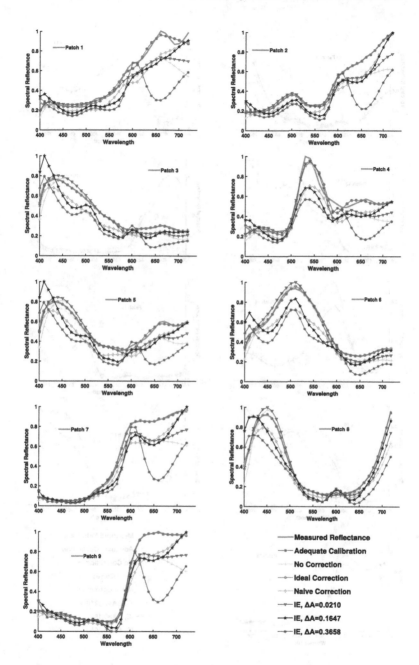

Fig. 2. First 9 reflectance patches from Macbeth ColorChecker. In each figure, there are curves of measured reflectance, spectral reconstruction from adequate calibration, no correction, ideal correction, naïve correction, reflectance reconstruction results after applying SAT from simulated illuminants having ΔA of 0.0210, 0.1647 and 0.3658, respectively.

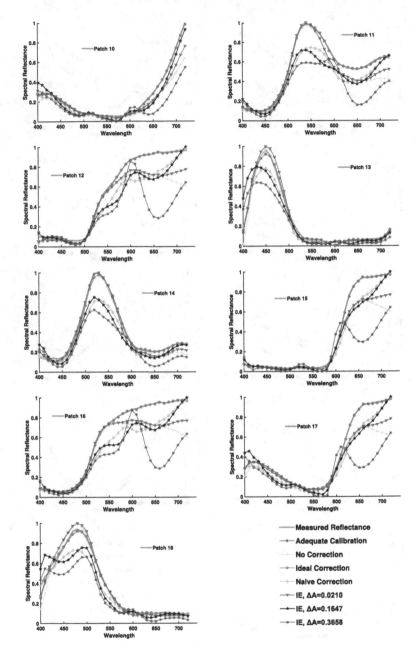

Fig. 3. Reflectance patches 10–18 from the Macbeth ColorChecker. In each figure, there are curves of measured reflectance, spectral reconstruction from adequate calibration, no correction, ideal correction, naïve correction, reflectance reconstruction results after applying SAT from simulated illuminants having ΔA of 0.0210, 0.1647 and 0.3658, respectively.

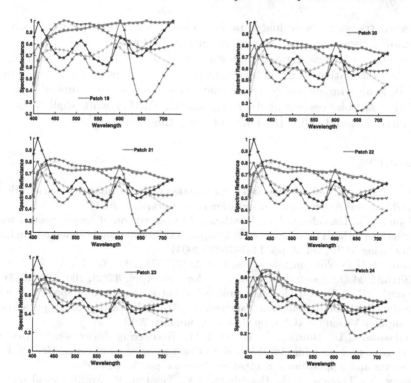

Fig. 4. Reflectance patches 19–24 from the Macbeth ColorChecker.

The dependence of SAT on the accuracy of illuminant estimation is evident when spectral reconstruction is performed. SAT being applied along with erroneous estimation of illuminant in the sensor domain causes error in the spectral reconstruction. It is interesting to note that even the naïve correction is able to perform well in comparison with the result of applying no correction, which makes the role of illuminant estimation as a significant factor for our proposed idea of multispectral constancy through SAT. SAT itself also needs to be investigated so that an efficient framework for the transformation of acquired spectral image into the illuminant free representation (multispectral constancy) can be achieved.

5   Conclusion

This work formalizes the concept of multispectral constancy, which permits multispectral image acquisition, independent of the illumination. Multispectral constancy is achieved via a spectral adaptation transform, which changes data representation from the actual sensor domain, towards a canonical one, where calibration applies. Simulation results show that a diagonal SAT permits to achieve similar reflectance reconstruction than when the samples are acquired under the illumination being used for calibration. However, when the spectral adaptation transform is evaluated based on an estimate of illumination, error in illuminant estimates makes the performance drop down significantly.

It is still to be investigated that what level of accuracy is required for the illuminant estimation to make this concept useful for computer vision applications. It is also important to recall that these results are compiled based on simulation of noiseless sensor data acquisition. Further work shall investigate these directions and define the limits of using this approach.

References

1. Hardeberg, J.Y.: Acquisition and Reproduction of Color Images: Colorimetric and Multispectral Approaches. Dissertation.com, Parkland (2001)
2. Connah, D., Hardeberg, J.Y., Westland, S.: Comparison of linear spectral reconstruction methods for multispectral imaging. In: International Conference on Image Processing, ICIP, vol. 3, pp. 1497–1500 (2004)
3. Foster, D.H.: Color constancy. Vis. Res. **51**(7), 674–700 (2011)
4. Fairchild, M.D.: Spectral adaptation. Color Res. Appl. **32**(2), 100–112 (2007)
5. Jiang, J., Liu, D., Gu, J., Süsstrunk, S.: What is the space of spectral sensitivity functions for digital color cameras? In: IEEE Workshop on Applications of Computer Vision (WACV), pp. 168–179, January 2013
6. Finlayson, G.D., Hordley, S., Hubel, P.M.: Recovering device sensitivities with quadratic programming. In: IS&T Sixth Color Imaging Conference: Color Science, Systems and Applications, Scottsdale, Arizona, pp. 90–95 (1998)
7. Wang, X., Thomas, J.-B., Hardeberg, J.Y., Gouton, P.: Multispectral imaging: narrow or wide band filters? J. Int. Colour Assoc. **12**, 44–51 (2014)
8. McCamy, C.S., Marcus, H., Davidson, J.G.: A color-rendition chart. J. Appl. Photograph. Eng. **2**(3), 95–99 (1976)
9. Thomas, J.-B.: Illuminant estimation from uncalibrated multispectral images. In: Colour and Visual Computing Symposium (CVCS), Gjøvik, Norway, pp. 1–6, August 2015
10. Land, E.H., McCann, J.J.: Lightness and retinex theory. J. Opt. Soc. Am. **61**, 1–11 (1971)
11. van de Weijer, J., Gevers, T.: Color constancy based on the Grey-edge hypothesis. In: IEEE International Conference on Image Processing, vol. 2, p. II–722-5, September 2005
12. van de Weijer, J., Gevers, T., Gijsenij, A.: Edge-based color constancy. IEEE Trans. Image Process. **16**, 2207–2214 (2007)
13. Khan, H.A., Thomas, J.-B., Hardeberg, J.Y., Laligant, O.: Illuminant estimation in multispectral imaging. Submitted to a Journal
14. von Kries, J.: Influence of adaptation on the effects produced by luminous stimuli. In: MacAdam, D.L. (ed.) Sources of Color Science, pp. 109–119 (1970)
15. Imai, F.H., Berns, R.S.: Spectral estimation using trichromatic digital cameras. In: International Symposium on Multispectral Imaging and Color Reproduction for Digital Archives, pp. 42–49. Chiba University, Chiba (1999)
16. Pratt, W.K., Mancill, C.E.: Spectral estimation techniques for the spectral calibration of a color image scanner. Appl. Opt. **15**, 73–75 (1976)
17. Hernández-Andrés, J., Romero, J., Lee, R.L.: Colorimetric and spectroradiometric characteristics of narrow-field-of-view clear skylight in granada, spain. J. Opt. Soc. Am. A **18**, 412–420 (2001)

State Estimation of the Performance of Gravity Tables Using Multispectral Image Analysis

Michael A.E. Hansen[1], Ananda S. Kannan[2], Jacob Lund[3], Peter Thorn[3], Srdjan Sasic[2], and Jens M. Carstensen[1,4(✉)]

[1] Videometer A/S, Hørsholm, Denmark
[2] Chalmers University of Technology, Göteborg, Sweden
[3] Westrup A/S, Slagelse, Denmark
[4] DTU Compute, Kongens Lyngby, Denmark
jmca@dtu.dk

Abstract. Gravity tables are important machinery that separate dense (healthy) grains from lighter (low yielding varieties) aiding in improving the overall quality of seed and grain processing. This paper aims at evaluating the operating states of such tables, which is a critical criterion required for the design and automation of the next generation of gravity separators. We present a method capable of detecting differences in grain densities, that as an elementary step forms the basis for a related optimization of gravity tables. The method is based on a multispectral imaging technology, capable of capturing differences in the surface chemistry of the kernels. The relevant micro-properties of the grains are estimated using a Canonical Discriminant Analysis (CDA) that segments the captured grains into individual kernels and we show that for wheat, our method correlates well with control measurements ($R^2 = 0.93$).

Keywords: CDA · Gravity tables · Multispectral imaging and state optimization

1 Introduction

Agriculture has been the key ingredient in the emergence and rise of human civilization as farming of domesticated plants/crops have fostered the modern world. Along with the improvement of modern farming techniques, advances in technology have aided in the growth and development of conventional grain and seed cleaning practices sch as gravity sorting. This fundamental technique aids in removing less dense material such as immature, low germination, and insect damaged seeds from the healthier seed lot. Typically this step is repeatedly undertaken in the final grain cleaning stages using equipment referred to as *gravity tables* [1]. These tables employ specific weight (measure of buoyancy or specific density) to separate material. The table has to be adjusted on a regular basis in order to perform optimally (every 15 min), and whereas gravity tables have improved in design and increased in capacity, the need for very skilled and

© Springer International Publishing AG 2017
P. Sharma and F.M. Bianchi (Eds.): SCIA 2017, Part II, LNCS 10270, pp. 471–480, 2017.
DOI: 10.1007/978-3-319-59129-2_40

Fig. 1. The Westrup KA-1200 gravity table (1), and the VideometerLab2.0 (2).

experienced manual labor for the essential task of monitoring and adjustment, has remained the same. Unfortunately, optimizing the table is difficult, due to grains being undistinguishable to the naked eye. Only operators with years of experience are capable of optimizing the settings of the table and maintain optimal operation. Hence, this is the next bottleneck that should be overcome in order to increase the overall productivity of grain and seed cleaning operations. So far, automation of these machinery have not been addressed, and this work represents a nascent foray into this domain by attempting to develop a self regulating gravity table that measures its current operating state using an advanced vision system. The first step in solving this problem is to prove that cameras are capable of measuring significant (and relevant) differences in the grain product profiles, and furthermore show that this difference actually correlates with grain density. It should be noted that the grains that are fed are pre-cleaned in several preceding operations and the final material processed on a gravity table is (assumed to be) homogeneous in shape,size and distribution. Hence, the table will sort material according to weight, and consequently the objective of this first study is to develop a suitable methodology for the estimation of kernel weights in batches using multispectral image analysis. We further demonstrate the feasibility of such a vision driven process optimization using *wheat* as a model crop.

In the following, we will introduce the instruments used in the study, followed by a description of the experimental setup, and the chosen strategy. We finally show the results, and illustrate how they can be used in the optimization process of the table parameters.

2   Instrumentation

Before presenting the methods and results, we give a short introduction to the apparatus used in this study.

2.1   Gravity Tables

Figure 1.(1) shows an industrial gravity table, and Fig. 2 illustrates the working principle [1]. Feed material, **X**, is introduced on a perforated deck, through which air is blowing. The table is tilted along two directions ("lengthwise" and "across" in Fig. 2). The deck is placed on springs, vibrating "lengthwise" with a constant frequency through an eccentric. As soon as material enters the deck, the air starts stratifying it vertically due to the inherent differences in specific weight (lighter on top of heavier). The "across"-angle is forces material to flow off the table during operation while the "lengthwise" tilt forces material to flow downwards (due to gravity). In addition, the vibrating action throws the bottom (heavy) material upwards, resulting in a horizontal separation of lower (heavy) from upper (lighter) material along the deck length. Finally, material exits the table as segregated fractions, Y.

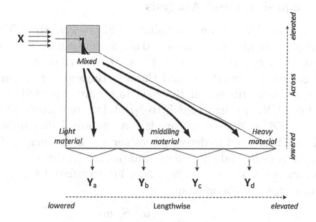

Fig. 2. Principle of operation of a gravity table.

2.2 Camera System

Images were captured using a VideometerLab vision system. The instrument captures images at 19 different wavelengths (from UV (365 nm) to NIR (970 nm)) with an image resolution of 2056×2056 pixels (44.5 μm per pixel and ≈306 Mb per image). The original technology behind the instrument was concieved at the Technical University of Denmark (DTU) and this system is now being developed and commercialized by Videometer A/S [2,4] (Fig. 1.(2)). The camera acquires images in which pixels contain a spectrum of (reflectance) values obtained at the different wavelengths. This information, is representative of the chemistry of an area above (and below) the surface. The slightest differences can be captured, through the use of multivariate statistical methods, capable of outperforming the human eye in both sensitivity, specificity and consistency. In this study, we use a VideometerLab (3.0) and the latest version of the accompanying software [2].

3   Materials and Methods

3.1 Gravity Table Setup

The table used in this study is a Westrup KA-1200 (Fig. 1.(1), having $N = 4$ outlets (labelled a, b, c, and d in Fig. 2). To simulate a production line, a hopper is placed next to the table, feeding the deck with material, recycled through a conveyor belt collecting (separated) material that leaves the table. This material is further transported to the side of the table, where a vacuum pump lifts the seeds back into the hopper. A 100 kilos of wheat from a grain cleaning plant is filled into the hopper and the table is started. We invited an expert to set the table, making sure it was running in an optimal state: $P^{opt} = [F^{opt}, \Gamma^{opt}, \theta_L^{opt}, \theta_T^{opt}] = [85, 40, 15, 8]$ (see Sect. 1). The parameter values are entered as percentage-of-maximum on the machine.

3.2   Image Acquisition and Analysis

All images are acquired the same way using a VideometerLab (v.3.0). For equal volumes (cups) of grain, images are acquired sequentially using the hopper, feeder and conveyor system described earlier. First, the instrument is calibrated for intensity and geometric distortions and these settings are kept constant for all subsequent acquisitions. Images of the reflectance are captured at 19 different wavelengths (from UV, through visual, to NIR), into one single high-resolution spectral image ($2056 \times 2056 \times 19$ pixels). A canonical discriminant analysis (CDA) model [5,6] is created to detect wheat kernels in the images. Pixel regions are labelled in the background and foreground (kernels), after which the between- and within covariance matrices, S_B and S_W, are calculated for the two classes and used to minimize the Rayleigh quotient

$$\min_{w} J(w) = \frac{w^T S_B w}{w^T S_W w}.$$

The solution, w, is a new projection (score) with maximum separation between foreground and background pixels. This is used in the segmentation that is done in two steps: (1) a pre-segmentation, where a threshold is applied to the score image, returning regions containing the kernels, (2) a watershed segmentation [7] is applied to the segmented regions in order to secure separate objects (if they are touching each other). From each kernel features are extracted relating to shape, intensity and texture, for each of the wavelengths. A total of 54 features were extracted from all kernel images.

3.3   Data Sampling

Two sets of samples were taken from the table: for the single kernel analysis, and for the density profiling (see Fig. 3). Few estimates of specific weight are generally accepted: The cup-method, measuring the weight of a fixed volume, recalculated to kg/m², and 2) The *1000 kernel weight*, calculating the average weight per kernel $\times$ 1000. Both methods give the same result, but during the course of development, we found that the cup-method was sensitive to variations in grain compaction within the cup (less important for increasing volumes). Therefore, we chose the 1000 kernel weight as the relevant metric, as it was robust and less sensitive towards sampling variation.

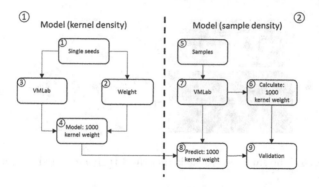

Fig. 3. Processing "pathway" of the analysis.

Single Kernel Modelling. A total of $4 \times 48 = 192$ wheat kernels were sampled randomly from the outlets of the table, and their weight found on an A&D HR-100AZ (102g/0.1 mg) scale, $\boldsymbol{M} = \{m_1, \ldots, m_{192}\}^T$. Images of the single kernels, were acquired on both sides (i.e. ventral and dorsal) using the VideometerLab and key features extracted. In the analysis, when creating a data set, we randomly select one side and use the representative features, such that we obtain a matrix of independent variables, $\boldsymbol{X}_M \in \mathbb{R}^{192 \times 54}$. Next, kernel weight was modelled using

$$\boldsymbol{M} = f(\boldsymbol{X}_M, \hat{\boldsymbol{\beta}}^*)$$

assumed to be linear, for simplicity. We applied stepwise regression [3] combined with a scoring schema. The scoring is done based on counting the number of times a feature has been selected, and for different models this is created randomly by picking kernel sides randomly (randomly picking only 90% of the kernels). We repeated this 1000 times, and choose the features that were selected for more than 30% of the times. The final generated regression model is used to estimate $\hat{\boldsymbol{\beta}}^*$ and this is further used in modelling the table density.

Table Density Profiling. Table parameters are varied individually, and 5 points are sampled $\pm 5\%$ around the optimal setting, P^{opt}. A total of 80 samples are taken (4 parameters $\times$ 5 parameter values $\times$ 4 outlets). We keep the load fixed (typically for production). Samples are collected in 0.147 L cups, taken at four positions equally distributed along the output side of the table. The weight of each sample (minus cup) is measured, $\boldsymbol{W} = \{w_1, \ldots, w_{80}\}^T$, and images of the kernels in each cup are acquired using the VideometerLab, with an attached feeder. Figures 4 and 7 show the resulting output for the varying parameters and the profile across the table.

4   Results

The kernels were analyzed as illustrated in Fig. 3. After the image analysis, features were exported as a comma separated (CSV) file into Matlab© (R2014a),

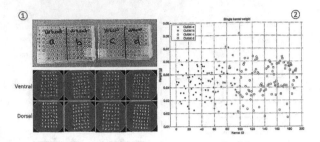

Fig. 4. The sampled wheat kernels and their measured weight.

from where the analysis was done. First, we create a model capable of estimating the weight of single kernels, based on visual characteristics only. Next, we use this model to estimate the weight of batches of kernels based on images of the individual kernels.

4.1   Single Kernel Modelling

Figures 4.(1) and 4.(2) shows the kernels picked for the single kernel weight model generation, and their measured weight. The average kernel weight was $45.15(\pm 11.90)$ mg. And although the kernels were sampled equally from each of the four outlets of the gravity table, no significant ($p = 0.05$) difference between

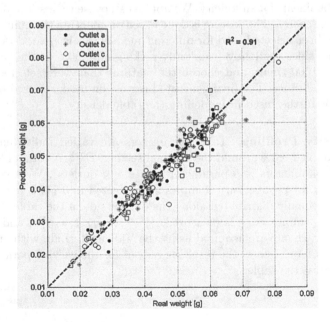

Fig. 5. Result after model generation and feature selection by stepwise regression (only 5 features).

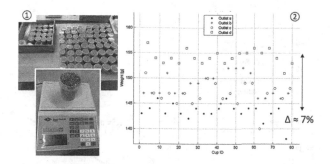

Fig. 6. Samples from the gravity table, and their measured weight. The difference between light (Outlet a) and heavy (Outlet d) material is 7% (if volume is assumed to be constant).

the weight of kernels could be detected (a and d are different at p = 0.11). In addition, no significant difference could be observed in the features from the different sides. A total of five out of 54 features were selected (with score): Area(70%), Roundness(37%), $Mean_{890\,nm}$(75%), $Mean_{940\,nm}$(90%), Volume(85%). Figure 5 shows the final result, with the predicted weight as function of the real kernel weight ($R^2 = 0.91$). The remaining features all were chosen less than 10% of the times.

4.2   Table Density Profiling

A total of 4964 images (1595 GB) were acquired ($\approx$20 images per cup) and 364 389 kernels were detected ($\approx$73 per image). Using the weight model, $\hat{\beta}^*$, we estimated the average weight of the kernels in each cup and multiplied it with a 1000 to obtain the thousand kernel weight. As a reference we calculated the 1000 kernel weight based on the measured cup weights. Figure 7 shows the correlation between measurements and the estimates based on image features alone, showing a high correlation ($R^2 = 0.93$). A slight bias (difference in the means) can be observed, coming from an inexact number of kernels. Some kernels were not visible in the camera field of view, or removed (half kernels at the border) during the segmentation process. A decrease in detected kernels relative to the actual, causes the estimated weight of the kernels to be less than the actual weight and the measured weight to be slightly higher. Figure 6 shows that we can use the estimated weight to optimize the process. If we had increased the fluidization conditions slightly, we would have obtained a much better separation between the ends (Outlets a and d) of the table. A similar trend is noted for the vibrational conditions as well, while changing the two angles would have had lower impacts. Figure 8 show the resulting output profiles from the table for the varying table settings. Figure 9 show the output profile for the four different sample points along the table.

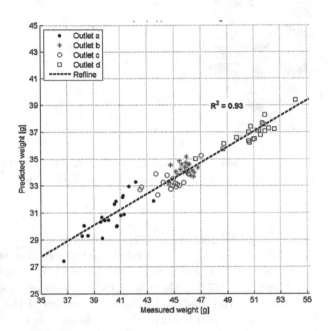

Fig. 7. Plot of the estimated 1000 kernel weight as function of the measured 1000 kernel weight.

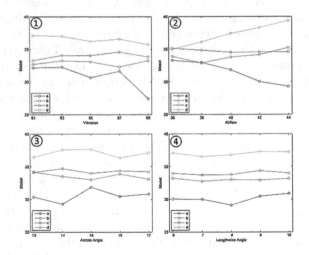

Fig. 8. Variation in the density output for varying parameters.

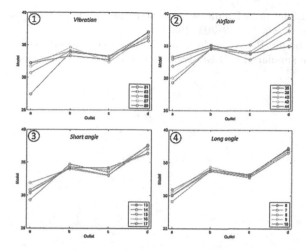

Fig. 9. Profile of density on the output.

5 Discussion

We have shown that the density profile of the output of gravity tables can be estimated using a multispectral camera setup coupled with simple CDA and regression modelling. As can be seen, there is a bias in the estimate of the weight due to missing seeds in the images that are outside the optical field of the camera (kernels being positioned at the edges of the image). This bias was noted to be constant and the measured metric can hence be utilized to quantify the relevant grain micro-properties (specific gravity). Such an analysis could be extended to other crop types such as maize, barley etc.

Acknowledgements. This project is supported by the Seventh Framework Programme of EU, Industry-Academia Partnerships and Pathways (IAPP) - Marie Curie Actions: Grant no. 324433 and the Innovation Fund Denmark under the SpectraSeed project (number 110-2012-1). The authors would also like to thank Jan Straby and Morten Seidenfaden from Westrup A/S for their help and support throughout this work.

References

1. Westrup machinery user manuals. http://www.westrup.com
2. VideometerLab user manual. http://www.videometer.com
3. Draper, N.R., Smith, H.: Applied Regression Analysis. Wiley, New Jersey (1998)
4. Carstensen, J.M., Hansen, J.F.: An apparatus and a method of recording an image of an object. Patent family EP1051660 (issued in 2003)
5. Rao, R.C.: The utilization of multiple measurements in problems of biological classification. J. Roy. Stat. Soc. Ser. B **10**(2), 159–203 (1948)

6. Fisher, A.R.: The use of multiple measurements in taxonomic problems. J. Roy. Stat. Soc. Ser. B **10**(2), 159–203 (1936)
7. Vincent, L., Soille, P.: Watersheds in digital spaces: an efficient algorithm based on immersion simulations. IEEE Trans. Pattern Anal. Mach. Intell. **13**(6), 583–598 (1991)

Author Index

Aanæs, Henrik   I-66, I-135, I-515, II-426
Abdollahzadeh, Ali   II-233
Acar, Erman   II-233
Alizadeh, Maryam   II-15
Allalou, Amin   II-257
Almstrup, Kristian   I-397
Alvén, Jennifer   II-28
Andersen Dahl, Vedrana   II-338
Anfinsen, Stian Normann   II-181
Anwer, Rao Muhammad   I-297
Arvanitis, Gerasimos   I-480
Ashmeade, Terri   II-350
Atkinson, Gary A.   II-123
Austvoll, Ivar   I-77, II-53
Ayyalasomayajula, Kalyan Ram   I-386

Badea, Mihai   I-337
Behrens, Thea   I-146
Bengtsson, Ewert   I-3
Beyerer, Jürgen   I-526, II-377
Bhogal, Amrit Pal Singh   I-184
Bianchi, Filippo Maria   I-419, I-431, II-181
Boesen, Lars   I-161
Bombrun, Maxime   II-257
Böttger, Tobias   I-54
Bours, Patrick   II-326
Branzan Albu, Alexandra   II-15
Brattland, Vegard   I-77
Breithaupt, Ralph   II-326
Breuß, Michael   I-502
Bruhn, Andrés   I-41
Brun, Anders   I-386
Bulatov, Dimitri   II-3
Busch, Christoph   I-325, II-448

Campos-Taberner, Manuel   II-205
Camps-Valls, Gustau   I-443, II-205
Carl, Jesper   I-161
Carstensen, Jens M.   II-471
Čech, Jan   II-389
Chowdhury, Manish   II-100
Christensen, Anders Nymark   II-109
Christensen, Lars Bager   I-350
Christophe, Francois   I-88

Chum, Ondřej   I-234
Clement, Jesper   I-66
Conradsen, Knut   II-109
Cote, Melissa   II-15
Cristea, Anca   II-159

Dahl, Anders B.   I-350
Dahl, Vedrana Andersen   II-109
Doulgeris, A.   II-136
Doulgeris, Anthony P.   II-159
Dragomir, Anca   I-407, II-362
Drengstig, Tormod   I-77

Eckerstorfer, M.   II-136
Eerola, Tuomas   II-245
Eftestøl, Trygve   II-53
Eilertsen, Gabriel   I-221
Einarsdottir, Hildur   I-350
Einarsson, Gudmundur   I-350
Eiriksson, Eythor R.   I-515
Eltoft, Torbjørn   II-159
Engan, Kjersti   I-362, II-53
Ersbøll, Bjarne K.   I-350
Eslami, Mohammad   I-273

Fagertun, Jens   I-374
Faion, Patrick   I-146
Fakotakis, Nikos   I-480
Favrelière, Hugues   I-550
Fischer, Thomas   I-502
Florea, Corneliu   I-337
Florea, Laura   I-337
Forssén, Per-Erik   I-221
Franc, Vojtěch   II-389
Franke, Uwe   I-98
Frimmel, Hans   II-76
Frintrop, Simone   I-260
Frisvad, Jeppe R.   I-515
Fröhlich, Björn   I-98

George, Sony   I-550
Giampieri, Mauro   I-285
Giuliani, Alessandro   I-285

Głowacka, Dorota   I-538
Goldgof, Dmitry   II-350
Gómez Betancur, Duván Alberto   II-76
Gouton, Pierre   II-401
Gupta, Anindya   I-407

Haario, Heikki   II-245
Haavik, Heidi   II-302
Hachaj, Tomasz   I-17
Hämmerle-Uhl, Jutta   I-184
Hannemose, Morten   II-426
Hansen, Bolette Dybkjær   II-302
Hansen, Michael A.E.   II-471
Haque, Mohammad A.   II-269
Hardeberg, Jon Yngve   I-550, II-459
Harder, Stine   II-438
Hasan, Ahmad   I-248
Hassanzadeh, Aidin   II-169
Heidemann, Gunther   I-146
Heikkilä, Janne   I-122
Hernández Hoyos, Marcela   II-76
Herrmann, Christian   II-377
Heun, Valentin   I-515
Hienonen, Petri   I-197
Hoeltgen, Laurent   I-502
Horvath, Peter   I-172
Hsu, Leighanne   I-457
Huttunen, Heikki   I-88

Imaizumi, Shoko   I-562

Jasch, Manuel   I-98
Jensen, Carina   I-161
Jensen, Janus N.   I-350
Jensen, Rasmus R.   I-66
Jensen, Sebastian Nesgaard   I-135
Jenssen, Robert   I-419, I-431, II-181
Jørgensen, Anders   I-374

Kaarna, Arto   II-169
Kahl, Fredrik   II-28, II-41
Kaipala, Jukka   II-221
Kälviäinen, Heikki   I-197, II-245
Kambhamettu, Chandra   I-457
Kampffmeyer, Michael   I-419, II-193
Kannala, Juho   I-122, II-88
Kannan, Ananda S.   II-471
Kasturi, Rangachar   II-350
Kauranne, Tuomo   II-169

Khan, Fahad Shahbaz   I-297
Khan, Haris Ahmad   II-459
Khawaja, Hassan A.   I-492
Khayeat, Ali Retha Hasoon   I-209
Kheiri, Peyman   I-146
Kiya, Hitoshi   I-562
Kjærulff, Søren   I-397
Klintström, Benjamin   II-65, II-100
Klintström, Eva   II-65, II-100
Koos, Krisztian   I-172
Koptyra, Katarzyna   I-17
Korsager, Anne Sofie   I-161
Krause, Oswin   I-397
Krumnack, Ulf   I-146

Laaksonen, Jorma   I-297
Landre, Victor   II-314
Längle, Thomas   I-526
Laparra, Valero   II-205
Lapray, Pierre-Jean   II-401
Larsen, Christian Thode   II-109
Larsen, Rasmus   II-109, II-147
Larsson, Måns   II-28, II-41
Laskar, Zakaria   II-88
Laugesen, Søren   II-438
Lauri, Mikko   I-260
Le Goïc, Gaëtan   I-550
Lehtola, Ville   I-88
Lei, Jun   I-468
Lensu, Lasse   I-197
Li, Guohui   I-468
Li, Shuohao   I-468
Lidayová, Kristína   II-76
Lindberg, Anne-Sofie Wessel   II-338
Lindblad, Joakim   I-407, II-257, II-362
Lindstroem, Rene   II-302
Livi, Lorenzo   I-419
Løkse, Sigurd   I-419, I-431
López, Miguel Bordallo   II-221
Luengo, David   II-205
Lund, Jacob   II-471
Luppino, Luigi Tommaso   II-181

Maes, Pattie   I-515
Mahoney, Andrew R.   I-457
Maiorino, Enrico   I-285
Malnes, E.   II-136
Mandrup, Camilla Maria   II-109
Mansouri, Alamin   I-550

Marstal, Kasper Korsholm II-438
Martini Jørgensen, Thomas II-338
Martino, Alessio I-285
Martino, Luca I-443, II-205
Matas, Jiří II-389
Maurer, Daniel I-41
Meidow, Jochen II-3
Meinich-Bache, Øyvind II-53
Melander, Markus I-197
Mercier, Gregoire II-181
Meyer, Johannes I-526
Micusik, Branislav I-29
Mikkonen, Tommi I-88
Moeslund, Thomas B. I-374, II-269
Molnár, József I-172
Moreno, Rodrigo II-65, II-100
Moser, Gabriele II-181
Moustakas, Konstantinos I-480
Mukundan, Arun I-234
Muñoz-Marí, Jordi II-205

Nasrollahi, Kamal II-269
Navaei, Pouria I-273
Nedergaard, Rasmus II-302
Neubert, Jeremiah I-248
Nielsen, Allan A. II-147
Nielsen, Jannik Boll II-426
Nielsen, Jon II-413
Nielsen, Mads I-397
Nielsen, Torben T. I-397
Nilsson, Mats II-257
Nissen, Malte S. I-397
Nøhr, Anne Krogh II-302
Nordeng, Ian E. I-248

Ogasawara, Takeshi I-562
Ogiela, Marek R. I-17
Olsen, Doug I-248
Olsen, Søren I. II-413
Orkisz, Maciej II-76
Østergaard, Lasse Riis I-161, II-302
Ostermann, Jörn I-313

Pai, Chih-Yun II-350
Pajdla, Tomas I-110
Panagiotopoulos, Theodore I-480
Partel, Gabriele II-257
Paulsen, Rasmus R. I-350, II-438
Pedersen, David B. I-515

Pedersen, Marius II-314
Peltonen, Sari II-233
Petersen, Martin Bæk II-109
Piekarczyk, Marcin I-17
Pilgaard, Louise Pedersen II-302
Pillet, Maurice I-550
Pitard, Gilles I-550
Plocharski, Maciej II-302
Pohl, Melanie II-3
Polic, Michal I-110
Pollefeys, Marc I-98
Pyykkö, Joel I-538

Qian, Xiaoyan II-257

Radow, Georg I-502
Rahtu, Esa II-290
Raja, Kiran II-448
Ramachandra, Raghavendra II-448
Ranefall, Petter II-257
Rasmussen, Jesper II-413
Rätsch, Matthias I-98
Rizzi, Antonello I-285
Rosin, Paul L. I-209
Ruoff, Peter I-77
Ruotsalainen, Ulla II-233

Saarakkala, Simo II-221, II-290
Salberg, Arnt-Børre I-431, II-193
Sampo, Jouni II-245
Sasic, Srdjan II-471
Schneider, Lukas I-98
Schöning, Julius I-146
Schuch, Patrick I-325
Schulz, Simon-Daniel I-325
Serpico, Sebastiano II-181
Sintorn, Ida-Maria I-407, II-362
Skovmand, Linda I-515
Skretting, Karl I-362
Sladoje, Nataša I-407, II-362
Smedby, Örjan II-65, II-76, II-100
Söllinger, Dominik I-184
Solorzano, Leslie II-257
Sorensen, Scott I-457
Sousedik, Ctirad II-326
Špetlík, Radim II-389
Steger, Carsten I-54
Stets, Jonathan D. I-66
Stoll, Michael I-41

Sun, Xianfang   I-209
Sun, Yu   II-350
Suurmets, Seidi   I-66
Suveer, Amit   I-407, II-362
Svendsen, Daniel H.   II-205

Thevenot, Jérôme   II-221, II-290
Thomas, Jean-Baptiste   II-401, II-459
Thorn, Peter   II-471
Tiulpin, Aleksei   II-290
Tolias, Giorgos   I-234
Torkamani-Azar, Farah   I-273
Treible, Wayne   I-457
Trier, Øivind Due   II-193
Trung, Pauline   I-184

Uhl, Andreas   I-184
Ulrich, Markus   I-54
Unger, Jonas   I-221

van de Weijer, Joost   I-297
Venkatesh, Sushma   II-448
Vertan, Constantin   I-337
Vicent, Jorge   I-443

Vickers, H.   II-136
Vogt, Karsten   I-313
Volz, Sebastian   I-41

Waaler, Dag   II-314
Wählby, Carolina   II-257
Wang, Chunliang   II-282
Wang, Tao   I-468
Wang, Xiaolong   I-457
Weber, Thomas   I-98
Wildenauer, Horst   I-29
Willersinn, Dieter   II-377
Wilm, Jakob   I-135

Xu, Shukui   I-468

Ylimäki, Markus   I-122

Zafari, Sahar   II-245
Zamzmi, Ghada   II-350
Zhang, Yuhang   II-41
Zitterbart, Daniel P.   I-457
Zsíros, László   II-426

Printed in the United States
By Bookmasters